physica status solidi (a)

VOL. 43 · NO. 2 · OCTOBER 1977

AKADEMIE-VERLAG · BERLIN

THEORIA CUM PRAXI

Classification Scheme

1. Structure of Crystalline Solids
   1.1 Perfectly Periodic Structure
   1.2 Solid-State Phase Transformations
   1.3 Alloys. Metallurgy
   1.4 Microstructure (Magnetic Domains See 18; Ferroelectric Domains See 14.4.1)
   1.5 Films
   1.6 Surfaces
2. Non-Crystalline State
3. Crystal Growth
4. Bonding Properties
5. Mössbauer Spectroscopy
6. Lattice Dynamics. Phonons
7. Acoustic Properties
8. Thermal Properties
9. Diffusion
10. Defect Properties (Irradiation Defects See 11)
    10.1 Metals
    10.2 Non-Metals
11. Irradiation Effects (X-Ray Diffraction Investigations See 1 and 10)
12. Mechanical Properties (Plastic Deformations See 10)
    12.1 Metals
    12.2 Non-Metals
13. Electron States
    13.1 Band Structure
    13.2 Fermi Surfaces
    13.3 Surface and Interface States
    13.4 Impurity and Defect States
    13.5 Elementary Excitations (Phonons See 6)
        13.5.1 Excitons
        13.5.2 Plasmons
        13.5.3 Polarons
        13.5.4 Magnons
14. Electrical Properties. Transport Phenomena
    14.1 Metals. Semi-Metals
    14.2 Superconductivity. Superconducting Materials and Devices
    14.3 Semiconductors
        14.3.1 Films
        14.3.2 Surfaces and Interfaces
        14.3.3 Device. Junctions (Contact Problems See 14.3.4)
        14.3.4 High-Field Phenomena, Space-Charge Effects, Inhomogeneities, Injected Carriers (Electroluminescence See 20.3; Junctions See 14.3.3)
    14.4 Dielectrics
        14.4.1 Ferroelectrics
15. Thermoelectric and Thermomagnetic Properties
16. Photoconductivity. Photovoltaic Effects
17. Emission of Electrons and Ions
    17.1 Field Emission Microscope Investigations
18. Magnetic Properties
    18.1 Paramagnetic Properties
    18.2 Ferromagnetic Properties
        18.2.1 Ferromagnetic Films
    18.3 Ferrimagnetic Properties
    18.4 Antiferromagnetic Properties

(Continued on cover three)

physica status solidi (a)

applied research

Volume 43 · Number 2 · Pages 371 to 710, K117 to K208, and A11 to A18

October 16, 1977

PSSA 43(2) 371—710, K117—K208, A11—A18 (1977)

AKADEMIE-VERLAG · BERLIN

Subscriptions and orders for single copies should be directed

in the DDR to the Postzeitungsvertrieb, to a book-shop or to the Akademie-Verlag, DDR-108 Berlin, Leipziger Straße 3 – 4;

in the other socialist countries to a book-shop for foreign-language literature or to the competent news-distributing agency;

in the BRD and BERLIN (WEST) to a book-shop or to the wholesale distributing agency Kunst und Wissen, Erich Bieber, 7 Stuttgart 1, Wilhelmstraße 4 – 6;

in ÖSTERREICH to the Globus-Buchvertrieb, 1201 Wien, Höchstädtplatz 3;

in the British Commonwealth and USA to Academic Press Inc. (London) Limited, 24 – 28 Oval Road, London N. W. 1 or to Academic Press Inc., 111, Fifth Avenue, New York 10003, USA;

in other countries to the international book- and journal-selling trade, to Buchexport, Volkseigener Außenhandelsbetrieb der Deutschen Demokratischen Republik, DDR-701 Leipzig, Postfach 160, or to the Akademie-Verlag, DDR-108 Berlin, Leipziger Straße 3 – 4.

Editorial Note:

"physica status solidi (a)" undertakes that an original paper *accepted for publication* before the 23rd of any month will be published within 50 days of this date unless the author requests a postponement. In special cases there may be some delay between *receipt* and *acceptance* of a paper due to the review and, if necessary, revision of the paper.

Schriftleiter und verantwortlich für den Inhalt: Professor Dr. Dr. h.c. P. Görlich, DDR 102-Berlin, Neue Schönhauser Str. 20 bzw. DDR-69 Jena, Schillbachstr. 24. Verlag: Akademie-Verlag, DDR-108 Berlin, Leipziger Straße 3 – 4; Fernruf: 22 36 221 und 22 36 229; Telex-Nr.: 114420; Postscheckkonto: Berlin 350 21; Bank: Staatsbank der DDR, Berlin, Kto.-Nr.: 6836-26-20712. Chefredakteur: Dr. H.-J. Hänsch. Redaktionskollegium: Prof. Dr. E. Gutsche, Dr. H.-J. Hänsch, Dr. H. Lange, Dr. S. Oberländer. Anschrift der Redaktion: DDR 102-Berlin, Neue Schönhauser Str. 20; Fernruf: 282 33 80. Veröffentlicht unter der Lizenz-Nr. 1620 des Presseamtes beim Vorsitzenden des Ministerrates der Deutschen Demokratischen Republik. Gesamtherstellung: VEB Druckerei „Thomas Müntzer", DDR-582 Bad Langensalza. Erscheinungsweise: Die Zeitschrift „physica status solidi (a)" erscheint jeweils am 16. eines jeden Monats. Jährlich erscheinen 6 Bände zu je 2 Heften. Bezugspreis eines Bandes 150,– M zuzüglich Versandspesen (Preis für die DDR 120,– M). Bestellnummer dieses Bandes: 1085/43.

AN (EDV) 20735

Contents

Short Notes

Systematic List

The Author Index of Volume 43 Begins on Page 711

(It will be delivered together with Volume 44, Number 1.)

Original Papers

phys. stat. sol. (a) **43**, 381 (1977)

Subject classification: 18.2

Institute of Physics, Czechoslovakian Academy of Sciences, Prague

Bubble Statics in Some Axially Symmetric Fields

By

I. Tomáš

Energy expressions are formulated corresponding to a bubble domain in an axially symmetric total field. A rather general magnetic system is described making it possible by a proper choice of constants to use the results in a number of physically and technically interesting situations. Multistability of bubbles can be achieved making them potentially usable as storage cells of random access memories.

Es werden Formeln für die Energie einer Bubble-Domäne in einem axial symmetrischen Gesamtfeld formuliert. Eine geeignete Konstantenwahl ermöglicht es, die Resultate des ziemlich allgemeinen magnetischen Systems in einer Reihe von Situationen, die physikalisch und technisch von Interesse sind, zu benutzen. Man kann eine Multistabilität von Bubble-Domänen erreichen, was eine potentielle Anwendungsmöglichkeit der multistabilen Bubble-Domänen in einer „Bubble Random Access Memory" zur Folge hat.

1. Introduction

Up to now the statics of a single bubble domain subjected to an axially symmetric total magnetic field has been described in two particular situations only. A single bubble in a plane-parallel plate located in a uniform bias field has been studied by Thiele [1] for the case of an infinite plate and by Szymczak and Wadas [2] for a finite circular plate. In both papers bubble dimensions versus bias field functions and stability regions have been determined. It has been found, moreover, that the equilibrium force equation may have at most one stable solution, or in other words there is at most one possible bubble diameter corresponding to a fixed bias field value.

If the profile of the total axially symmetric field is more complicated than in the above-mentioned cases, the bubble behaviour may be qualitatively changed. Under special conditions the bubble domain may acquire one of two or more different stable diameters corresponding to a fixed bias field magnitude. This means that the force equation may have more than one stable solution. The proper shape of the total field may be brought about by stray fields of axially symmetric magnetic structures in the vicinity of the bubble. Formulas expressing the total energy of the bubble domain in a number of such situations are given in the next section.

2. Energy Expressions

Assuming uniform bias field and isotropic wall energy the symmetry of the total field will be that of the demagnetizing field. Various shapes and magnitudes of stray fields having rotational symmetry can be realized for instance by the magnetic system sketched in Fig. 1.

Let us have two plane-parallel circular magnetic plates denoted as plate 1 and 2, having thicknesses h and dh, respectively. The plates are parallel to each other, the distance between them being gh. (All lengths are expressed as multiples of the thickness of plate 1.) Magnetization is allowed only to be perpendicular to the plates. Differently magnetized regions are separated by rigid boundaries with diameters bjh and ch

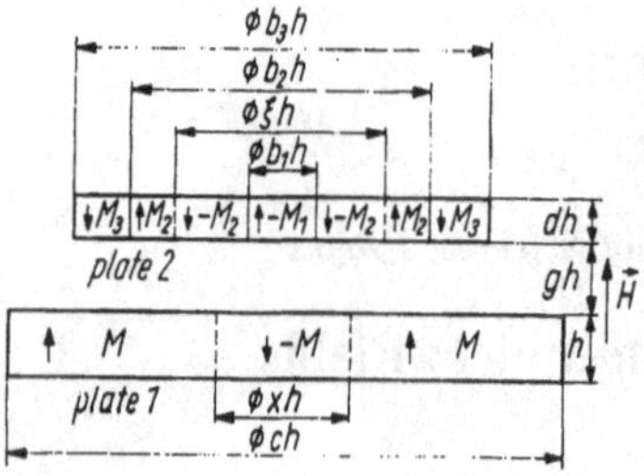

Fig. 1. Schematic cross-section through the considered magnetic system consisting of plate 1 and 2 magnetized perpendicularly to the plates surfaces. Differently magnetized annuli are separated by rigid boundaries with diameters ch, b_1h, b_2h, ... or mobile 180° domain walls with diameters xh and ξh (see text)

($j = 1, 2, \ldots, m$) and/or by movable 180° domain walls with diameters xh and ξh. During the radial outward motion of these domain walls the original direction of magnetization changes within the volume encircled by the wall to the opposite one. Defining $b_0 = 0$, γ_x and γ_j are energy densities of the 180° domain walls having diameters xh and ξh ($b_{j-1} \leqq \xi \leqq b_j$) in the plates 1 and 2, respectively. The absolute value of the magnetization of the plate 1 is M. In the plate 2, the magnetization of an annulus with the inner diameter $b_{j-1}h$ and the outer diameter b_jh is M_j. All the magnetic system is located in an uniform bias field $\boldsymbol{H}$ the positive direction of which is identical with the direction of magnetization of the plate 1 outside the volume encircled by the domain wall.

The total energy of such a system can be understood as being formed by three components. We can write as usually

$$\varepsilon_T = \varepsilon_D + \varepsilon_H + \varepsilon_W\,, \tag{1}$$

where ε_D is the demagnetizing energy, ε_H the energy of the system in the bias field H, and ε_W the domain walls energy. (We take account only of the mobile 180° domain walls with diameters xh and/or ξh.)

The demagnetizing energy may be easily expressed if the system of Fig. 1 is understood as a superposition of uniformly charged discs located in four different planes — bottom and top surfaces of the plate 1 and bottom and top surfaces of the plate 2. A scheme of such a superposition of charged discs for the top surface of the plate 2 is shown in Fig. 2. It can be seen from Fig. 2, that the magnetostatic surface charge in the j-th annulus — between diameters $b_{j-1}h$ and b_jh — and also the perpendicular magnetization of this annulus can be expressed as

$$M_j = M \sum_{i=j}^{m} k_i\,, \tag{2}$$

where k_i is the reduced magnetostatic surface charge of the i-th circular disc.

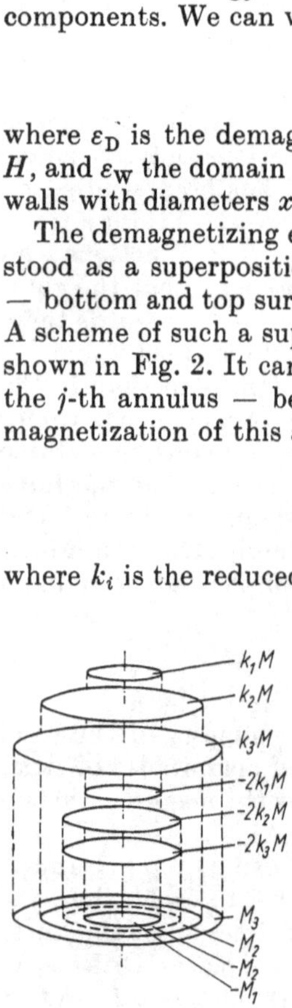

Fig. 2. Superposition of the charged circular discs creating together the charged top surface of the plate 2 in Fig. 1. $M_3 = k_3M$, $M_2 = (k_2 + k_3)\,M$, $-M_2 = (-k_2 - k_3)\,M$, $-M_1 = (-k_1 - k_2 - k_3)\,M$

Employing the classical calculation of potentials and energies of charged circular discs described for instance in [3] or [4] the demagnetizing energy of the system in Fig. 1 can be expressed as follows, for ξ in the interval:

$$b_{j-1} \leqq \xi \leqq b_j; \quad j = 1, 2, \dots, m, \quad b_0 = 0, \tag{3}$$

$$\varepsilon_D = \frac{\pi^2 M^2 h^3}{2} \int_0^\infty [Z^2(x, c) + Z^2(\xi, b_i) - Z^2(x, c)\, e^{-2t} - Z^2(\xi, b_i)\, e^{-2td} -$$
$$- Z(\xi, b_i)\, Z(x, c)\, e^{-2tg} + Z(\xi, b_i)\, Z(x, c)\, e^{-2t(g+1)} +$$
$$+ Z(\xi, b_i)\, Z(x, c)\, e^{-2t(d+g)} - Z(\xi, b_i)\, Z(x, c)\, e^{-2t(d+g+1)}] \frac{dt}{t^2}, \tag{4}$$

where[1])

$$Z(x, c) = -cJ_1(ct) + 2xJ_1(xt), \tag{5}$$

$$Z(\xi, b_i) = \sum_{i=1}^{m} \operatorname{sgn}(j - i - 1)\, k_i b_i J_1(b_i t) + 2\xi J_1(\xi t) \sum_{i=j}^{m} k_i. \tag{6}$$

In (5), (6) and below in (10) to (13) J_0 and J_1 are Bessel functions of the zeroth and first order and we define

$$\left.\begin{aligned} \operatorname{sgn}(j - i - 1) &= +1; \quad i \leqq j - 1, \\ &= -1; \quad i > j - 1. \end{aligned}\right\} \tag{7}$$

The individual terms in the expression for the demagnetizing energy of the system in Fig. 1 can be identified as follows. The first and second terms in the argument of the integral (4) are given by contributions of the magnetostatic self-energies of both surfaces of the plate 1 and both surfaces of the plate 2, respectively. The next two terms are contributions of the interaction energies between the charges at the bottom and the top surfaces of each plate. Thus the third term corresponds to the interaction of the bottom and the top surfaces of plate 1 and the fourth one expresses the same for plate 2. The last four terms are given by the interaction energy among the surfaces of different plates. Thus the fifth one represents interaction between the top surface of plate 1 and the bottom surface of plate 2, the sixth one interaction between the bottom surfaces of plates 1 and 2, the seventh one interaction between the top surfaces of plates 1 and 2, and the last (eighth) term expresses interaction between the bottom surface of plate 1 and the top surface of plate 2.

With the aid of superposition of magnetized cylinders the energy of the described magnetic system in the uniform bias field $\boldsymbol{H}$ parallel to the axis of symmetry can be for ξ in the range (3) expressed as

$$\varepsilon_H = \frac{\pi M H h^3}{4} \left[-c^2 + 2x^2 + d \sum_{i=1}^{m} \operatorname{sgn}(j - i - 1)\, b_i^2 k_i + 2\xi^2 d \sum_{i=j}^{m} k_i\right]. \tag{8}$$

The first two terms in the brackets of (8) correspond to the contribution of plate 1 and the last two terms to the contribution of plate 2.

The energy of the mobile 180° domain walls can be for ξ from (3) written as

$$\varepsilon_W = \pi h^2 [x\gamma_x + \xi\gamma_f d], \tag{9}$$

where the first term is the energy of the domain wall in the plate 1 (diameter xh) and the second one is the energy of the domain wall in the plate 2 (diameter ξh).

[1]) For $m = j = 1$, $k_1 = 1$, $b_1 = c$, $\xi = x$ (6) is evidently identical to (5). In spite of that the use of different symbols $Z(x, c)$ and $Z(\xi, b_i)$ has been chosen for the sake of clarity of the description.

For the description of the static behaviour of a bubble domain with diameter xh and/or ξh in the considered system, the first and second derivatives of the total energy ε_T with respect to the bubble diameter must be known. The derivatives of the components ε_H and ε_W are elementary. When differentiating the demagnetizing component ε_D the order of differentiation and integration may be interchanged and the following expressions used:

$$\frac{\partial}{\partial x} Z(x, c) = 2xtJ_0(xt) , \tag{10}$$

$$\frac{\partial}{\partial \xi} Z(\xi, b_i) = 2\xi tJ_0(\xi t) \sum_{i=j}^{m} k_i , \tag{11}$$

$$\frac{\partial^2}{\partial x^2} Z(x, c) = 2t[J_0(xt) - xtJ_1(xt)] , \tag{12}$$

$$\frac{\partial^2}{\partial \xi^2} Z(\xi, b_i) = 2t[J_0(\xi t) - \xi tJ_1(\xi t)] \sum_{i=j}^{m} k_i . \tag{13}$$

There are no problems connected with the computation of the ε_H and ε_W components of the total energy and with their derivatives. In computing the demagnetizing energy ε_D and its derivatives the formulas found in [5] can be used. The integrals of Bessel functions are thus expressed by means of complete elliptical integrals. Programs for their computation have been published in [6].

3. Conclusion

The presented expressions for the energy of the axially symmetric magnetic system of Fig. 1 make it possible to solve the bubble domain statics in a number of concrete situations. Besides the above-mentioned bubble domain in infinite ($d = 0$, $c = \infty$) [1] and finite ($d = 0$) [2] plates the results of the present paper can be used in more complicated cases. At least several possible situations that seem to be physically interesting and comparatively easily realizable are sketched in Fig. 3 together with the corresponding values of some constants[2]):

a) Bubble domain in plate 1 influenced by the presence of a magnetically hard plate 2 with magnetization kM and domain structure in the pattern of concentric annuli ($\xi = 0$; $k_i = (-1)^i\, 2k$; $i = 1, 2, 3$; $k_4 = k$) (see Fig. 3a).

b) Bubble domain in plate 1 influenced by the presence of discrete annuli from a magnetically hard material with magnetization kM ($\xi = 0$; $k_i = (-1)^i\, k$; $i = 1, 2, 3$) (see Fig. 3b).

c) Bubble domain in a plate with empty annular grooves ($g = 0$; $\xi = x$; $\gamma_1 = \gamma_3 = 0$; $\gamma_2 = \gamma_4 = \gamma_x$; $k_i = (-1)^i$; $i = 1, 2, 3, 4$) (see Fig. 3c).

d) Bubble domain in a plate around an empty cylindrical hole ($g = \infty$; $d = 1$; $b_1 \leqq \xi \leqq b_2$; $k_i = (-1)^i$; $i = 1, 2$) (see Fig. 3d).

e) Bubble domain in a plate around a cylindrical hole partially filled by a cylinder with magnetization kM ($g = \infty$; $d = 1$; $b_2 \leqq \xi \leqq b_3$; $k_1 = -k$; $k_i = (-1)^{i+1}$; $i = 2, 3$) (see Fig. 3e).

f) Two coaxial bubble domains in two different plates see also [7] ($k_1 = k$) (see Fig. 3f).

---

[2]) The set of constants is not unique for each case. The same situations can be described in several alternative ways.

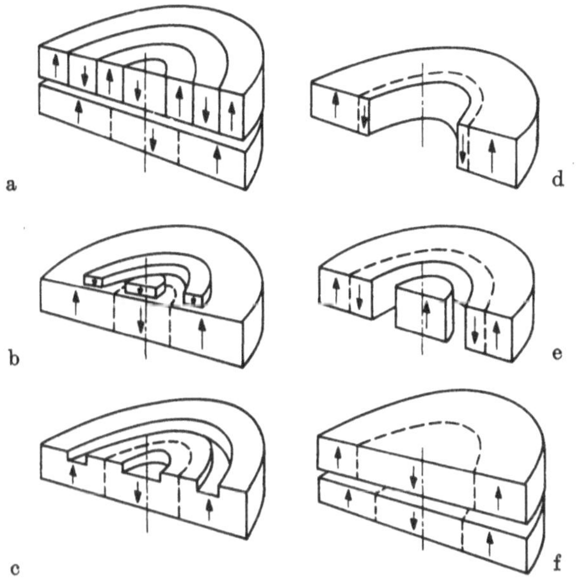

Fig. 3. Examples of bubbles in some situations which can be described by energy formulas (4), (8), and (9). Further descriptions see in the text

By a proper choice of the material and geometrical parameters multistability of the bubble domain can be achieved in cases a) to e). Such bubbles have been suggested as potential storage elements. It was shown by Tomáš [8] that storage elements based on the principle of bubble domain multistability could be used in random access memories with both destructive and non-destructive read-out.

Acknowledgements

The author would like to thank J. Kaczér and V. Kamberský for carefully reading the manuscript and valuable remarks.

References

[1] A. A. Thiele, Bell Syst. tech. J. **48**, 3287 (1969).
[2] R. A. Szymczak and R. S. Wadas, IEEE Trans. Magnetics **7**, 361 (1971).
[3] J. Kaczér and Z. Klem, phys. stat. sol. (a) **35**, 235 (1976).
[4] D. J. Craik, J. Phys. D **7**, 1566 (1974).
[5] G. Eason, B. Noble, and I. N. Sneddon, Phil. Trans. Roy. Soc. (London) **247**, 529 (1955).
[6] R. Bulirsch, Numerische Mathematik **13**, 305 (1969).
[7] W. F. Druyvesteyn, D. L. A. Tjaden, and J. W. F. Dorleijn, Philips Res. Rep. **27**, 7 (1972).
[8] I. Tomáš, Magnetic Bubble Domains in Some Axially Symmetric Fields, to be presented at INTERMAG 1977; IEEE Trans. Magnetics, September 1977, to be published.

(Received June 13, 1977)

phys. stat. sol. (a) 43, 387 (1977)

Subject classification: 14.2

Faculty of Physics, Babeş-Bolyai University, Cluj-Napoca[1])

Superconductivity in Nb_3Ge Films above 21.9 K

By

G. ILONCA

Niobium–germanium films which remain superconducting up to 21.9 K. are prepared by electron beam evaporation technique, using co-deposition both of the elements Nb and Ge on hot sapphire substrates. The high T_c's of these films are attributed to the formation of a more perfect stoichiometric Nb_3Ge compound than previously obtainable. The dependences of T_c, resistivity, and structure on chemical composition for Nb–Ge films are studied.

Niob–Germanium-Schichten, die supraleitend bleiben bis 21,9 K, werden mittels Elektronenstrahlverdampfungstechnik hergestellt, wobei gleichzeitiges Aufbringen der Elemente Nb und Ge auf ein heißes Saphir-Substrat verwendet wird. Die hohen T_c-Werte dieser Schichten werden der Bildung einer im Vergleich zu früher perfekteren stöchiometrischen Nb_3Ge-Verbindung zugeschrieben. Die Abhängigkeit von T_c, des Widerstands und der Struktur von der chemischen Zusammensetzung der Nb–Ge-Schichten wird untersucht.

1. Introduction

Most high-temperature superconductors exhibit some form of structural instability. These instabilities, as now evidenced, are indeed probably the cause of the high T_c values, a conjecture which has lead to new efforts to obtain unstable or metastable phases in the search for better superconductors. A search for metastable phases in about 40 alloys prepared by sputtering and electron beam evaporation onto hot substrates [1] has been carried out.

At the present time, most known high-transition-temperature ($T_c \leqq 23$ K) superconductors are A-15(β-W)structure compounds in which one of the constituent elements is either niobium or vanadium. This fact explains the continued interest in the study of niobium or vanadium alloys which contain, or might be expected to contain, the A-15(β-W)phase. Gavaler et al. [2, 3] and Testardi et al. [4] obtained T_c onsets of ≈ 23 K in films of Nb–Ge sputtered onto hot substrates at high argon pressure.

The question how sensitive the T_c of Nb–Ge is to the composition near the exact stoichiometric ratio 3:1 has attracted considerable comment but, as yet, no quantitative answers. Carpenter and Searcey [5] were the first to synthesize and identify A-15 structure Nb–Ge, and Geller [6] was the first to recognize that the compound normally grown to became Nb_3Ge was, in fact, nonstoichiometric. Later, Matthias et al. [7] splat-cooled the approximate Nb_3Ge and obtained a very broad superconducting transition beginning at ≈ 17 K. The high T_c was assumed to result from the nearly exact stoichiometric Nb_3Ge obtained by the rapid-quench preparation compared to the bulk compound where the Nb/Ge ratio is 3.3 [8] to 4 [9].

The major content of this paper is devoted to the identification of the chemical composition and crystal structure of high-T_c Nb–Ge evaporated films.

[1]) Cluj-Napoca, cod. 3400, Roumania.

2. Experimental Technique

2.1 Preparation of the samples

The Nb_3Ge films for the present study were prepared by a deposition process using the electron beam evaporation technique, which has been described by Hammond [10].

The substrate is 6.0 cm in length, with sources approximately 21.6 cm below the substrate. The films have been deposited onto polished sapphire substrates, the temperature of the substrate varied between 600 and 800 °C. The elements were co-evaporated using one source for niobium and another one for Ge.

The evaporation rates were controlled by monitors (ERM) with the sensors in Bayard-Alpert configuration. The film thickness was of the order 2 μm.

2.2 T_c measurements

Superconducting transition temperatures were determined by resistance measurements and magnetic susceptibility measurements using a standard Schawlow-Devlin [11] resonance method. Standard four-probe ac resistance measurements were made using pressure contacts. The temperatures were measured to an accuracy of $\approx \pm 0.3$ K with calibrated germanium resistors. The experimental set-up was described by Lebeau and Pinel [12].

2.3 X-ray measurements

X-ray diffraction data for the superconducting films were generally obtained by use of an X-ray diffractometer (XRD) and CuK_α radiation.

2.4 Chemical composition measurements

The variation of niobium and germanium concentrations was determined by electron microprobe analysis. The films were thicker than the infinite thickness required for these elements; so a correction for film thickness was not necessary. The I_0 values of Nb and Ge were determined by measuring the intensities of $Nb(L_\alpha)Ge(K_\alpha)$ from pure Nb and Ge. For a quantitative analysis, the ratios of the net X-ray intensities from the samples to those obtained from pure elements (I/I_0) were calculated. Absorption corrections were calculated by the method of Philbert [13] modified by Duncumb and Shields [14]; Heinrich's [15] absorption coefficients were used for both the elements.

3. Results

3.1 X-ray results

Table 1 contains the critical temperatures, depositing conditions, and lattice parameters of the A-15 phase obtained by the electron beam evaporation technique. The average lattice constants were computed from the diffraction lines for a film thickness of 2 μm between 10 and 125°.

The lattice parameters were found to decrease with decreasing Nb/Ge ratio. For a Nb/Ge ratio ≈ 3, where the T_c values are highest, $a \approx (5.14 \pm 0.03)$ Å, which is considerably less than the value ≈ 5.17 Å obtained for bulk A-15 Nb–Ge. This result is in general agreement with the previous findings of Gavaler et al. [3] and Testardi et al. [4]. The large lattice parameter of the bulk material was the prime indicator that its composition was nonstoichiometric [6]. Note, however, that films prepared by this method of the composition Nb/Ge ≈ 4 have, in fact, considerably smaller lattice parameters than those of the bulk. The pronounced difference between bulk and film

Table 1

Summary of critical temperature and X-ray data for Nb–Ge films deposited by electron beam evaporation technique

evaporating conditions	Nb/Ge ratio	T_c (K) (start—end)	a (Å) for the A-15 phase	remarks
$T_s = 750$ °C $t \approx 2$ μm $p \approx 2 \times 10^{-6}$ Torr	1.97	18.5—13.1	5.11 ± 0.03	A-15 W extra lines attributed to Nb
$T_s = 750$ °C $t \approx 2$ μm $p \approx 2 \times 10^{-6}$ Torr	2.02	13.2—5.6	5.12 ± 0.03	D8m structure W extra lines attributed to A-15 phase
$T_s = 750$ °C $t \approx 2$ μm $p \approx 2 \times 10^{-6}$ Torr	2.2	19.3—16.1	5.13 ± 0.02	two phases: A-15 + Nb_5Ge_3
$T_s = 750$ °C $t \approx 2$ μm $p \approx 2 \times 10^{-6}$ Torr	2.8	21.0—19.8	5.14 ± 0.02	A-15 no extra lines observed
$T_s = 750$ °C $t \approx 2$ μm $p \approx 2 \times 10^{-6}$ Torr	2.98	21.9—20.2	5.14 ± 0.03	A-15 no extra lines observed
$T_s = 760$ °C $t \approx 2$ μm $p \approx 2 \times 10^{-6}$ Torr	3.15	19.8—17.7	5.15 ± 0.02	A-15 two VW extra lines attributed to Nb_5Ge_3
$T_s = 750$ °C $t \approx 2$ μm $p \approx 2 \times 10^{-6}$ Torr	3.3	20.1—19.2	5.14 ± 0.03	A-15 one VW extra line attributed to Nb
$T_s = 780$ °C $t \approx 2$ μm $p \approx 2 \times 10^{-6}$ Torr	4.15	11.4—9.8	5.14 ± 0.02	A-15 one W extra line attributed to Nb_5Ge_3
$T_s = 750$ °C $t \approx 2$ μm $p \approx 2 \times 10^{-6}$ Torr	5.1	5.2—3.6	5.15 ± 0.02	A-15 two VW extra lines attributed to Nb_5Ge_3

W weak; VW very weak.
T_s is the substrate temperature and t the approximate film thickness.

Nb–Ge at the same composition is more clearly demonstrated in the superconducting T_c. The very weak second-phase lines observed in these films can be associated with Nb_5Ge_3 and Nb.

3.2 Composition dependence of T_c

An important question concerning the superconductivity of Nb–Ge films deposited by the electron beam technique is how critically do the high T_c values depend upon achieving exact stoichiometry. Fig. 1 shows the variation of T_c with Nb/Ge ratio as determined from microprobe analysis.

The data show that the highest T_c values fall into the compositional range Nb/Ge $\approx$ $\approx$ 2.7 to 3.2; they do not show the critical dependence on composition which might be expected to be attributed to perfect stoichiometry.

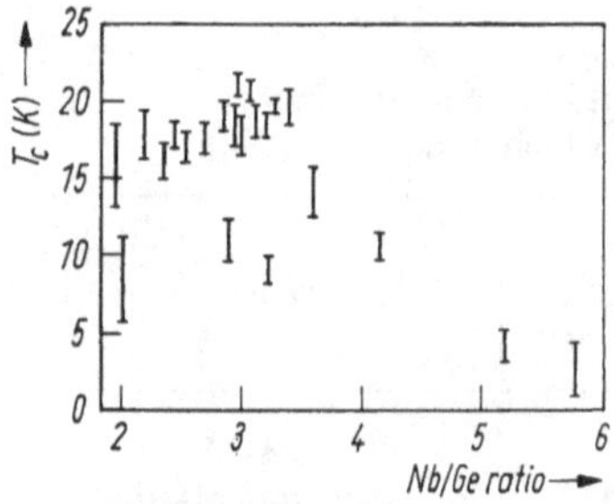

Fig. 1. Variation of T_c with Nb/Ge ratio

It is important to note that over most of this range (Nb–Ge $\approx$ 2.2 to 4.15) X-ray diffraction showed only very weak second phases. Thus the dependence of T_c on composition represents predominantly the single A-15 phase behaviour.

3.3 Composition dependence of substrate temperature (T_s) and electrical resistivity

The relative measurements are given in Fig. 2. It shows how the optimum value of T_s varied with chemical composition over the range of Nb/Ge 2.2 to 5.5. A minimum in the optimum T_s occurs for Nb/Ge in the interval $\approx$ 2.9 to 3.1, with T_s rising only slowly on either side of this range. The lowest optimum T_s value occurs in the same composition range which yields the highest T_c values. Since the X-rays show that over most of this range the predominant phase is A-15, we conclude that compositions yielding the best T_c require formation at the lowest temperatures.

The electrical resistivities of films deposited at the optimum T_s over the composition range studied generally are in the range $\approx$ 110 to 140 μΩ cm at 300 K and $\approx$ 40 to 120 μΩ cm at 25 K.

The resistance ratio $\varrho(300\text{ K})/\varrho(25\text{ K})$ at optimum T_s is also shown as a function of composition in Fig. 2. We find a maximum of the resistance ratio for Nb/Ge $\approx$ 2.7 to 3.2. Note that considerably away from 3:1, there is a disordered solution rather than a mixture of ordered phases.

3.4 Correlation of resistance ratio and T_c

Fig. 3 presents a sampling of T_c versus resistance ratio data for about 50 films and is representative, in fact, of almost all samples obtained in the present study. The data points are given here irrespective of substrate temperature, evaporation rate,

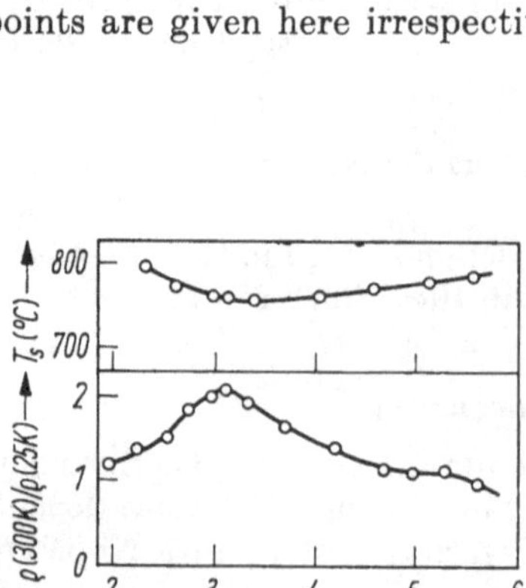

Fig. 2. Variation of optimum T_s and resistance ratio with Nb/Ge content

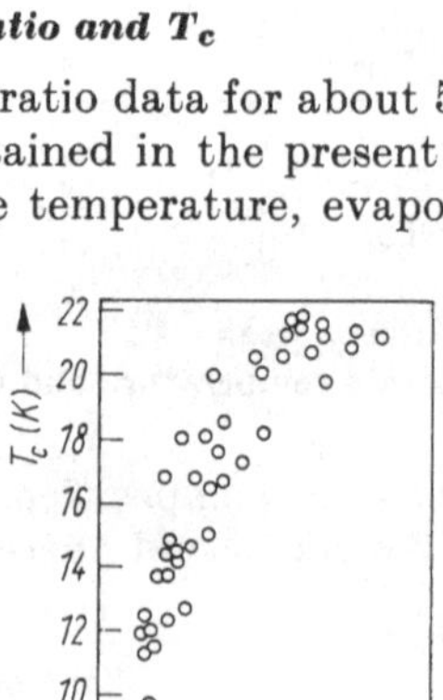

Fig. 3. Correlation of electrical resistance ratio with T_c

film thickness, crystal structure, and even chemical composition in the range Nb/Ge 2.2 to ≈ 5.5. The only restriction in choosing points for this plot was that the width of the superconducting transition is less than ≈ 2 to 3 K, thereby insuring that those films which were grossly inhomogeneous, by this criterion, are excluded.

Note that within the scatter there is a reasonable correlation between T_c and resistance ratio. Resistance ratios less than 1 generally occur in films which are not superconducting. However, increases in resistance ratio up to 2 to 3 are generally accompanied by superconductivity between 18 and 21.9 K.

The number of samples with T_c higher than 21 K was only eight.

3.5 *Variation of properties and structure with deposition temperature*

In the interval of deposition temperature between 620 and 760 °C large changes occur in the superconducting T_c, electrical resistivity, resistance ratio, and X-ray structure. Fig. 4 shows how these properties vary for seven films deposited simultaneously in seven runs, where the average chemical composition Nb/Ge was 3.2 to 2.9.

X-ray diffraction studies showed an amorphous structure at $T_s = 620$ K which gave way to a predominant A-15 structure at the higher T_s values. This is in agreement with the previously reported results of Testardi et al. [4] and Gavaler et al. [3].

From here we can note that the variation in the Nb/Ge ratio is too small to account for the T_c behaviour. We conclude that crystallographic and/or other microscopic changes are primarily responsible for the different physical behaviour.

4. Conclusion

This study allows one to see that the formation of A-15 Nb_3Ge can occur only at low temperatures. For Nb–Al and Nb–Ga a similar shift of the A-15 phase towards the 3:1 composition has been shown in the works of Hammond [16] and Webb and Engelhardt [17]. The kinetics of this reaction may well be too slow in the bulk, but greatly enhanced in the films where surface diffusion can control the growth. Thus, both the

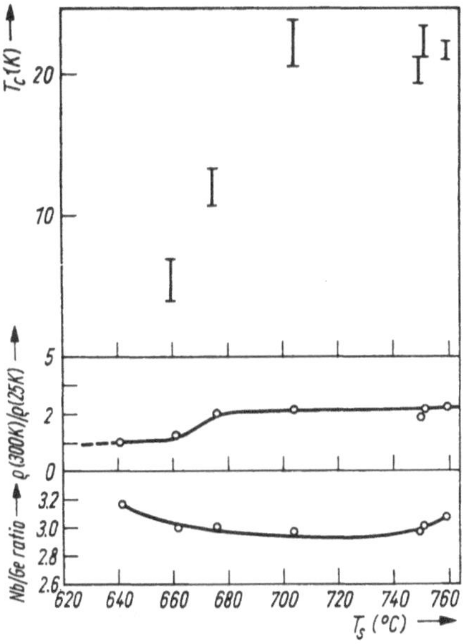

Fig. 4. Variations of T_c, electrical resistivity ratio, and Nb/Ge ratio with deposition temperature (T_s)

low formation temperature and the surface reactions may be important in the film samples.

This investigation established, too, that it is possible to obtain superconducting materials with a higher T_c by electron beam technique in a poor vacuum (about 2×10^{-6} Torr). This technique is able to make a large quantity of superconducting compounds, in short time, without special conditions analogous to that of the sputtering technique [2 to 4].

References

[1] L. R. TESTARDI, J. H. WERNICK, W. A. ROYER, D. D. BACON, and A. R. STORM, J. appl. Phys. **45**, 446 (1974).
[2] J. R. GAVALER, Appl. Phys. Letters **23**, 480 (1973).
[3] J. R. GAVALER, M. A. JANOCKO, and C. K. JONES, J. appl. Phys. **45**, 7 (1974).
[4] L. R. TESTARDI, J. H. WERNICK, and W. A. ROYER, Solid State Commun. **15**, 1 (1974).
[5] J. H. CARPENTER and A.W. SEARCEY, J. Amer. Chem. Soc. **78**, 2079 (1956).
[6] S. GELLER, Acta cryst. **9**, 885 (1956).
[7] B. T. MATTHIAS, T. H. GEBALLE, R. H. WILLENS, E. CORENZWIT, and G. W. HULL, JR., Phys. Rev. **139**, A1501 (1965).
[8] B. T. MATTHIAS, T. H. GEBALLE, L. D. LONGINOTTI, E. CORENZWIT, G. W. HULL, JR., R. H. WILLENS, and J. P. MAITA, Science (Washington) **156**, 645 (1967).
[9] Gmelins Handbuch der Anorganischen Chemie, Teil B2: Niob, Verlag Chemie GmbH, Weinheim 1971 (p. 201).
[10] R. H. HAMMOND, IEEE Trans. Magnetics **11**, 201 (1975).
[11] A. L. SCHAWLOW and G. E. DEVLIN, Phys. Rev. **113**, 120 (1959).
[12] C. LEBEAU and J. PINEL, J. Phys. E **4**, 857 (1971).
[13] J. PHILBERT, L'analyse quantitative en microanalyse par sonde électronique, IRSID Publ., Series B, No. 51, 1965.
[14] P. DUNCUMB and P. K. SHIELDS, Effect of Critical Excitation Potential on the Absorption Correction in X-Ray Microanalysis, Tube Investment Res. Labor., Tech. Rep. No. 181 (1968).
[15] K. F. J. HEINRICH, in: The Electron Microprobe, Ed. T. MCKINLEY, K. HEINRICH, and D. WHITTEY, Wiley, New York 1966 (p. 284).
[16] R. H. HAMMOND, Proc. Applied Superconductivity Conf., Chicago 1974, published in: IEEE Trans. Magnetics **11**, 201 (1975).
[17] G. W. WEBB and J. J. ENGELHARDT, Proc. Applied Superconductivity Conf., Chicago 1974, published in: IEEE Trans. Magnetics **11**, 208 (1975).

(Received June 24, 1977)

phys. stat. sol. (a) 43, 393 (1977)

Subject classification: 1.3 and 18.2; 21.1; 21.1.1

Department of Electronics, Faculty of Engineering, Hokkaido University, Sapporo

Effects of Annealing on Magnetostrictive Vibrations of Co–Fe Alloys

By

K. MUKASA and M. MAEDA

The free decay of vibration with time after removal of the exciting force and the amplitude of vibration are measured for 70% Co–Fe alloys produced by cold-rolling and annealing for about 3 h at 150 to 750 °C. It is found that the changes in the decay and the amplitude of vibrations are dependent on the method of such annealing. The most suitable condition in this experiment for the use of magnetostrictive vibrator is annealing at about 400 °C and for 3 h. The results of above measurements are discussed.

An 70% Co–Fe-Legierungen, die durch Kaltwalzen und Tempern für ungefähr 3 h zwischen 150 und 750 °C hergestellt wurden, werden die Schwingungsamplitude und das zeitliche Abklingen der magnetostriktiven Schwingung nach Aufhebung der anregenden Kraft gemessen. Es wird gefunden, daß das Abklingen und die Schwingungsamplitude von der Temper-Methode abhängig sind. Die beste Bedingung für den Einsatz als magnetostriktiven Schwinger in diesem Experiment ist ein Drei-Stunden-Tempern bei etwa 400 °C. Die Ergebnisse der Messungen werden diskutiert.

1. Introduction

The viscosity of liquids or liquid metals, for instance fused solder, was measured using the viscometer [1] which employs a magnetostrictive strip as a probe. In these measurements, the free decay of the vibration with time after removal of the exciting force in accordance with the viscosity of the liquid is detected. The magnetostrictive material is required to satisfy a certain condition: it can be used at a temperature above about 200 °C and negligibly small internal friction or internal loss is necessary. The latter is contrary to the requirement for high damping alloys. In this experiment 70% Co–Fe alloys were chosen as the material of the strip in view of the above requirements. In this paper we describe the results of the experiment on the effect of annealing on the amplitude and the decay of vibrations. The most suitable annealing condition for the use of a magnetostrictive vibrator is also shown.

2. Experimental Procedures

The alloy was made from iron and cobalt of high commercial purity, melted in a high-frequency induction furnace, cast as a rod 35 mm in diameter and cut with thickness of 1 mm. This was cold-rolled to 0.3 mm, corresponding to a nearly 70% reduction, the specimens cut to a length of 5 cm and a width of 3 mm and annealed for about 3 h at 150 to 750 °C. In the present study, a 70% Co–30% Fe alloy is used.

The apparatus used for measuring the magnetic loss in ferromagnetic metals consists of three parts as the schematic diagram of the experimental arrangement in Fig. 1 shows: a power source, a coil for excitation and detection and a subsequent amplifier. A pulse current is passed through the coil by the SCR from charged condenser by the dc power supply. The coil is wound with about 400 turns on a quartz tube 6 mm in diameter, placed around the specimen and creates fields up to about 140 Oe. On the

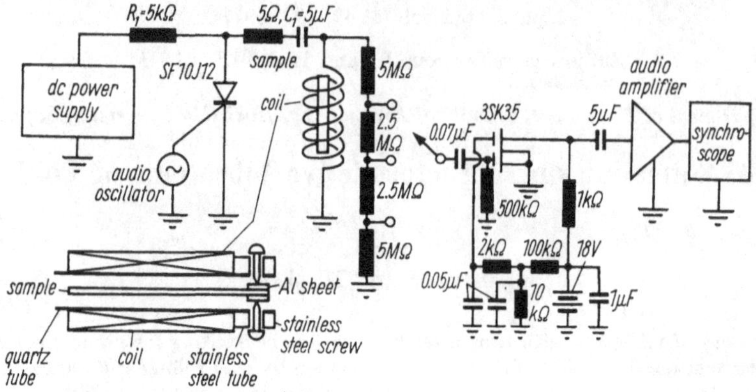

Fig. 1. Experimental set-up used for measuring the magnetic loss

other hand, the charge flows slowly through the resistor R_1 shown in Fig. 1, charging the capacitor C_1, so the amplitude of its current will be far below the discharging limit of the SCR. The pulse is derived at 70 Hz from the audio oscillator. The sample plate is with one end fixed at a stainless steel tube with two stainless steel screws and the other end free. The amplitude of vibrations and the decay of amplitude with time after removal of the exciting force corresponding to the dissipative properties of materials are detected by the same coil. This output is applied to the resistors of higher value than the impedance of the coil and then to the FET, so as not to distort the signal, and is amplified to a level suitable for the oscilloscope input requirements by the transistor amplifier.

Magnetization curve and magnetic hysteresis curve are measured using a magnetic fluxmeter, magnetostrictive constant and Young's modulus using strain gauges, and electric resistance using the potentiometric dc method.

3. Results and Discussion

Fig. 2 shows the curve of decay of amplitude with time for the specimen annealed for about 3 h at 400 °C. Changes in loss, Q^{-1}, and amplitude with annealing temperature,

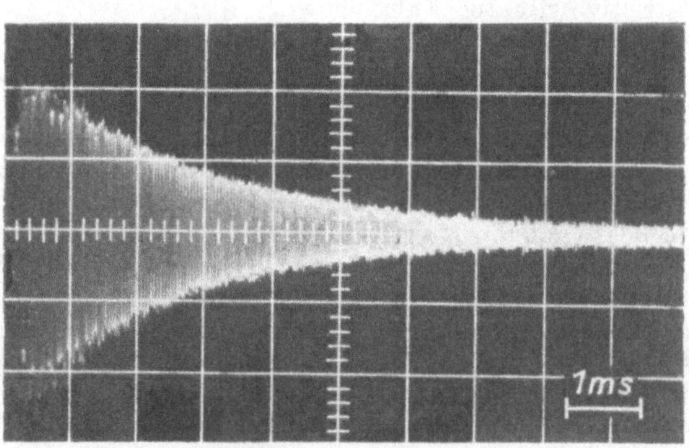

Fig. 2. The curve of decay of amplitude with time for the specimen annealed for about 3 h at 400°C

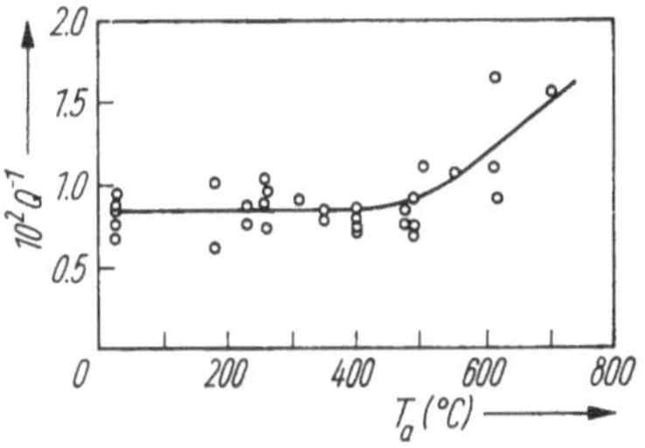

Fig. 3. Changes in loss, Q^{-1}, with annealing temperature, T_a

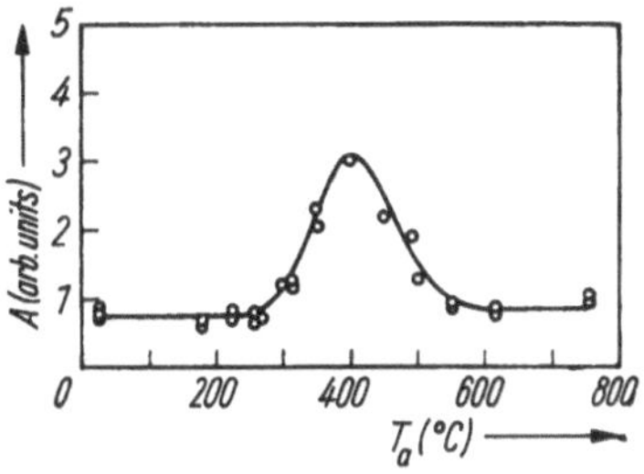

Fig. 4. Changes in amplitude, A, with annealing temperature, T_a

T_a, are shown in Fig. 3 and 4, respectively. The specimen exhibits an increase in loss, Q^{-1}, above about 500 °C and a peak in amplitude at about 400 °C.

The typical structure observed using scanning electron microscopy in the specimen annealed at 400, 490, 615, and 755 °C is shown in Fig. 5. The recrystallized grains are found above 615 °C, so we can estimate that there is the stage of recrystallization at temperatures above about 610 °C.

The measured differential thermal analysis (DTA) curve is shown in Fig. 6. In this experiment a heating rate of 20 °C/min was employed. The recrystallization peak is obtained at about 610 °C, that is confirmed by SEM as shown in Fig. 5. It is assumed that below about 400 °C there is recovery. The endothermic change at about 550 °C seems to be the transformation from order to disorder reported by Yokoyama [2]. The recrystallization peak seems to be lowered by this change. It seems that the exothermic peak corresponds to the transformation from disorder to order in a cold-worked material [3]. The change corresponding to the Curie point is observed near 900 °C.

The coercive force H_c can be obtained from the hysteresis curves as shown in Fig. 7. The value of the coercive force decreases above about 200 °C. In Fig. 8 the magnetostriction, λ, is plotted against the magnetization, I, for samples annealed at different annealing temperature. Magnetostriction increases in magnitude with increase in annealing temperature above about 260 °C. The curves for the samples annealed below about 260 °C show a similar tendency. So it is considered that there is a difference in the magnetization process, and comparing with the change of coercive force in Fig. 7, it is easier to magnetize the sample annealed above 260 °C than that below 260 °C.

We consider the behaviour in Fig. 3 on the basis of some experimental results. There are two possible magnetic losses associated with the problem. The first kind is due to the hysteresis loss and the second kind to the eddy current loss. Fig. 7 shows the effect of annealing temperature on the hysteresis. It is clear from the figure mentioned above that the hysteresis loss decreases with annealing temperature. The experimental results cannot be explained by this kind of loss.

The annealing temperature in which Q^{-1} increases as shown in Fig. 3 is inside the range in which the sample is easy to magnetize as shown by the above results. The decrease below 10% is observed in the electric resistivity for the sample above about 500 °C. The strain dependence of loss is shown in Fig. 9 for the sample annealed at 615 °C. The dependence is constant in the range of strain. A similar tendency is observed for the sample annealed at 400 °C.

The value of Q^{-1} depends on the amplitude of applied field or strain in the case of hysteresis loss, but does not in the case of eddy current loss [4]. So it is considered that the experimental results as shown in Fig. 3 can be explained by the eddy current loss.

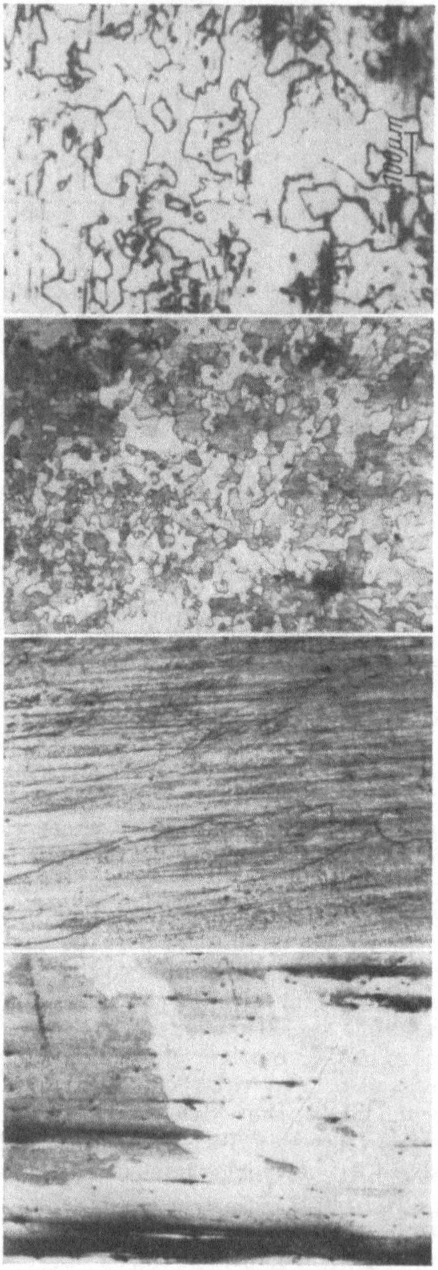

Fig. 5. The typical structure observed in the scanning electron microscope. a) T_a = 400, b) 490, c) 615, d) 755 °C

Next we consider the behaviour in Fig. 4. The electromechanical coupling factor, K, is obtained by using the equation

$$K = \sqrt{\frac{\Gamma^2}{E}} = \left(\frac{\partial\lambda}{\partial I}\right)\sqrt{\chi E}\,,$$

where χ, E, λ, I, and Γ are magnetic susceptibility, Young's modulus, magnetostriction, magnetization, and $\Gamma = \partial\lambda/\partial I$. The measured values of E are constant in the

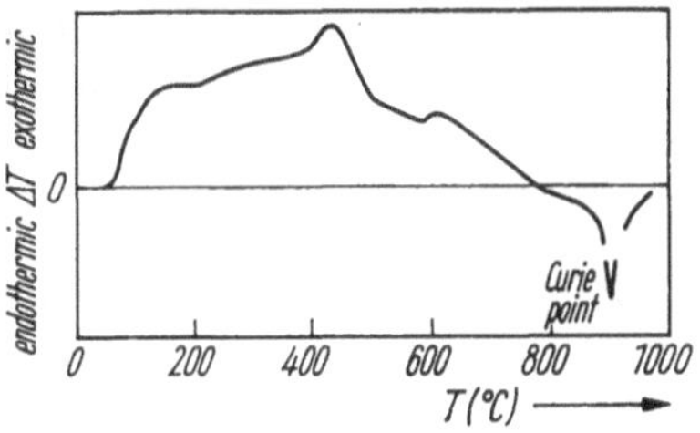

Fig. 6. Differential thermal analysis curve

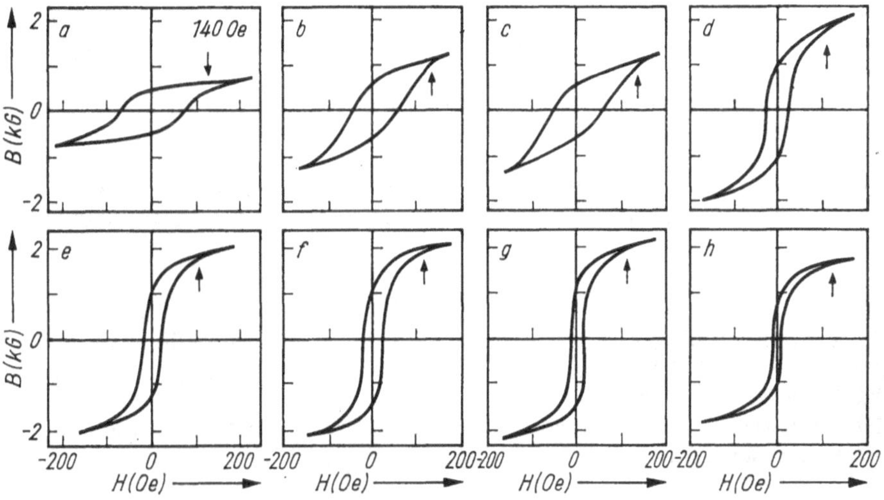

Fig. 7. Magnetic hysteresis curves measured with a magnetic fluxmeter. a) Cold-rolled, b) $T_a =$ $= 255$, c) 260, d) 400, e) 490, f) 550, g) 615, h) 755 °C

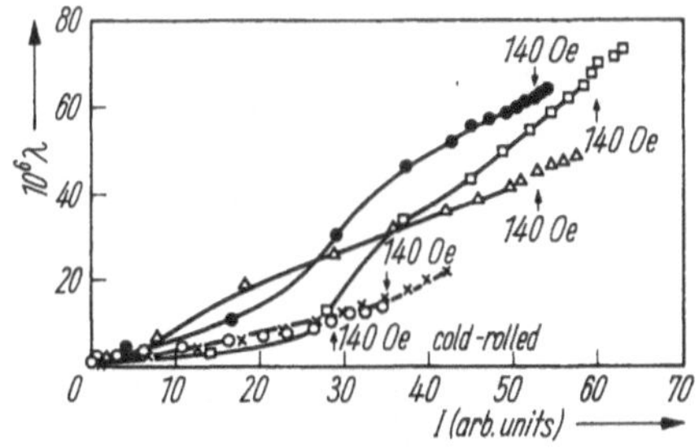

Fig. 8. Magnetostriction, λ, plotted against the magnetization, I. ○ cold-rolled × $T_a = 260$, △ 400, □ 755, ● 490 °C

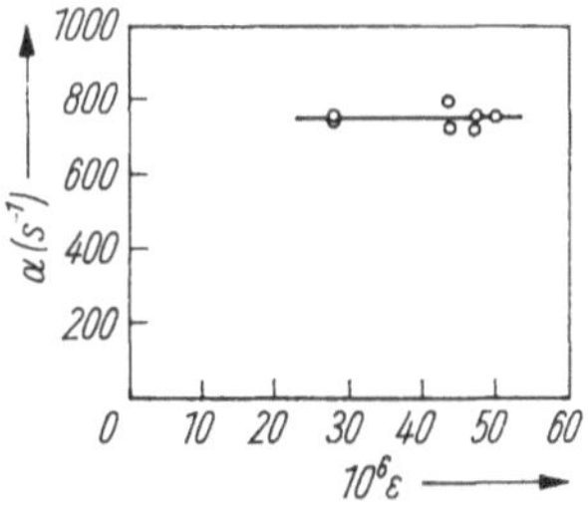

Fig. 9. The strain dependence of attenuation coefficient, α, for the sample annealed at 615 °C

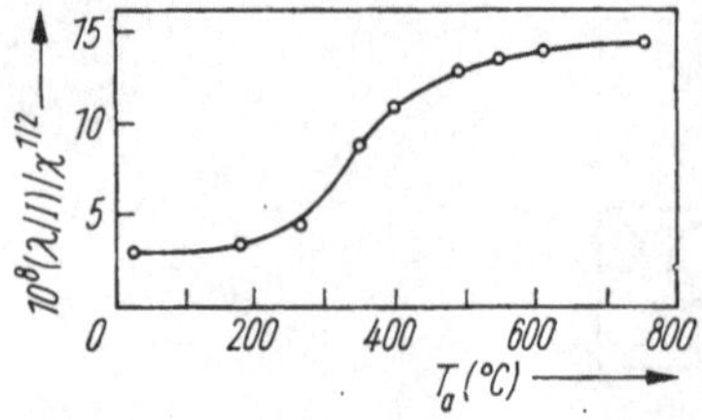

Fig. 10. The relation between the value of $(\lambda/I)\sqrt{\chi}$ and the annealing temperature

range of annealing temperatures used. $\partial\lambda/\partial I$ is approximated by the measured value of λ/I. So the relation between the value of $(\lambda/I)\sqrt{\chi}$ and the annealing temperature is shown in Fig. 10, which explains the increase below 400 °C in Fig. 4, but does not the decrease above 400 °C. It was found in the discussions about coercive force and magnetostriction that in the range above 400 °C it is easier to magnetize, and it was observed that the value of loss, Q^{-1}, increases above about 500 °C as shown in Fig. 3. It is assumed that the energy of vibration above 400 °C is more dissipated during every half cycle in exciting the mode by appling the magnetic field than below 400 °C. So the amplitude of vibration may decrease above 400 °C. It can be considered that this effect would be expressed in terms of the electromechanical coupling factor represented by a complex number.

From the experimental results described above, we conclude that the most suitable condition in this experiment for the use as magnetostrictive vibrator was annealing at about 400 °C for 3 h.

Acknowledgements

The authors would like to thank Asst. Prof. K. Sugimoto for his valuable advice to the experiment and thank Mr. I. Ishida for providing the magnetic fluxmeter and for his cooperation in the measurement.

This work was supported in part by a Grant-in-Aid for Scientific Research from the Ministry of Education.

References

[1] W. ROTH and S. R. RICH, J. appl. Phys. 24, 940 (1953).
[2] T. YOKOYAMA, J. Japan Inst. Metals 17, 263 (1953) (in Japanese).
[3] W. B. HUTCHINSON, F. M. C. BESAG, and C. V. HONESS, Acta metall. 21, 1685 (1973).
[4] R. M. BOZORTH, Ferromagnetism, Van Nostrand Co., 1951.
S. CHIKAZUMI, Physics of Magnetism, John Wiley & Sons, New York 1959.

(Received May 6, 1977)

phys. stat. sol. (a) 43, 399 (1977)

Subject classification: 1

Department of Physics, Erevan State University

X-Ray Spectrometers with Diffractional Focussing

By

G. M. ALADZHADZHYAN, P. A. BEZIRGANYAN, O. S. SEMERDZHYAN, and D. M. VARDANYAN

New types of X-ray spectrometres, based on the diffractional focussing phenomena are theoretically considered and proposed for (n; −n) and (n; n) arrangements of the crystals. Schemes are given for obtaining the spectrum in the spectrometres. The spectrum of the $MoK_{\alpha_{1,2}}$ and $MoK_{\beta_{1,3}}$ doublets are experimentally obtained. In the (n; −n) arrangement, for the width of the MoK_{α_1} line a value of 0.25 X-units is obtained. In the (n; n) arrangement values of 0.24 X-units and 0.23 X-units are obtained for widths of MoK_{α_1} and MoK_{β_1}, respectively.

Теоретически рассмотрены и предложены новые типы рентгеновских спектрометров, основанные на эффекте дифракционной фокусировки, при (n; −n) и (n; n) расположении кристаллов. Разработаны и приведены схемы опытов для получения спектров. Экспериментально получены спектры $MoK_{\alpha_{1,2}}$ и $MoK_{\beta_{1,3}}$ дуплетов. В расположении (n; −n) для ширины MoK_{α_1} линии получили значение 0,25 X-units. В расположении (n; n) для ширины MoK_{α_1} получили 0,24 X-units, а для MoK_{β_1} 0,23 X-units.

1. Introduction

For the first time the effect of diffractional focussing in the approximation of geometrical optics was shown in [1], when considering the problems of imaging stacking faults. In [2] the intensity at the focus is calculated in the case of successive diffraction of a narrow X-ray beam by a two-plate system. In [3] it is noted that the cases of focussing considered in [1, 2] are particular cases of a more general phenomenon which includes a wide variety of problems.

As it is shown in [2], after passing through a system consisting of two plates with equal thickness, the doubly reflected beam images the initial form of the slit, with a precision of a few microns. However, except the method of obtaining the focussing effect considered in [1 to 3], this effect can also be obtained in the case of (n; n) arrangement of crystals, which has the essential advantage with respect to (n; −n) arrangement, that it has as it were a dispersion.

In this paper a new case of diffractional focussing is considered, which is applied to the construction of an X-ray spectrometer with higher resolution. X-ray spectra are investigated, applying the phenomenon of diffractional focussing to systems of (n; −n) and (n; n) arrangements.

The comparison of the results for the widths of $MoK_{\alpha_{1,2}}$ and $MoK_{\beta_{1,3}}$ obtained by the spectrometers with (n; −n) and (n; n) arrangements shows that in the case of (n; −n) arrangements the width of the initial slit and the dynamical background in the region of the Borrmann delta affect the natural width. In the case of (n; n) arrangement, the obtained values practically are equal to the values of the natural widths of these lines.

2. General Theory of Diffractional Focussing of a System Composed of Two Identical Crystals in (n; —n) and (n; n) Arrangements in Laue Case

In [2] a particular case of a two-plate system is investigated, with (n; —n) arrangement and symmetrical reflection. Let us consider the general case of asymmetry for both arrangements, (n; —n) and (n; n).

The general solutions of the Takagi equations [5] with the Laplace integral transformation are written as follows [4]:

$$\psi_h(x, z_1) = -\frac{ikc\chi_h}{2\gamma_h} \int\limits_{c_0-i\infty}^{c_0+i\infty} dp\, \Phi_0^i(p) \frac{\exp(px - paz_1)}{\sqrt{\beta^2 - p^2b^2}} \times$$

$$\times \left[\exp\left(i\sqrt{\beta^2 - p^2b^2}\, z_1\right) - \exp\left(-i\sqrt{\beta^2 - p^2b^2}\, z_1\right)\right]; \tag{1}$$

$$\Phi_0^i(p) = \int\limits_0^\infty \exp(-px)\, \psi_0(x, 0)\, dx; \qquad \psi_0(x; 0) = \psi_0^i(x) \exp\left[-2\pi i(\boldsymbol{K}_0^i - \boldsymbol{K}_0)\, \boldsymbol{r}\right],$$

where $\psi_0^i(x)$ is the quasiamplitude, $\boldsymbol{K}_0^i$ the wave vector of the incident wave packet,

$$\beta = \frac{\pi kc(\chi_h\chi_{\bar{h}})^{1/2}}{(\gamma_0\gamma_h)^{1/2}}, \quad b = \frac{\sin 2\theta_B}{2\gamma_0\gamma_h}, \quad a = \frac{\sin(\alpha_0 - \alpha_h)}{2\gamma_0\gamma_h},$$

$\boldsymbol{K}_0$ is the wave vector of the transmitted beam. The directions of the x- and z_1-axes and the angles α_0 and α_h are described in Fig. 1. χ_h and $\chi_{\bar{h}}$ are the Fourier transforms of the crystal polarizability, θ_B is the Bragg angle, c the polarization factor, γ_0 and γ_h are the direction cosines of incident and reflected waves under the exact Bragg angle, k is the wave number of the incident wave in vacuum. As the vectors $\boldsymbol{K}_0$ and $\boldsymbol{K}_h$ originate at the Lorentz point, the parameter of misorientation is $\beta_h = 0$ [5].

a) Let us consider the (n; —n) arrangement (Fig. 2a). From the continuity of the wave functions on the exit surface of the first crystal we have

$$\psi_h^i(x) = \exp\left[-2\pi i(\boldsymbol{K}_h^i\boldsymbol{r})\right] = \psi_h(x, z_1 = d) \exp\left[-2\pi i\boldsymbol{K}_h\boldsymbol{r}\right],$$

where ψ_h^i is the quasiamplitude and $\boldsymbol{K}_h^i$ the wave vector of the wave emerging from the first crystal. Choosing the wave vector $\boldsymbol{K}_h^i$ in such a way that $(\boldsymbol{K}_{h_1} - \boldsymbol{K}_h^i) = \boldsymbol{n}A$ (the condition of phase continuity), where $\boldsymbol{n}$ is the normal to the surface $z_1 = d$ (A = const), we obtain

$$\psi_h^i(x) = \psi_h(x, z_1 = d).$$

This wave packet with an amplitude ψ_h^i is incident on the entrance surface of the second crystal (Fig. 2a). From the continuity condition of the wave functions on

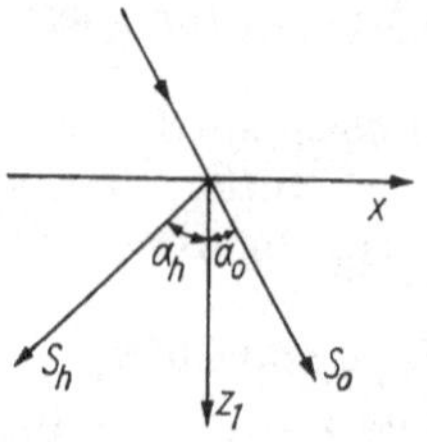

Fig. 1. Directions of the x- and z_1-axes and definition of the angles α_0 and α_h

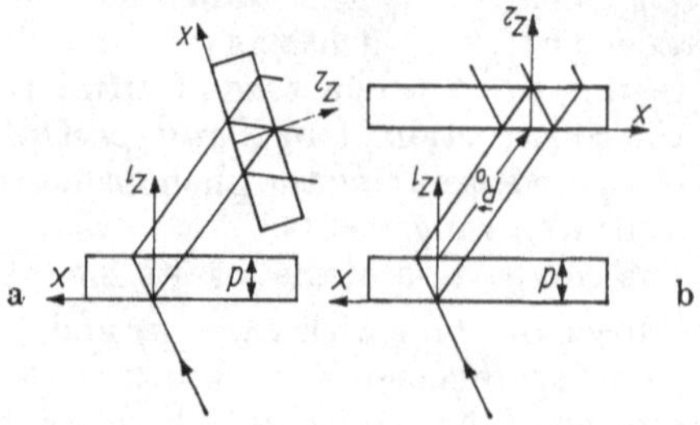

Fig. 2. Schemes for obtaining the focussing in a) (n; —n), b) (n; n) arrangement of crystals

this surface, taking into consideration the phase continuity, for the boundary conditions we have

$$\psi_h^{II}(x, z_2 = 0) = \psi_h^i(x, d) ,$$

where ψ_h^{II} is the amplitude of the twice diffracted wave. From (1) for ψ_h^{II} we obtain

$$\psi_h^{II}(x, z_2) = -\frac{ikc\chi_{\bar{h}}}{2\gamma_0} \int\limits_{c_0 - i\infty}^{c_0 + i\infty} dp\, \Phi_h^i(p) \frac{\exp(px - paz_2)}{\sqrt{\beta^2 - p^2 b^2}} \times$$
$$\times \left[\exp\left(i\sqrt{\beta^2 - p^2b^2}\, z_2\right) - \exp\left(-i\sqrt{\beta^2 - p^2b^2}\, z_2\right)\right], \qquad (2)$$

where

$$\Phi_h^i(p) = \int\limits_0^\infty \exp(-px)\, \psi_h^i(x, d)\, dx .$$

In (2) $\exp[-2\pi i(\boldsymbol{K}_h^{II} - \boldsymbol{K}_h^i)\, R_0]$ is a constant phase factor.

In the case of incidence of an infinitely narrow beam onto the first crystal, we have

$$\psi_0^i(x) = \delta(x) , \qquad \Phi_0^i(p) = 1 ,$$

$$\psi_h^i(x, z_1 = d) = \frac{\pi kc\chi_h\gamma_0}{\sin 2\theta_B} J_0\left(\frac{2\pi kc(\chi_h\chi_{\bar{h}})^{1/2}(\gamma_0\gamma_h)^{1/2}}{\sin 2\theta_B} \sqrt{(d \operatorname{tg} \alpha_0 - x)(d \operatorname{tg} \alpha_h + x)}\right) \times$$
$$\times\, \theta(d \operatorname{tg} \alpha_0 - x)\, \theta(d \operatorname{tg} \alpha_h + x) ,$$

where $J_0(x)$ is a Bessel function of zeroth order, $\theta(x) = 1$ at $x > 0$ and $\theta(x) = 0$ at $x < 0$.

Applying the convolution theorem for the Laplace transformation of the amplitude of the twice diffracted beam we have

$$\psi_h^{II}(x, z_2) = -\frac{\beta^2}{b^2} \int\limits_{-\infty}^{\infty} J_0\left(\frac{\beta}{b}\sqrt{(d \operatorname{tg} \alpha_0 - x)(d \operatorname{tg} \alpha_h + x)}\right) \times$$
$$\times\, J_0\left(\frac{\beta}{b}\sqrt{[z_2 \operatorname{tg} \alpha_0 - (x - x')][z_2 \operatorname{tg} \alpha_h + (x - x')]}\right) \times$$
$$\times\, \theta(d \operatorname{tg} \alpha_0 - x)\, \theta(d \operatorname{tg} \alpha_h + x)\, \theta[z_2 \operatorname{tg} \alpha_h + (x - x')]\, \theta[z_2 \operatorname{tg} \alpha_0 - (x - x')]\, dx' .$$

Taking this integral at $z_2 = d$ we obtain focussing around the point $x = 0$

$$\psi_h^{II}(x, d) \approx -\frac{\beta}{b} \exp\left(-\frac{\beta}{b}|x|\right). \qquad (3)$$

In the remaining parts of the twice diffracted beam the intensity is equal to

$$\left|\frac{\beta}{b|x|}\sqrt{4b^2d^2 - x^2}\, J_1\left(\frac{\beta}{b}\sqrt{4b^2d^2 - x^2}\right)\right|^2$$

(J_1 is a Bessel function of first order) which is less than the intensity of the beam near the focus point $x = 0$ by approximately two orders of magnitude. From (3) is seen that the best focussing is obtained in the symmetrical case.

Actually the width at half maximum (WHM) of the intensity for the central peak in the asymmetrical case is greater and the height is less than in the symmetrical case.

b) Let us consider the (n; n) arrangement (Fig. 2b). From the continuity condition of the wave functions on the entrance surface of the second crystal we have for the boundary conditions.

$$\psi_{\mathrm{h}}^{\mathrm{II}}(x, z_2 = 0) = \psi_{\mathrm{h}}^{\mathrm{i}}(x, d) \exp\left[-2\pi i(\boldsymbol{K}_{\mathrm{h}}^{\mathrm{II}} - \boldsymbol{K}_{\mathrm{h}}^{\mathrm{i}})\,\boldsymbol{r}\right].$$

In this case the phase factor $\exp(i\alpha x)$ appears under the integral, where α is the relative misorientation parameter of the two crystals. However, if the second crystal is set under the exact Bragg angle, we obtain

$$(\boldsymbol{K}_{\mathrm{h}}^{\mathrm{II}} - \boldsymbol{K}_{\mathrm{h}}^{\mathrm{i}})\,\boldsymbol{r} = (\boldsymbol{K}_{\mathrm{h}}^{\mathrm{II}} - \boldsymbol{K}_{\mathrm{h}}^{\mathrm{i}})\,(\boldsymbol{R}_0 + \boldsymbol{r}'),$$

where $\boldsymbol{r}'$ is the radius vector of a point on the $z_2 = 0$ surface. If the second crystal is set at the exact angle of reflection, the origin of the wave vector $\boldsymbol{K}_{\mathrm{h}}^{\mathrm{i}}$ is located at the Laue point La (Fig. 3), and the origin of the wave vector $\boldsymbol{K}_{\mathrm{h}}^{\mathrm{II}}$ is located at the Lorentz point $\mathrm{Lo^{II}}$, so that $\mathrm{LaLo^{II}}$ is perpendicular to the entrance surface of the second crystal:

$$(\boldsymbol{K}_{\mathrm{h}}^{\mathrm{II}} - \boldsymbol{K}_{\mathrm{h}}^{\mathrm{i}})\,(\boldsymbol{R}_0 + \boldsymbol{r}') = (\boldsymbol{K}_{\mathrm{h}}^{\mathrm{II}} - \boldsymbol{K}_{\mathrm{h}}^{\mathrm{i}})\,\boldsymbol{R}_0\,. \tag{4}$$

Taking into account (4), the determination of $\psi_{\mathrm{h}}^{\mathrm{II}}$ leads to the same result as in the first case, in (n; —n) arrangement. It is not difficult to see that the same condition (4) would be obtained in the asymmetrical case.

Thus, the problem of diffractional focussing for the general cases of asymmetry in (n; n) and (n; —n) arrangements of crystals shows that

1. diffractional focussing is also obtained in the case of (n; n) arrangement,
2. as in the case of (n; —n), also in (n; n) arrangement a better focussing is obtained in case of symmetrical reflections.

3. Spectrometers with Diffractional Focussing

As we have noticed, by diffractional focussing the intensity is concentrated in a small spatial region. Apparently, this effect can be used for the construction of X-ray spectrometers of high resolution.

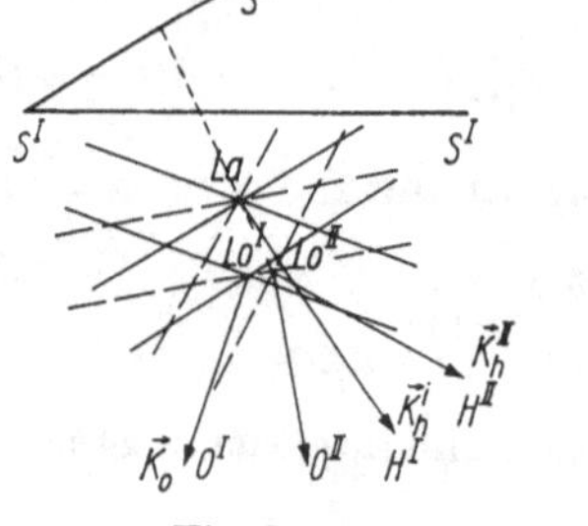

Fig. 3

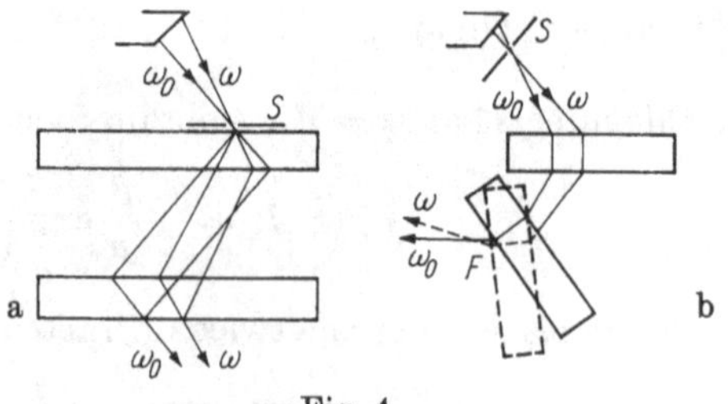

Fig. 4

Fig. 3. Laue (La) and Lorentz (Lo) points and directions of the wave vectors $\boldsymbol{K}_{\mathrm{h}}^{\mathrm{i}}$, $\boldsymbol{K}_{\mathrm{h}}^{\mathrm{II}}$, $\boldsymbol{K}_0$ in the reciprocal lattice. $\mathrm{SS^I}$ and $\mathrm{SS^{II}}$ are the traces of the entrance surfaces of the first and the second crystals, respectively

Fig. 4. Schemes of the experiments for obtaining the spectrum in a) (n; —n), b) (n; n) arrangement of the crystals. Distance between the slit and the first axis is 200 mm

Indeed, maintaining the conditions of focussing for two identical crystals in Laue reflection, we can obtain a higher resolution, which is the main purpose of the second part of this study.

3.1 A two-crystal spectrometer with (n; —n) arrangement of crystals in the Laue case

The scheme of the experiment for obtaining a spectrum is shown in Fig. 4a. An X-ray beam from a wide-focus source is incident on the crystal analyser of the spectrometer through a narrow slit, which is placed on the entrance surface of this crystal. The crystal analyser is a two-plate diffraction lens. Each line of the spectrum on the exit surface of the crystal analyser being focussed forms a thin line which is parallel to the slit S. As a result, on a photographic plate which is set perpendicular to the emerging beam the spectrum of the primary X-ray beam is formed.

Two non-coherent lines of the spectrum, which differ from each other by their wavelength difference, $d\lambda$ are separated on the photoplate by a distance dx.

From the Bragg condition we obtain [6]

$$dx = (D - 2M)\frac{d\lambda}{\lambda} \operatorname{tg} \theta_B (1 + \operatorname{tg}^2 \theta_B),$$

where D is the distance form the slit to the photoplate and M is the distance between the plates of the analyser.

As to each point of the dispersion curve of the primary quasi-monochromatic wave on the exit of the diffraction lens corresponds a certain spacial distribution $\psi_h^{II} = -\frac{\beta}{b}\exp\left(-\frac{\beta}{b}|x|\right)$, to find out the real shape of the distribution of the intensity of the quasicoherent parts of the spectrum on the photoplate we must takethe integral $\psi_h^{II}(\omega)$ along the dispersion contour of the spectral line.

The dispersion contour of the primary wave, on account of the damping of the source by $\exp(-\gamma t)$, is determined by the expression

$$\exp[i(\omega_0 - \gamma)t]\,\theta(t) = \frac{1}{2\pi i}\int_{-\infty}^{\infty} \frac{\exp(i\omega t)}{\omega - (\omega_0 + i\gamma)}\,d\omega .$$

Taking this into account, for the intensity on the exit from the two-plate system we obtain

$$|E_h|^2 = \left|\frac{1}{2\pi i}\int_{-\infty}^{\infty} \psi_h^{II}(\omega)\,\frac{\exp(i\omega t)}{\omega - (\omega_0 + i\gamma)}\,d\omega\right|^2 . \tag{5}$$

We assume a wave with dispersion and central frequency ω_0 to be focussed at the point $x = 0$. From (3) it is easy to derive

$$\psi_h^{II}(\omega) = \frac{B}{\sqrt{\omega^2 - \omega_0^2 \sin^2 \theta_0}}\exp\left(-\frac{B\,|x|}{\sqrt{\omega^2 - \omega_0^2\sin^2\theta_0}}\right),$$

where

$$B = \pi k_0 c_0 \omega_0 (\chi_{h0}\chi_{\bar{h}0})^{1/2} .$$

Here the parameters corresponding to the frequency ω_0 are denoted by the index 0. As it is seen from the above expression, in case of $|x| \leq \Lambda$ (Λ being the extinction depth) $\psi_h^{II}(\omega)$ is much less dependent on ω than the function $1/[\omega - (\omega_0 + i\gamma)]$. $\psi_h^{II}(\omega)$

has an essential input in the case of $|x| > 3\Lambda$. But in these points the intensity of the focus is practically equal to zero. Excluding the practically meaningless case of $\theta_B \approx \approx \pi/2$, when the special point $\omega = \pm\omega_0 \sin\theta_0$ of the function $\psi_h^{II}(\omega)$ remains outside the boundaries of integration, one can take the function $\psi_h^{II}(\omega)$ out of the integral (5). This means that the coherent lines of the spectrum are added on the exit surface of the crystal analyser non-coherently. On the photoplate, the obtained size corresponding to the WHM 2γ increases by Δ, where Δ is the WHM of the line of the focus.

For a Si interferometer in the case of the $MoK_{\alpha_1}(220)$ reflection $\Delta = 1.4\,\mu m$. Choosing $D - 2M = 400$ mm, to the natural width 2γ of MoK_{α_1} corresponds $dx \approx 58\,\mu m$, that is to say diffractional widening Δ practically does not have any affect on the natural shape of the spectral line.

In our experiment the dimensions of the projection of the focus of the X-ray source are $0.4 \times 0.4\,mm^2$. The width of the slit $S = 5\,\mu m$.

We have used MoK_α and MoK_β radiation. The two-plate system is prepared from a dislocation-free Si single crystal. The large surface of the crystal is perpendicular to the growth axis [111]; ($2\bar{2}0$) reflections are investigated. After the mechanical and chemical treatment of the specimen, the thicknesses of the plates were equal to each other and were $(450 \pm 3)\,\mu m$. Photoplates MK with a resolution of $3.6\,\mu m$ are used.

As the difference of the thicknesses of the plates and the misorientation lead to an increase of the WHM of the focus lines [2], we had to check the quality of the focussing effect of the system. For this purpose we photographed a topograph of the double-reflected beam with monochromatic radiation.

The spectrum of the $MoK_{\alpha_{1,2}}$ doublet, taken by this method, is shown in Fig. 5. Distinct maxima of K_{α_1} and K_{α_2} lines are accompanied by the background, which is the result of an overlapping of the dynamic background

$$\frac{\beta}{b\,|x|}\sqrt{4b^2d^2 - x^2}\,J_1\left(\frac{\beta}{b}\sqrt{4b^2d^2 - x^2}\right)$$

in the region of the Borrmann delta. For the WHM of the MoK_{α_1} line a value of 0.25 X-units is obtained. Note, that for this WHM a spectrometer with a bent crystal gives a value of 0.26 to 0.28 X-units [7] and a double-crystal spectrometer in the Bragg case gives a value of 0.24 X-units [8].

The WHM of a spectral line is strictly dependent on the slit width S. As it is shown above, the distance on the photoplate, corresponding to the natural width of the spectral line, increases by $S + \Delta$. Hence, with the increase of the slit width r the resolution of the spectrometer quickly decreases.

3.2 A Two-crystal spectrometer with (n; n) arrangement, in the Laue case

Though a spectrometer of diffractional focussing in the case of (n; −n) has a high resolving power, it has also some disadvantages. These disadvantages are 1. a complicated "mathematical" calculation for the subtraction of the shape of the background to obtain the real shape of the dispersion distribution and 2. the absence of

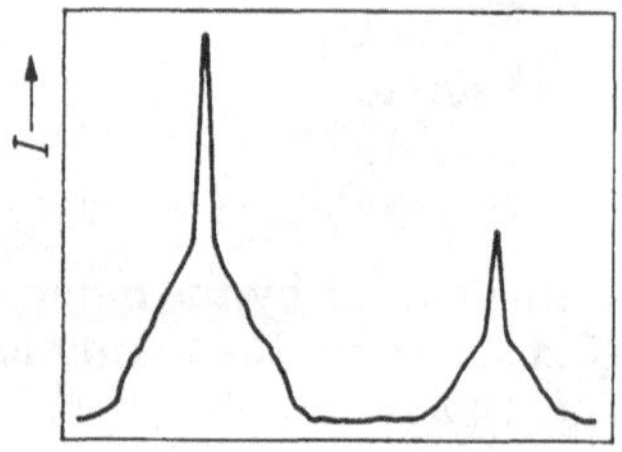

Fig. 5. Microphotometrical record of the intensity of the $MoK_{\alpha_{1,2}}$ doublet obtained by the scheme Fig. 4

dispersion, as a result of which one cannot obtain a resolution equal to unity in the case of (n; n) arrangement.

The scheme of the experiment is shown in Fig. 4b. From a narrow slit S a divergent X-ray beam is incident on the first crystal, which is adjusted on the first axis of a two-axis goniometer. Waves of different length are reflected under different angles from the first crystal and hit the second crystal which is adjusted on the second axis of the goniometer. Let a ray of frequency ω_0 be reflected from the two crystals under an angle θ_0 and pass through both axes of the goniometer. Let another ray of frequency ω be reflected from the first crystal under the Bragg angle θ_1, but not pass through the first axis. For this ray to be reflected fom the second crystal, the latter must be rotated until the reflecting planes make an angle θ_1 with the incident ray. The rays reflected under angles θ_1 and θ_0, if the second crystal is rocked, will intersect at a point F. Denoting the distance from the slit to the first axis as ξ and from the first axis to the second one as η, the distance from the second axis to the point F will be

$$R = \frac{\xi - \eta}{3} - \frac{4}{3} \frac{(\xi - \eta)\, \Delta\theta \sin 2\theta_0}{1 - 4 \sin^2 \theta_0}, \qquad (6)$$

where $\Delta\theta = \theta_0 - \theta_1$. In obtaining (6) we neglected the terms containing $(\Delta\theta)^2$.

The consideration of the conditions of focussing in the case of a misorientation of the crystals is of great significance. If the second crystal is set under an exact reflection position for a wave of frequency ω_0, then it will be in a misoriented position for a wave of frequency ω. The intensity of focussing quickly decreases by the increase of the misorientation. For instance, for a $MoK_\alpha(220)$ reflection the rotation of the reflecting planes by 0.04″ decreases the intensity at the focus practically to zero [2]. In the Bragg case reflection the value of this rotation angle is equal to the width of the Darwin table, which is larger by about two orders of magnitude.

This is one of the important advantages of a diffractional focussing spectrometer.

In our experiment the window of the detector is arranged at a distance $(\xi - \eta)/3 = 10$ mm.

The dimensions of the X-ray source, the slit width, and the crystals are the same as in the case of (n; —n) arrangement. The spectra were obtained by rocking the second crystal around the second axis. The scheme of the experiment gives a possibility to get rid of the dynamic background (cf. Section 3.1). Except the line of the focus, for each position of the second crystal a dynamic background also exists in the region of the Borrmann delta for the specific wavelength. So, as during rocking, the second crystal beams of different wavelengths pass through the point F with an error of the order of $(\Delta\theta)^2$, choosing the width of the window of the detector

$$F_W = S + \Delta + \frac{4}{3} \frac{(\xi - \eta)\, (\Delta\theta)_m^2 \sin 2\theta_0}{1 - 4 \sin^2 \theta_0}, \qquad (7)$$

one can mechanically get rid of the influence of the background in the region $4d \operatorname{tg} \theta_0$ of the Borrmann delta. In (7) $(\Delta\theta)_m$ is the angular region of rocking of the second crystal.

For our specimen $4d \operatorname{tg} \theta = 360\ \mu m$. For the $MoK_{\alpha_{1,2}}$ doublet the last term in (7) is equal to $8 \times 10^{-2}\ \mu m$, and for the $MoK_{\beta_{1,3}}$ doublet $5.6 \times 10^{-6}\ \mu m$, so that the slit width of the detector was chosen $F_W = 20\ \mu m$. In this case for the WHM of the MoK_{α_1} line we have a value of $0.24U$. The spectrometer clearly resolved the lines K_{β_1} and K_{β_3}, which are very difficultly resolved by usual spectrometres. In Fig. 6 the spectrum of the $MoK_{\beta_{1,3}}$ doublet is shown, where the WHM of the K_{β_1} line is $0.23U$.

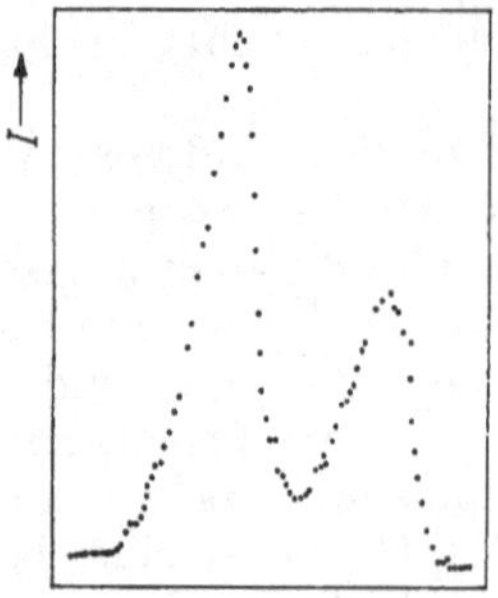

Fig. 6. Spectrum of the $MoK_{\beta_{1,3}}$ doublet obtained in (n; n) arrangement of the crystals with diffractional focussing

The resolving power of the sepctrometer can be increased, if the primary beam is collimated by narrow slits. To obtain the spectrum, one must rotate the crystals around their axes with a velocity $\dot{\theta}$, rotate the second axis of the goniometer around the first axis with a velocity $2\dot{\theta}$, and at the same time rotate the detector around the second axis with a velocity $\dot{\theta}$. In this case the slit width must be chosen as $F_W = S + \Delta$.

4. Conclusion

Thus, we succeeded in applying the phenomenon of diffractional focussing to the construction of a spectrometer with high resolution.

For the spectral width of the MoK_{α_1} line in (n; n) arrangement 0.24 X-units is obtained, while a usual two-crystal Bragg spectrometer in (n; n) arrangement gives 0.29 X-units.

This is due to the spatial width of a few ten microns of the reflected beam in case of Bragg reflection from a crystal of an infinitely narrow primary beam. In case of a spectrometer with diffractional focussing the emerging beam has a width of the order of 1 μm.

Note that diffractional focussing can be obtained also from different reflecting planes, or from the same planes with different orders of reflections. In this case, it is clear that the thicknesses of the plates must be different, but chosen in such a way, as to satisfy the conditions of focussing.

Acknowledgements

We wish to thank A. Rostomyan and K. Truni for their advices of great value in carrying out the present research.

References

[1] A. AUTHIER, A. D. MILNE, and M. SAUVAGE, phys. stat. sol. **26**, 469 (1968).
[2] V. L. INDENBOM, I. SH. SLOBODETSKII, and K. G. TRUNI, Zh. eksper. teor. Fiz. **66**, 1110 (1974).
V. L. INDENBOM, E. V. SUVOROV, and I. SH. SLOBODETAKII, Zh. eksper. teor. Fiz. **71**, 359 (1976).
[3] P. A. BEZIRGANYAN, Fiz. tverd. Tela **5**, 21 (1976).
[4] K. G. TRUNI, D. M. VARDANYAN, L. V. LEVONYAN, and P. A. BEZIRGANYAN, Fiz. tverd. Tela **1**, 20 (1973)
[5] S. TAKAGI, Acta cryst. **15**, 1131 (1962).
S. TAKAGI, J. Phys. Soc. Japan **26**, 1239 (1969).
[6] P. A. BEZIRGANYAN, G. M. ALADZHADZHYAN, and K. G. TRUNI, Izv. Akad. Nauk Armenian SSR **10**, 449 (1975).
V. L. INDENBOM, G. M. ALADZHADZHYAN, Dokl. Akad. Nauk SSSR **227**, 827 (1976).
[7] B. G. GOKHAL, Ann. Phys. (France) **7**, 852 (1952).
[8] L. G. PARRAT, Rev. mod. Phys. **31**, 616 (1959).

(Received April 21, 1977)

phys. stat. sol. (a) 43, 407 (1977)

Subject classification: 13.4 and 14.3; 10.2; 22.1.2

Institute of Solid State Physics, Academy of Sciences of the USSR, Chernogolovka

Investigation of the Energy Spectrum and Kinetic Phenomena in Dislocated Si Crystals (I)

By

V. A. Grazhulis, V. V. Kveder, and V. Yu. Mukhina

An investigation is made of the dc conductivity and the Hall effect in heavily dislocated Si crystals doped with boron (10^{13} cm^{-3}) and phosphorus (2×10^{14} cm^{-3}). In such crystals two levels are shown to exist, namely, donor-acceptor level with $E_1 = (0.42 \pm 0.03)$ eV above the valence band and acceptor level with $E_2 = (0.43 \pm 0.03)$ eV below the conduction band. Both E_1- and E_2-levels are assumed to be due to dislocations and to correspond to states of electrons localized in cores of dislocations with edge components.

Настоящая работа посвящена исследованию проводимости и эффекта Холла сильно деформированных кристаллов Si, с высокой плотностью дислокаций, легированных бором (10^{13} cm^{-3}) и фосфором (2×10^{14} cm^{-3}). Как показано, в таких кристаллах существует два уровня, а именно, донорно-акцепторный уровень $E_1 = (0{,}42 \pm 0{,}03)$ eV выше валентной зоны и акцепторный уровень $E_2 = (0{,}43 \pm \pm 0{,}03)$ eV ниже зоны проводимости. Предполагается, что как уровень E_1, так и уровень E_2 обусловлены дислокациями и соответствуют состояниям электронов, локализованных в ядрах дислокаций с краевыми компонентами.

1. Introduction

Despite the numerous experimental data published [1 to 20], up to now there has been no common idea on the effects of dislocations on the energy spectrum and kinetic properties of covalent semiconductor crystals. Therefore, investigations of the behaviour of these crystals with dislocations are still urgent.

Si single crystals should be perspective in this respect since the ESR signal due to dislocations observed in these crystals provides additional important information on electronic states on dislocations [11 to 14]. Apart from this, because of the recently revealed high microwave conductivity along dislocations in Si [15], investigations of the energy spectrum and kinetic phenomena in the said crystals become a still more burning problem, because this conductivity, probably, demonstrates some features of one-dimensionality — the mechanism of this conductivity is to be elucidated. To solve this problem one has, first of all, to know the structure of the energy spectrum of electrons on dislocations and some other dislocation parameters important for the physics of one-dimensional systems, especially for understanding the conductivity mechanism in such systems.

In the present work, an attempt was made to investigate dislocated Si single crystals by employing different experimental measurements, namely, dc conductivity and Hall effect measurements, microwave conductivity measurements, and measurements of photoconductivity and magnetoresistance in the microwave range. The same specimens were investigated by different methods in order to obtain a fuller information on their properties. Crystals of n-type containing 2×10^{14} cm^{-3} phosphorus and p-type containing $\approx 10^{13}$ cm^{-3} boron were used.

The crystals were grown by zone melting in vacuum. The specimens were prepared by a standard technique described in [10]. The dislocations were introduced by compression of the crystals along [110] at 700 °C in argon atmosphere. The deformation curves of the investigated specimens are given in Fig. 1.

In the present paper we report on the results obtained only from the dc Hall effect and dc conductivity measurements. The results obtained in the microwave range are forwarded to be published as Part II [23] of this paper.

2. Experimental Results

In the Hall effect measurements the magnetic field was 10 to 20 kOe, the Hall factor was assumed to be unity.

Crystals with high dislocation density were predominantly studied: the number of dangling bonds in dislocation cores, N_D, for these crystals greatly exceeded the number of chemical donors, N_d, or acceptors, N_a. The magnitude of N_D reached 6×10^{16} cm^{-3}. The concentration of dangling bonds was estimated by measuring the integral intensity of the dislocation ESR spectrum [11 to 14] at $T = 77$ K within an accuracy of $\approx 40\%$. The accuracy was sufficient for the purposes of the present work. Note, that the conventional electron microscope method that can be, of course, used for these crystals allows one to determine dislocation density and, hence, the number of dangling bonds only to the order of magnitude.

The first item to be discussed here is that of the properties of plastically deformed p-type Si single crystals.

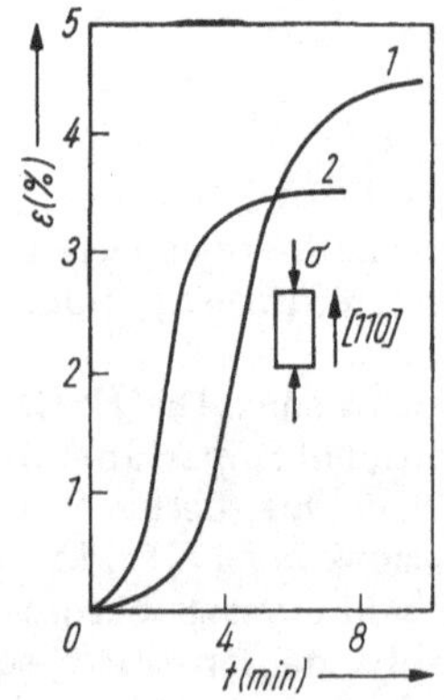

Fig. 1

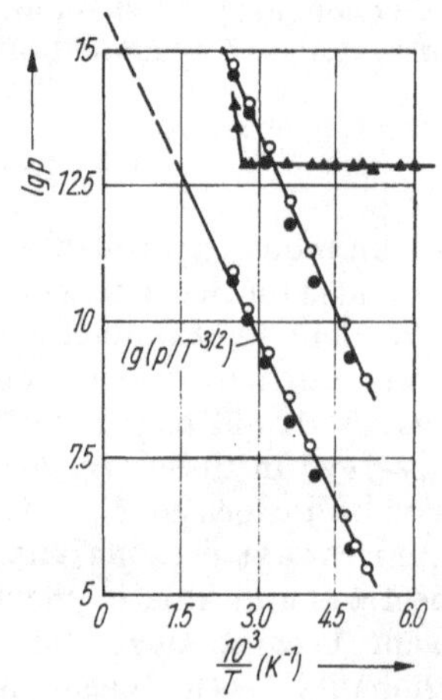

Fig. 2

Fig. 1. Typical deformation curves of the specimen ($\varepsilon = \Delta l/l$ relative length change) at 700 °C. t is the time of deformation, σ the deformation stress; during deformation σ is constant. (1) crystals doped with phosphorus (2×10^{14} cm^{-3}), $\sigma = 94$ N/mm^2; (2) crystals doped with boron (10^{13} cm^{-3}), $\sigma = 113.8$ N/mm^2

Fig. 2. Free hole concentration p as a function of inverse temperature. ▲ original p-type crystal doped with boron ($\approx 10^{13}$ cm^{-3}); ○ deformed specimen, $\varepsilon = 3.5\%$ (see Fig. 1), $N_D \approx 4 \times 10^{16}$ cm^{-3}; ● specimen with $\varepsilon = 3.5\%$ subjected to additional annealing after deformation for 1 h at 500 °C in an argon atmosphere. The lower straight line shows the value lg $(p/T^{3/2})$ as a function of the inverse temperature. The slope gives the activation energy 0.42 eV

Fig. 2 presents the temperature dependence of free hole concentration for the original specimen and for a strongly deformed crystal. As is seen from the figure, a new energy level (that may be related to dislocations) with the activation energy (0.42 ± 0.03) eV is observed in deformed crystals. The hole mobility μ (Fig. 3) is decreased with increasing dislocation density and is practically temperature-independent within the investigated temperature range in deformed crystals.

Annealing the crystal at 500 °C for 1 h does not result in a marked change in the activation energy and hole concentration (Fig. 2). In more detail, the effect of annealing and deformation temperatures on the properties of Si single crystals was studied elsewhere [13, 16, 17]. One of the major conclusions of these works is that at deformation temperatures $T \lessapprox 700$ °C the properties of heavily deformed Si crystals are predominantly determined by "pure" dislocations, and at $T \gtrapprox 800$ °C agglomerates of point defects and impurities on dislocations may considerably contribute to crystal properties.

Note, that when recording the ESR spectrum of our deformed specimens, a signal from dislocations alone was observed. No signal from other centres was present. This fact and also annealing data allow us to assume that in our case the predominant paramagnetic and electrically active centres are dislocations, and that dislocations are mainly responsible for the electronic properties of deformed crystals.

Actually, the results of Fig. 2 are peculiar for dislocations and may be interpreted as follows.

In the range of "low" temperatures, $10^3/T \gtrapprox 3.3$, dislocations act as donors and compensate chemical acceptors; this results in a decrease of the free hole concentration as compared with the original crystal. In this state dislocations are positively charged. At increasing T, due to thermally activated transitions of electrons from the valence band to dislocations, the positive charge of the latter is decreased and at $10^3/T \approx 3.3$ it equals zero.

In the $10^3/T \lessapprox 3.3$ range dislocations accept some additional electrons (become negatively charged) which means they act as acceptors increasing the hole concentration in the valence band as compared with the concentration in undeformed crystals. Consequently, we observe donor–acceptor activity of dislocations in p-type crystals.

In the whole temperature range the free hole concentration (within the accuracy of measurements) is quite well discribed by $p = N_v \exp(E_1/kT)$, where N_v is the effective number of states in the valence band, E_1 the donor–acceptor dislocation level; the energy is counted from the top of the valence band. (Let us note, that in this case the Fermi level $E_F \approx E_1$; some additional information concerning the position of E_F will be given below.) In fact, in the effective mass approximation we have [22] $N_v \approx \approx 4.83 \times 10^{15}(m_v/m_0)^{3/2}\, T^{3/2}$, therefore extrapolation to the ordinate axis of the dependence $\lg(p/T^{3/2})$ must yield the value $\lg(N_v/T^{3/2})$. If the split-off valence subband is ignored, the effective density-of-states mass m_v for Si equals 0.6 [22] and $\lg(N_v/T^{3/2}) \approx 15.3$. In Fig. 2 the dependence $\lg(p/T^{3/2}) = F(1/T)$ is also plotted. The point of intersection of the line $F(1/T)$ with the ordinate axis corresponds to the

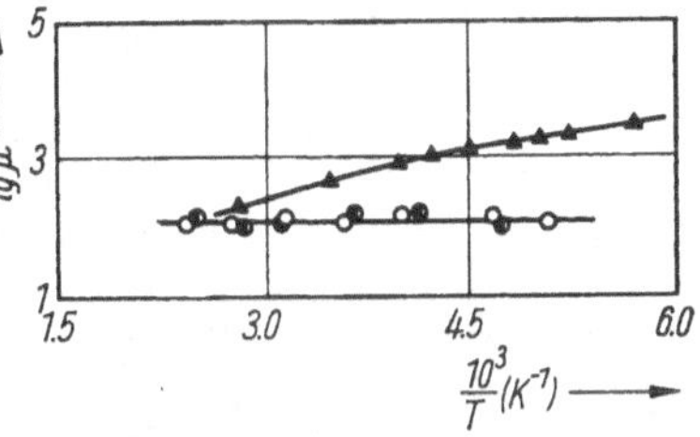

Fig. 3. Mobility μ of free holes as a function of inverse temperature in logarithmic scale. ▲ original crystal doped with boron ($\approx 10^{13}$ cm^{-3}); ○ deformed specimen, $\varepsilon = 3.5\%$ (see Fig. 1); ● deformed specimen ($\varepsilon = 3.5\%$) subjected to additional annealing (see Fig. 2)

value lg $(N_v/T^{3/2}) \approx 15.5$ which satisfactorily agrees with the value 15.3. The slope of lg $(p/T^{3/2})$ gives the activation energy $E_1 = 0.42$ eV. Note, that the upper straight line in Fig. 2 has a somewhat larger slope corresponding to 0.44 eV [18].

Thus, in the case of heavy concentration of dislocation dangling bonds, $N_D \gg N_a$, in the range $T \lesseqgtr 400$ K we obtain the usual Boltzmann statistics and one deep donor–acceptor dislocation level E_1 independent of the number of dislocations and temperature. The latter is due to the fact that throughout the investigated range the number of electrons or holes accepted by dislocations is small as compared with the number of dangling bonds. Therefore the linear charge density (or the occupation ratio f [1]) is small and the effect of the Coulomb interaction between the trapped carriers is negligible [1]. In fact, we have $f \lesseqgtr 10^{-2}$ for all T (see Fig. 3).

Let us come back to the question concerning the position of E_F. The Fermi level position can be obtained from the condition of electrical neutrality of the crystal. It can easily be shown that, in our case, when $N_D \gg N_a$ and $f \ll 1$, the Fermi level E_F practically coincides with E_1 throughout the investigated temperature range. Due to this fact, the free hole concentration (that, in general, is given as $p = N_v \exp [(E_v - E_F)/kT]$, we assume $E_v = 0$) is determined by the effective density of states in the valence band and is independent of the number of dangling bonds. The experiments also showed that as long as $N_D \gg N_a$, the free hole concentration p may be considered, within logarithmic accuracy, independent of the dislocation density, that is of the N_D-value.

Now, let us consider the behaviour of Si crystals doped with a donor impurity. On the basis of the available data predominantly obtained in Ge, one may expect that dislocations will act as acceptors. We did observe this effect and, besides, some additional important peculiarities in the behaviour of dislocations in n-Si crystals which, to our knowledge, were neither observed in Si [2 to 4], nor in Ge [5 to 7].

Fig. 4 presents the temperature dependences of the free carrier concentrations for n- and p-type crystals after a heavy dislocation density has been introduced, so that $N_D \gg N_a, N_d$. The dangling bond concentration for these crystals is approximately equal, its value is (4 to 6) $\times 10^{16}$ cm^{-3}. As seen from Fig. 4, all crystals turned out to be p-type after introducing the dislocations, the free hole concentration being similarly dependent on temperature: $p = N_v \exp (E_1/kT)$, where $E_1 = (0.42 + 0.03)$ eV, as it was previously.

The n–p transition of the crystals appears because of the fact that chemical donors turn out to be fully compensated due to the acceptor behaviour of some levels with

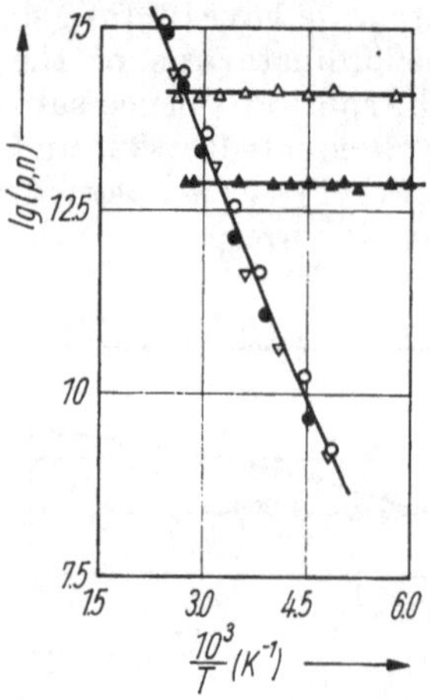

Fig. 4. Free carrier concentration as a function of inverse temperature for the original crystals and crystals with high dislocation density. ▲ free hole concentration for the original crystal with B content $\approx 10^{13}$ cm^{-3}; ○ deformed specimen with content $\approx 10^{13}$ cm^{-3}, $\varepsilon = 3.5\%$, $N_D \approx$ (4 to 6) 10^{16} cm^{-3}; ● the same specimen after annealing (see Fig. 3); △ free electron concentration of the original crystal with phosphorus content 2×10^{14} cm^{-3}; ▽ free hole concentration in deformed specimen doped with phosphorus, $\varepsilon = 4.1\%$, $N_D \approx$ (4 to 6) $\times 10^{16}$ cm^{-3}. The specimen was transformed to p-type

energy E_1. The number of acceptor-type levels (N_{D_1}) in our case is greater than the number of chemical donors and therefore these levels are still able to accept additional electrons from the valence band.

The results of Fig. 4 suggest the following conclusion: whatever the doping type, at $N_D \gg N_a, N_d$ and $N_{D_1} > N_d$ (low occupation ratios f) dislocations demonstrate donor–acceptor activity with the energy level E_1. In this case the crystals turn out to be always p-type ($E_F \approx E_1$) and the free hole concentration is independent of dislocation density; it is only determined by the effective density of states in the valence band and by the E_1-level position.

Let us consider, additionally, the behaviour of n-specimens at different dislocation densities in the case of phosphorous-doped Si.

We have investigated specimens with $\varepsilon = 1.0, 1.4, 1.6, 1.8, 1.9, 2.5, 4.1\%$ (see Fig. 1). It was established that during deformation of phosphorus-doped crytals the specimens become p-type at $\varepsilon \gtrapprox 1.8$ and remain n-type at $\lessapprox 1.8\%$.

The specimens with $\varepsilon \approx 1.8\%$ are "critical" that is the Hall effect is small and difficult to measure. (Note, that in these specimens we have revealed the effect of electric conductivity oscillations arising at certain threshold voltages [21].)

Fig. 5 presents the temperature dependence of electron concentration in the conduction band for two crystals containing an amount of dislocations sufficient to fully compensate 2×10^{14} donors, but insufficient for an n–p transition in these crystals.

As is seen from the figure, such dislocated crystals retain the conductivity type, but the free electron concentration is sharply decreased as compared with the original

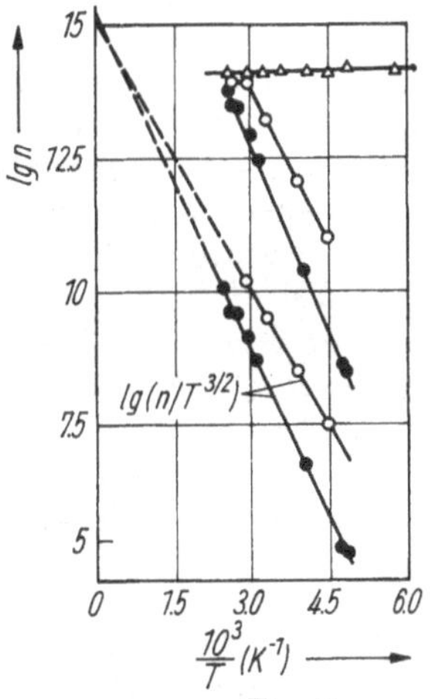

Fig. 5

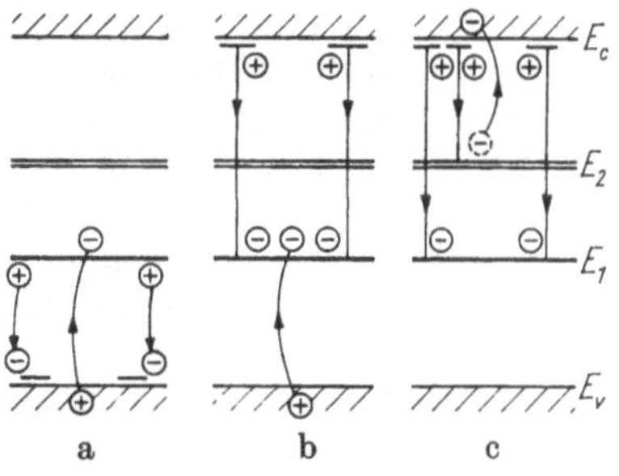

Fig. 6

Fig. 5. Free electron concentration as a function of inverse temperature. △ original crystal with phosphorus content 2×10^{14} cm^{-3}; ○ deformed specimen with $\varepsilon = 1.4\%$; ● specimen with $\varepsilon = 1.6\%$ (see Fig. 1, curve 1)

Fig. 6. Energy level scheme for dislocated Si crystals: a) Crystals containing shallow acceptor chemical impurity; arrow upwards shows electron transition from the valence band onto dislocation due to thermal activation; $E_F \approx E_1$, $f_2 \approx 0$. b) Crystal doped with donor impurity but transformed to p-type due to the acceptor behaviour of E_1-levels of the dislocations; arrows downwards show trapping of donor electrons, arrow upwards shows thermal transition of electrons from valence band to E_1-levels; $E_F \approx E_1$, $f_2 \approx 0$. c) Crystal doped with donor impurity which is fully compensated due to capturing the electrons on E_1 and E_2 levels; arrows downwards show acceptor compensation, arrow upwards shows thermal transition of electrons tc conduction band; $E_F \approx E_2$, $f_1 = f_{1\,max}$, $f_2 \neq 0$

crystal, and is exponentially dependent on temperature in the low-temperature range, $n = N^* \exp(-E_2/kT)$. The activation energy is somewhat decreased with decreasing dislocation density. For a specimen with $\varepsilon = 1.6\%$ we have $E_2 \approx 0.43$ eV, for a specimen with $\varepsilon = 1.4\%$, $E_2 \approx 0.4$ eV. It should be emphasized that the energy is now counted from the bottom of the conduction band and that the relation between E_F and E_2 is more complicated in this case than that between E_F and E_1 in p-type crystals. Due to this fact the pre-exponential factor N^* markedly differs from N_c, the effective density of states in the conduction band. In more detail, this problem will be discussed later.

Now, which are the occupation ratio f for these crystals and the Coulomb interaction in this case? The estimates show that for these specimens $f \lessapprox 3 \times 10^{-2}$, therefore it can be said that a principally new acceptor level E_2 is observed in these crystals. This level is not due to the upward shift of the E_1 acceptor level because of the Coulomb interaction. The latter is too weak to explain the position of E_2. In fact, since small occupation ratios are deals with, the energy of the Coulomb interaction can be estimated (to the order of magnitude) as the energy of two neighbouring electrons on dislocations. Therefore $E_{\text{Coul}} \sim e^2 f/\varkappa a$, where a is the lattice constant ($a \approx \approx 5$ Å) and $\varkappa$ the electric permittivity ($\varkappa \approx 10$). Accordingly, at $f \approx 10^{-2}$ one can obtain $E_{\text{Coul}} \approx 10^{-2}$ eV and therefore, $E_{\text{Coul}} \ll E_1$ and $E_{\text{Coul}} \ll E_2 - E_1$; we have $E_2 - E_1 \approx 0.27$ eV. The order of magnitude of E_{Coul} is just the one needed to explain a small difference in the energies E_2 for the specimens with $\varepsilon = 1.4$ and 1.6%. So, the level system shown in Fig. 6 should be introduced. Using this level scheme two occupation ratios for electrons can be determined: f_1 the occupation ratio for the E_1-level, and f_2 the occupation ratio for the E_2-level. By assuming $f_1 = \Delta n_1/N_D$ and $f_2 = \Delta n_2/N_D$ (Δn_1, Δn_2 are the numbers of electrons captured on the levels E_1 and E_2, respectively) we obtain that the total occupation ratio for dislocations is $f = f_1 + f_2$.

One of the most important results of the present work is the following: the magnitude of f_1 for electrons has a limited value $f_{1\,\text{max}} \ll f_{\text{max}}$, where f_{max} is the maximum total occupation ratio for electrons captured by dislocations which, due to the Coulomb interaction, is usually of the order of 0.1 to 0.2 [1, 13].

Actually, for the sample with $\varepsilon = 1.4\%$ (Fig. 6) the curve $\lg n(1/T)$ has a bent at $10^3/T \approx 3.0$. The "plateau" corresponds to the case, when the E_2 acceptor level is empty and the free electron concentration is $n_0 - \Delta n_1$, where n_0 is the concentration of free electrons in the original crystal. The values of Δn_1 and $f_1 = \Delta n_1/N_D$ can easily be determined from Fig. 6. The f_1-value, in this case, corresponds to $f_{1\,\text{max}}$ the latter turns out to be $\approx 2 \times 10^{-2}$.

Apart from this, by comparing the total occupation ratios f for the specimens with $\varepsilon < 1.8$ and $> 1.8\%$, we arrive at the conclusion that the n–p transition with increasing dislocation density occurs at $f \approx 10^{-2}$, that is at $f \approx f_{1\,\text{max}}$. So, for the crystals close to the "critical" deformation, one can assume $f_1 \approx f_{1\,\text{max}} \approx 2 \times 10^{-2}$, and $f_2 \approx 0$. Due to the low value of $f_{1\,\text{max}}$ the n–p transition of Si crystals occurs at high dislocation densities only, when $N_D \approx 10^2 N_d$. The latter circumstance sharply distinguishes dislocations from chemical acceptors for which "inversion" occurs at $N_a = N_d$.

Generally speaking, the above behaviour of the E_1-levels should result in the fact that in p-crystals with increasing T($T \gtrapprox 300$ K) there must be a certain "plateau" in the dependence $\lg p(1/T)$. However, the temperatures at which this "plateau" can be noticeable are directly adjacent to the temperature range where the intrinsic conductivity already dominates (see Fig. 2), therefore no "plateau" is seen on the curve $\lg p(1/T)$. According to data published in [5] one can expect that for Ge crystals $E_1 \approx 0.1$ eV and, therefore, intrinsic conductivity must occur only after the E_1-level has been saturated by captured electrons. As a result the bending of the line $\lg p(1/T)$

is assumed to be noticeable in this case. In [5] carried out for Ge crystals the bending of the curve $\lg p(1/T)$ was really observed in the range of elevated temperature (the author of this work, however, discussed this phenomenon only in terms of Coulomb interaction between the trapped electrons).

Let us come back to the connection between E_F and E_2 for the samples under discussion.

The dependence of E_F on E_2 and on other crystal parameters can be found from the condition of electric neutrality that, at sufficiently low temperature, is given as $N_D^- = N_d^+ = N_d$, where N_D^- is the number of electrons accepted on dislocations, N_d^+ is the number of ionized donor impurity. We have ignored holes captured by dislocations, free electrons, and free holes (low-T approximation). If N_{D1}, N_{D2} are the effective numbers of states which may be occupied by captured electrons, corresponding to levels E_1 and E_2, then

$$N_D^- = \frac{N_{D1}}{\{\exp[(E_1 - E_F)/kT] + 1\}} + \frac{N_{D2}}{\{\exp[(E_2 - E_F)/kT] + 1\}}.$$

The energies E_1 and E_2 may be considered independent of T in the range of low T (Fig. 4, 5), the factors of spin degeneracy [22] are not important for our consideration. Note, that $N_{D1} = f_{1\,max} N_D$ and N_{D2} is supposed to be of the order of N_D. Since, in our case the occupation ratio for the states with the energy E_2 is small (see Fig. 5), we have $(E_2 - E_F)/kT \gg 1$. On the other hand, since the states with energy E_1 have to be filled (low T) we have $(E_F - E_1)/kT \gg 1$. This means the level E_F is in the upper half of the forbidden band (n-type crystal). So, we obtain $N_D^- \approx N_{D1} + N_{D2} \times \exp[(E_F - E_2)/kT]$. This allows one to obtain easily the expression for E_F from the condition of electric neutrality: $E_F \approx E_2 - kT \ln[N_{D2}/(N_d - N_{D1})]$. Accordingly, $n = N_c \exp[(E_F - E_c)/kT] = N^* \exp(-E_2/kT)$, where $N^* = N_c(N_d - N_{D1})/N_{D2}$. Here, E_c is the energy of the lower edge of the conduction band, we assume $E_c \equiv 0$. Thus, the free electron concentration is described by the exponential with the activation energy equal to E_2. The pre-exponential factor differs from N_c, namely, $N^* < N_c$, since the inequality $N_{D2} > N_d > N_{D1}$ must be fulfilled according to definition. This conclusion is in qualitative agreement with the results of Fig. 5.

The ε-dependence of the Fermi level E_F is of special interest. The value E_F can directly be determined from the experimental data. Fig. 7 presents the dependence $E_F(\varepsilon)$ at low temperatures. (At these temperatures, $E_F \approx E_1$ for p-type crystals, $E_F \approx E_2$ for n-type crystals.) As is seen from the figure, E_F undergoes a "jump" at $\varepsilon \approx 1.8\%$. Since N_D monotonously increases with ε (up to $\varepsilon \approx 5\%$ [11, 13]), we again arrive at the conclusion that a jump of E_F upwarps with decreasing ε p–n transition of the crystal) is associated with the occupation of the new level (E_2) and not associated with the shift of the E_1-level due to the Coulomb interaction.

So, we may conclude that the energy level scheme is that shown in Fig. 6. Now, a few words about the dc conductivity, σ_0, of dislocated crystals.

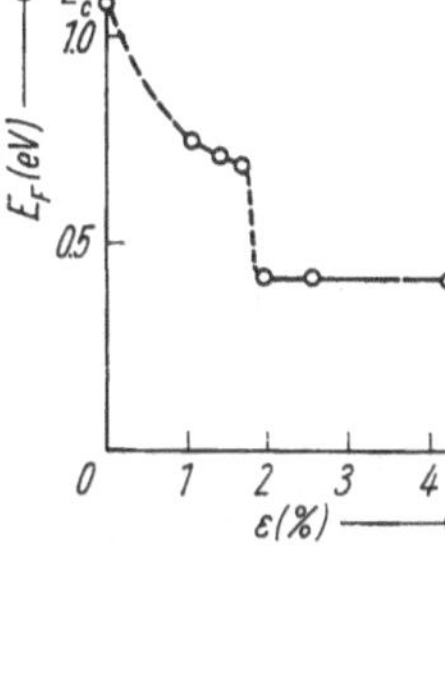

Fig. 7. Dependence of the Fermi level on deformation for specimens with phosphorus content 2×10^{14} cm^{-3}. The "jump" of E_F at $\varepsilon \approx 1.8\%$ is seen. The energy of the gap, $E_c - E_v$, is assumed to be 1.1 eV

The conductivity of these crystals depends exponentially on temperature: $\sigma_0 \sim \sim \exp(-E_D/kT)$. The activation energy E_0 for p-type crystals corresponds to E_1, for n-type to E_2. In the case of phosphorus-doped crystals E_0 essentially increases (as compared with E_1-, E_2-values) for "critical" specimens, when the n–p transition occurs. For such specimens we have obtained $E_0 \approx 0.67$ eV. This is, probably, due to the exponential dependence of the mobility on temperature [10]. One of the most important results obtained when investigating the conductivity is, however, the following: no visible dc conductivity along dislocations occurred in the investigated crystals. This result shall be discussed in Part II [23] of this paper.

3. Conclusion

In this paper main attention is paid to crystals with high dislocation density, i.e. with small occupation ratios.

In this case, an interpretation of the results is markedly alleviated since the energy of the Coulomb interaction is appreciably smaller than the characteristic energies for the dislocation spectrum, which gives rise to the fact that the concentration of free holes or electrons in the range of sufficiently low temperatures ($T \lesseqgtr 300$ K) is exponentially dependent on temperature; the power of the exponential curve for p-type crystals is determined by E_1 and for n-type crystals by E_2. Due to the small magnitude of f, the obtained values of E_1 and E_2 are likely to strictly correspond to the case $f = 0$.

It should also be noted that the occurrence of the acceptor levels in the upper half of the gap of plastically deformed silicon crystals was observed earlier [2, 3]. However, there was some spread of energies of these levels and some uncertainty in the interpretation of this phenomenon. Evidently, a spread in values of E_2 may be due, first of all, to the Coulomb interaction. We have carried out some control experiments using phosphorus-doped silicon crystals (2×10^{14} atoms/cm^3) containing lower dislocation densities (as compared with those treated above) for which the Coulomb interaction can be rather strong. As a result, the following "spectrum" of E_2-values was obtained: 0.43, 0.4, 0.35, 0.3 eV. So, a certain spread in the values of the dislocation acceptor levels for n-type Si, existing in literature, can be, generally speaking, attributed to the Coulomb interaction, though there might be other reasons, too [1].

As to the E_1-level position the following is to be pointed out: the value $E_1 = = (0.42 \pm 0.03)$ eV obtained in the present work satisfactorily agrees with the value of the activation energy for the conductivity of plastically deformed p-type silicon crystals obtained eralier in [18] (no measurements of the Hall effect were carried out in [18]).

The nature of levels E_1 and E_2 and their behaviour will be discussed in Part II [23] of this paper. By the way, some new arguments will be given for the fact that E_1-states (both acceptor and donor type) and E_2-states are really localized on dislocations.

Acknowledgements

The authors express their gratitude to Prof. Yu. A. Osipyan, Prof. V. I. Nikitenko. and Prof. V. B. Shikin for helpful discussions.

References

[1] W. BARDSLEY, Progr. Semicond. **4**, 155 (1960).
[2] H. WEBER, W. SCHRÖTER, and P. HAASEN, Helv. phys. Acta **41**, 1255 (1968).
[3] W. SCHRÖTER, and R. LABUSCH, phys. stat. sol. **36**, 539 (1969).
[4] R. H. GLAENZER and A. G. JORDAN, Solid state Electronics **12**, 247 (1969).
[5] W. SCHRÖTER, phys. stat. sol. **21**, 211 (1967).

[6] R. Labusch and R. Schettler, phys. stat. sol. (a) **9**, 455 (1972).
[7] P. Haasen and W. Schröter, IV. Internat. Summer School Defects, Zakopane, June 1973 (p. 231).
[8] Z. Gołacki, T. Figielski, and M. Jastrzębska, phys. stat. sol. **11**, K35 (1965).
[9] T. Wosiński and T. Figielski, phys. stat. sol. (b) **71**, K73 (1975).
[10] V. A. Grazhulis, V. Yu. Mukhina, Yu. A. Osipyan, and S. A. Shevchenko, Soviet Phys. — J. exper. theor. Phys. **41**, 1076 (1975).
[11] V. A. Grazhulis and Yu. A. Osipyan, Materialy soveshchaniya po defektam struktury v poluprovodnikakh, Part I, Novosibirsk, 1969 (p. 374).
[12] F. D. Wohler, H. Alexander, and W. Sander, J. Phys. Chem. Solids **31**, 1381 (1970).
[13] V. A. Grazhulis and Yu. A. Osipyan, Zh. eksper. teor. Fiz. **60**, 1150 (1971).
[14] S. V. Broude, V. A. Grazhulis, V. V. Kveder, and Yu. A. Osipyan, Soviet Phys. — J. exper. theor. Phys. **39**, 721 (1974).
[15] V. A. Grazhulis, V. V. Kveder, V. Yu. Mukhina, and Yu. A. Osipyan, Zh. eksper. teor. Fiz., Pisma **24**, 164 (1976).
[16] V. G. Eremenko, V. I. Nikitenko, and E. B. Yakimov, to be published.
[17] V. A. Grazhulis and Yu. A. Osipyan, Zh. eksper. teor. Fiz. **58**, 1259 (1970).
[18] V. G. Eremenko, V. I. Nikitenko, and E. B. Yakimov, Zh. eksper. Teor. Fiz. **67**, 1148 (1974).
[19] M. C. Collet, J. Electrochem. Soc. **117**, 259 (1970).
[20] K. K. Ju, A. G. Jordan, and R. L. Longini, J. appl. Phys. **38**, 572 (1967).
[21] V. A. Grazhulis, V. V. Kveder, and V. Yu. Mukhina, Fiz. tverd. Tela **19**, 585 (1977).
[22] P. S. Kireev, Fizika poluprovodnikov, Vysshaya Shkola, Moskva 1969.
[23] V. A. Grazhulis, V. V. Kveder, and V. Yu. Mukhina, phys. stat. sol. (a) **44**, (1977).

(Received June 6, 1977)

phys. stat. sol. (a) 43, 417 (1977)

Subject classification: 19; 11; 22.1.2

Department of Physics, University of California, Los Angeles (a) and Technology Research Center, Japan Petroleum Development Corporation, Tokyo[1]) *(b)*

An ESR Study of Annealing Characteristics for Heavily Irradiated Silicon[2])

By

H. NAKATA[3]) (a) and H. MORISHIMA (b)

When silicon substrates are irradiated by 2.5×10^{16} $^{31}P^+$ ions/cm^2 at 190 keV in the ⟨111⟩ direction, the ESR spectrum showed two resonances ($g = 2.0057$ and $g = 2.0026$). The former represents the amorphous state while the latter is a signal which cannot be seen at lower doses. They have different annealing characteristics between 100 and 800 °C. The composition of two independent Lorentzian spectra designed for these g-values can successfully reproduce the experimental spectrum observed after annealing at high temperatures.

Wenn Siliziumsubstrate mit $2{,}5 \times 10^{16}$ $^{31}P^+$-Ionen/cm^2 bei 190 keV in ⟨111⟩-Richtung bestrahlt werden, zeigt das ESR-Spektrum zwei Resonanzlinien ($g = 2{,}0057$ und $g = 2{,}0026$). Die erstere rührt vom amorphen Zustand her, während die letztere ein Signal ist, das bei niedrigeren Dosen nicht beobachtbar ist. Sie zeigen unterschiedliche Temperungscharakteristiken zwischen 100 und 800 °C. Die Zusammensetzung von zwei unabhängigen Lorentzlinien mit diesen g-Werten kann das experimentelle Spektrum, das nach der Temperung bei hohen Temperaturen beobachtet wird, erfolgreich reproduzieren.

1. Introduction

When semiconductor materials are irradiated by low-energy heavy ions, several kinds of unpaired electrons are generated at high doses [1 to 10]. In this case, on-channel rather than off-channel irradiation is expected to reveal fundamental phenomena about the interaction of ions with atoms in a material since the former process is simple.

In our previous ESR study [11], a structure was found in the relation between the number of unpaired electrons and the dose in the case of on-channel irradiation. Another remarkable result was that the ESR spectrum has two peaks ($g = 2.0057$ and $g = 2.0026$) at 2.5×10^{16} ions/cm^2 whereas there was only one peak at lower doses except for very small peaks.

As shown in our previous paper, the resonance at $g = 2.0057 \pm 0.0002$ with $\Delta H_{msl} = 9$ to 12 G at room temperature belongs to the amorphous state of silicon [12 to 16]. The other one will be discussed in Section 5. Normally it is accepted that an irradiated substrate is locally amorphous at low doses ($\leqq 1 \times 10^{14}$ ions/cm^2), uniformly amorphous at intermediate doses ($\leqq 1 \times 10^{16}$ ions/cm^2) to a considerable extent, and uniformly amorphous at high doses ($\geqq 1 \times 10^{16}$ ions/cm^2).

The purpose of the present study is to understand the behaviour of the ESR spectrum after the substrate irradiated at 2.5×10^{16} ions/cm^2 is annealed. This paper is called ref. [11] in our previous paper [11].

---

[1]) Mori-18 Bldg. Akefune-cho, Minato-ku, Tokyo, Japan.

[2]) Work supported in part by Japan Society.

[3]) Permanent address: Minamiuemachi, Kishiwada, Osaka, Japan.

2. Experimental Method

A p-type, ⟨111⟩ Si substrate was irradiated at 26 °C by 190 keV $^{31}P^+$ ions with a dose of 2.5×10^{16} ions/cm² in the ⟨111⟩ on-channel direction (hereafter called Sample 1625). Details are given elsewhere [11].

Small pieces of this substrate were annealed separately at 100 to 800 °C in dry nitrogen gas for 30 min. Such dependences as sample size, ambient temperature, observation angle, and magnetic field modulation width were carefully checked.

3. Results

Fig. 1 shows ESR spectra for Sample 1625, observed at room temperature after 30 min annealing. Conditions for ESR were the same as those in our previous study, except the present RF power of 25.6 mW and the modulation width of 1 G. These spectra are not normalized in the ordinate of Fig. 1. Hereafter, resonances at $g = 2.0057$ and $g = 2.0026$ are called Signals L and R, respectively.

4. Spectral Analysis

Wave analysis [17, 18] is useful to extract a fundamental relation between R and L from the experimental spectrum in the annealing process.

In the solid sample, the electron spin feels slightly differing dipolar fields produced by surrounding spins. If the spin–lattice relaxation is slower than the spin–spin one, all the spin packets never lead to thermal equilibrium at the same time (inhomogeneous broadening); the absorption curve becomes Gaussian.

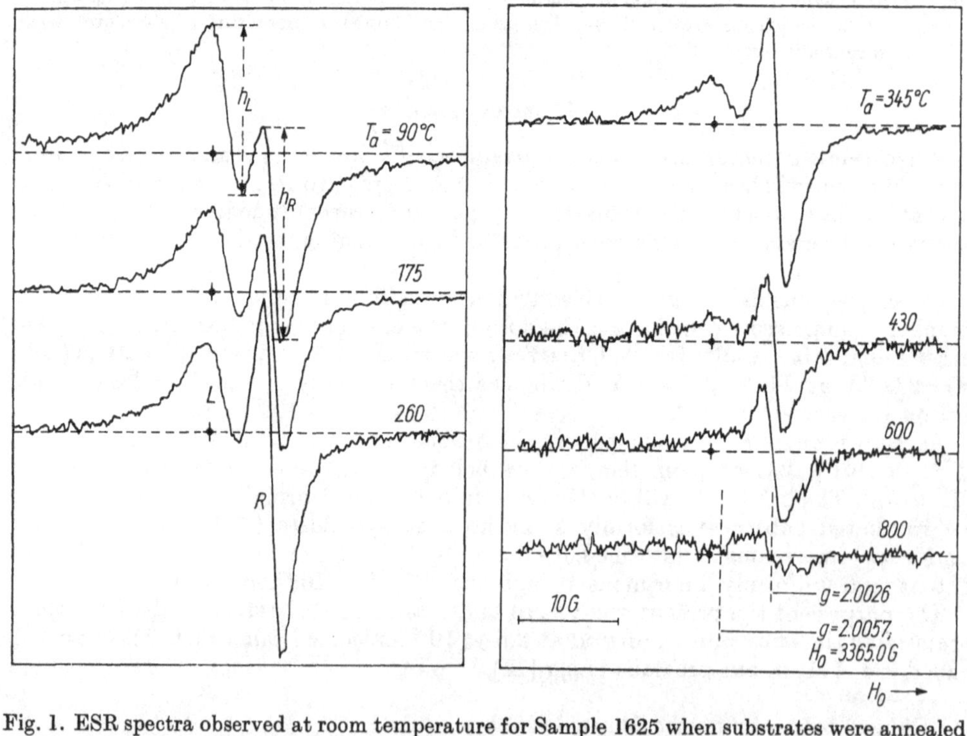

Fig. 1. ESR spectra observed at room temperature for Sample 1625 when substrates were annealed separately at 100 to 800 °C in dry nitrogen gas for 30 min (modulation width 1 G). No normalization is applied to the spectra in this figure. (T_a is the annealing temperature)

The ESR spectrum of a paramagnetic impurity in a solid is determined by the Gaussian distribution of neighboring spin magnetic moments, i.e. Gaussian lineshape. However, even in the case of solids, the exchange energy may be completely large at a close spin–spin distance. Electron exchange will possibly take place, making a Lorentzian lineshape just as in the case of a liquid. Beside the rapid spin–spin exchange process, both the rapid spin–lattice relaxation process and the rapid Brownian motion of the molecule, which depend on temperature, may also induce a Lorentzian lineshape (homogeneous broadening).

Because of a long tail in our spectrum, the Lorentzian function was more suitable. (The Gaussian parametrization was a poor representation of the data.) It was assumed that this spectrum was composed of two independent Lorentzian functions of different maximum slope widths and amplitudes, as marked in the figures as *A* and *B* for L and R, respectively (see Fig. 2).

ΔH_{msl} was fixed at 2:1 (or 4.14:2.07 G) for *A* versus *B* at any annealing temperature. Some of the results are shown in Fig. 2. Thus, each of these experimental spectra can be approximately decomposed into two sets of perfectly independent Lorentzian functions. The agreement is better at higher temperatures. There is almost no deviation above 260 °C.

At lower temperatures the amplitude of the composite spectrum disagrees more or less with that of the experimental one around the central part of the abscissa. An addition of a "third" small Lorentzian curve to the two-elemental composite spectrum in the vicinity of that fiducial point of the abscissa will eliminate the deviation.

5. Discussion

As far as L is concerned, the Si state before annealing is considered uniformly amorphous to a considerable extent, according to the theory that defects which do not overlap anneal rapidly above 200 °C, while overlapping defects anneal above 500 °C.

Fig. 3 shows that the annealing temperature dependence is monotonic for L, while there is a peak in the intermediate stage for R. Amorphous centers are almost removed after the substrate was annealed above 500 °C. This phenomenon represents a transition from amorphization to recrystallization in an original crystalline substrate.

Let us consider some problems about our results. The line width is not Gaussian but Lorentzian. Then one may review the system of spins by saying that the spin–spin distance is short in the two kinds (L and R) of unpaired electrons and the spin–spin exchange interaction energy is larger than the width of the distribution of dipolar fields. The result that the two kinds of radicals for A and B have independent resonant linewidth suggests that the exchange interaction for one kind of spin is independent of that for the other kind of spin; therefore both kinds of spins must be localized in separated regions.

This interpretation is not inconsistent with our three-step mechanism [11] that heavy irradiation induces amorphous centers in the shallow region and $g = 2.0026$ centers in the deeper region. If both kinds of unpaired electrons were mixed and mutually adjacent in the substrate, there would occur an exchange interaction more or less between *A* and *B*; then the lineshape for *A* or *B* would be deformed from the perfect Lorentzian function due to spectral coherence. The actual linewidth and resonant position for *A* or *B* are always fixed, independently of the number of spins or the ratio of *A*/*B* at any annealing temperature.

These phenomena and interpretations may be quite reasonably applied even if the sample should have a carbon compound on the surface i.e., when an oil film is built up by a poor diffusion pump. In that case the ions cause radiation damage in the film

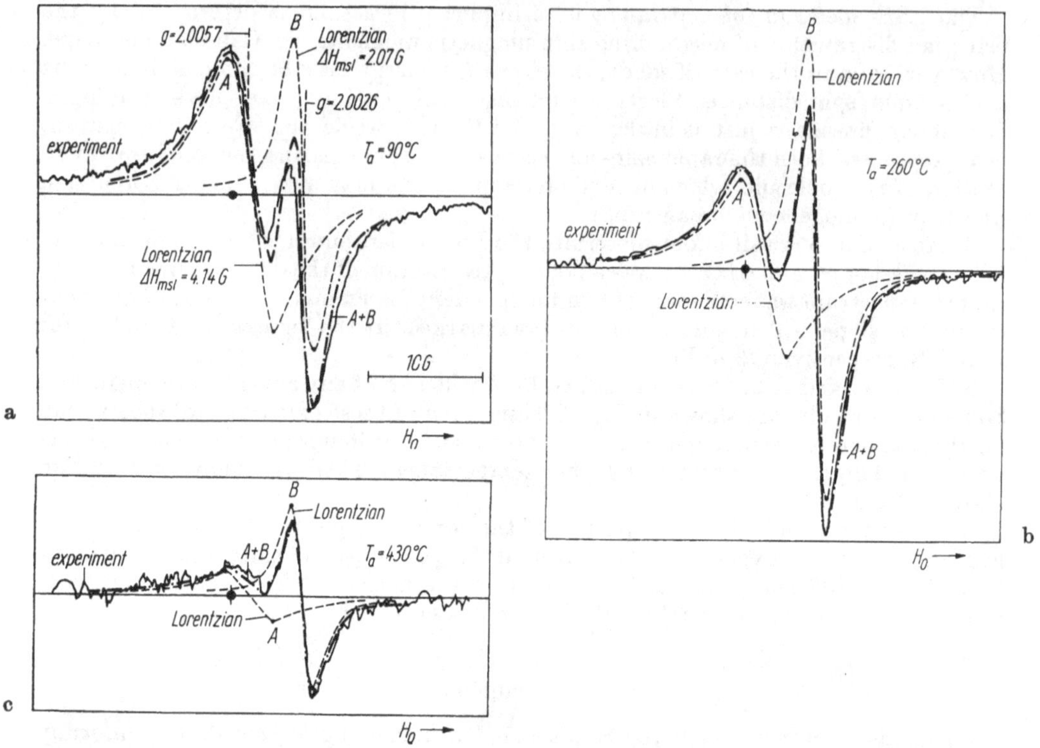

Fig. 2. Lorentzian composition and decomposition of ESR spectra measured at room temperature after 30 min annealing at a) $T_a = 90$, b) 260, and c) 430 °C. (The modulation width is 1 G.) d) Dependence of the relative intensity of the differential Lorentzian curve on annealing temperature, measured at room temperature. This is calculated by the first derivative of the Lorentzian function $L'(X) = -16h_0X(X^2+3)^{-2}$, where h_0 is the height in the differential curve; $X = 2x/\Delta H_{msl}$ is the abscissa measured from the origin (center) in units of $\Delta H_{msl}/2$. (1) $T_a = 90$, (2) 175, (3) 260, (4) 345, (5) 430 °C. e) Dependence of the relative intensity of the Lorentzian absorption curve on annealing temperature, measured at room temperature. This is obtained from the Lorentzian function $L(X) = 4h_0\,\Delta H_{msl}(X^2+3)^{-1}$, where $\Delta H_{1/2} = \sqrt{3}\Delta H_{msl}$. (1) $T_a = 90$, (2) 175, (3) 260, (4) 345, (5) 430 °C

and they give the ESR signal at $g = 2.0026$ [19 to 23]. One ($g = 2.0057$) may be in the Si substrate and the other ($g = 2.0026$) may be in the C film. We can neglect SiO_2.

To check this possibility, some pieces of the irradiated Si substrate were put in triclene ($ClCH{=}CCl_2$) for 40 min, 24 h, and 50 d at room temperature. As a result of this test, it turned out that there was no change in B/A within the experimental error ($\pm 3\%$) before and after this chemical treatment. It follows that what can be solved by trichloroethylene has never been produced on the surface of the substrate. Although the time required for an irradiation of 2.5×10^6 ions/cm² was approximately 2 h, it was not affected by other (e.g., C, O, etc.) ions than P ions.

In order to recognize that the resonant lineshape was determined by the spin–spin exchange interaction, ESR measurements were performed at 77 K. The Lorentzian shape was supported again. In the vicinity of the fiducial point of the abscissa, the disagreement was more pronounced at 77 K than at room temperature due to un-

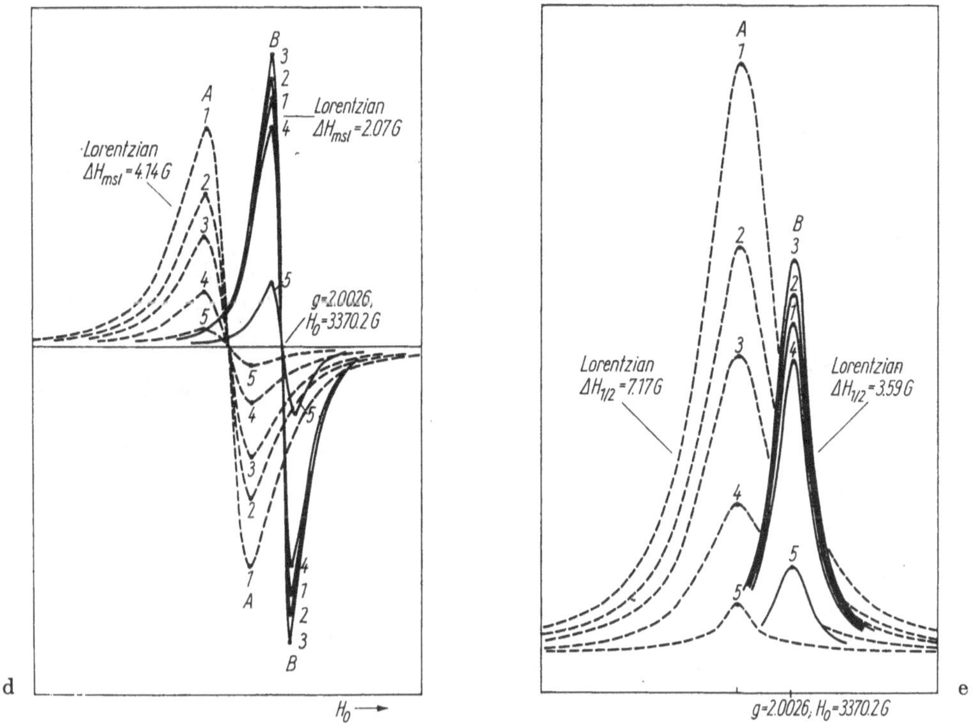

Fig. 2

known causes. The linewidth was not changed. Therefore, it is neither motional narrowing nor spin–lattice relaxation mechanism but the spin–spin exchange interaction that dominates the resonant lineshape. The first and second among these events depend on temperature; the spin–spin relaxation mechanism does not depend on it but the spin–spin distance.

The spin–spin relaxation time (T_2) derived from the Lorentzian spectrum

$$T_2 = \left(\frac{2\hbar}{\sqrt{3}}\right) g^{-1} \mu_B^{-1} (\Delta H_{msl})^{-1}$$

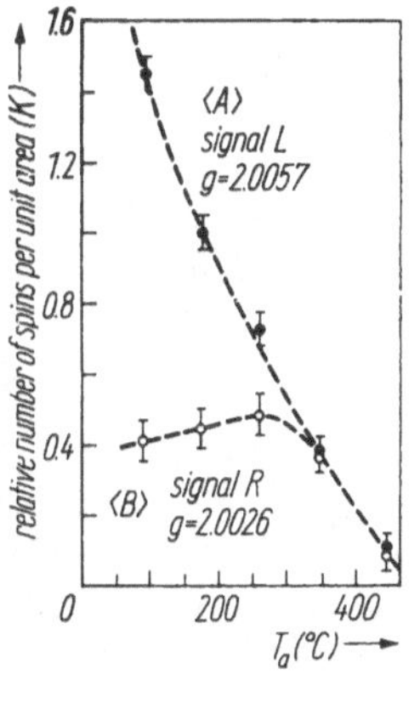

Fig. 3. A plot of the relative number of spins per unit area computed by the equation $K = (2\pi/\sqrt{3})\, h_0 (\Delta H_{msl})^2$ as a function of annealing temperature. Dashed curves are drawn by eye through data points

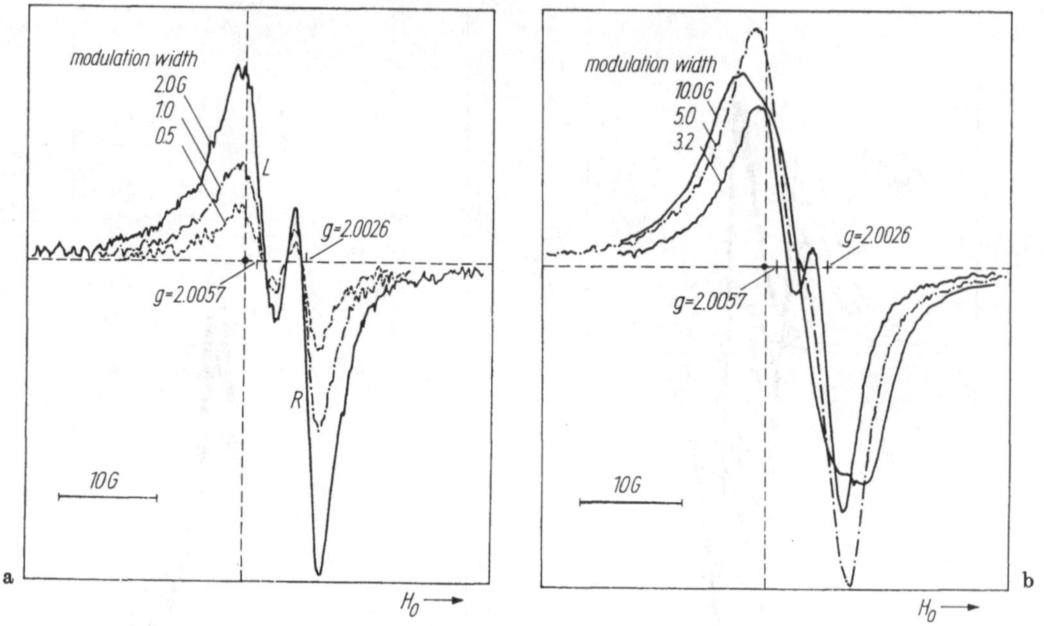

Fig. 4. Spectral variation of Sample 1625 with the magnetic field modulation width ranging from 0.5 to 10 G, observed at room temperature, before annealing

yields 1.6×10^{-8} and 3.2×10^{-8} s for A and B, respectively; where $\hbar$ is Planck's constant divided by 2π, g is the g-value for unpaired electrons, and μ_B is the Bohr magneton. The quantity T_2 is also correlated with the exchange interaction energy which is inversely proportional to the inter-spin distance. The difference in the linewidth (ΔH_{msl}) between A and B suggests the difference in the unpaired electron–electron distance between A and B. The wider linewidth (A) is correlated with the smaller interaction energy and it stems from the longer spin–spin distance.

Regardless of monocrystalline samples, our previous study shows no angular dependence at room temperature. It is partly due to the possibility that amorphous centers are extremely prevalent in comparison with other centers, and spectral anisotropy, if any, is smoothed out by the spin–spin exchange interaction.

Uniform removal or recovery of centers induced by annealing will enlarge the average distance between unpaired electrons and widen the observed exchange narrowed width. Since the data are far from this, it seems that the progress of recovery is local and the local average concentration of spins in small regions is kept constant during annealing.

Finally, we show the clear dependence of the spectrum on the magnetic field modulation width in Fig. 4.

6. Conclusions

In this study of heavy irradiation, a few conclusions can be drawn:

1. Two kinds of unpaired electrons are generated at 2.5×10^{16} ions/cm^2.
2. They have perfectly Lorentzian shapes with different maximum slope widths.
3. They show different annealing characteristics.
4. Neither of them is affected by triclene.

References

[1] D. F. Daly and K. A. Pickar, Appl. Phys. Letters **15**, 267 (1969).
[2] K. L. Brower, F. L. Vook, and J. A. Borders, Appl. Phys. Letters **15**, 208 (1970), **16**, 108 (1970).
[3] K. L. Brower, and J. A. Borderes, Appl. Phys. Letters **16**, 169 (1970).
[4] B. L. Crowder, R. S. Title, M. H. Brodsky, and G. D. Pettit, Appl. Phys. Letters **16**, 205 (1970).
[5] S. Hasegawa, R. Kontani, and T. Shimizu, Japan. J. appl. Phys. **10**, 655 (1971).
[6] F. F. Morehead, B. L. Crowder, and R. S. Title, J. appl. Phys. **43**, 1112 (1972).
[7] K. L. Brower and W. Beezhold, J. appl. Phys. **43**, 3499 (1972).
[8] S. Hasegawa, K. Ichida, and T. Shimizu, Japan. J. appl. Phys. **12**, 1181 (1973).
[9] K. Murakami, K. Masuda, K. Gamo, and S. Namba, Japan. J. appl. Phys. **12**, 1307 (1973).
[10] K. C. Chu, W. R. Hurren, E. Hale, and J. Reigle, J. appl. Phys. **44**, 4243 (1973).
[11] H. Nakata and H. Morishima, phys. stat. sol. (a) **33**, 633 (1976).
[12] D. Haneman, Phys. Rev. **170**, 705 (1968).
[13] D. Haneman, M. F. Chung, and A. Taloni, Phys. Rev. **170**, 719 (1968).
[14] M. H. Brodsky and R. S. Title, Phys. Rev. Letters **23**, 581 (1969).
[15] D. Kaplan, D. Lépine, Y. Petroff, and P. Thirry, Phys. Rev. Letters **35**, 1376 (1975).
[16] B. P. Lemke and D. Haneman, Phys. Rev. Letters **35**, 1379 (1975).
[17] R. S. Alger, Electron Paramagnetic Resonance, Interscience Publishers, New York 1972 (p. 1).
[18] J. E. Wertz and J. R. Bolton, Electron Spin Resonance, McGraw-Hill, New York 1972 (p. 1).
[19] D. J. Miller and D. Haneman, Surface Sci. **19**, 45 (1970).
[20] D. J. Miller and D. Haneman, Surface Sci. **24**, 639 (1971).
[21] D. J. Miller and D. Haneman, Phys. Rev. B **3**, 2918 (1971).
[22] D. Haneman, private communication (1975).
[23] E. Hale, private communication (1975).

(Received April 7, 1977)

phys. stat. sol. (a) 43, 425 (1977)

Subject classification: 1.4 and 10.1; 1.2; 21.1

Centre de Recherche sur les Mécanismes de la Croissance Cristalline, C.N.R.S., Centre Scientifique de St-Jérôme, Marseille[1])

X-Ray Topographic Investigation of Dislocations in Titanium Single Crystals Grown by Recrystallization

By

C. JOURDAN and J. GASTALDI

Dislocations in as grown titanium single crystals, produced by recrystallization after phase transformation under high vacuum (5×10^{-10} Torr), are studied by X-ray topography. All kinds of line defects are observed: edge, screw, and helical dislocations, prismatic and glissile loops, sub-boundaries and non-identified defects called "black dots". Their Burgers' vectors are of $\boldsymbol{a}$ or $\boldsymbol{c} + \boldsymbol{a}$ type. It is established that residual impurities, polishing conditions, and the number of cycle around the phase transformation point $\alpha \rightleftarrows \beta$, did not increase the number of defects. Particularly, X-ray topography enables to determine that "black dots" did not appear during polishing but are present in the crystals before.

Nous avons étudié les dislocations contenues dans les monocristaux de titane obtenus par recristallisation par la méthode du changement de phase et recuit sous ultra-vide. De nombreux types de défauts ont été trouvés: dislocations coins, vis et hélicoídales, boucles glissiles et prismatiques, sous-joints et des défauts non identifiés que nous avons appelés "points noirs". Les vecteurs de Burgers de ces défauts sont du type $\boldsymbol{a}$ et $\boldsymbol{c} + \boldsymbol{a}$; nous montrons que les impuretés résiduelles, les conditions de polissage et le nombre de cycles autour du point des transformation $\alpha \rightleftarrows \beta$ n'augmentent pas le nombre de défauts. D'autre part la topographie montre que les "points noirs" n'apparaissent pas au cours du polissage.

1. Introduction

Till now, dislocations in titanium single crystals have been studied by etch-pitting [1, 2] and by electron microscopy [3 to 5]. No X-ray topography has been made because of the difficulties to grow crystals of this metal with a low density of dislocations. Only recently we were successful in our attempt to overcome this hindrance [6].

It is well known, nowadays, that X-ray topography is more accurate than etch-pitting to characterize dislocations and is more suitable than electron microscopy to observe directly these defects in bulk materials (for example: using a 1 mm thick sample for aluminium and a some 1/10 mm thick sample for titanium). Because of that, X-ray topography is an excellent tool for studying, without damages, defects occurring during the growth or during the deformation of crystals.

This paper deals with the advantages of this method to study dislocations in highly perfect titanium crystals (i.e. having 10^2 to 10^3 cm/cm^3 of dislocations) prepared using the procedure developed in our laboratory [6]. We have compared our results with those obtained using electron microscopy and we have also established that the crystal quality varies with the metal purity, the number of cycles and the way of polishing.

---

[1]) 13397 Marseille, Cedex 4, France.

2. Experimental Procedure

The procedure used for the growth of titanium crystals has been detailed previously. We shall remind briefly that these crystals are produced under high vacuum ($5 \times \times 10^{-10}$ Torr), by recrystallization at a temperature slightly lower than the phase transformation point temperature $\alpha \rightleftarrows \beta$, after some cycles around this transformation point. The crystals are then cooled down very slowly to room temperature (3 °C/h).

We studied titanium crystals presenting different degrees of purity: very impure titanium crystals ("éponge refondue"), titanium crystals containing few metallic elements (M.R.C.), and titanium crystals out of which, both metallic and metalloidic elements had been removed ("Van Arkel").

We also changed the number of cycles around the transformation point and used two types of polishing baths to test the eventuality of titanium hydride formation during the polishing procedure.

After each experiment, crystals were systematically observed by X-ray topography using MoK_α radiation. The characterization of dislocations (line direction, Burgers vector, ...) was achieved with the aid of the 0002 reflection and if that was not possible (planes 0002 nearly parallel to the large faces of the flat samples of $30 \times 10 \times 0.2$ mm^3), to remove all kinds of ambiguity in the determinations, we used up to ten reflections.

3. Results

3.1 Influence on crystal quality of metal purity, number of cycles, polishing conditions

We prepared a great number of titanium crystals and we studied systematically by X-ray topography about twenty of them. Thus we are now able to assure that it is possible to prepare by our method crystals of different orientations (Fig. 1), highly

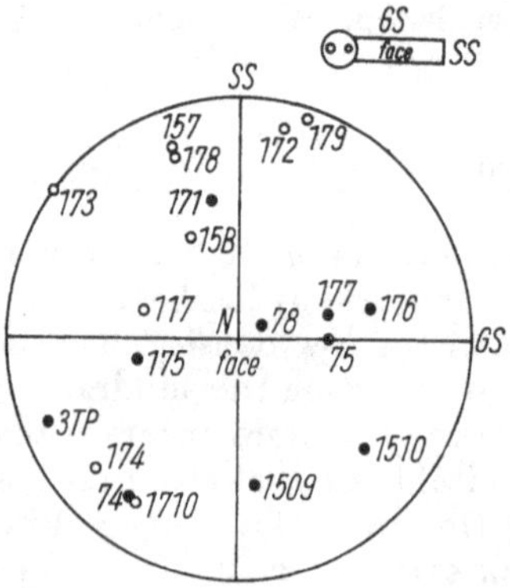

Fig. 1. Orientations of the various purity titanium crystals studied ○ basal plane of "Van Arkel" titanium crystals, ⊙ basal plane of "M.R.C." titanium crystals, ● basal plane of "éponge refondue" titanium crystals. GS great side direction of sample, SS small side direction of sample, N normal to the sample face

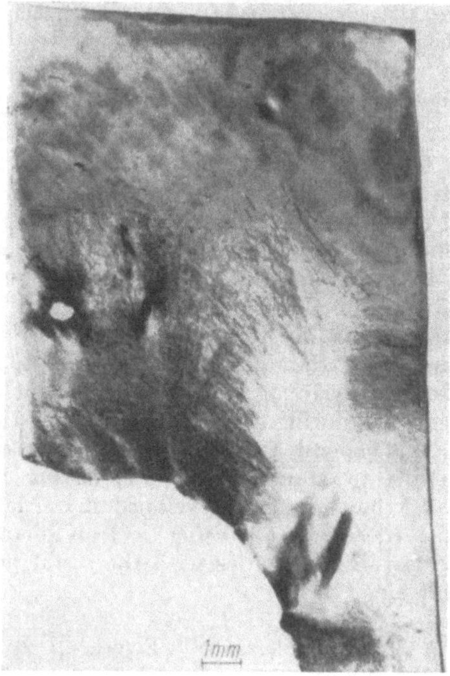

Fig. 2. X-ray topograph of the titanium crystal 176 ("éponge refondue") (see orientation Fig. 1). Reflection $\bar{1}101$

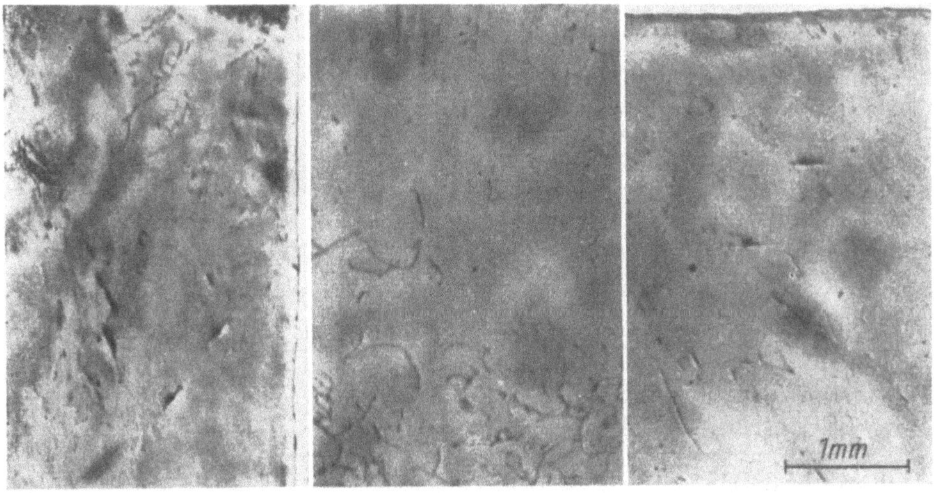

a b c

Fig. 3. X-ray topographs of various purity titanium crystals. a) "Van Arkel", reflection $\bar{1}10\bar{1}$; b) "M.R.C.", reflection $\bar{1}100$; c) "éponge refondue", reflection $\bar{1}101$

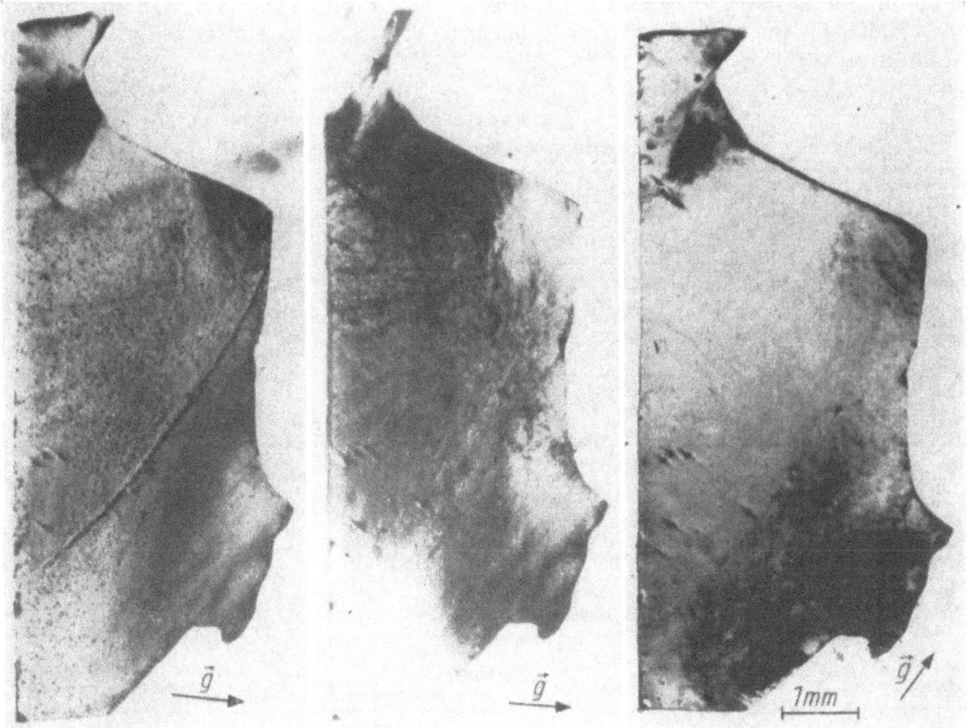

a b c

Fig. 4. X-ray topographs of the titanium crystals 173 ("Van Arkel") (see orientation Fig. 1). a) Before chemical polishing, reflection $1\bar{1}0\bar{1}$; b) after chemical polishing, reflection $1\bar{1}0\bar{1}$; c) after chemical polishing, reflection 0002 perfect, relatively large ($20 \times 10 \times 0.2$ mm^3) (Fig. 2), whatever the purity of the titanium used was (Fig. 3).

When, before annealing crystals in the α-phase, the number of cycles was increased, crystals larger than after a few cycles, but of the same quality, were obtained.

The chemical or electrochemical polishing, carried out when the crystals has been taken out of the furnace and before X-ray examination, does not cause additional defects. But it removes surface defects and it allows an improved visibility of volume defects on X-ray topographs. The "black dots" that can be seen on these topographs do not appear during the polishing as they exist before it, they are related with phenomena occurring during the crystal formation (Fig. 4).

3.2 *Dislocations analysis*

In the crystals that had been investigated, curved or linear single dislocations, loops in rows and concentric loops, small angle grain boundaries and unidentified defects, previously called "black dots", were observed.

The Burgers vectors of these dislocations are mostly of the $\boldsymbol{a}$ type ($\frac{1}{3}\langle 11\bar{2}0\rangle$); the contrast vanishes on 0002 pictures (Fig. 4). Nevertheless, this rule cannot be always used. For some crystals, mainly those prepared from impure titanium ("éponge refondue" and "M.R.C."), dislocations can be seen on images obtained by diffraction on planes parallel to the basal plane (Fig. 5). If the visibility criterion $\boldsymbol{g} \cdot \boldsymbol{b} \times \boldsymbol{n} \neq 0$ ($\boldsymbol{g}$ diffraction vector, $\boldsymbol{b}$ Burgers vector, $\boldsymbol{n}$ dislocation line direction), using different $\boldsymbol{g}$, is applied, it is possible to assign to these defects $\boldsymbol{c} + \boldsymbol{a}$ type ($1/3\,\langle 11\bar{2}3\rangle$) Burgers vectors. This is not very usual, but it confirms what had been observed in deformed titanium by electron microscopy [4, 7].

Table 1

Impurity contents of the three kinds of titanium used

	atomic P.P.M.										
	H	C	O	Al	V	Cr	Fe	Ni	Cu	As	Zr
éponge refondue	non analysed	280	900 to 1800	10 to 200	15 to 60	40 to 60	150 to 250	50 to 70	1 to 3	2 to 3	10 to 40
M.R.C.	190	440	750 to 1800	11	3	9 to 15	9 to 17	8	<5	0.2 to 2	3
"Van Arkel"	undetected			0.6 to 3.5	0.6 to 14	0.02 to 7	2 to 64	0.07 to 13	undetected	2	0.4 to 2

Table 2

Composition of the titanium poslishing baths used

	baths	operating temperatures
chemical polishing	alternate soaking into: 1st bath 50% H_2SO_4 50% HF; 2nd bath HNO_3 dilute	20 °C
electrolytical polishing	100 ml $C_2H_4OHCOOH$, 70 ml HF, 40 ml H_2SO_4	3 to 4 °C

Table 3

Dislocation character inside a titanium crystal 175 ("éponge refondue") (see Fig. 1) marked on Fig. 5. $\boldsymbol{b}$ Burgers vector, $\boldsymbol{l}$ dislocation direction

	$\boldsymbol{b}$	$\boldsymbol{l}$ near by	$\boldsymbol{b} \times \boldsymbol{l}$	character
dislocations 1	$1/3\,[\bar{2}110]\,\boldsymbol{a}$	$[01\bar{1}\bar{2}]$	$(01\bar{1}1)$	edge
subboundary 2	$1/3\,[\bar{2}110]\,\boldsymbol{a}$	$[01\bar{1}\bar{2}]$	$(01\bar{1}1)$	tilt
dislocations 3	$1/3\,[\bar{2}110]\,\boldsymbol{a}$	$[\bar{2}110]$		screw-helical
dislocations 4	$1/3\,[\bar{1}2\bar{1}0]\,\boldsymbol{a}$	$[\bar{1}2\bar{1}3]$	$(10\bar{1}0)$	mixed
dislocations 5	$1/3\,[11\bar{2}\bar{3}]\,\boldsymbol{c}+\boldsymbol{a}$	$[\bar{2}110]$	$(01\bar{1}1)$	quasi-edge
subboundary 6	$1/3\,[11\bar{2}0]\,\boldsymbol{a}$	$[\bar{1}2\bar{1}\bar{3}]$	$(\bar{1}101)$	mixed

3.2.1 Single dislocations

The study of the character of single dislocations shows that they can be of helical, screw, edge, or mixed type and that their slip planes can be of $\{1\bar{1}00\}$ or $\{1\bar{1}01\}$ type (Table 3) (Fig. 5). Some $\boldsymbol{a}$ type Burgers vector dislocations have the same slip planes as some $\boldsymbol{c}+\boldsymbol{a}$ Burgers vector dislocations. The concentric loops and the loops in rows are always associated to "black dots" (Fig. 6).

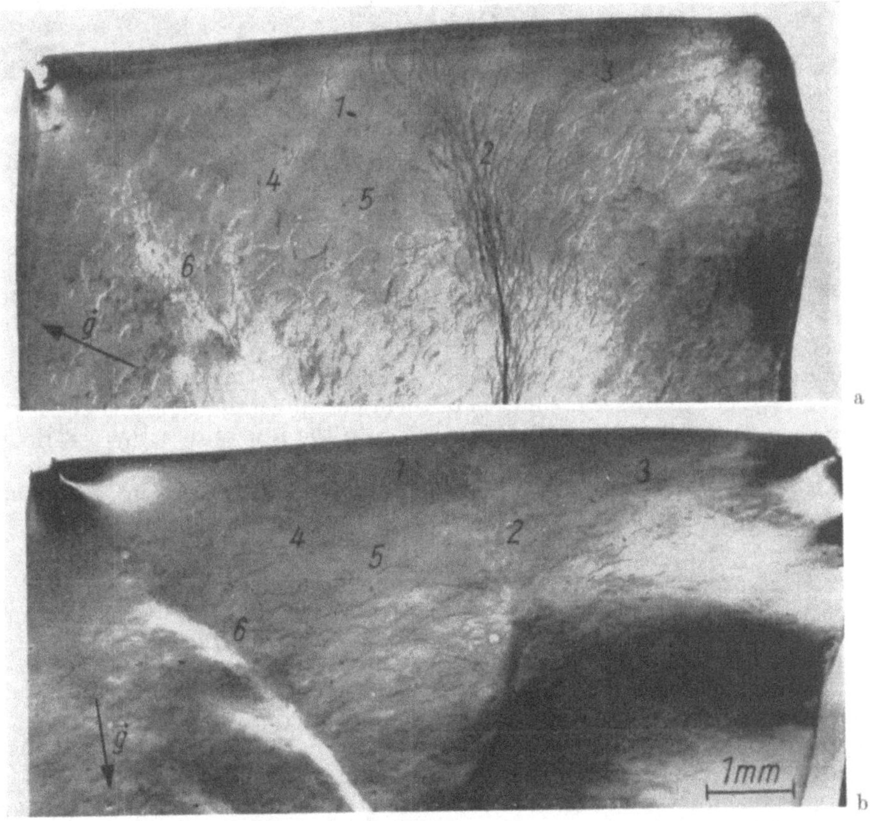

Fig. 5. X-ray topographs of the titanium crystal 175 ("éponge refondue") (see orientation Fig. 1). a) Reflection $\bar{1}10\bar{1}$; b) reflection 0002

a 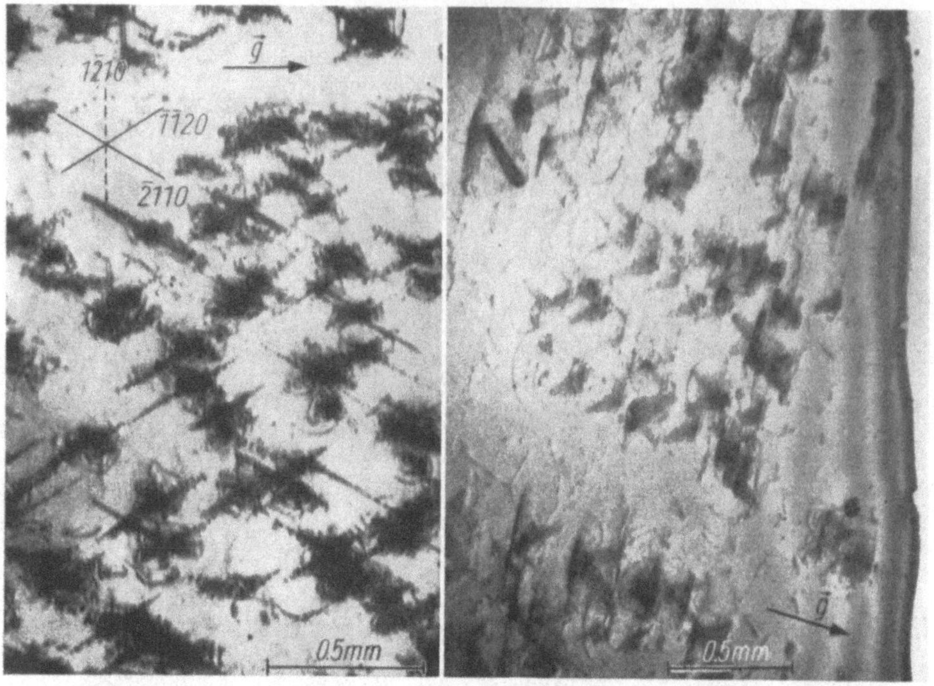 b

Fig. 6. X-ray topographs of titanium crystals with "black dots". a) "M.R.C." titanium crystal 78 (see orientation Fig. 1), reflection $\bar{1}010$. Loops in rows should be noticed, one is missing because it has a $\frac{1}{3}[1\bar{2}10]$ Burgers vector and that $\boldsymbol{g} \cdot \boldsymbol{b} = 0$; b) "Van Arkel" titanium crystal 117 (see orientation Fig. 1), reflection $10\bar{1}1$. Loops in rows and glissile loops are visible

3.2.2 Loops in rows

They consist of loops which roughly have equal size and which sometimes are normal to the row axes. Thus, if we look at Fig. 6, we may conclude that they are made of loops having all the same orientation. In fact, a careful examination of the various reflections shows that this is not true. On Fig. 7, three reflections of the same part of a crystal can be seen. On each of them one direction $\langle 11\bar{2}0\rangle$ is missing and it can be seen very clearly that there is no loop in the row of direction $\langle 1\bar{2}10\rangle$ on the $\bar{1}010$ reflection. This row consists of dislocation loops having a $1\bar{2}10$ Burgers vector. The same phenomenon can be seen on the $0\bar{1}10$ and $\bar{1}101$ reflections, where loops in rows of axes $\langle\bar{2}110\rangle$ and $\langle\bar{1}\bar{1}20\rangle$ are lacking, which gives us the direction of the Burgers vector in their loops. Besides that, if we consider the outlines on Fig. 7, it is obvious on Fig. 7a that the visible loops in the direction $\langle\bar{1}\bar{1}20\rangle$ are located on a $(\bar{1}010)$ plane, but on Fig. 7b, in the same direction $\langle\bar{1}\bar{1}20\rangle$, the loops are on a $(0\bar{1}10)$ plane. This is a general result: along the three directions $\langle 11\bar{2}0\rangle$ loops in rows are composed of two loop families. Accordingly, for a precise direction, loops are located on the two prismatic planes situated in its neighbourhood. If the contrast conditions are taken in account, it is easy to understand why only one loop family can be seen on the X-ray topograph. It is known that for a visible dislocation the contrast is maximum when the diffraction vector $\boldsymbol{g}$ is normal to the line direction. Strong contrasts are, of course, always prominent on X-ray topographs. If one has this in mind when watching Fig. 7, one can verify that the most visible loop family is the one normal to the vector $\boldsymbol{g}$. When the number of loops in a row is not important both families can be seen. This

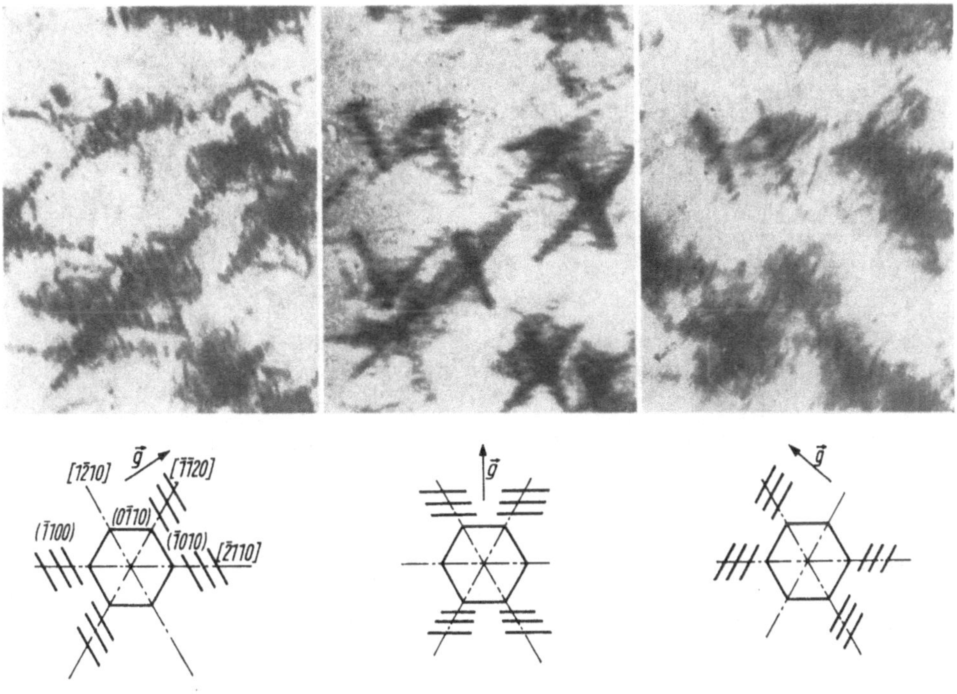

a b c

Fig. 7. Characteristics of loops in rows. a) Reflection $\bar{1}010$; b) reflection $0\bar{1}10$; c) reflection $\bar{1}101$, outline of rows

is the case on Fig. 8 where for each direction $\langle 11\bar{2}0\rangle$ both possible directions have been mentioned. On Fig. 8a, which is a $\bar{1}010$ reflection topograph, it can be seen that loops labelled 1 and 2 which are perpendicular to the vector $\boldsymbol{g}$ are darker than loops 1′ and 2′.

Before closing about these loops in rows, an additional remark should be made. Loops in $\{01\bar{1}0\}$ planes are not in a minimum energy configuration. They should turn and locate on $\{1\bar{2}10\}$ planes which are normal to the row direction. The energy decrease due to the line shortening being very weak, this process will take place only when the row is located in a slightly disturbed region (Fig. 9). In the parts of

Fig. 8. Detail of a row with a small number of loops a) outline, b) reflection $\bar{1}010$

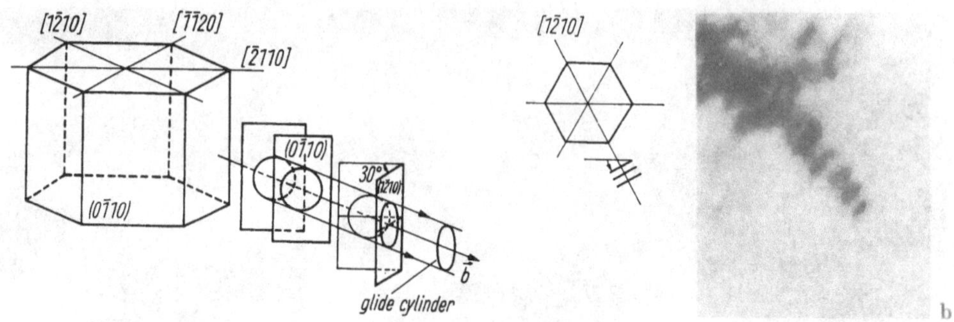

a b

Fig. 9. Detail of a row and loops rotations seen in a plane normal to the row axis. a) Outline, b) X-ray topograph with its outline

the crystals having a high "black dots" density, local stresses in and between the rows will stop this movement and keep the loops in a prismatic plane.

3.2.3 Concentric loops

These loops have a much simpler geometry. They are located in the basal plane and have an ***a*** type Burgers vector, so they are glissile loops. They are probably generated by a Frank-Read mechanism. Fig. 6, where several concentric loops, or what remains of them, can be seen, is a good representation of this mechanism.

3.2.4 Small-angle grain boundaries

Big titanium crystals are often surrounded by smaller highly disoriented crystals ($\theta > 10°$). The resulting high-angle grain boundaries do not give visible stresses on topographs. Moreover, the thinness of the sample (slightly above the method resolution: 100 μm) preludes dislocation observation as it has been done in aluminium [8]. Nevertheless, it may occur that the neighbouring crystals, surrounding or included, are slightly disoriented ($1' < \theta < 10°$). In that case, strong contrast fading in the big crystal (Fig. 1) can be seen near the boundaries on the topographs. This is the characteristic picture of a small-angle grain boundary, it destroys the crystal quality and one has to get rid of it if one is interested in the crystal perfection. Sometimes, dislocation tangles making a very small-angle grain boundary might be observed ($\theta < 1'$: below X-ray rocking curve width) (Fig. 5, mark 2).

3.2.5. "Black dots"

About "black dots" only a few things can be said. Their chemical analysis is not the purpose of this paper. We are carrying it out now. It can be noticed that their number varies much not according to the metal purity but to the batch number. Thus we made "éponge refondue" titanium crystals with few "black dots" (Fig. 5a) and "Van Arkel" titanium crystals presenting a lot of these "dots".

4. Discussion

4.1 Preparation of titanium crystals with a good crystal perfection

The good quality of titanium crystals of all purities ("éponge refondue", "M.R.C.", "Van Arkel") we prepared (Fig. 3) shows that purity itself is not a crucial parameter when recrystallization is made under high vacuum. The results are in agreement with those obtained on aluminium [9]. They mean that dislocation sources, mainly those of vacancy nature, are only a little if not at all activated. This phenomenon can be explained by the fact that vacancy supersaturation, in crystals grown by recrystallisation under high vacuum ($P \leqq 10^{-8}$ Torr) and cooled down slowly (3 °C/h) to room temperature, is never very high. It is certainly that which could be calculated for a perfect crystal, cooled in the same conditions and the external surfaces of which are clean [10].

We have already mentioned that, generally speaking, crystals were larger when we carried out several cycles around the phase transformation point (4 to 5 cycles). This is a peculiarity of this preparation method which uses the energy of phase transformation $\beta \rightarrow \alpha$ as volume energy to grow the α-grain. What is its mechanism? One should think that for each cycle the passage through the phase transformation point ($\alpha \rightleftarrows \beta$) will destroy the prior cycle effect and whatever our standpoint could be, dimension or perfection, there is no advantage to multiply the number of cycles. Accordingly, no differences were ever observed in the density and the configuration of defects between, for example, the first and the fourth cycle. But it has been found that the crystal size was then increasing. This could be considered as an odd contradiction as it is known that during the transformation of α-phase (h.c.) into β-phase (c.c.) some crystallographic relations are satisfied

$$(0001)\ \alpha \parallel \{110\}\ \beta$$

and

$$\langle 11\bar{2}0 \rangle\ \alpha \parallel \langle 111 \rangle\ \beta\,.$$

Consequently, one β grain can give twelve α grains but one α grain can be transformed into six β grains. In the best case, i.e. when, at high temperature, only one β grain is formed after several $\alpha \rightleftarrows \beta$ passages, many α phase crystals should be observed. Considering the previous relation, this should be the smallest number of crystals to be obtained. These results should be explained through a grain consuming process like grain growth which would follow the phase transformation. During α-phase annealing the most stable α crystal would grow at the expense of its neighbours. A kinematical study of this phenomenon is certainly interesting, its results will be published in a further paper.

X-ray topographs before and after sample polishing (chemical or electrolytical polishing), show us that "black dots" do not appear during this treatment (Fig. 4). This observation casts a doubt on Bailey's conclusion [11] about precipitation formation in zirconium annealed under $P < 10^{-5}$ Torr. It is not sure that the precipitates he has observed by electron microscopy were nucleated during electrochemical thinning. Electron micrographs made only after the procedure do not prove this fact. Our experiment (Fig. 4) shows that we must not forget another possibility, if precipitates are hydrides, they can be simply associated with difficulties of hydrogen to evolve from crystals, even under high vacuum annealing (5×10^{-10} Torr in our case). X-ray topography is, in this case, more relevant than electron microscopy to determine the actual moment of the precipitate formation.

Rather perfect titanium crystals can be produced by recrystallization under high vacuum ($P \leqq 10^{-8}$ Torr) even if the metal is impure ("éponge refondue"). Phase transformation does not inhibit the preparation of such crystals. However, to prepare crystals of an excellent quality, the identification and the expulsion of "black dots" still have to be done.

4.2 c + a Burgers vector dislocations

Such defects have been observed by electron microscopy in impure deformed titanium [4, 7]. The peculiarity of our observations (Fig. 5) is that we did them obtained in crystals by recrystallization without applying external stresses. Dislocations of this type can be seen in impure titanium and near grain boundaries. They had never been observed in pure titanium ("Van Arkel"). It could be thought that they are formed by crystals meeting at high temperature, when the resulting stresses cannot initiate usual gliding systems. For the same reason when impurities hinder a prismatic gliding and favour a secondary one, these dislocations are also observed. Accordingly we can see in Table 1 that the mains impurities in impure titanium ("éponge refondue", "M.R.C.") are hydrogen, carbon, and oxygen. These impurities may cause the hindrance of prismatic gliding systems. If we remember that our crystals are obtained at 850 °C, we shall find that our results are in agreement with those of Paton and Backofen [12] who stated that $\boldsymbol{c} + \boldsymbol{a}$ Burgers vector dislocations are more numerous at a high-temperature deformation.

4.3 Dislocation in rows

Dislocations in rows are spreading from "black dots". The chemical composition of these dots has not yet been determined, but it is sure that their image looks like the one of spherical stress concentration around a particle (hydride precipitates, hydrogen bubbles, ...). Jones and Mitchell [13] stated that for such a system (glass balls in AgCl) dislocation loops can be nucleated and emitted on their glide cylinder. Lally and Partridge used this mechanism to explain with considerations of anisotropic gliding the presence of dislocation loops around hydrogen bubbles included in magnesium [14]. Although gliding in titanium is not identical to gliding in magnesium when crystals are submitted to external stresses, we may consider that in the case of spherical internal stresses, these metals have, especially at high temperatures, the same behaviour. Accordingly, it could be thought that all gliding systems in titanium $|\{1\bar{1}00\}\langle 11\bar{2}0\rangle$, $(0001)\langle 11\bar{2}0\rangle$, $\{1\bar{1}01\}\langle 11\bar{2}0\rangle|$ will be activated. The presence of both glissile basal loops and prismatic loops in our crystals can be explained in this way (Fig. 7).

5. Conclusion

This X-ray topographic work on dislocations in titanium crystals, prepared by recrystallization after phase transformation $\alpha \rightleftarrows \beta$ and under $P \approx 5 \times 10^{-10}$ Torr shows that:

1) Metal impurities and polishing procedure do not increase the number of defects. It is the same for the number of cycles around the phase transformation point $\alpha \rightleftarrows \beta$ that can be then multiplied to obtain larger crystals.

2) These crystals sometimes contain dislocations of $\boldsymbol{c} + \boldsymbol{a}$ Burgers vectors (those of less pure titanium) related to their growth and ascribed to high-temperature deformation (two grains meeting). They contain also gliding dislocations and prismatic dislocations located around a stress centre. A chemical study of these stress centres ("black dots") must be realized because they look like those found in zirconium by electron microscopy [15] but it is not certain that they are hydrides. Moreover, they alter the crystals the quality of which would be excellent without them.

Acknowledgements

This work was supported by C.N.R.S. — R.C.P. N° 244 entitled "Mechanic Properties of Titanium and its Alloys". Thanks are due to Mr. J. P. Langeron, manager of this R.C.P., for his assistance and to Mr. F. L. Rigaud, Institut de Petroléochimie, for his help in writing this paper in English.

References

[1] T. R. Cass, Trans. AIME **239**, 1864 (1967).
[2] T. Tanaka and H. Conrad, Acta metall. **19**, 1001 (1971).
[3] P. G. Partridge, Metallurg. Rev. **118**, 169 (1968).
[4] T. R. Cass, The Science, Technology, and Application of Titanium, Pergamon Press, 1970 (p. 459).
[5] A. Akhtar and E. Teghtsoonian, J. Crystal Growth **28**, 227 (1975).
[6] C. Jourdan, D. Rome-Talbot, and J. Gastaldi, Phil. Mag. **26**, 1053 (1972).
[7] J. C. Williams and M. J. Blackburn, phys. stat. sol. **25**, K1 (1968).
[8] J. Gastaldi and C. Jourdan, J. Crystal Growth **35**, 17 (1976).
[9] J. Gastaldi, P. Marzo, and C. Jourdan, J. Crystal Growth **18**, 77 (1973).
[10] J. Gastaldi, Thèse Université d'Aix – Marseille III, 1975.
[11] J. E. Bailey, Acta metall. **11**, 267 (1963).
[12] N. E. Paton and W. A. Backofen, Metallurg. Trans. **1**, 2839 (1970).
[13] D. A. Jones and J. W. Mitchell, Phil Mag. **3**, 1 (1958).
[14] J. S. Lally and P. G. Partridge, Phil. Mag. **13**, 9 (1966).
[15] J. P. Langeron, Thèse Université de Paris, 1965.

(Received June 30, 1977)

phys. stat. sol. (a) 43, 437 (1977)

Subject classification: 5 and 18.2; 21.1.1; 21.4

Institut für Angewandte Physik (a) and Institut für Experimentalphysik (b) der Technischen Universität Wien

Hyperfine Interactions in $RE_2(Fe, Ni)_{17}$

RE = Y, Gd, Dy

By

W. STEINER (a), R. HAFERL (b), and H. WELSER (a)

Mössbauer measurements are performed on $RE_2(Fe_{1-x}Ni_x)_{17}$, RE = Y, Gd, Dy, compounds. The spectra are analyzed using 7 (8) independent subspectra for the samples with the Th_2Zn_{17} (Th_2Ni_{17}) structure type. The mean hyperfine fields obtained are nearly independent of the RE atoms. Up to $x \approx 0.2$ a small increase in the hyperfine splitting and a constant value for further increasing x is found. This is in line with the magnetic measurements which can be explained by a filling up of the majority spin subband in analogy with the binary (Fe, Ni) alloys at low Ni concentrations ($x \lesssim 0.15$). From the concentration dependence of the quadrupole splitting no change of the easy axis caused by the Ni substitution can be observed and from the isomer shift charge transfers may be excluded.

An den Verbindungen $RE_2(Fe_{1-x}Ni_x)_{17}$, RE = Y, Gd, Dy werden Mößbauermessungen durchgeführt. Die Spektren werden mit 7 (8) unabhängigen Subspektren für die Proben mit Th_2Zn_{17} (Th_2Ni_{17}) Struktur analysiert. Die erhaltenen mittleren Hyperfeinfelder sind nahezu unabhängig von den RE Atomen. Bis $x \approx 0,2$ wird ein geringer Anstieg und für höhere x ein konstantes Hyperfeinfeld beobachtet. Dies ist in Übereinstimmung mit den magnetischen Messungen, die durch eine teilweise Auffüllung des Majoritäts-Spinbandes, in Analogie zum binären (Fe, Ni) System bei kleinen Ni-Konzentrationen ($x \lesssim 0,15$), erklärt werden. Aus der Konzentrationsabhängigkeit der Quadrupolaufspaltung kann auf keine durch die Ni-Substitution verursachte Drehung der leichten Richtung geschlossen werden. Die Isomerieverschiebungen weisen darauf hin, daß Ladungstransfer weitgehendst auszuschließen ist.

1. Introduction

A strong dependence of both the crystallographic and the magnetic properties upon the 3d component was found for intermetallic compounds with a composition near RE_2M_{17} (RE = Y, Gd, Dy, M = Fe, Ni). As a result of the different atomic radii for RE_2M_{17} the rhombohedral and the hexagonal structure type was detected ([1] and references scited herein), whereas for RE_2Ni_{17} only the hexagonal type could be observed [2 to 5]. Complete solid solubility [6, 7] as well as a change of the easy axis of magnetization depending on the Co concentration [8 to 11] was found for $RE_2(Fe, Co)_{17}$. On the other hand mixed crystals, only up to $x = 0.5$, 0.4, and 0.3 for RE = Y, Gd, and Dy, respectively, were formed for $RE_2(Fe_{1-x}Ni_x)_{17}$ and further for all these compounds an increasing excess of RE with increasing x was necessary to prepare single phased samples [12, 13]. $RE_2(Fe, Ni)_{17}$ with RE = Y, Dy exhibit the hexagonal structure type, whereas for RE = Gd a change from the rhombohedral to the hexagonal structure for $x > 0.2$ occurs. The strong dependence of the structure type formed can be observed from the c/a ratio, because for the sample with the boundary concentration it is approximately the same as that for the corresponding RE_2Ni_{17} compound [13].

The magnetic measurements show an antiparallel coupling of the RE and the 3d moments. The mixed crystals $Y_2(Fe, Ni)_{17}$ are ferromagnetic. The additional conduction electrons contributed by the RE atoms seems to have only a small influence on the mean magnetic moment in RE_2Fe_{17} and in samples for which a substitution of Fe by Ni was possible. Because of the analogous concentration dependence of the mean 3d moment in the binary (Fe, Ni) alloys and the ternary $RE_2(Fe, Ni)_{17}$ compounds a similar shape of the density of states near the Fermi energy for the 3d band can be assumed. This is supported by the results obtained for $RE_2(Fe, Co)_{17}$ ([14] and references cited therein). Up to $x \approx 0.15$, one of the unfilled 3d subbands, proposed for RE_2Fe_{17}, will be filled up. The linear decrease of the mean moment for further increasing x leads to the assumption of a constant Fe moment of approximately 2.5 and a Ni moment of 0.6 μ_B [13].

To allow further examinations of these assumptions Mössbauer measurements have been performed on these mixed crystal series. For $Gd_2(Fe, Ni)_{17}$ the measurements also have been carried out at 80 K, to get a practicable low T/T_C ratio. Gd has been chosen to eliminate the influence of the crystal field on the concentration dependence of the mean 3d moment.

2. Experimental Results

The sample preparation and characterization has been described in [12, 13]. The experimental equipment was specified in [1].

An antiparallel coupling of the RE and Fe moments has been detected for RE_2Fe_{17}, RE = Gd, Dy [15 to 17]. For this case the calculated dipole fields lead to no further possibility to distinguish between the non-equivalent Fe sites and the same number of subspectra is obtained as for Y_2Fe_{17} [1]. From magnetic measurements it can be assumed that a RE^{3+} ion is formed. Therefore, assuming a point charge model, the same results will also be expected for the crystalline electric field gradient, as for Y_2Fe_{17}. Following [1], the obtained spectra have been analysed with seven (eight) independent subspectra having a fixed intensity ratio of 1:0.5:1:0.5:0.5:0.25:0.5 (1:0.5:0.5:0.5:0.5:0.5:0.25:0.5) for samples with the Th_2Zn_{17}(Th_2Ni_{17}) structure type. Due to the low resolution of the measured spectra, this analysis will always carry some degree of arbitrariness. Furthermore for both Gd_2Fe_{17} and Dy_2Fe_{17} the same considerations have to be performed concerning a non-spherical charge cloud and a covalent bond superimposed upon the metallic bond, as for Y_2Fe_{17} [1].

To reduce the number of free parameters two assumptions have been introduced:

(i) The RE excess, which probably leads to a statistical substitution of some of the 3d by RE atoms at the c(f) sites [1, 18] has been neglected and an ideal Th_2Zn_{17} (Th_2Ni_{17}) structure type was proposed.

(ii) The changes for a given Fe site caused by the Ni substitution have been neglected. This seems to be valid with regard to the entire shape of the measured spectra, excluding those of $Y_2(Fe_{0.5}Ni_{0.5})_{17}$, which is the Ni-rich terminal composition containing 38.5% RE excess. But the increasing linewidth with increasing Ni content of approximately 25% indicates that this assumption will be doubtful at high Ni concentrations.

The main difference between the measured spectra at 300 K obtained for the hexagonal and the rhombohedral samples is given in the shape of the fifth line at positive velocities (Fig. 1). For both structure types the lowest Fe–Fe distance is given for the f- and c-sites, respectively (Wyckoff notation). Analogous to Y_2Fe_{17} the largest hyperfine field has been attributed to those sites and this will not be altered by Ni substitution. Although evidence for a preferred substitution of 3d atoms on one of the Fe sites has been determined for $Y_2(Fe, Co)_{17}$ [19] and $RE_2(Fe, Ni)_{17}$, RE = Er, Tm [20], no definite prediction could be obtained from the measured spectra. This is mainly due to the large number of subspectra needed for the analysis

Fig. 1. (1) Rhombohedral $Gd_2(Fe_{0.8}Ni_{0.2})_{17}$, (2) hexagonal $Gd_2(Fe_{0.7}Ni_{0.3})_{17}$, (3) hexagonal $Dy_2(Fe_{0.8}Ni_{0.2})_{17}$, and (4) hexagonal $Y_2(Fe_{0.8}Ni_{0.2})_{17}$ at 300 K

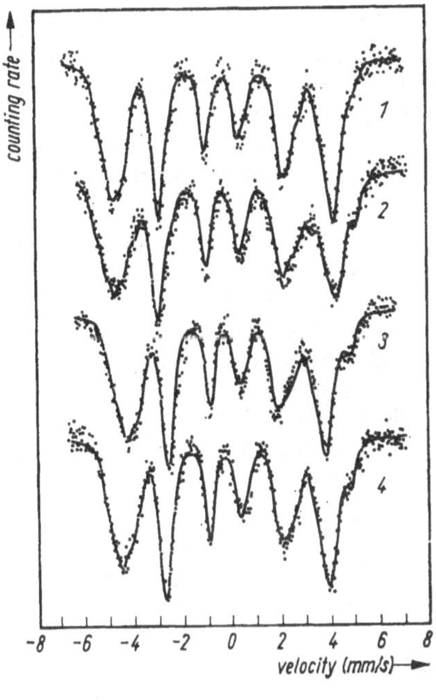

and the low resolution of the measured spectra. From this, some uncertainties result for the line positions but, contrary to [20, 21], the intensity of the spectra corresponding to the c-sites agrees with the ideal occupation number as well as with a statistical substitution. At 80 K the hyperfine field obtained for the c-sites and for $x > 0.2$ fort he f-sites, remains constant for $Gd_2(Fe, Ni)_{17}$. This is in agreement with $RE_2(Fe_{0.88}Ni_{0.12})_{17}$, RE = Tm, Er, where for an easy basal plane a concentration-independent hyperfine field at 0 K has been detected, too [20].

3. Discussion

Since a coordination of the subspectra to the different Fe sites is only possible with a large amount of uncertainty, the concentration dependence of the weighted mean values of the hyperfine parameters have been calculated [1]. The values of the hyperfine fields determined for RE_2Fe_{17}, RE = Gd, Dy, at 300 K are slightly higher than those obtained from [22], whereas good agreement is found for the mean value of Gd_2Fe_{17} at 80 K [23]. At room temperature for $x < 0.2$ the hyperfine fields for

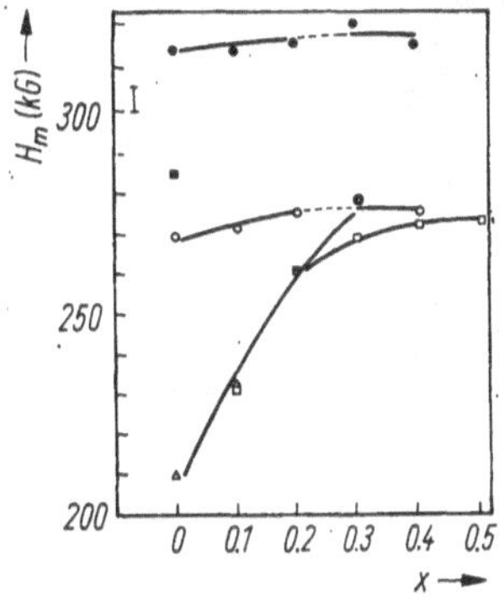

Fig. 2. Hyperfine fields of $Y_2(Fe, Ni)_{17}$ (□ 300, ■ 80 K), $Gd_2(Fe, Ni)_{17}$ (○ 300, ● 80K), and $Dy_2(Fe, Ni)_{17}$ (△ 300 K)

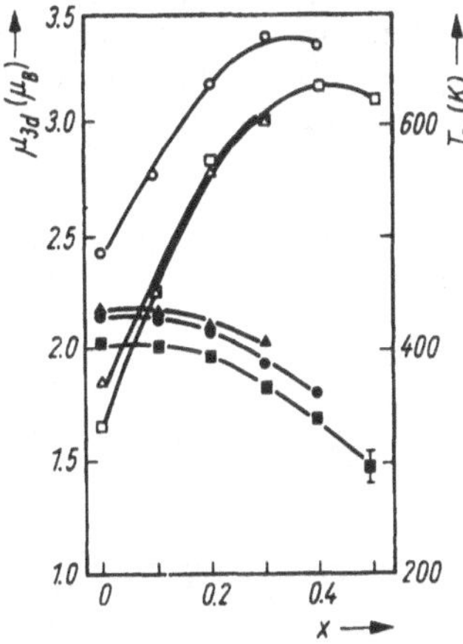

Fig. 3. Mean 3d moments (full symbols) and Curie temperatures (open symbols) of $RE_2(Fe, Ni)_{17}$, RE = Y (□), Gd (○), and Dy (△)

$RE_2(Fe, Ni)_{17}$, RE = Y, Dy, are appreciably smaller than those of the corresponding $Gd_2(Fe, Ni)_{17}$ compound (Fig. 2). This is mainly due to the large discrepancies of the T/T_C ratios. For $Dy_2(Fe_{0.7}Ni_{0.3})_{17}$ $T/T_C = 0.5$ and for $Gd_2(Fe_{0.7}Ni_{0.3})_{17}$ $T/T_C = 0.44$ (Fig. 3). Therefore the hyperfine fields are the same in view of the overall error. For $Gd_2(Fe, Ni)_{17}$ and $Dy_2(Fe, Ni)_{17}$ the mean 3d moments are in accord with each other. The small deviations might be due to influences of the crystalline electric field acting on the Dy moment, which have been neglected for the calculation of the mean 3d moments [13]. The values obtained for $Y_2(Fe, Ni)_{17}$ are about $0.1\mu_B$ smaller than those for $RE_2(Fe, Ni)_{27}$, RE = Gd, Dy. Since Y does not contribute to the mean magnetic moment and furthermore since the interactomic distances are nearly the same for $Dy_2(Fe, Ni)_{17}$ and $Y_2(Fe, Ni)_{17}$, this small contribution must be due to the influence of RE atoms on the mean 3d moment. However, the influence on the Fe hyperfine field is also small (Fig. 2).

The analogy of the concentration dependence of the mean 3d moment of $RE_2(Fe,Ni)_{17}$ with those obtained for both the binary (Fe, Ni) [24] and (Fe, Co) [25] alloys as well as for $RE_2(Fe, Co)_{17}$ [14] leads to the assumption of a similar shape of the density of states for the 3d electrons near the Fermi energy. Two unfilled bands and a minimum in the density of states of the minority band near the Fermi level were proposed [26]. According to the increasing mean moment, the mean hyperfine field should also increase with increasing x. In view of the overall error for $Gd_2(Fe, Ni)_{17}$ up to $x \approx 0.2$ an increasing (and for $x > 0.2$ a constant) hyperfine field is observed at 80 K (Fig. 2). This is in agreement with $RE_2(Fe, Ni)_{17}$, RE = Er, Tm, if an easy basal plane plane is observed [20]. For $x \geqq 0.2$ a linear decrease of the mean 3d moment was detected, leading to the assumption of a constant Fe moment of 2.5 and a Ni moment of $0.6\mu_B$ [13]. The results obtained from the hyperfine fields are in line with this interpretation, and the increasing Curie temperatures measured in this concentration range (Fig. 3) indicate a strong dependence of the 3d–3d exchange energy on the interatomic distances along the crystallographic c-direction.

From the overall shape of the Mössbauer spectra as well as from the concentration dependence of the mean quadrupole splitting an easy axis situated in the basal plane has to be assumed (Fig. 4). Contrary to $RE_2(Fe, Co)_{17}$, RE = Y, Gd [9, 11, 19], no change of the easy axis caused by the Ni substitution could be observed. Furthermore for high Co concentrations an appreciable dependence of the hyperfine splitting upon the Co content could be detected [14], indicating that the above given explanation might only be valid for low Co or Ni concentrations. There a relatively small disturbance can be assumed for the band structure, because all compounds with analogous

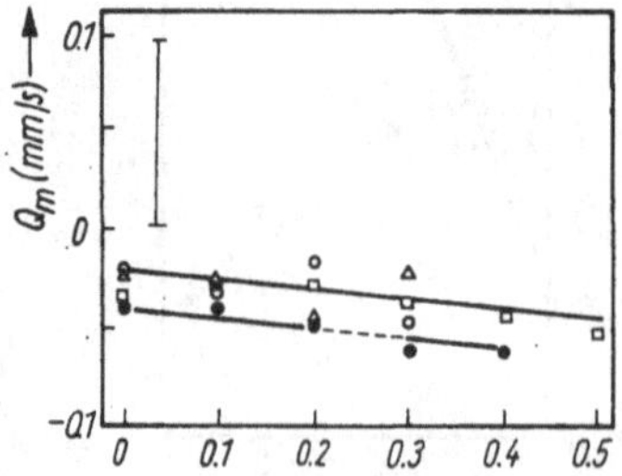

Fig. 4. Mean quadrupole splitting of $RE_2(Fe, Ni)_{17}$, RE = Y (□ 300, ■ 80 K), Gd (○ 300, ● 80 K), and Dy (△ 300 K)

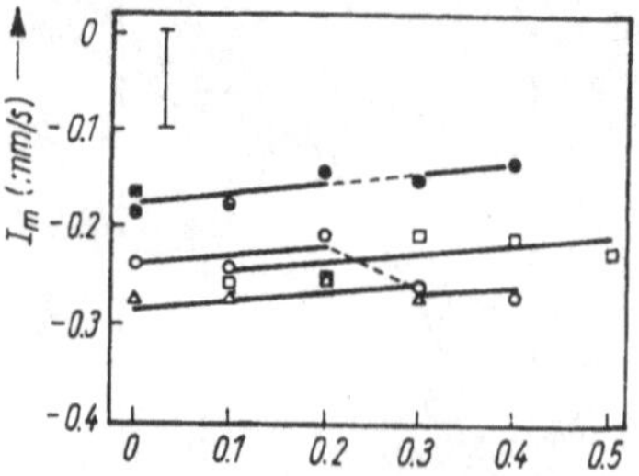

Fig. 5. Mean isomer shift of $RE_2(Fe, Ni)_{17}$, RE = Y (□ 300, ■ 80 K), Gd (○ 300, ● 80 K), and Dy (△ 300 K)

structure types, such as $RE(Fe, Co)_2$ [27, 28], and $RE(Fe, Co)_3$ [29], show a similar behaviour of the mean 3d moments and the mean hyperfine fields in this concentration region. For $Dy_2(Fe, Ni)_{17}$ and $Dy_2(Fe, Co)_{17}$ an easy basal plane was always observed. The anisotropy of the Dy ions dominates that of the (Fe, Ni) and (Fe, Co), respectively. The mean quadrupole splitting is the same for both structure types Th_2Zn_{17} and Th_2Ni_{17}, indicating their similarity, and is almost independent of the Ni concentration, though a drastic decrease of the c-parameter can be observed with increasing x. The temperature dependence of the quadrupole splitting is small and that of the isomer shift is about the same as for Y_2Fe_{17} (Fig. 4, 5).

Since a RE^{3+} ion will always be obtained, the number of conduction electrons stays constant for all series $RE_2(Fe, Ni)_{17}$. A slightly increasing isomer shift resulting from the substituted Ni atoms could be detected, and in spite of the overall erro charge transfers may be excluded (Fig. 5).

4. Conclusion

Up to $x \approx 0.2$ a small increase in the hyperfine splitting and a constant value for further increasing x is found for $RE_2(Fe, Ni)_{17}$ compounds. The mean 3d moment decreases linearly with increasing x for $x > 0.2$. This indicates a constant Fe moment for high Ni concentrations. For low Ni content ($x \leqq 0.15$) a filling up of one of the unfilled 3d subbands, proposed for RE_2Fe_{17}, may explain the concentration dependence of the mean 3d moment and the mean hyperfine field. Because a similar behaviour could be observed for all other structurally related compounds $RE(Fe, Co)_2$, $RE(Fe, Co)_3$, and $RE_2(Fe, Co)_{17}$ a similar shape of the density of states seems to be possible for these compounds for low contents of Co or Ni. Nevertheless it appears to be impossible to decide whether this partial filling or the decrease of a portion of the negative exchange will be the correct explanation. No change of the easy axis caused by the Ni substitutions could be observed. From the concentration dependence of the isomer shift charge transfers can be excluded with reasonable certainty.

References

[1] W. Steiner and R. Haferl, phys. stat. sol. (a) **42**, 739 (1977).
[2] K. H. J. Buschow, J. less-common Metals **11**, 204 (1966).
[3] J. Laforest, R. Lemaire, D. Paccard, and R. Pauthenet, C. R. Acad. Sci. (France) **B 264**, 676 (1967).
[4] P. D. Carfagna and W. E. Wallace, J. appl. Phys. **39**, 5259 (1968).
[5] D. Givord, F. Givord, R. Lemaire, W. J. James, and J. S. Shah, J. less-common Metals **29**, 389 (1972).
[6] W. Steiner, B. K. Choi, G. Hilscher, and H. Kirchmayr, Proc. Internat. Conf. Magnetism, Vol. IV, Moscow 1973 (p. 63).
[7] W. Steiner, H. R. Kirchmayr, and W. Springs, Z. angew. Phys. **32**, 146 (1971).
[8] A. E. Ray and K. J. Strnat, AFML-TR-71-210 (1971).
[9] R. S. Perkins and H. Nagel, Physica (Utrecht) **80 B**, 143 (1975).
[10] K. J. Strnat, IEEE Trans. Magnetics **8**, 511 (1972).
[11] R. Grössinger, W. Steiner, F. Culetto, and H. Kirchmayr, Physica (Utrecht) **86—88 B**, 210 (1977).
[12] W. Steiner and H. Hrubec, J. less-common Metals **41**, 165 (1975).
[13] W. Steiner, A. Planck, and G. Weihs, J. less-common Metals **45**, 143 (1976).
[14] W. Steiner, R. Haferl, and B. K. Choi, phys. stat. sol. (a) **43**, 111 (1977).
[15] D. Givord and R. Lemaire, IEEE Trans. Magnetics **10**, 109 (1974).
[16] D. Givord, R. Lemaire, W. J. James, J. M. Moreau, and J. S. Shah, IEEE Trans. Magnetics **7**, 657 (1971).
[17] K. H. J. Buschow, phys. stat. sol. (a) **7**, 199 (1971).

[18] F. Givord and R. Lemaire, J. less-common Metals **21**, 463 (1970).
[19] J. Deportes, D. Givord, R. Lemaire, H. Nagai, and Y. T. Yang, J. less-common Metals **44**, 273 (1976).
[20] P. C. M. Gubbens and K. H. J. Buschow, phys. stat. sol. (a) **34**, 729 (1976).
[21] P. C. M. Gubbens, A. M. van der Kraan, and K. H. J. Buschow, Solid State Commun. **19**, 355 (1976).
[22] L. M. Levinson, E. Rosenberg, A. Shaulov, S. Shtrikman, and K. Strnat, J. appl. Phys. **41**, 910 (1970).
[23] P. C. M. Gubbens, J. J. van Loef, and K. H. J. Buschow, J. Physique **35**, C6 (1974).
[24] C. E. Johnson, M. S. Ridout, T. E. Cranshaw, and P. E. Madsen, Phys. Rev. Letters **6**, 450 (1961).
[25] B. De Mayo, D. W. Forester, and S. Spooner, J. appl. Phys. **41**, 1319 (1970).
[26] N. F. Mott, Adv. Phys. **13**, 325 (1964).
[27] A. P. Guimaraes and D. St. P. Bunbury, J. Phys. F **3**, 885 (1973).
[28] A. M. van der Kraan and P. C. M. Gubbens, J. Physique **35**, C6-469 (1974).
[29] S. K. Arif, D. St. P. Bunbury, and G. J. Bowden, J. Phys. F **5**, 1792 (1975).

(Received March 31, 1977)

Note added in proof:

In [14] on page 115 in the sentence "Therefore common 3d bands have been assumed and the *two* additional 3d electrons. . ." (7th and 8th line from the top) the "two" has to be deleted.

phys. stat. sol. (a) 43, 443 (1977)

Subject classification: 1.2; 1.3; 8; 21.1

Institut für Festkörperforschung der Kernforschungsanlage Jülich[1])

A Differential Thermal Analysis Study of the Vanadium–Hydrogen System

By

T. SCHOBER and A. CARL

Results of a differential thermal analysis (DTA) study of vanadium hydrides are presented. A new VH-phase diagram is obtained containing a hitherto unknown low-temperature phase. The existence and extension of two-phase regions is shown. The V–H solubility limit as expressed by $c = c_0 \exp(-\Delta H/kT)$ is determined to: $\Delta H = 0.153$ eV and $c_0 = 880\%$. Some metallographic results are also included.

Die Ergebnisse einer Differential-Thermoanalyse (DTA)-Untersuchung an Vanadium-Hydriden werden dargestellt. Es wird ein neues VH-Phasendiagramm mit einer bisher unbekannten Tieftemperaturphase erhalten. Für die VH-Löslichkeitsgrenze, ausgedrückt durch $c = c_0 \exp(-\Delta H/kT)$, wird $\Delta H = 0{,}153$ eV und $c_0 = 880\%$ gefunden. Es werden auch einige metallographische Ergebnisse wiedergegeben.

1. Introduction

Early studies of the V–H system include the work of Maeland [1], Arons et al. [2], Asano and Hirabayashi [3], Wanagel et al. [4 to 6], and Westlake et al. [7]. The V–D system is explicitly excluded here since there are large differences compared to the V–H system [2, 8 to 10, 18]. Significant progress occurred very recently through the resistivity and NMR work of Fukai and Kazama [11] and the calorimetric work of Asano et al. [12]. In these investigations, the ranges of existence of a new high-temperature phase (ε) and of the fully ordered low-temperature phase δ [3, 5, 7, 11, 12] were established. Further already known phases of the V–H system were: the α-phase — a disordered b.c.c. solution of H in V, the β-phase — an ordered monoclinic phase with octahedral occupancy [2, 12 to 14], and the dihydride phase γ of composition VH_2 [1, 15]. It is the purpose of the present work to present results of an extensive differential thermal analysis (DTA) study on VH alloys. It emphasizes the question of two-phase regions and their extension between the various phases. The results of Asano et al. [12] and Fukai and Kazama [11] are critically examined.

2. Experimental

High-purity V ingots were spark-cut to small cubes of about 20 to 30 mg.[2]) The hydrogen charging was accomplished either by the electrolytic technique or by gas-phase charging in standard equipment (see reference [16], Chapter 5). Very high concentration master alloys were prepared in a special high-pressure apparatus with UHV components which is schematically shown in Fig. 1. Alloys with slightly lower

[1]) 5170 Jülich, BRD.

[2]) Electro-refined V (U.S. Bureau of Mines, Boulder City, Nevada) was used. The interstitial impurity levels were: C < 86, N < 10, O < 70 wt-ppm. We deliberately avoided the use of powder samples since there is ample evidence in our opinion that such samples are (i) inhomogeneous in concentration and (ii) rather impure due to a pick-up of oxygen and nitrogen during charging.

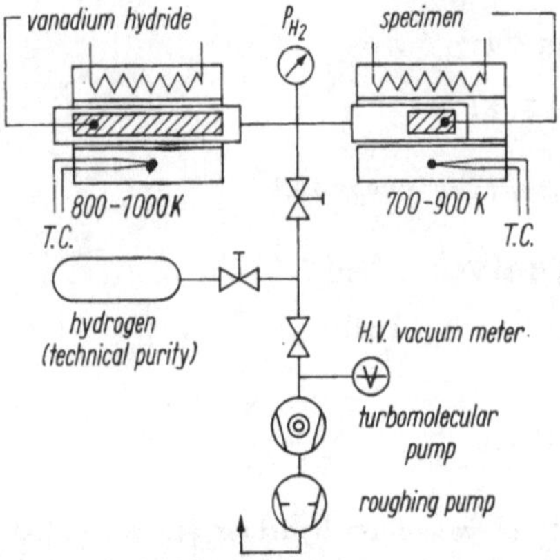

Fig. 1. High-pressure hydrogen charging apparatus. Only UHV components were used. Hydrogen of technical purity is dissolved in a first step in a vanadium rod (left-hand side). Highest-purity hydrogen at pressures up to 150 bar is then obtained through the thermal decomposition of the vanadium hydride. Simultaneously, the sample at the right-hand side is gas phase charged with hydrogen. Obviously, a thorough pump-out is required between the two cycles

concentrations were achieved through controlled-vacuum hot extraction in a hydrogen gas analyzer (cf. [16], Chapter 5). In the same analyzer, the H content was determined after the experiment to a precision of about $\pm 0.2\%$.[3]) The DTA equipment used was a Mettler TA2000(B) apparatus. In order to detect the onset and the end of a phase transformation (for instance, the extension of a two-phase region), the scanning speeds were held so low that

$$\tau_r \ll \Delta t\,. \tag{1}$$

(Here, τ_r is the thermal relaxation time of the DTA system, and Δt the time spent in the vicinity of the phase transformation.) This requirement resulted in scanning rates from 0.1 to 0.5 K min^{-1}. In contrast, high heating rates (≈ 5 K min^{-1}) were used to detect low-enthalpy phase changes. Here, $\tau_r \approx \Delta t$ and no two-phase regions could be determined. Several important alloys were also given a metallographic polish and observed at room temperature and on a heating stage up to temperatures of 200 °C. (The procedure was presented in [17].)

3. Results

Concerning the VH solvus at low concentration, it is commonly expressed by

$$c = c_0 \exp\left(-\frac{\Delta H}{kT}\right). \tag{2}$$

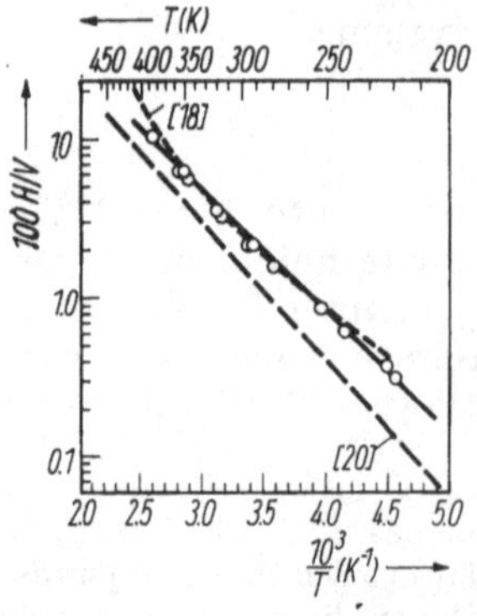

Fig. 2. VH solvus as obtained from present DTA measurements. Dashed lines: results by Westlake and Ockers [18] and Chang and Wert [20]

[3]) The concentration is indicated in this work in % as given by: (atom ratio H/V) · 100.

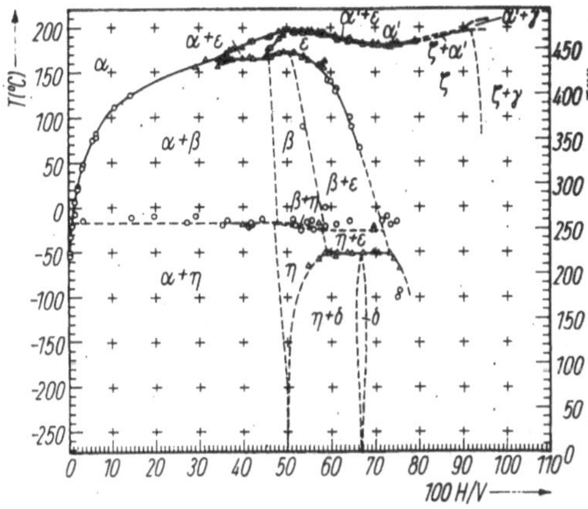

Fig. 3. VH phase diagram based on present DTA data. Open circles: heating rate 5 Kmin^{-1}; triangles: heating rate (0.2 to 0.5) Kmin^{-1}. Note the following transition temperatures: eutectoid reaction $\varepsilon \rightarrow \alpha + \beta$ at (164 ± 1) °C, decomposition of β at (173 ± 1) °C, decomposition of ε at (197 ± 1) °C, peritectoid reaction $\delta \rightarrow \eta + \varepsilon$ at −(51 ± 1) °C

Here, c is the atom ratio V/H in % and ΔH the enthalpy. The results of our DTA measurements are shown in Fig. 2. Included in Fig. 2 is the resistivity work of Westlake and Ockers [18] which is seen to be in very good agreement with our results. From our data we obtain $\Delta H = 0.153$ eV and $c_0 = 880\%$. Another recent resistivity solvus study was presented by Heller and Wipf [19] who obtained $\Delta H = 0.13$ eV and $c_0 = 360\%$. These values are in fair agreement with this work when the uncertainties and experimental difficulties are considered. Finally, we note the solvus work of Chang and Wert [20] which again is in fair agreement with the above values (see Fig. 2). Resistivity and internal friction were utilized in the latter study.

The phase diagram based on our DTA data is shown in Fig. 3 (for exact transition temperatures see figure caption). Fig. 4 shows the enlarged high-temperature region above 30% and 100 °C. It shows in more detail the extension of the (α'–ε) two-phase region which was about 2 to 3 K for concentrations above 50%. — Typical DTA curves are shown in Fig. 5a and b as measured with the precise slow heating rate of 0.5 K min^{-1}. We bear in mind that the DTA signal is essentially proportional to the heat flow from or to the specimen during the transition. In single-phase regions, there is to a first approximation no heat flow, and the DTA signal follows the base line. In contrast, heat flow is expected in two-phase regions due to the heat of transformation (provided the adjacent phase boundaries are not vertical), and the DTA signal will deviate from the base line.

As far as the decomposition of phase δ is concerned, a round maximum was proposed in recent studies [11, 12]. Such a solution is very difficult to incorporate into a phase diagram, since it requires a single-phase region above the decomposition temperatures. We list below evidence which is in support of a normal peritectoid reaction of δ decomposing at −51 °C into η and ε.

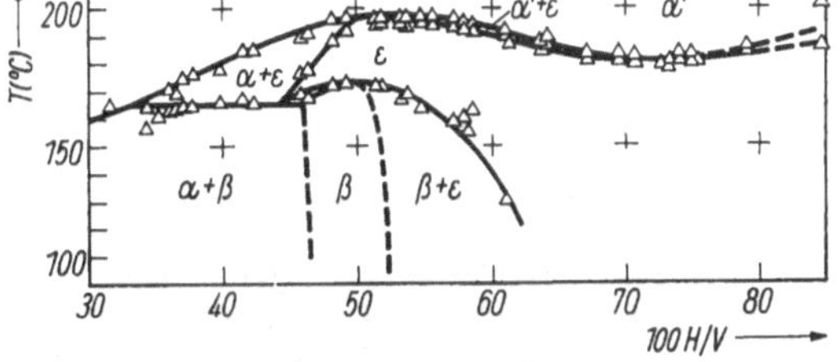

Fig. 4. Enlarged section of Fig. 3: partial VH phase diagram. Note the extension of the ($\alpha + \varepsilon$), ($\alpha' + \varepsilon$), and ($\varepsilon + \beta$) phase regions

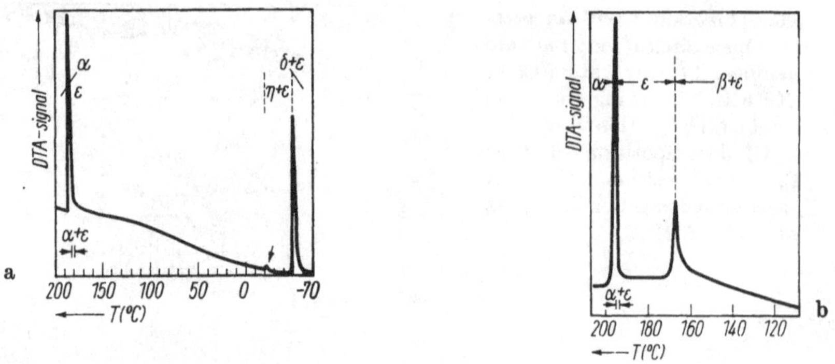

Fig. 5. Typical DTA curves of VH alloys taken during heating cycles at 0.5 Kmin^{-1}. The peak areas are approximately proportional to the transition enthalpies. a) $c = 69.5\%$; at -51 °C the peritectoid $\delta \rightarrow \eta + \varepsilon$ transition is complete. Note small peak at -20 °C where $\eta \rightarrow \beta$. Around 180 °C, ε transforms to α'. b) $c \approx 53\%$. $(\beta + \varepsilon)$ two-phase region persists up to about 170 °C, followed by a single-phase ε-region (note horizontal base line). The $\varepsilon \rightarrow \varepsilon + \alpha' \rightarrow \alpha'$ transition starts at 193 °C and is terminated at 195 °C

1. A few cooling runs between 59 and 73% demonstrated that δ-phase DTA peaks started in all cases at the same temperature of -51 °C.

2. Specimens which are not in the immediate vicinity of 66.6% display DTA curves as shown in Fig. 6a for a 59% alloy. Heat of reaction is registered for temperatures far below $T_p = -51$ °C. T_p is obtained here as the temperature above which no heat of reaction is observed anymore. Oppositely, specimens in the immediate vicinity of 66.6% show DTA curves as in Fig. 6b. Here, the heat flow starts only slightly below T_p. T_p is obtained by an extrapolation of the peak flank onto the base line. (The paths in the phase diagrams are schematically inserted in Fig. 6b.) It is also

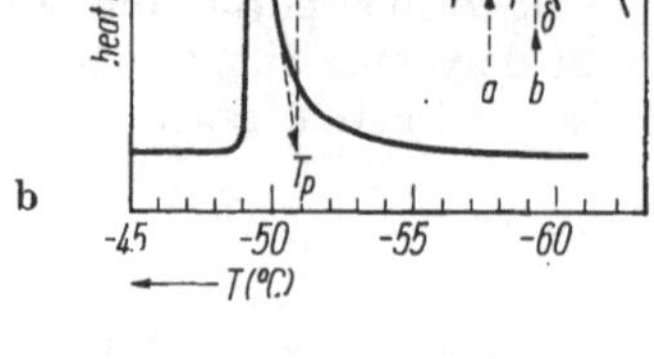

Fig. 6

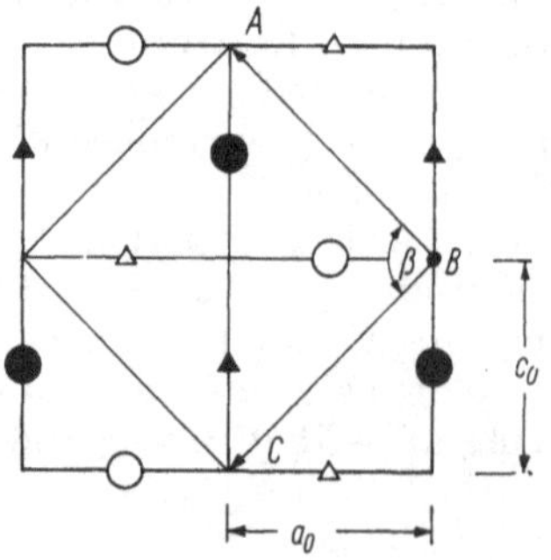

Fig. 7

Fig. 6. DTA curves of the $\delta \rightarrow \eta + \varepsilon$ transformation. a) $c_H = 59\%$. No take-up of heat is observable above T_p. b) $c_H = 67.1\%$. As a good approximation T_p gives the starting point of the take-up of heat. (Single phase δ transforms to η and ε). Heating cycle 0.5 K/min

Fig. 7. The structure of the VH β-phase after [2, 4, 6, 9, 12, 14]. Hydrogen occupies octahedral O_{z1} sites. (In phase ε, the O_{z1}- and O_{z2} sites are filled with equal probability [14]). O_{z1} site: ● H at $z = 0.1$, ○ H at $z = 1/2$; O_{z2} site: ▲ $z = 0.1$, △ $z = 1/2$ $(c_0/a_0 \approx 1.1)$

Fig. 8. Polarized-light photograph of a 20% VH alloy. Visible are thin plates on (10 3 3) planes arising from gas phase charging. Note the internal domain structure in the plates. The additional irregular patches with a domain contrast are artifacts. They were shown to be surface hydrides induced by wet grinding and polishing

evident from the above considerations that the locus of the maxima of the DTA curves alone as a function of concentration goes over a maximum. Those maxima are not a true measure for T_p, and this may be the reason for the discrepancy between this and previous work [11, 12].

A rather common observation in high-concentration VH alloys were small, very sharp peaks in the DTA curves at high temperatures. They were even observed in single-phase α-regions. We attribute them to sudden bursts of hydrogen into the atmosphere caused by cracks in the oxide layers. There cracks may be caused by plastic deformation processes during or after phase transitions. Specimens above 70% were quite vulnerable to such hydrogen loss. This may explain some of the experimental difficulties in measuring the $\varepsilon \rightarrow \alpha$ transition temperatures above 70%.

A new feature of the phase diagram in Fig. 3 is the occurrence of phase η. The $\eta \rightarrow \beta$ transition was only visible through very small DTA peaks. The involved enthalpies were two or three orders of magnitude lower than in the $\beta \rightarrow \alpha$ transition and were just at the limit of detection of our very sensitive DTA apparatus. The structure of η probably is very similar to the β-phase (see Fig. 7), but has a higher degree of order.

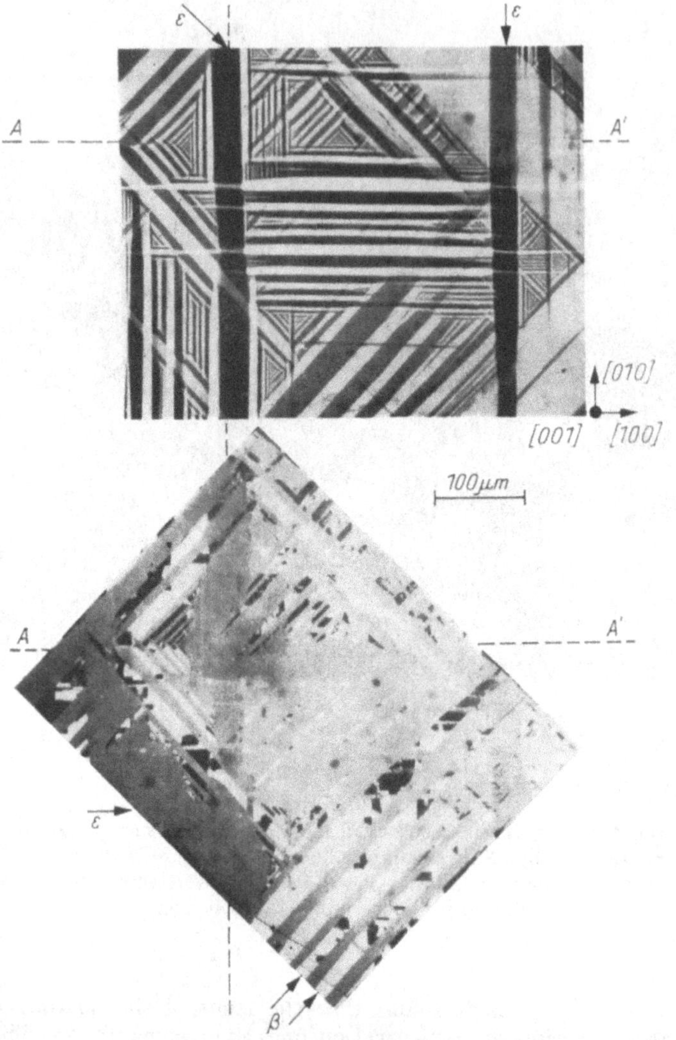

a

b

Fig. 9. Polarized-light photographs of the same area of a 56% VH alloy at room temperature. The vertical as well as the horizontal axes AA′ coincide in both photographs. a) Domain patterns of β and ε are simultaneously visible. The angle between the patterns is 45 °C. The vertical domains parallel to ⟨100⟩ are domains of phase ε. b) Same area as in a), but in a different state of polarization. ε-phase domains are now invisible. The full domain structure of β is now visible

The metallographic morphologies of phases in the α + β phase region is shown in Fig. 8. β precipitates in form of thin plates on (10 3 3) planes [6]. The polarized-light micrograph shows clearly the internal domain arrangement of the β-phase. We also obtained definite metallographic evidence for the existence of a (β + ε) two-phase region (see Fig. 3). Polarized-light observations of alloys around 55 to 60% revealed at room temperature two distinct sets of domain patterns which were shown to be attributable to β and ε, respectively. An example is presented in Fig. 9a and b. Here, identical areas are shown for two different states of polarization. In Fig. 9a, we see two interpenetrating domain patterns of ε and β. In Fig. 9b, the state of polarization was changed in order to make ε-phase domains invisible. In Fig. 9b only the β-phase domains remain in contrast. Very conclusively, heating stage experiments under polarized light revealed the disappearance of β-type domains at around 170 °C.

4. Discussion

Comparison of Fig. 3 with the diagrams in references [11, 12] shows that: a) the existence and extension of two-phase regions was established in the present work, b) the discrepancy of 20 °K between the disordering temperatures in [11, 12] of V_2H alloys was resolved (a value of (197 ± 1) °C was found in this study), c) δ displays a peritectoid transition to η and β.

As proposed by Asano et al. [14] the difference between phases β and ε is occupancy in O_{z1} sites in phase β, and equal occupancy in O_{z1} and O_{z2} sites in phase ε (see Fig. 7). The superstructure VH (here labelled ζ) is achieved by filling of the empty sites of phase ε [14]. It is very probable that ζ (with $c \approx 100\%$) decomposes according to the peritectoid reaction $\zeta \rightarrow \alpha' + \gamma$. This reaction was tentatively included in Fig. 3.

The remaining uncertainties in the diagram are the structure of phase η and the low-temperature phase boundaries from 75 to 100%. We expect here further ordered phases.

Acknowledgements

Discussions with Prof. H. Wenzl and Dr. J.-M. Welter and Dr. J. Hauck are acknowledged. Technical assistance was provided by F. Birmans. The high-purity vanadium was obtained through the kind cooperation of Dr. G. Hörz, Stuttgart.

References

[1] A. J. Maeland, J. phys. Chem. **68**, 2197 (1964).
[2] R. R. Arons, H. G. Bohn, and H. Lütgemeier, J. Phys. Chem. Solids **35**, 207 (1974).
[3] H. Asano and M. Hirabayashi, phys. stat. sol. (a) **16**, 69 (1973).
[4] J. Wanagel, S. L. Sass, and B. W. Batterman, phys. stat. sol. (a) **10**, 49 (1972).
[5] J. Wanagel, S. L. Sass, and B. W. Batterman, phys. stat. sol. (a) **11**, 767 (1972).
[6] J. Wanagel, S. L. Sass, and B. Batterman, Met. Trans. **5**, 105 (1974).
[7] D. G. Westlake, S. T. Ockers, M. H. Mueller, and K. D. Anderson, Met. Trans. **3**, 1709 (1972).
[8] H. Asano and M. Hirabayashi, phys. stat. sol. (a) **15**, 267 (1973).
[9] D. G. Westlake, M. H. Mueller, and H. W. Knott, J. appl. Cryst. **6**, 206 (1973).
[10] K. I. Hardcastle and T. R. P. Gibb, J. phys. Chem. **76**, 927 (1972).
[11] Y. Fukai and S. Kazama, Scripta metall. **9**, 1073 (1975).
[12] H. Asano, Y. Abe, and M. Hirabayashi, Acta metall. **24**, 95 (1976).
[13] V. A. Somenkov, I. R. Entin, A. Yu. Chervyakov, S. Sh. Shilshtein, and A. A. Chertkov, Soviet Phys. — Solid State **13**, 2178 (1972).
[14] H. Asano, Y. Abe, and M. Hirabayashi, J. Phys. Soc. Japan **41**, 974 (1976).
[15] J. J. Reilly and R. H. Wiswall, Inorg. Chem. **9**, 1678 (1970).
[16] T. Schober and H. Wenzl, in: Hydrogen in Metals, Ed. G. Alefeld and J. Völkl, Springer-Verlag, Berlin 1977 (to be published).
[17] T. Schober and U. Linke, Metallography **9**, 309 (1976).
[18] D. G. Westlake and S. T. Ockers, Met. Trans. **4**, 1355 (1973); **6A**, 399 (1975).
[19] R. Heller and H. Wipf, phys. stat. sol. (a) **33**, 525 (1976).
[20] H. Y. Chang and C. A. Wert, Acta metall. **21**, 1233 (1973).

(Received May 24, 1977)

phys. stat. sol. (a) **43**, 451 (1977)

Subject classification: 2 and 9; 14.3.4; 22

Zentralinstitut für Elektronenphysik der Akademie der Wissenschaften der DDR, Berlin

Determination of the Charge of Migrating Impurities from Electrodiffusion Profiles: the System a-As_2Se_3:Cu

By

E. NEBAUER, P. SÜPTITZ, and I. WILLERT

By means of a precision radiotracer-sectioning technique, diffusion and electrodiffusion profiles of copper in amorphous As_2Se_3 are measured between 120 and 180 °C. Fickian diffusion behaviour is found. From the $D(T)$-dependence an interstitial-like mechanism is proposed. Using a δ-like Cu source the electrodiffusion profiles (EDP) show at low electric fields of $\approx 10^2$ V/cm nearly ideal Gaussian shape from which the diffusion coefficient D (being equal to the "field-free" value) and the mobility μ could be determined with high accuracy. Via the Nernst-Einstein relation, the charge q of the migrating Cu species could be determined to $(+\ 1.0 \pm 0.1)\ e$ for the whole temperature range ascertaining that the experimental q's are true ones. At electric fields of $\gtrsim 10^3$ V/cm strong asymmetries in the EDP's occur which still remain unexplained.

Vermittels einer Präzisions-Radiotracer-Sektionstechnik werden Diffusions- und Elektrodiffusionsprofile von Kupfer in amorphem As_2Se_3 zwischen 120 und 180 °C gemessen. Hinsichtlich der Diffusion wird Ficksches Verhalten gefunden, aus dem $D(T)$-Verhalten wird ein „interstitial-like"-Mechanismus gefolgert. Unter Verwendung einer δ-förmigen Cu-Quelle weisen die Driftdiffusionsprofile bei kleinen elektrischen Feldstärken ($\approx 10^2$ V/cm) eine nahezu ideale Gauß-Form auf, aus der der Diffusionskoeffizient D (der mit dem feldfreien Wert übereinstimmt) und die Beweglichkeit μ sehr genau bestimmt werden können. Damit kann unter Verwendung der Nernst-Einstein-Relation die Ladung q der wandernden Spezies im gesamten Temperaturbereich zu $(+\ 1{,}0 \pm 0{,}1)\ e$ bestimmt werden. Es wurde sichergestellt, daß die experimentellen q-Werte wahre q's sind. Bei elektrischen Feldern $\gtrsim 10^3$ V/cm treten starke Asymmetrien in den Driftprofilen auf, die zur Zeit noch nicht verstanden werden.

1. Introduction

More than twenty years have passed since the first detailed investigations of the influence of a steady electric field on the migration of foreign atoms in a crystal [1]. The experimental method — electrodiffusion or electromigration[1]) — can be used under such conditions that drifted concentration profiles of Gaussian shape may be expected. This has the unique advantage to allow the simultaneous determination of the diffusion coefficient and the mobility. Both quantities inserted in the Nernst-Einstein relation give a reliable value of the effective electric charge q of the migrating species. Though such experiments are relatively simple in principle, they have been applied in a very different scale to the various groups of solids. In metals a considerable number of electromigration profiles have been measured and q-analysed (see e.g. [2]). Studies of thin metallic layers lead to an understanding of the grain-boundary electromigration mechanism and its influence on the failure in devices (see e.g. [3]). Concerning ionic crystals the amount of relevant papers is small though the determination of

[1]) The notion "electromigration" is mostly used if remarkable concentrations of electrons (holes) influence the ion drift, i.e. in metals and low-resistivity semiconductors. In ionic crystals and in high-resistivity (amorphous) semiconductors only the direct external field–ion interaction exists, in this case "electrodiffusion" is more appropriate.

the true ion charges is often possible in a straightforward manner, for a good example see [4]. Studies in crystalline semiconductors mostly suffer from the non-Fickian behaviour of the diffusion profiles due to complicated volume mechanisms and/or dislocation contributions. Further problems result from the influence of the electron/hole and/or vacancy "wind" on the measured (effective) charge. Hence, the determination of atomic mobilities based on drifted concentration peaks is exceptional, for an example (Cu in GaAs) see [5]. With other techniques, like the sheet resistance/sectioning procedure [6], p–n techniques [7] and the C_+/C_- method [8] applied to Ge, Si, and III–V compounds, at least the sign of the charge of migrated dopants and sometimes also the charge number could be determined. The C_+/C_- method with the somewhat problematic Joule heating technique was practised especially by the Boltaks group in Leningrad (see e.g. [9]), for a study with external heating of the sample see [10].

First investigations of the atomic transport of Ib elements in amorphous As_2Se_3 (a-As_2Se_3) revealed a fairly regular behaviour [11 to 13], though the results on the charge of the dopants remained contradictory possibly due to experimental differences. In this paper we show for the system a-As_2Se_3:Cu that with improved experimental conditions ideal (or nearly ideal) drifted Gaussian profiles could be obtained which allow the determination of the charge of the migrating dopants with a satisfying exactness. Independently of the special system we discuss the conditions for successful electrodiffusion measurements.

2. Experimental Arrangement Following from Simple Electrodiffusion Formulas

For the determination of the parameters of the electrodiffusion — mobility μ and diffusion coefficient D — a great variety of special experimental methods is known arising from the numerous realizable initial and boundary conditions for the corresponding differential equation [14]. In our opinion the variant of drifted Gauss concentration profiles resulting from δ-like diffusant sources yields the best simultaneous μ and D-values because a very precise measurement of the position of the peak of the electrodiffusion profile (EDP) — giving μ — and an accurate wing fitting — giving D — are possible. Starting from the diffusion equation with (external) electric field term in its simplest form [8, 14]

$$\frac{\partial C_+}{\partial t} = D \frac{\partial^2 C_+}{\partial x^2} - \mu E \frac{\partial C_+}{\partial x} \tag{1}$$

(C_+ concentration of the diffusant in field direction, E electric dc field, x coordinate, t time) one obtains the well-known proportionality

$$C_+(x, t) \sim \exp\left[-\frac{(x - \mu Et)^2}{4Dt}\right] \tag{2}$$

provided the source is δ-like and the sample length L is very large compared with $(Dt)^{1/2}$ and μEt. Hence, the concentration profile is a Gaussian shifted by $x_{max} \equiv x(C_{max}) = \mu Et$ from the origin $x = 0$. The mobility follows from

$$\mu = \frac{x_{max}}{Et}. \tag{3}$$

Applying these formulae the following conditions must be observed:

a) Adequate geometry, especially one-dimensionality of the diffusion profile and avoiding of reflections of the front and/or back flank of the EDP at the surfaces;

b) use of sufficiently thin diffusant layers deposited onto one face of the sample or in sandwich geometry so that the condition $C_{max} \ll S$ (S solubility) is fulfilled for all relevant x and t and no chemical interaction with the annealing atmosphere takes place;

c) x-independent electric field strength, i.e. homogeneity in the bulk as well as near the electrodes: no conductivity modifications by the diffusant or by the electrode material;

d) C-independent volume diffusion coefficients and absence of short-circuit diffusion paths, i.e. overall Fickian behaviour;

e) constant ion mobility;

f) homogeneous sample temperature realized by external heating and working in such regions of the current(I)–voltage(U)-characteristics that no remarkable Joule heating is produced [10];

g) a precision technique for sectioning and counting allowing the measurement of drift profiles at least over two orders of magnitude in C.

We realized these requirements in the following way:

a) The measurements were made on a-As_2Se_3 bulk samples of rectangular shape having some millimeters in all dimensions, they are prepared from "standard" melt-grown ingots [15] by a microtome technique. The annealings were made mainly in a temperature range of $\approx$ 120 to 180 °C (T measurement by Pt resistance thermometer) in air or in an evacuated ampoule with such time/field strength combinations that $4(Dt)^{1/2} \lessapprox x_{max} \ll L$ holds, i.e. $t \gtrapprox 20(kT)^2/(DE^2)\, q^2$.

b) Metallic copper containing the radioactive isotope Cu-64 as tracer was evaporated onto *one* of the microtomed main planes[2]) using a mask; the effective thickness of the layers was $\lessapprox$ 1 Å. The evaporated side of the sample was used later as anode.

c) As in our earlier investigations [12, 13] both main faces were covered with colloidal graphite electrodes which act approximately as ohmic contacts. By current measurements on undoped and (via electrodiffusion) homogeneously Cu-doped samples it was secured that at the annealing temperatures the diffused copper does not remarkably modify the conductivity properties of a-As_2Se_3. Hence, E can be determined "geometrically" by U/d (U dc voltage, d electrode distance).

d) and e) The fulfilment of these conditions will be shown in Section 4.

f) The annealings were made in the T plateau of tube-like furnaces at such electric powers $N = IU$ that the sample overtemperature τ due to Joule heating was $\lessapprox$ 0.1 K[3]) and thus negligible in our experiments. At temperatures of 180 (120) °C the ohmic region of $I(U)$ extends to some hundred (few thousand) V/cm. Generally we used field strengths of (some) 10^2 up to 1×10^3 V/cm.

g) Our measuring technique of the concentration profiles was improved: Instead of the residual activity/etching technique used earlier, a layer activity/microtome method was applied using a "Low Background Anticoincidence β Counter Assembly" (Nuclear Enterprises Ltd.) and a microtome type K (Jung AG) with a diamond knife and a pneumatic sample holder (step width 1 μm). The detection limit was about 10^{15} Cu atoms/cm³. The masses of the single sections were determined with a Sartorius electronic microbalance. The absolute gauging of the half-life-corrected concentration profiles was performed by Cu-64 etalons.

---

[2]) The sandwich method was not used as standard procedure because the $x = 0$ inhomogeneities disturb the drift profiles.

[3]) For the estimation the approximate formula $\tau = IU/(2Fh)$ (F area of the main faces, h heat transfer number) from [16] was used.

3. Experimental Results

At first a series of "field-free" diffusion profiles in the range ≈ 100 to 200 °C was measured using δ-like as well as quasi-infinite Cu sources. These concentration profiles had the expected dependence ($C \sim \exp(-x^2/4Dt)$ or $\sim \operatorname{erfc}(x/2\sqrt{Dt})$) over more than three orders of magnitude in C. No $D(C)$ effect and no tails (i.e. diffusion contributions of short-circuit paths) appear. Fig. 1a shows some results for both boundary conditions with the layer activity detected, Fig. 1b demonstrates the erfc-dependence of the residual activity $A(x)$ for quasi-infinitely thin starting layers.

All determined D-values are shown in an Arrhenius plot (Fig. 2) giving the following relation for $D(T)$

$$D(T) = 5 \times 10^{-2} \exp(-0.79 \text{ eV}/kT) \text{ cm}^2/\text{s} \tag{4}$$

with an error of about ±0.03 eV in the activation enthalpy.

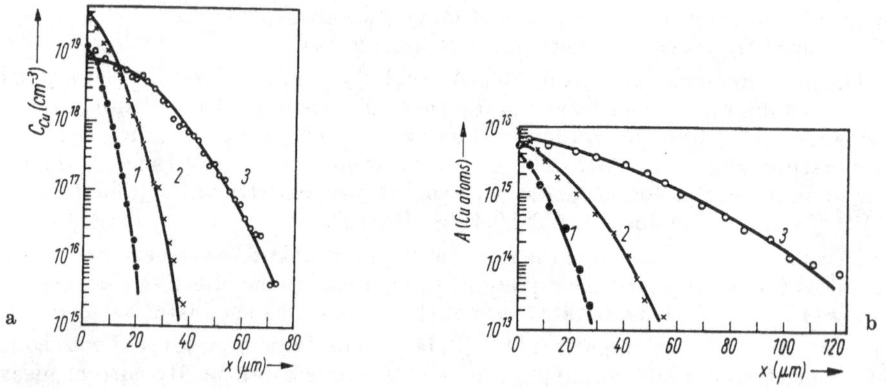

Fig. 1. Field-free diffusion profiles. a) $C(x)$ profiles measured with the layer activity/microtome technique. (1) $T = 98.3$ °C, $t = 1.92 \times 10^5$ s, quasi-infinite Cu source, $D = 8.5 \times 10^{-13}$ cm²/s (erfc fitting), $S \approx 1.1 \times 10^{19}$ cm^{-3}; (2) $T = 141.6$ °C, $t = 3.24 \times 10^4$ s, quasi-infinite Cu source, $D = 1.2 \times 10^{-11}$ cm²/s (erfc fitting), $S \approx 4 \times 10^{19}$ cm^{-3}; (3) $T = 160.2$ °C, $t = 5.76 \times 10^4$ s, δ-like Cu source, $D = 3.0 \times 10^{-11}$ cm²/s ($\exp(-x^2)$-fitting). b) $A(x)$ profiles measured by the residual activity/etching technique. (1) $T = 121.1$ °C, $t = 1.37 \times 10^5$ s, $D = 3.3 \times 10^{-12}$ cm²/s; (2) $T = 159.6$ °C, $t = 6.21 \times 10^4$ s, $D = 2.5 \times 10^{-11}$ cm²/s; (3) $T = 191.9$ °C, $t = 5.26 \times 10^4$ s, $D = 1.8 \times 10^{-10}$ cm²/s. All experiments with δ-like Cu sources, accordingly the D-values from erfc fitting

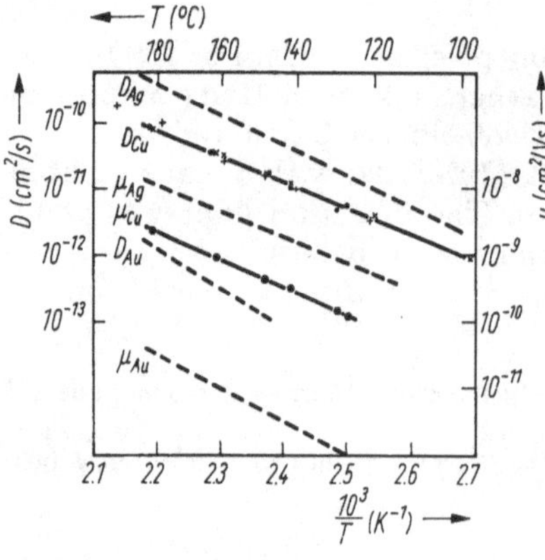

Fig. 2. Temperature dependence of the diffusion coefficient and of the mobility of copper and, for comparison, of silver [12, 15] and gold [13, 15] in a-As_2Se_3. Symbols: Field-free experiments: + from etching measurements, × from microtome technique; ● from (low field) electrodiffusion profiles

With the stationary Cu source some preliminary solubility determinations were made by extrapolating $C \sim \operatorname{erfc} x$ to $x = 0$. The values of S lie in the range 10^{19} to 10^{20} cm^{-3} for $T \approx 100$ to 170 °C. The enthalpy in the $S(T)$ dependence is $\approx (0.5 \pm 0.15)$ eV.

The electrodiffusion experiments yielded distinctly marked concentration peaks displaced from the origin. The parameter combinations were chosen in such a way that C_{max} was not higher than 10^{17} to 10^{18} Cu atoms/cm^3 still allowing the measurement of the EDP's over at least two orders of magnitude in C. Unlike our earlier experiments on a-As_2Se_3:Ag;Au [12, 13] mainly electric dc fields of (some) 10^2 V/cm were applied now. Under such "low field conditions" we found in a part of the experiments symmetrical EDP's which could be well fitted by Gaussians (Fig. 3a), disregarding the surface region. The D-values determined from both flanks agree very well (within $\approx \pm 10\%$) with the corresponding D's from field-free experiments. More characteristic of the low field EDP's is a weak asymmetry of the profiles, where the front (right) flank could be fitted with the corresponding field-free diffusion coefficient, but the back (left) flank with a (about 50%) larger effective D-value, two examples are shown in Fig. 3b and c. The D's resulting from the fitting of the symmetrical EDP's and from the front flank of the slightly asymmetric EDP's are included in Fig. 2, too. At present it is not possible to allot the occurrence of the weak asymmetry to such experimental facts like differing peak concentration, distance from origin, and sample history.

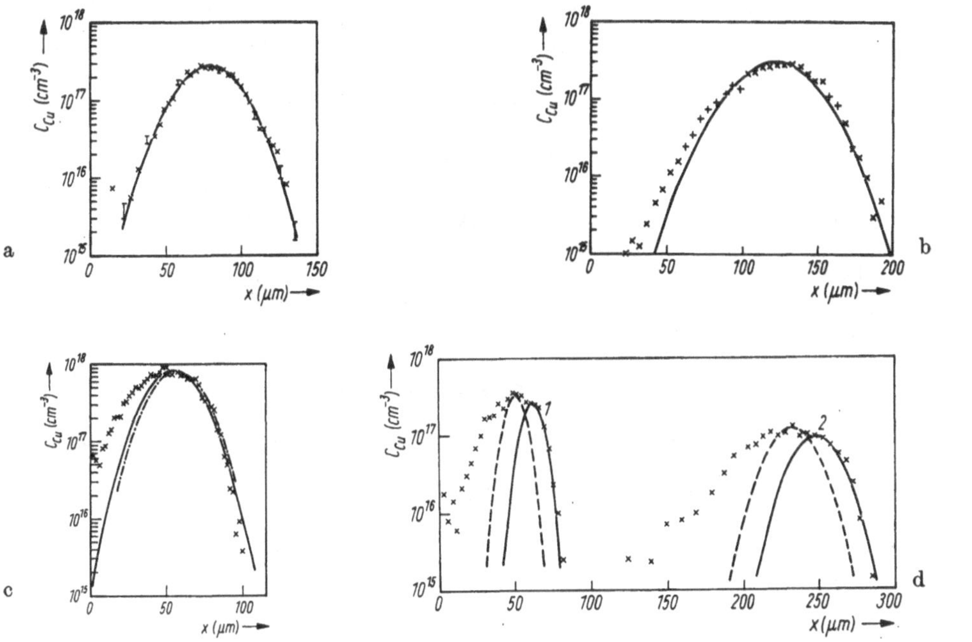

Fig. 3. Electrodiffusion profiles. a) Symmetrical profile ($T = 182.7$ °C, $t = 1.87 \times 10^4$ s, $E = 190$ V/cm), b) and c) slightly asymmetric profiles ($T = 162.4$ °C, $t = 7.6 \times 10^4$ s, $E = 182$ V/cm and $T = 141.6$ °C, $t = 9.35 \times 10^4$ s, $E = 181$ V/cm), d) strongly asymmetric profiles: (1) $T = 148.6$ °C, $t = 1.08 \times 10^4$ s, $E = 1025$ V/cm; (2) $T = 148.6$ °C, $t = 5.4 \times 10^4$ s, $E = 1000$ V/cm. The solid lines are theoretical curves after equation (2) (fitting of the front side or of the whole profile). The bars in a) indicate the errors (from counting and weighing). The dashed-dotted curve in c) is calculated from equation (2) in [13], considering the semi-infiniteness of the samples. The dashed curves in d) are fits with the field-free D's in the C maximum

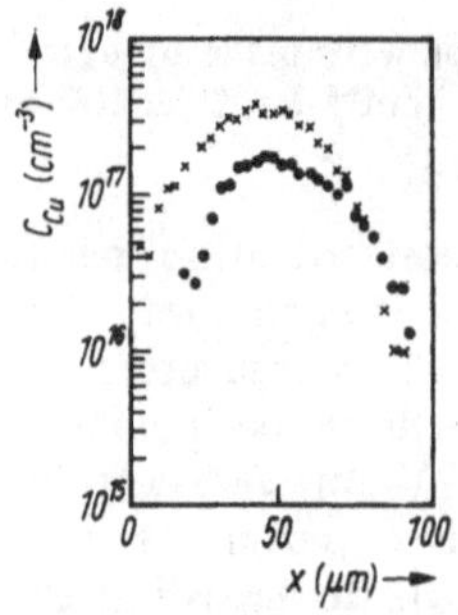

Fig. 4. Comparison between electrodiffusion profiles of a Cu (diffusion-) doped As_2Se_3 sample (●) and an undoped sample (×) ($T = 148.8$ °C, $t = 1.03 \times 10^5$ s, $E = 111$ V/cm)

Clear deviations from Gaussian curves appear at fields in the order of magnitude of 10^3 V/cm independent of the annealing time. The electrodiffusion profiles become strongly asymmetric (Fig. 3d) but the front flank (below $\approx 1/4\, C_{max}$) can be described again with the D-values of field-free experiments. From the back flank an effective D-value may be evaluated which is increased by about half an order of magnitude.

The determination of the mobility μ from x_{max} (equation (3)) is unambiguous for EDP's obtained at low E's (uncertainty $\approx 5\%$). The T dependence (see Fig. 2) is given by

$$\mu(T) = 3 \exp(-0.82\ \text{eV}/kT)\ \text{cm}^2/\text{Vs} \tag{5}$$

with an error of the activation enthalpy of about $\pm$ 0.03 eV, i.e. the enthalpies of D and μ are equal within the experimental uncertainty. The mobility values calculated from x_{max} of the EDP's obtained at high E's seem to be increased by 20 to 50%. For all EDP's belonging to the same temperature but different (E, t) combinations the proportionality $x_{max} \sim Et$ is sufficiently fulfilled.

A further experimental result should be mentioned: At the same (T, t, E) combination EDP's were measured on an undoped sample and a sample which was inhomogeneously predoped with Cu by diffusion over the EDP range ($\approx 10^{20}$ cm^{-3} near $x = 0$, $\approx 10^{19}$ cm^{-3} at the right flank). The result is shown in Fig. 4. In contrast to the usual situation the profile of the doped sample has a slight asymmetry with the lower D on the *back* side. This is plausible because a slight decrease of D at Cu doping was already observed in the case of Ag and Au diffusion in a-As_2Se_3 [15].

4. Discussion

One of the most important results is the fact that in amorphous As_2Se_3 the electrodiffusion profiles of copper have an ideal or nearly ideal Gaussian shape under appropriate conditions (low electric fields etc.). On the basis of the (μ, D) pairs calculated from such Gaussians (Fig. 2) at different temperatures the charge q of the copper species can be determined by the Nernst-Einstein relation, assuming that it can be used in the simplest form

$$\frac{\mu}{D} = \frac{q}{kT} = \frac{eZ}{kT} \tag{6}$$

(Z charge number). Inserting μ from (3) yields

$$Z = \frac{x_{max} kT}{eD(U/d)\, t}\,. \tag{6a}$$

Considering the uncertainties of the measured quantities entering (6a) the total error for Z is about 10%. The experimental results are shown in Fig. 5. It can be concluded

Fig. 5. Charge numbers Z at different temperatures, evaluated by means of equation (6a). The bars indicate the scattering of the results

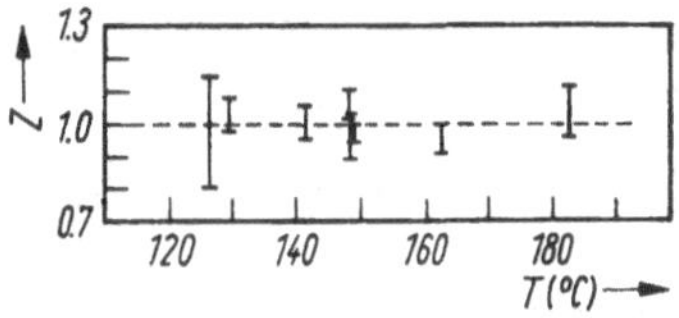

that Z is temperature-independent and equal to unity in the available T range. The determined charges are not effective and/or apparent charges but true ones, because contributions resulting from charge carrier and/or vacancy "winds" are absent. After Fiks' formula [17]

$$q_{\text{eff}} = q_{\text{true}}(1 + pl_{\text{h}}\Sigma_{\text{h}}) \tag{7}$$

(p hole concentration, l_{h} mean free path, Σ_{h} cross-section of the impurities for holes) the term $pl_{\text{h}}\Sigma_{\text{h}}$ ($\approx 10^{12} \times 10^{-7} \times 10^{-13}$) is $\ll 1$ and thus the hole wind is negligible. Vacancy flow terms [18] should be not essential in amorphous material. Therefore it follows that the copper species are singly positively charged in the moment of jumping.

Concerning the shape of the electrodiffusion profiles at present no plausible arguments for the asymmetry of the EDP's can be given. The higher left-hand D's cannot be caused by

- the influence of the semi-infinite geometry. The exact solution presented in [13], being slightly asymmetric, cannot explain the measured profiles, as is demonstrated in Fig. 3c;
- the simultaneous migration of different species. In principle it is possible to construct the measured asymmetric profiles from two (or more) Gaussian profiles. But the species with the smaller mobility should then have higher D's, i.e. they should have — according to $\mu \sim ZD$ — the smaller charge number. This is not very probable, the more so because the field-free profiles show only one component (uniform D);
- matrix modifications by the drifting Cu species. Corresponding calculations have shown that the C_{max}-values at each time are below $\approx 5 \times 10^{18}$ cm^{-3} for $x > 1$ μm. Related test experiments on Cu pre-drifted samples yielded the same EDP.

Finally one may ask whether the asymmetry properties of the drifting Cu species "packet" perhaps can be explained by a "dispersive" transport theory with an entire spectrum of transition rates — to our knowledge no theoretical investigations have been published up to now.

Concerning the Cu diffusion behaviour the measured overall erfc (x) or exp ($-x^2$) concentration distributions suggested — similar as in the cases of the systems a-As_2Se_3:Ag;Au [15], but unlike a-As_2S_3:Ag [19] — the model of a uniform migration process, probably an interstitial-like mechanism. The main argument follows from the smallness of the pre-exponential factor in (4) ($D_0 \approx 5 \times 10^{-2}$ cm^2/s) (cf. [20]) and of the activation enthalpy.

Further Fig. 2 shows the surprising fact that in the investigated T interval the sequence of the diffusion coefficients of the Ib group elements is $D_{\text{Ag}} > D_{\text{Cu}} > D_{\text{Au}}$ (the same holds for the μ's) which is different from that following from geometrical arguments. However, deviations from the $D_{\text{Cu}} > D_{\text{Ag}} > D_{\text{Au}}$-behaviour are also known for crystalline semiconductors (e.g. Si) but are not well explained up to now.

In the end the enthalpy H_{inc} necessary for the incorporation of a Cu atom in the As_2Se_3 matrix should be determined. Assuming that the Cu atoms in the external phase Cu_{ext} interact with quasi-interstitials $V_i^\times$ according to $Cu_{ext} + V^\times \leftrightharpoons Cu_i^\times$ [4]), the concentration of these Cu interstitials should be equal to the solubility

$$S \equiv [Cu_i^\times] \sim \exp(-H_{inc}/kT)\, a_{Cu} \sim \exp(-H_{inc}/kT) \frac{p_{Cu}}{p_{Cu}^{(0)}} \exp(-(H_v - H_v^{(0)})/kT) \tag{8}$$

(a_{Cu} activity of Cu_{ext}, $p_{Cu}^{(0)}$ saturation vapour pressure of pure Cu). Unfortunately, no information concerning the As–Se–Cu ternary phase diagram is available, thus the $a_{Cu}(T)$ dependence cannot be determined. But from other (known) systems a value of about 0.1 eV for $H_v - H_v^{(0)}$ can be estimated giving $H_{inc} \approx H_{solub}^{exp} + 0.1$ eV ≈ 0.6 ($\pm$ 0.2) eV.

5. Final Remarks

It was shown that among the noncrystalline (high resistivity) semiconductors like As chalcogenides one can find systems which, firstly, allow a sufficiently accurate determination of the charge of the migrating impurities and, secondly, have the property that Z expected from the main valency of the drifting dopant is really observed. But it seems to be rash to transfer the results to neighbouring systems. Recent experiments on the system a-As_2S_3 : Ag [21] have shown that these EDP's are far from Gaussian shape for all relevant conditions. Obviously a variety in the transport and charge behaviour of the Ib elements in a-As_2S_3 and a-As_2Se_3 exists.

References

[1] M. CHEMLA, Thesis, University Paris, 1954.
M. CHEMLA, Ann. Phys. (France) ser. 13, 1, 959 (1956).
[2] N. V. DOAN, J. Phys. Chem. Solids 32, 2135 (1971).
[3] F. M. D'HEURLE, Proc. IEEE 59, 1409 (1971).
[4] V. C. NELSON and R. J. FRIAUF, J. Phys. Chem. Solids 31, 825 (1970).
[5] R. N. HALL and J. H. RACETTE, J. appl. Phys. 35, 379 (1964).
[6] D. H. LAANANEN and W. B. NOWAK, Rev. sci. Instrum. 40, 422 (1969).
[7] C. S. FULLER and J. C. SEVERIENS, Phys. Rev. 96, 21 (1954).
[8] P. SH. MALKOVICH, Fiz. tverd. Tela 2, 2784 (1960).
[9] M. K. BAKHADYRKHANOV, B. I. BOLTAKS, and G. S. KULIKOV, Fiz. tverd. Tela 14, 1671 (1972).
[10] E. NEBAUER and F. QUEDNAU, phys. stat. sol. (a) 26, 225 (1974).
[11] V. KH. BIKTIMIROVA, B. I. BOLTAKS, Z. U. BORISOVA, T. D. DZHAFAROV, and A. A. OBRAZTSOV, Fiz. Tekh. Poluprov. 8, 2174 (1974).
B. I. BOLTAKS, Z. U. BORISOVA, V. KH. BIKTIMIROVA, T. D. DZHAFAROV, A. A. OBRAZTSOV, and E. A. SHENYAVSKAYA, phys. stat. sol. (a) 30, 731 (1975).
[12] E. A. LEBEDEV, P. SÜPTITZ, and I. WILLERT, phys. stat. sol. (a) 28, 461 (1975).
[13] P. SÜPTITZ, J. TELTOW, E. A. LEBEDEV, and I. WILLERT, phys. stat. sol. (a) 31, 31 (1975).
[14] B. I. BOLTAKS, Diffuziya v poluprovodnikakh, Fizmatgiz, Moscow 1961.
[15] P. SÜPTITZ and I. WILLERT, phys. stat. sol. (a) 28, 223 (1975).
[16] E. NEBAUER and E. JAHNE, phys. stat. sol. 8, 881 (1965).
[17] V. B. FIKS, Ion Conductivity in Metals and Semiconductors, Izd. Nauka, Moscow 1969 (in Russian).
[18] J. R. MANNING, Diffusion Kinetics for Atoms in Crystals, Van Nostrand Co., Inc., Princeton 1968.
[19] A. BUROFF, E. NEBAUER, P. SÜPTITZ, and I. WILLERT, phys. stat. sol. (a) 40, K195 (1977).
[20] P. G. SHEWMON, Diffusion in Solids, McGraw-Hill Publ. Co., New York 1963.
[21] A. BUROFF, E. NEBAUER, and P. SÜPTITZ, to be published.

(Received August 10, 1977)

[4]) This is in our opinion not in contrast to the above result of singly positively charged migrating species. Because of the conductivity properties a quasi-chemical reaction of the type $Cu_{ext} + V_i^\times \rightarrow Cu_i^{\cdot} + e'$ cannot be assumed.

phys. stat. sol. (a) **43**, 459 (1977)

Subject classification: 1.4; 10.1; 21.1

Laboratoire Structure et Propriétés de l'Etat Solide, associé au C.N.R.S. n° 234, Université des Sciences et Techniques de Lille, Villeneuve D'Ascq[1])

On the Nature of Deformation Twins in Ordered Ni_3V

By

G. VANDERSCHAEVE and T. SARRAZIN

Electron microscopy observations on deformed specimens of ordered Ni_3V show that deformation proceeds mainly by propagation of stacking faults and twinning. Dark field experiments proove that these deformation twins are ordered; it is concluded that deformation in this alloy is controlled by gliding of those partial dislocations leading to a stacking fault without any first neighbours wrong bonds, as previously expected on a theoretical basis.

Les observations en microscopie électronique de l'alliage ordonné Ni_3V montrent que la déformation s'effectue principalement par propagation de fautes d'empilement et par maclage. Les observations en champ sombre montrent que ces macles sont ordonnées; la déformation dans cet alliage est donc contrôlée par le glissement des dislocations partielles créant une faute d'empilement sans mauvaises liaisons entre premiers voisins.

1. Introduction

The structure of ordered Ni_3V is of the DO_{22} type. It can be described as an example of a long-period ordered alloy (with periodic antiphase boundaries every M (100) atomic planes) with $M = 1$. In preceding studies [1 to 3] on deformation modes of long-period ordered alloys a model has been developed in which the deformation proceeds by propagation of those dislocations leading to a stacking fault without any wrong first neighbour bonds; so-called geometrical faults [2]. Following this model the long-range order should thus be preserved during the deformation. On the other hand, Buynova et al. [4 to 6] introduce the concept of planes with "disturbed long-range order". Accordingly the crossing of gliding dislocations over periodic antiphase boundaries would cause the development in their slip planes of new segments of antiphase boundary. Since the periodic antiphase boundaries are distributed over intervals of a few lattice parameters, disturbance of long-range order should occur at the very initial stage of deformation. The present study is an attempt to decide between the two models. It shows conclusively that in Ni_3V, deformation proceeds by slip of partial dislocations, resulting in a profuse twinning of the crystal: the twins do preserve the DO_{22} order, demonstrating that the long-range order is not disturbed during deformation.

2. Experimental

Polycrystalline (grain size ≈ 0.1 mm) ordered (domain size ≈ 0.5 μm) samples have been deformed in compression at room temperature up to approximately 5%. The elastic limit is as high as 120 kp/mm². Thin foils suitable for electron microscopy have been prepared by jet electropolishing using an aceto-perchloric bath at 70 V.

---

[1]) B.P. 36, 59650 Villeneuve D'Ascq, France.

3. Crystallography

3.1 Structural considerations

The (111)[2]) close-packed plane is depicted on Fig. 1 for the long period along [100]. The stacking sequence of the DO_{22} structure involves six such planes and can be written as

$$\ldots A_1B_1C_1A_2B_2C_2A_1B_1C_1 \ldots .$$

In order to determine the nature of the observed deformation twins, we have to describe the new structures obtained by successive shears on the (111) planes. The shear vectors considered here are of the type either $\frac{1}{6}\langle 112\rangle$ or $\frac{1}{3}\langle 112\rangle$. Two of these shear vectors, $\frac{1}{6}[\bar{2}11]$ and $\frac{1}{3}[2\bar{1}\bar{1}]$, (these two vectors being crystallographically equivalent modulo a lattice translation $\frac{1}{2}[\bar{2}11]$) induce a "true" twin, i.e. the new stacking sequence is

$$\ldots A_1B_1C_1A_2B_2C_2\vdots A_1C_2B_2A_2C_1B_1A_1 \ldots .$$

This is a mirror structure of the preceding one, the long-range order is not disturbed by these successive shear translations. Notice that these translations correspond to Shockley partials which introduce a stacking fault without order fault [2].

All the four other shear translations introduce first neighbour wrong V–V bonds between successive (111) layers; $\frac{1}{6}[1\bar{2}1]$ or $\frac{1}{6}[11\bar{2}]$, introduce a stacking sequence of twelve planes, and $\frac{1}{3}[\bar{1}\bar{1}2]$ or $\frac{1}{3}[\bar{1}2\bar{1}]$, a sequence of three planes

$$\ldots A_1B_1C_1A_2B_2C_2\vdots A_1C_1B_2A_1C_1B_2 \ldots .$$

instead of the sixfold sequence of the DO_{22} structure.

3.2 Diffraction

In long-period ordered alloys the occurrence of periodic antiphase boundaries every $M(100)$ planes, for example, induces those diffraction spots (hkl) with h, k, l mixed and $k + l$ odd to be extinguished, but there remain rows of spots with indexes $\frac{1}{2M}\,kl$, $\frac{3}{2M}\,kl$, ..., $(k + l$ odd). So in the DO_{22} structure, due to the occurrence of a longperiod superstructure with $M = 1$, new spots with indexes $\langle h \pm \frac{1}{2}\,k\,l\rangle^*$ $(k + l$ odd) are expected and these spots are very useful to determine the direction of the long period within domains: an index $h \pm \frac{1}{2}$ indicates that the long period is along [100] and so on.

Structure factors have been calculated for all the possible sheared structures. From the occurrence of forbidden reflections, the following situations are expected.

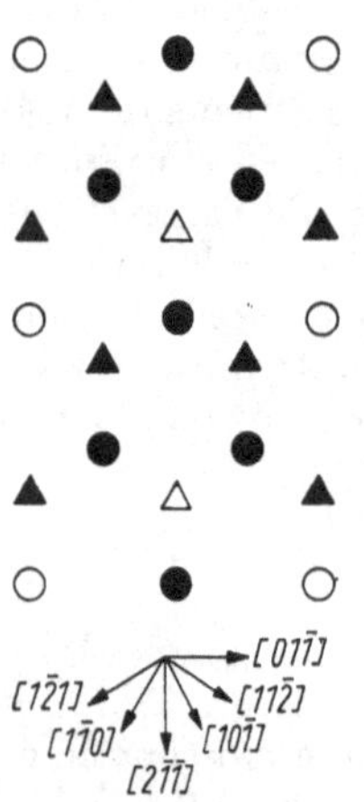

Fig. 1. (111) plane of Ni_3V with long period along [100]. A_1 plane: ● Ni, ○ V; B_1 plane: ▲ Ni, △ V

[2]) For the sake of simplicity we use throughout this paper disordered face-centred cubic notations in indexing crystallographic planes and directions.

3.2.1 True twin

In this case, the twin reciprocal lattice corresponds to a mirror image of the matrix reciprocal lattice; the indexes of the "twinned" reflections can be obtained by the usual way [7]. The following superstructure reflections are considered.

(i) $[100]^*$ is not forbidden, its index in the matrix lattice is $\frac{1}{3}[\bar{1}22]^*_M$.

(ii) the reflexions $\langle\frac{1}{2}\,10\rangle^*$ are of special interest; we have

$$[\tfrac{1}{2}\,01]^*_T \equiv [\tfrac{1}{2}\,10]^*_M, \qquad [\tfrac{1}{2}\,10]^*_T \equiv [\tfrac{1}{2}\,01]^*_M,$$

$$[-\tfrac{1}{2}\,10]^*_T \equiv [\tfrac{5}{6} - \tfrac{2}{3}\,\tfrac{1}{3}]^*_M, \quad [-\tfrac{1}{2}\,01]^*_T \equiv [\tfrac{5}{6}\,\tfrac{1}{3} - \tfrac{2}{3}]^*_M .$$

Notice that all these reflections are due to a long period along [100]. So, in a dark field image taken with a $[\frac{1}{2}\,01]^*_M$ or $[\frac{1}{2}\,10]^*_M$ reflection, both twins and matrix are in contrast; but in a dark field image with a $[-\frac{1}{2}\,01]^*_M$ or $[-\frac{1}{2}\,10]^*_M$ reflection, twins are out of contrast.

3.2.2 Other twinned structures

For the structures deduced from the DO_{22} structure by successive shears either $\frac{1}{6}[1\bar{2}1]$ or $\frac{1}{6}[11\bar{2}]$ (stacking sequence of twelve planes) it can be demonstrated that
(i) the reflection $\frac{1}{3}[\bar{1}22]^*_M \equiv [100]^*_T$ is forbidden.
(ii) the reflections corresponding to $\langle\frac{1}{2}\,10\rangle^*_T$ are also forbidden. So in a dark field image taken with this reflection the twins are out of contrast.

For the structure deduced by sucessive shears either $\frac{1}{3}[\bar{1}\bar{1}2]$ or $\frac{1}{3}[\bar{1}2\bar{1}]$ (stacking sequence of three planes), calculations indicate that
(i) the reflection $\frac{1}{3}[\bar{1}22]^*_M \equiv [100]^*_T$ is not forbidden.
(ii) the reflections corresponding to $\langle\frac{1}{2}\,10\rangle^*$ are forbidden (this is consistent with a stacking sequence of only three planes, i.e. the new unit cell is cubic and reflections with fractional indexes are not allowed). The twins are therefore out of contrast in dark field images taken with these reflections.

Of course these calculations are also valid for a superperiod along either [010] or [001] by changing the order of the indexes in an appropriate way. So we have the possibility by studying dark field images with different reflections to determine the nature of the deformation twins observed by electron microscopy. We show below that the observed twins are in contrast for dark field images taken with reflections like $[\frac{1}{2}\,01]^*_M$, while being out of contrast with reflections like $[-\frac{1}{2}\,01]^*_M$. So the long-range order is not destroyed during deformation.

4. Results

4.1 Electron microscopic observations

Fig. 2 to 5 are electron microscopic pictures of a deformed crystal. The three possible orientations of the long period are present. According to Tanner [8], the domain boundaries are found of the type {110}. We can observe (Fig. 2) that deformation proceeds by propagation of twins and stacking faults often stopped at domain boundaries. From Fig. 2b, where they are seen end-on, the fault plane is (111). In the dark field picture of Fig. 2c, taken with $[002]^*_T$, it can be seen that some twins are extended over many domains.

Fig. 3, 4, and 5 are dark field pictures shown in order to determine the nature of twins in [100], [001], and [010] domains, respectively. As indicated above it is possible to know the orientation of the long period by dark field images with $\langle\frac{1}{2}\,0\,1\rangle^*$ reflections. From all these pictures we conclude that the long-range order is not disturbed in the twins.

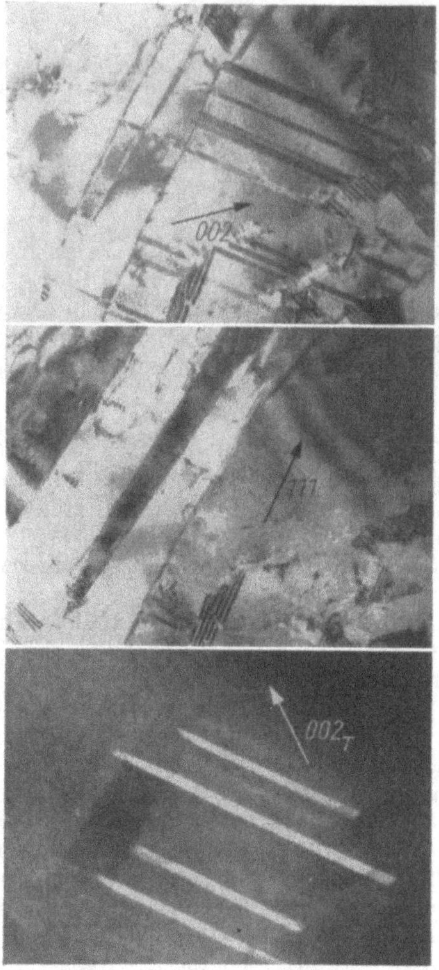

a
b
c

Fig. 2. Deformation twins in ordered Ni_3V. (Magnification, here and in Fig. 3 to 6 22000 ×)

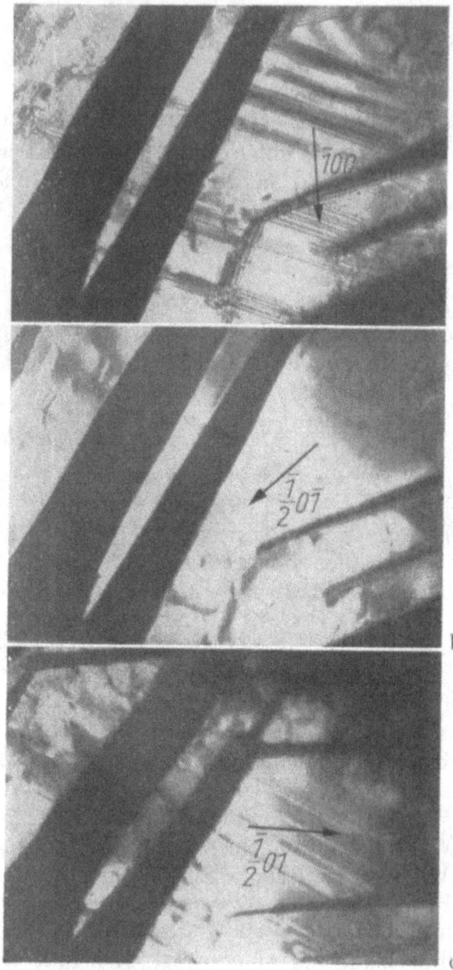

a
b
c

Fig. 3. Analysis of twins in (100) domain. The twins are out of contrast in a) $\boldsymbol{g} = [100]^*$ and c) $\boldsymbol{g} = [-\frac{1}{2}\,01]^*$, and in contrast in b) $\boldsymbol{g} = [-\frac{1}{2}\,0\bar{1}]^*$ indicating that they are ordered

It can be seen indeed that the twins are

— out of contrast for reflections like $[100]^*_M$ (Fig. 3a, 4a),
— in contrast for reflections like $[\frac{1}{2}\,01]^*_M$ (Fig. 3b, 4b, 5),
— out of contrast for reflections like $[-\frac{1}{2}\,01]^*_M$ (Fig. 3c, 4c).

It follows that deformation twins are ordered, in the sense that this mode of twinning does not introduce wrong first neighbour bonds.

The Burgers vector of the partial dislocations within the twin has been determined by the conventional extinction method (Fig. 6). First we note that a lot of dislocations are stopped at the domain boundaries (Fig. 6a) indicating that these boundaries are obstacles to the glide of dislocations. Except an isolated one, all the dislocations within the twin are out of contrast with $\boldsymbol{g} = [02\bar{2}]^*$ (Fig. 6b) and $\boldsymbol{g} = [1\bar{1}3]^*$ and the Burgers vector is therefore parallel to $[2\bar{1}\bar{1}]$. As indicated above $\frac{1}{6}[\bar{2}11]$ and $\frac{1}{3}[2\bar{1}\bar{1}]$

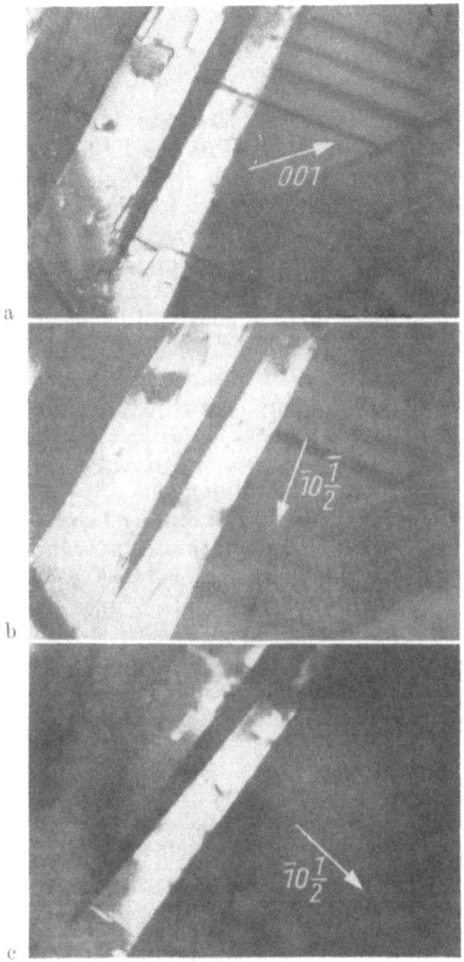

Fig. 4. Analysis of twins in (001) domain. The twins are out of contrast in a) $\boldsymbol{g} = [001]^*$ and c) $\boldsymbol{g} = [\bar{1}0\,\frac{1}{2}]^*$ but in contrast in b) $\boldsymbol{g} = [\bar{1}0 - \frac{1}{2}]^*$, indicating that they are ordered

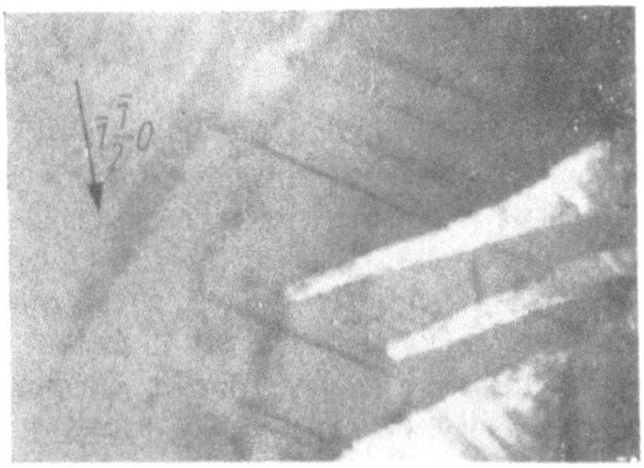

Fig. 5. The twin is in contrast with $\boldsymbol{g} = [1\,\frac{\bar{1}}{2}\,0]^*$ in the (010) domain, indicating that it is ordered

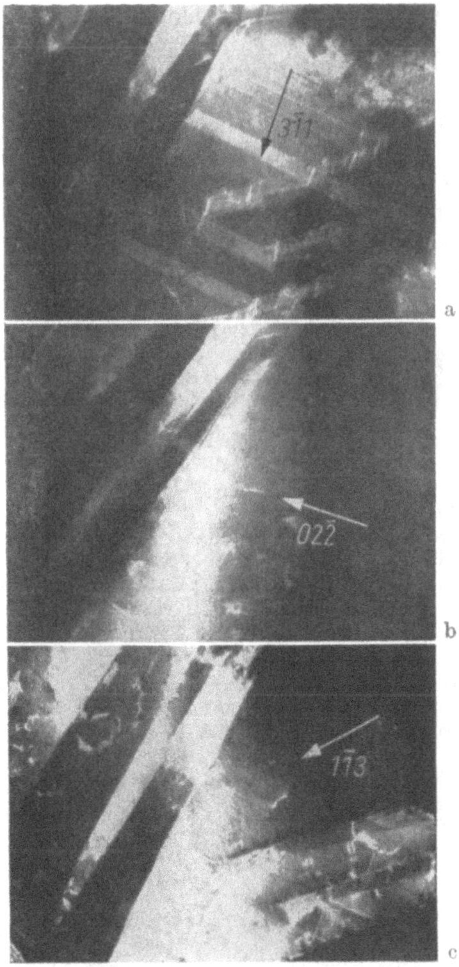

Fig. 6. Analysis of the Burgers vector of twin dislocations. Notice the presence of numerous dislocations in domain boundaries

are the shear vectors (modulo a lattice translation) creating stacking faults without order fault in a [100] domain, which is consistent with the occurrence of "true" twins by deformation.

4.2 Discussion

From all the preceding experimental evidences, it is clear that the concept of "planes with disturbed long-range order" is not applicable in this case. First, the twins are ordered. Second, the Burgers vector of twin dislocations is the one which introduces a geometrical fault without wrong first neighbour bonds. Another experimental indication is the fact that dislocations are stopped at domain boundaries: If the long-range order was really disturbed in glide planes, the domain boundaries would not act as obstacles to the propagation of dislocations.

We have also to explain why some twins extend over many domains. In preceding studies on propagation of dislocations in long-period ordered alloys, we determined the possible Burgers vectors of partial dislocations leading to a geometrical fault. It happens that $\frac{1}{6}[2\bar{1}\bar{1}]$ or $\frac{1}{3}[\bar{2}11]$ dislocations cannot propagate in a [010] or [001] domain without creating wrong first neighbour bonds; only $\frac{2}{3}\langle 211\rangle$ dislocations can glide in all domains, without disturbing long-range order. So there is a strong reason to think that those twins which extend experimentally over many domains are led by $\frac{2}{3}\langle 211\rangle$ dislocations propagating through the crystal. Of course another mechanism is on the other hand conceivable, requiring production and migration of a vacancy every atomic row in the slip plane [9]. For example the shear vector $\delta B = \frac{1}{6}[\bar{2}11]$ in a [100] domain has to be turned into $\delta C = \frac{1}{6}[11\bar{2}]$ for propagation in a [001] domain. This requires in the slip plane an additional displacement of $BC = \frac{1}{2}[10\bar{1}]$ which could be obtained by convenient vacancy jumps along every atomic row. However, such a vacancy production and migration process seems rather unlikely here, owing to the low ratio of deformation temperature to the melting point of this alloy ($T < 0.2\,T_m$).

Acknowledgement

The authors wish to thank Prof. B. Escaig for stimulating discussions.

References

[1] G. Vanderschaeve, G. Coulon, and B. Escaig, phys. stat. sol. (a) **9**, 541 (1972).
[2] G. Vanderschaeve and B. Escaig, phys. stat. sol. (a) **20**, 309 (1973).
[3] G. Vanderschaeve, J. Phys. **35**, C7–47 (1974).
[4] L. N. Buynova, V. I. Syutkina, O. D. Shashkov, and E. S. Yakovleva, Fiz. Metallovi i Metallovedenie **29**, 1221 (1970).
[5] L. N. Buynova, V. I. Syutkina, O. D. Shashkov, and E. S. Yakovleva, Fiz. Metallov i Metallovedenie **33**, 1195 (1972).
[6] L. N. Buynova, V. I. Syutkina, O. D. Shaskov, and E. S. Yakovleva, Fiz. Metallov i Metallovedenie **34**, 561 (1972).
[7] P. B. Hirsch, A. Howie, R. B. Nicholson, D. W. Pashley, and M. J. Whelan, Electron Microscopy of Thin Crystals, Butterworths London 1965 (p. 141).
[8] L. E. Tanner, phys. stat. sol. **30**, 685 (1968).
[9] A. Guimier, Thesis University of Nancy, 1972.

(Received May 31, 1977)

phys. stat. sol. (a) 43, 465 (1977)

Subject classification: 20.3; 13.4; 22.9

Physikalisches Institut der Universität Bayreuth[1])

Fluorescence of Localized States in a Partially Polymerized Diacetylene Single Crystal

By

H. Eichele and M. Schwoerer

Partially polymerized crystals of 2,4-Hexadyne-1,6-diol-bis-(p-toluene sulfonate) (PTS) show fluorescence. Excitation at 4.2 K yields two emission bands at 20486 and 20802 cm^{-1} (accompagnied by vibrational side bands) and displaced bands, whose wavenumbers are $\bar{\nu}_{fluorescence} = \bar{\nu}_{excitation} - \bar{\nu}_i$ ($\bar{\nu}_1 = 988.7\ cm^{-1}$, $\bar{\nu}_2 = 1227\ cm^{-1}$, $\bar{\nu}_3 = 1513\ cm^{-1}$, and $\bar{\nu}_4 = 2085\ cm^{-1}$). The $\bar{\nu}_i$ are assigned to the vibronic energies of the excited state of the fluorescing center.

Teilpolymerisierte Kristalle von 2,4-Hexadiine–1,6-Diol-bis-(p-Toluol-Sulfonat) (PTS) fluoreszieren. Bei 4,2 K führt die Anregung zu zwei Emissionsbanden bei 20486 und 20802 cm^{-1} (begleitet von vibronischen Seitenbanden) und verschobenen Banden mit Wellenzahlen $\bar{\nu}_{Fluoreszenz} = \bar{\nu}_{Anregung} - \bar{\nu}_i$ ($\bar{\nu}_1 = 988,7\ cm^{-1}$, $\bar{\nu}_2 = 1227\ cm^{-1}$, $\bar{\nu}_3 = 1513\ cm^{-1}$ und $\bar{\nu}_4 = 2085\ cm^{-1}$). Die $\bar{\nu}_i$ werden vibronischen Energien des angeregten Zustandes des fluoreszierenden Zentrums zugeschrieben.

1. Introduction

Single crystals of suitably substituted diacetylenes polymerize topochemically yielding large polymer crystals [1]. Polymerization is induced thermally by annealing or photochemically by irradiation with UV or higher-energy radiation. In partially polymerized crystals stable intermediates during polymerization have been observed by absorption spectroscopy at low temperatures [2]. ESR investigations revealed triplet carbenes as intermediates in partially thermally polymerized 2,4 hexadiyne-1,6-diol-bis-(p-toluene sulfonate) (PTS) [3]. Recently Bloor et al. [4] reported additional optical absorption and fluorescence bands in rapidly grown, highly defective samples of PTS, which is by far the most investigated polydiacetylene crystal.

In order to get further spectroscopic evidence about the intermediates in PTS both the luminescence spectrum and the excitation spectrum were investigated. They are treated in the present paper.

2. Experimental

The monomer was prepared by Wegner's method [1]. Single and polycrystalline (2 to 5 crystallites) samples were grown from acetone or dichloromethane solution and cleaved to approximately 25 mm^2 area and 1 mm thickness. Partial thermal polymerization occurred during crystal growth. The samples were stored at −25 °C prior to measurement. In unpolarized light the colour of the investigated crystals ranged from transparent pink to deep red.

Powder samples were obtained by pulverizing either single or polycrystals. The samples were cooled to 1.5, 4.2, and 77 K by immersion in liquid helium or liquid nitrogen. The fluorescence was observed perpendicular to the unfocused excitation beam of an argon ion laser or a cw dye laser (beam diameter approximately 1.7 mm).

[1]) D858 Bayreuth, BRD.

The argon ion laser has been operated in single line mode (except UV) for direct excitation of the fluorescence or as a pump source for coumarin 2 and coumarin 47 laser dyes. The dye laser was tuned from 440 to 480 nm in increments of approximately 0.25 nm.

The fluorescence spectra were recorded using 5 mW excitation power, stabilized with a light regulator.

3. Spectra

Fluorescence with a lifetime shorter than 100 μs has been observed by excitation with wavelengths in the range 488 to 440 nm and with the UV lines (363.8 to 333.6 nm) of the argon laser. This fluorescence has a quantum energy of approximately 2.5 eV and lies above the strong absorption edge of the polymer (2 eV). Fig. 1 gives a comparison of five spectra obtained at 4.2 K from a polycrystalline sample with different excitation wavelengths. The three lower spectra show broad fluorescence bands between 20200 and 20900 cm^{-1} followed by vibronic side-bands at lower energies.

The shapes and the relative intensities of these bands strongly depend on the excitation wavelengths. This is clearly demonstrated in Fig. 1: In the second spectrum from the top, the exciting 488 nm argon line, which is resonant with the fluorescence band, enhances the vibronic sidebands.

No fluorescence is observed by irradiation with the argon lines from 496.5 to 514.5 nm, which lie below the fluorescence emission (top spectrum).

Fig. 2a shows part of the spectrum of one single crystal at 4.2 K at different excitation energies. The change of the relative intensities of the emission bands is clearly demonstrated.

The shape, relative intensities, and spectral positions of the fluorescence will be discussed in detail below.

When the crystal was irradiated at 4.2 K with the frequency doubled 514.5 nm line, no fluorescence could be observed. Additionally the irradiated part of the crystal surface darkened at 4.2 K. This could be observed directly in the cryostat. No fluorescence was observed on subsequent tuning the laser frequency to lower energies. But the other parts of the same crystal surface, which were not irradiated with such shortwave light, did further show fluorescence. Such a behaviour has not been observed with irradiation wavelengths above 333.6 nm. There is no significant difference be-

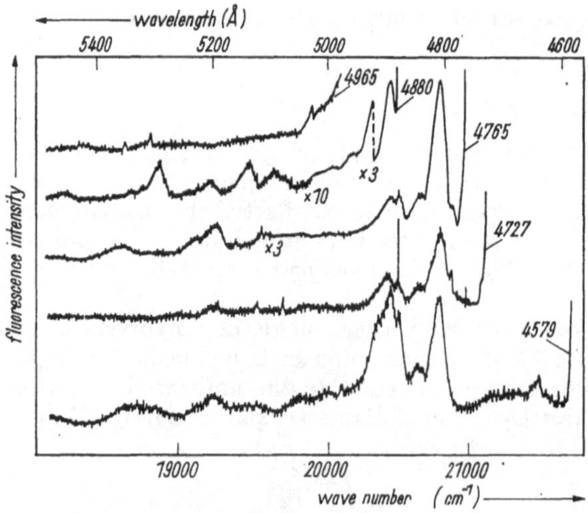

Fig. 1. Fluorescence spectra obtained fromone polycrystallines ample at 4.2 K using five different excitation wavelengths (parameter in Å). The fluorescence strongly depends on teh excitation wavelength. No fluores cence occurs with excitation wave lengths of 4965 Å or longer. Resolution: 2 cm^{-1}

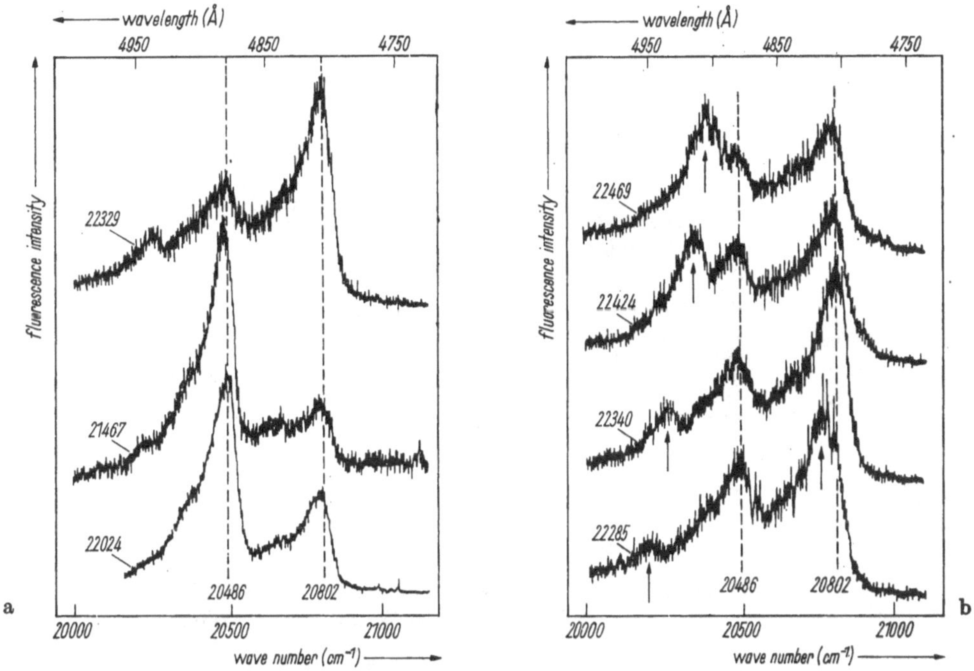

Fig. 2. Fluorescence spectra of one single crystal (4.2 K) for different excitation light wave numbers (parameter in cm^{-1}; resolution 2 cm^{-1}): a) demonstrating the resonant enhancement of one (two upper curves) or both (lower curve) of the two fixed-position bands; b) demonstrating the fixed-position bands and the displaced bands (indicated by arrows)

tween spectra obtained from single crystals, polycrystals, or powder, except the bands from powder samples are broader (factor 1.5 to 2). The main bands of all recorded spectra are located between 20200 and 20900 cm^{-1}.

The fluorescence intensity depends on the location of the laser spot on the surface of the sample (except when powder samples were used), on the colour of the samples (i.e. on the degree of polymerization), and on the use of single crystals or polycrystals.

The strongest emission was observed from a sample consisting of four crystallites with a common (100) surface, when the boundary between the crystallites was irradiated. There is no significant increase in intensity, when the (poly)-crystalline material is powdered. Pink samples show stronger fluorescence than deep red ones.

The fluorescence bands are unsymmetrical with a sharp edge on the high-energy and a broad edge on the low-energy side (Fig. 2). The linewidths are 100 cm^{-1} or larger. They do not change on cooling from 4.2 to 1.5 K, but are considerably broader at 77 K.

In Table 1 the vibronic frequencies obtained from polycrystalline and powder samples as well as from a single-crystalline sample are listed. These values are averages from many spectra with specified standard deviations and statistical errors. They are compared to resonant Raman data from Bloor et al. [5].

4. Fixed and Displaced Bands

In order to investigate the dependence of the relative fluorescence intensities on the excitation wavelengths, many spectra were recorded with small increments of the exciting wavelength using one single-crystalline sample.

Table 1
Vibronic frequencies observed in the fluorescence spectra. Data are average values from different recordings

fluorescence data (partially polymerized material, this work)			resonance Raman data (fully polymerized material [5])
average value (cm^{-1})	statistical error (cm^{-1})	standard deviation (cm^{-1})	(cm^{-1})
956 (960)	2.5 (4.5)	11 (9)	955 ± 2
1213 (1215)	1.9 (11.9)	7.4 (26)	1205 ± 2
1529 (1537)	2.2 (6)	9 (17)	1487 ± 2
2092 (2089)	6.3 (9)	14 (12.7)	2088 ± 2

Without brackets: one single-crystalline sample excited with different wavelengths. In brackets: different polycrystalline and powder samples.

In Fig. 2b four recordings demonstrate the behaviour of the fluorescence. The excitation wave number increases from the bottom to the top curve as indicated. Two bands have fixed positions at 20486 and 20802 cm^{-1} and only vary in intensity. Other bands, which are indicated by arrows, are displaced proportional to the changes in the excitation wave number. They also vary in intensity, getting strong near the fixed-position bands and weak aside them.

Fig. 3 shows the spectral positions of the observed intensity maxima versus the wave number of the exciting light. Two fluorescence bands have fixed positions at 20486 and 20802 cm^{-1} as indicated by the two horizontal lines. Four additional lines connect the spectral positions of the displaced bands obeying the equation

$$\bar{\nu}_{\text{fluor}} = \bar{\nu}_{\text{exc}} - \bar{\nu}_i \,. \tag{1}$$

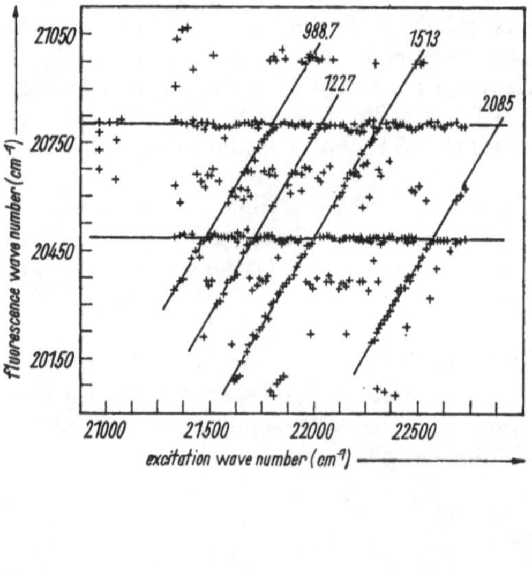

Fig. 3. Spectral positions of observed fluorescence maxima versus excitation wave number. The figure shows the two fixed-position bands (horizontal lines) and four displaced bands (oblique lines) according to $\bar{\nu}_{\text{fluor}} = \bar{\nu}_{\text{exc}} - \bar{\nu}_i$. The frequencies $\bar{\nu}_i$ are written to the corresponding oblique lines (parameter in cm^{-1})

These lines are least-square fits to the experimental data according to (1). Only the frequencies $\bar{\nu}_i$ ($i = 1, 2, 3, 4$) were adjusted:

$$\bar{\nu}_1 = (988.7 \pm 2)\ \text{cm}^{-1}\,,$$
$$\bar{\nu}_2 = (1227 \pm 2.5)\ \text{cm}^{-1}\,,$$
$$\bar{\nu}_3 = (1513 \pm 2.5)\ \text{cm}^{-1}\,,$$
$$\bar{\nu}_4 = (2085 \pm 4)\ \text{cm}^{-1}\,.$$

The specified error is the calculated error of the fit parameter ν_i inherent to the least-square fit.

At the crossing points of the horizontal and oblique lines the fixed-position band seems to shift. But this shift is due to an overlap of a displaced band and a fixed-position band, when the displaced band crosses the fixed-position band.

5. Excitation Mechanism

In the following analysis of the experimental data we assign the observed frequencies $\bar{\nu}_1$, $\bar{\nu}_2$, $\bar{\nu}_3$, and $\bar{\nu}_4$ to the vibronic frequencies of the excited state of the fluorescing centre. In the corresponding energy level scheme (Fig. 4) we adopt the 20486 and 20802 cm^{-1} levels as independent vibronic ground-state levels. Each of these levels has the same sequence of vibronically excited states.

Fluorescence excitation light of approximately 21467 cm^{-1} is resonant with the 988.7 cm^{-1} vibronic state of the 20486 cm^{-1} level. Because of fast vibronic relaxation this yields a selective population of the lower one of the two fixed-position bands. Therefore we expect this band to be strongly enhanced. On the other hand, excitation with light of 22329 cm^{-1} is in resonance with the 1513 cm^{-1} vibronically excited state of the upper fixed-position band (20802 cm^{-1}). In this case we expect enhancement of this emission. The two upper curves in Fig. 2a indeed show this behaviour.

22024 cm^{-1} excitation is resonant with both the 1227 cm^{-1} vibronic state of the upper fixed-position band (20802 cm^{-1}) and with the 1513 cm^{-1} state of the lower fixed-position band (20486 cm^{-1}). This should enhance both bands, which is really observed (see Fig. 2a, lower curve).

In order to understand the appearance of displaced bands, we tried to simulate the unsymmetrical fluorescence bandshape. It turned out that it is impossible to fit the bandshape with one unsymmetrical Gaussian curve. However, a sequence of Gaussian subbands, summed up, is able to simulate the low-energy tail of the observed fluorescence band. This is schematically sketched in Fig. 5a, where 16 symmetrical Gaussian subbands of equal width build up one broad unsymmetrical band.

If we assume each of the subbands to have the same sequence of vibronically excited states as the fixed position bands, we can selectively enhance one of the subbands by the mechanism described

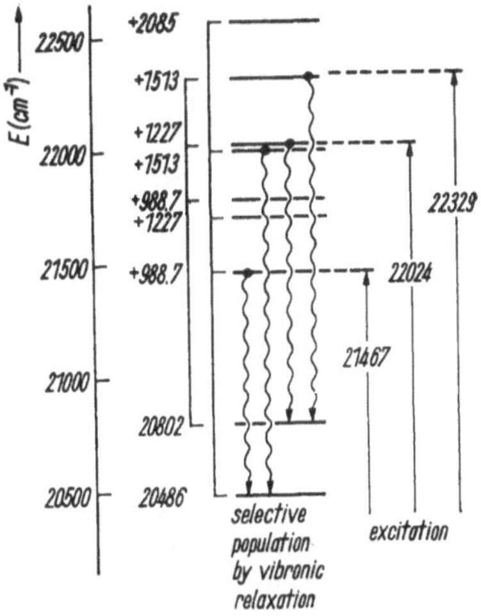

Fig. 4. Energy level scheme of the excited state of the fluorescing centre. The figure shows three resonant excitations, which cause different populations of the vibronic ground-state levels

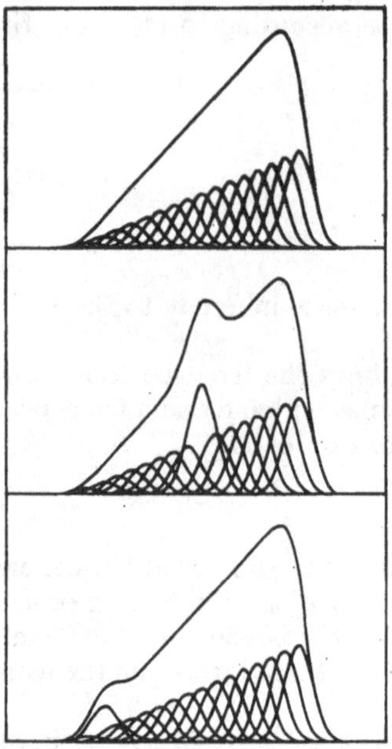

Fig. 5. Schematic representation of the displaced bands. a) 16 symmetric Gaussian subbands build up the unsymmetric fluorescence band (no enhancement of subbands). b) Enhancement of one subband by a factor of two gives an additional fluorescence peak. c) Changing the excitation wave number selects another subband, which is shifted exactly by the change in the excitation wave number and enhanced by a factor of three

above. In Fig. 5b one subband is enhanced in amplitude a factor of two, while leaving all the others unchanged. This yields an additional peak within the original fluorescence band.

When the laser frequency is tuned, for instance to lower energies, we select another subband which is now enhanced, whilst the other one reduces in amplitude (Fig. 5c). Therefore we observe displaced bands by means of selective enhancement of fluorescing subbands. It is noteworthy that this effect requires an enhancement of the subbands by at least a factor of two. The fixed-position bands do indeed show intensity variation by factors greater than two when the laser frequency is tuned.

6. Discussion

From our data, as well as from the data reported by Bloor et al. [4], it seems reasonable to assign the observed fluorescence to localized states connected to the polymer chain. Localization is understood in two senses:

First, the fluorescing centres are geometrically localized, i.e. the fluorescence intensity depends on the localization of the laser spot on the crystal surface. The strongest emission is observed in polycrystalline samples when the boundaries between crystallites are irradiated.

Secondly, the fluorescing centres are energetically isolated from the perfect chain, but nevertheless connected to a defect of the polymer chain.

The reasons are as follows:

(i) The vibronic frequencies obtained from the analysis of the excitation mechanism are nearly identical to those occurring in the fluorescence spectra. This indicates that the fluorescing centre itself absorbs the exciting light, i.e. no energy transfer from another absorbing centre to the fluorescing one is required.

(ii) The polymer molecule shows strong absorption ranging from approximately 16000 cm^{-1} to the UV (i.e. the band edge is approximately 0.5 eV lower than the fluorescence energy). Despite this fact fluorescence is observed. Therefore we conclude that there is no efficient energy transfer from the fluorescing centre to the polymer chain, which would quench the fluorescence.

(iii) The existence of displaced bands shows that it is possible to strongly enhance one subband of the fluorescence emission without affecting its neighbours. This is evidence that spectral diffusion is absent.

(iv) A comparison of the vibronic frequencies obtained from our fluorescence spectra with resonance Raman results [5, 6] suggests that the fluorescing centres are connected to the polymer chain.

The resonance Raman data show that the frequencies are sensitive to the degree of polymerization and strains of the polymer molecule: The 2088 cm^{-1} vibration in fully polymerized crystals lowers to approximately 2013 cm^{-1} in 0.1% polymer in monomer matrix, but remains at 2088 cm^{-1} when the polymer is extracted. The 1487 cm^{-1} vibration lowers to 1455 cm^{-1} in 0.1% polymer in monomer matrix, but remains constant in the extracted polymer. A similar behaviour shows the 1205 cm^{-1} vibration.

This demonstrates that the vibronic frequencies are extremely sensitive to strains in the material (as explained by Bloor et al. [6]). Therefore, despite some differences, we adopt the frequencies obtained from fluorescence as well as from excitation data to be due to vibrations characteristic to the polymer chain.

The largest difference shows the C = C stretching vibration (1529 cm^{-1} from fluorescence, 1487 cm^{-1} from resonance Raman data). This indicates that the fluorescing centre is accompanied by a significant perturbation of the C=C bond.

A 1531 cm^{-1} vibration has been observed with samples at low conversion after extraction of the polymer [5, 6] and has been associated to kinks in the chain introduced during extraction.

(v) The vibronic structure of the excitation as well as the fluorescence emission undoubtedly rules out this luminescence to be due to impurities in the samples. Oligomers can also be excluded. A distribution of chain lengths will cause a broad spectral distribution of fluorescing bands [7] towards higher and lower energies (the low-energy limit is approximately 16000 cm^{-1} for the infinite chain). This is not observed.

From the arguments given above we conclude that the fluorescing centres are localized states connected to the polymer chain. Nevertheless, these states must be different from the band states of the polymer chain, which are singlet states. Because the C≡C stretching vibration (2092 cm^{-1}) is that of a relaxed (strain-free) molecule, the fluorescing centre must be localized at a strain-free part of the polymer molecule which may possibly be a chain end.

It has been suggested that active chain ends, which are responsible for the polymerization reaction, could be trapped in dislocations [4]. If this is really true, the fluorescing centre could be a trapped ground-state triplet carbene at the end of a polymer molecule which has been observed during thermal polymerization of PTS [3]. In this case absorption and emission of the localized centre involve allowed T–T transitions. It is expected to have a similar but not completely identical vibronic structure as the perfect polymer chain.

7. Conclusions

Partially polymerized PTS single crystals show fluorescence with a quantum energy significantly higher than the absorption edge of the perfect polymer chain. We showed that this fluorescence is due to localized states which are connected to the polymer chain and gave some arguments for these states to be trapped active chain ends (triplet carbenes). The electronic structure of these states is unknown so far. Further investigations are intended to show whether these states are connected to the triplet states.

Acknowledgements

We thank D. Bloor for useful discussions in the early state of the work and R. Huber for preparing the samples.

References

[1] G. WEGNER, Makromol. Chem. **145**, 85 (1971).
[2] U. WALTER, Diploma Work, Mainz 1974.
[3] H. EICHELE, M. SCHWOERER, R. HUBER, and D. BLOOR, Chem. Phys. Letters **42**, 342 (1976).
[4] D. BLOOR, D. N. BATCHELDER, and F. H. PRESTON, phys. stat. sol. (a) **40**, 279 (1977).
[5] D. BLOOR, F. H. PRESTON, D. J. ANDO, and D. N. BATCHELDER, in: Structural Studies of Macromolecules by Spectroscopic Methods, Ed. K. J. IVIN, Wiley, New York 1976 (p. 91).
[6] D. BLOOR, L. KOSKI, G. C. STEVENS, F. H. PRESTON, and D. J. ANDO, J. Mater. Sci. **10**, 1678 (1975).
[7] L. SALEM, The Molecular Orbital Theory of Conjugated Systems, W. A. Benjamin, Inc., New York/Amsterdam 1966 (p. 365).

(Received June 6, 1977)

phys. stat. sol. (a) **43**, 473 (1977)

Subject classification: 1.1; 22.1.2

Laboratoire de Minéralogie-Cristallographie, associé au C.N.R.S., Université Pierre et Marie Curie, Paris[1]) *(a)*
and Institut für Angewandte Physik der Technischen Hochschule Wien (b)

Synchrotron Radiation Topographic Evidence for the Interaction of Plane-Polarized X-Ray Photons with Perfect Crystals

By

M. SAUVAGE (a), J. F. PETROFF (a), and P. SKALICKY (b)

Evidence for the coherent splitting of a plane-polarized X-ray wave inclined with respect to the plane of incidence during Bragg diffraction in a perfect crystal is obtained, using synchrotron radiation white beam topography. This new tool makes feasible further experiments on symmetry dependent dielectric properties of crystals in the X-ray range.

Es werden Hinweise für die kohärente Aufspaltung einer bezüglich der Einfallsebene geneigten planpolarisierten Röntgenwelle während der Bragg-Beugung in einem perfekten Kristall durch Topographie mit weißem Strahl der Synchrotronstrahlung erhalten. Diese neue Methode ermöglicht weitere Experimente an symmetrieabhängigen dielektrischen Eigenschaften von Kristallen im Röntgenbereich.

1. Introduction

As is well known, due to the tensorial character of their dielectric constant in the optical frequency range, crystals show orientation dependent properties such as birefringence and optical activity for electromagnetic waves. However, in the X-ray frequency range, a controversy is found in the literature: according to Moliere's theory [1] birefringence should occur in simple transmission of X-rays, the effects being detectable close to absorption edge frequencies only while a computation by Hojo et al. [2] demonstrates that the probability for a plane-polarized X-ray photon to be scattered into a different polarization state through interaction with the electrons of an atom is zero as long as hydrogen-like wave functions are used. Due to the lack of experimental data, it is not possible to decide for any of these statements in the case of simple transmission. However, with crystals adjusted for Bragg reflexion and perfect enough to fulfil the requirements of dynamical theory, polarization effects independent of crystal symmetry and implying the existence of two principal polarization states over which any incident wave is projected, have been observed.

2. Basic Theory and Previous Experiments

2.1 Dispersion surface and crystal waves

The dynamical theory (Ewald [3], Laue [4]) establishes that in the case of two reciprocal lattice points closed to the Ewald sphere, the origin 0 and the node H, the extremities of the wave vectors of the waves propagating in the crystal, are located on a dispersion surface whose section in the plane of incidence is approximated by two sets of hyperbolae, one for each polarization state (Fig. 1). The usual continuity conditions enable one to determine the eight crystal waves excited by a single incident plane wave:

[1]) 4 place Jussieu, 75230 Paris, Cedex 05, France.

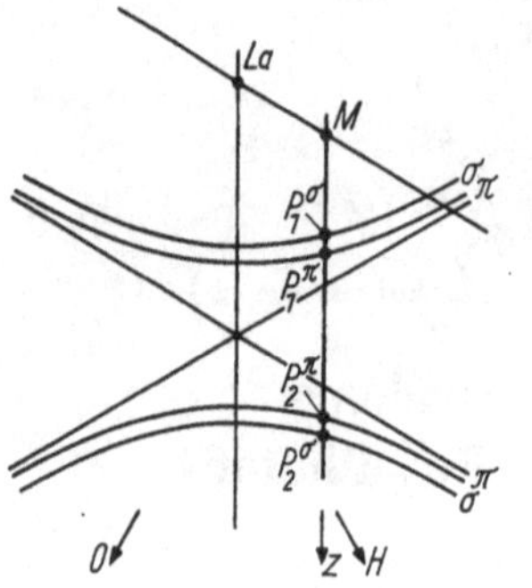

Fig. 1. Dispersion surface for both polarization directions. Tie-point of the crystal wave-fields excited by an incident plane wave (OM) in the symmetrical Laue case (normal Mz to the entrance surface parallel to the reflecting plane)

transmitted waves	$\overrightarrow{OP_1^\sigma}$, $\overrightarrow{OP_1^\pi}$, $\overrightarrow{OP_2^\sigma}$, $\overrightarrow{OP_2^\pi}$,
reflected waves	$\overrightarrow{HP_1^\sigma}$, $\overrightarrow{HP_1^\pi}$, $\overrightarrow{HP_2^\sigma}$, $\overrightarrow{HP_2^\pi}$.

σ and π stand for polarization normal and parallel to the plane of incidence, respectively. The discussion will be restricted to reflected waves which are more often studied experimentally. The crystal waves belonging to the same state of polarization but to different branches of the dispersion surface are coherent for any type of incident wave; their interferences during propagation give rise to the well-known "pendellösung fringes" which provide a most sensitive way of detecting polarization effects in crystals as will be developed in the next section.

2.2 Previous experiments

When the incident wave is unpolarized, the σ and π components are incoherent, each one excites two coherent waves whose resulting amplitude is a periodic function of the crystal thickness with periods

$$\Lambda_\sigma \propto \frac{1}{|P_1^\sigma P_2^\sigma|}, \qquad \Lambda_\pi \propto \frac{1}{|P_1^\pi P_2^\pi|}.$$

The σ and π systems add up incoherently and the fringe pattern observed with a wedge shaped crystal shows fading areas corresponding to a beat between both systems of slightly different periods (Fig. 2). Such beat was first observed and explained by Hart and Lang [5] and Hattori et al. [6].

Up to now, very few experiments have been performed with plane polarized X-rays due to the high loss of intensity in the polarizing process. Hart and Lang [5] used a germanium crystal adjusted for 333 reflexion of Cu$K\alpha_1$ as a polarizer and excited

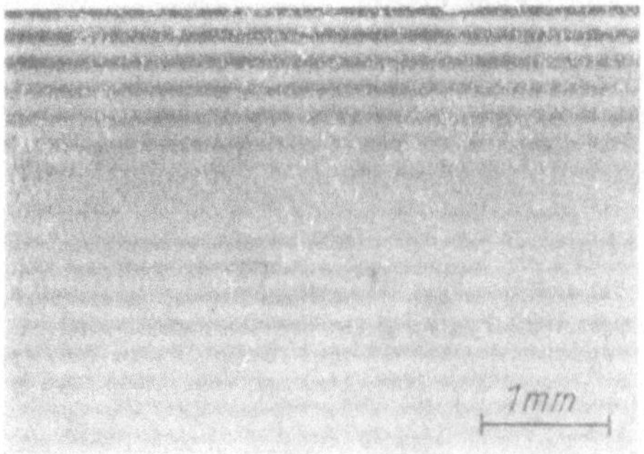

Fig. 2. Classical traverse topograph of a silicon wedge. 220 reflexion, Mo$K\alpha_1$. Pendellösung fringes with unpolarized incident radiation, first fading after seven fringes

only the σ branches of the dispersion surface thus producing a fringe system without fading in a silicon wedge.

Another experiment has been performed by Skalicky and Malgrange [7] on a two-crystal spectrometer where the first crystal followed by a narrow slit delivered a plane-polarized pseudo plane wave since the π component was nearly absorbed. The second crystal, a wedge shaped silicon, was mounted on a scanning device and adjusted at peak intensity (zero departure from Bragg incidence). The reflected intensity was measured as a function of crystal thickness in two experimental situations:

(i) Polarization normal to the plane of incidence.

As in Hart and Lang's experiment the σ state only is excited and the intensity record shows a monotonic decrease of the oscillation contrast.

(ii) Polarization at 45° with respect to the plane of incidence.

As the crystal is traversed in front of the beam, the reflected intensity shows sequences with vanishing oscillations which are interpreted by the authors in the following way: the coherent splitting of the incident wave into σ and π components whose wave vectors differ by

$$\Delta \boldsymbol{k}_1 = P_1^\sigma P_1^\pi \quad \text{and} \quad \Delta \boldsymbol{k}_2 = P_2^\sigma P_2^\pi$$

gives rise to a modification of the wave field polarization state during propagation: wave fields 1 and 2 become elliptically polarized with the particular cases of linear and circular polarization for definite crystal thicknesses and their interference reproduces the experimental features of the fringe pattern.

This experiment was the first performed with the polarization inclined with respect to the plane of incidence, it gave a definite proof of the physical existence of principal polarization states associated with Bragg diffraction and demonstrated the possibility of producing elliptically polarized X-ray waves. It has to be emphasized that the nature of the diffracting crystal and particularly its symmetry properties do not interfere with the effect under investigation here.

3. Present Experiment

The present experiment confirms the result of Skalicky and Malgrange [7]. It has been performed at Lure, (Orsay, France), with the synchrotron beam delivered by the D.C.I. storage ring of the Laboratoire de l'Accélérateur Linéaire. Due to the small angular divergence of the X-ray synchrotron beam, white beam topography is a most suitable technique to image crystal defects with high speed and high resolution as is demonstrated by the important amount of work already performed at DESY (Tuomi et al. [8]) and at Daresbury (UK) (Hart [9]; Tanner et al. [10]). However, no experiment taking advantage of the nearly pure linear polarization state of the synchrotron beam had been performed. In the present work, the silicon wedge has been adjusted for 220 reflexion of $\lambda \approx 0.9$ Å with polarization normal to the plane of incidence. The beam topograph presented in Fig. 3a was observed after 2 min exposure on a L4 50 μm nuclear plate, it shows an integrated intensity fringe pattern with a monotonic contrast decrease characteristic of a pure linear polarization state (σ state). If π components were present, the first fading should be observed after 4.3 fringes, its absence is an evidence of the predominance of the horizontal polarization component in the synchrotron beam which is indeed expected at short wavelengths.

The 220 reciprocal vector of the crystal has been then rotated around the incident beam axis in order that 220 reflexion should occur with a plane of incidence inclined

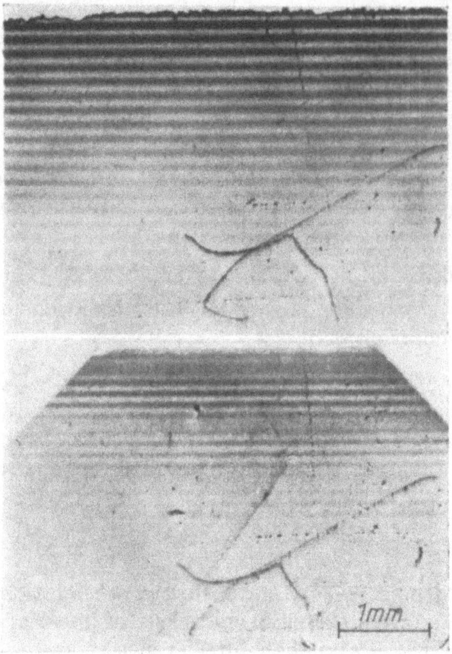

a

b

Fig. 3. Synchrotron radiation topographs of a silicon wedge, 2 min exposure time on Ilford L4 50 μm nuclear plates, D.C.I. storage ring operated at 1.72 GeV, 80 mA: a) 220 reflexion, $\lambda \approx 0.9$ Å. Polarization normal to the plane of incidence. No periodic fading in the fringe pattern. b) 220 reflexion, $\lambda \approx 1.1$ Å. Polarization at 45° to the plane of incidence. Periodic fadings separated by 5.2 fringes $\rightarrow \theta_B \approx 17.2°$

at 45° to the beam polarization, for a similar wavelength ($\lambda \approx 1.1$ Å). The topograph shown in Fig. 3b and obtained with the same exposure time of 2 min displays clearly fading areas in the integrated fringe pattern which can be explained in the same way as in the plane-wave experiment by Skalicky and Malgrange.

A quantitative comparison particularly concerning fringe positions and interfringe distances is difficult since white beam topography settings do not provide accurate measurements of the crystal angular position: in the present experiment, the best way of evaluating the selected wavelength is the ratio ϱ between the mean fringe distance Λ_m and the separation Λ_b between two fadings which depends only on 2θ,

$$\varrho = \frac{\Lambda_m}{\Lambda_b} = \frac{2(1 - |\cos 2\theta|)}{1 + |\cos 2\theta|}.$$

It is worth noticing that both topographs 3a and 3b have been recorded with the same exposure time of 2 min, which means that no important intensity loss occurs when working with polarization inclined with respect to the plane of incidence. This was definitely not the case with the two-crystal setting: when both crystals were adjusted with their planes of incidence parallel, an effective counting time of 30 min was necessary to record the information corresponding to topograph 3a, the time spent on readjusting the crystal after each of the 60 steps used to scan the whole fringe pattern is not evaluated here; on the other hand, when the planes of incidence were inclined at 45° to each other, *240 h* of counting time were necessary to bring the information provided by topograph 3b in the present work.

4. Conclusion

The purpose of the present experiment was not to produce new physical results since the existence of elliptically polarized X-ray waves in crystals had already been demonstrated by Skalicky and Malgrange [7]. Its major aim was to show that a new field of experiments is now open: very small effects such as birefringence and rotatory power in the X-ray range were impossible to detect as long as polarized X-rays were produced with a dramatic intensity loss, in very restrictive geometrical settings and for a limited number of characteristic lines. None of these drawbacks is encountered with synchrotron radiation which is a *white*, *plane-polarized*, and *highly intense* X-ray beam and consequently symmetry dependent polarization properties of crystals in the X-ray range will now be investigated.

Acknowledgements

The authors wish to express their thanks to the members of the Laboratoire de l'Accélérateur Linéaire and in particular to Dr. P. Marin and his group for having made possible the use of the D.C.I. synchrotron radiation.

References

[1] G. Moliere, Ann. Phys. (Leipzig) **35**, 272 (1939).
[2] A. Hojo, Y. H. Ohtsuki, and S. Yanagawa, J. Phys. Soc. Japan **21**, 2082 (1966).
[3] P. P. Ewald, Ann. Phys. (Leipzig) **54**, 519 (1917).
[4] M. von Laue, Röntgenstrahl Interferenzen, Frankfurt/Main (1960).
[5] M. Hart and A. R. Lang, Acta cryst. **19**, 73 (1965).
[6] H. Hattori, H. Kuriyama, and N. Kato, J. Phys. Soc. Japan **20**, 1047 (1965).
[7] P. Skalicky and C. Malgrange, Acta cryst. **A28**, 501 (1972).
[8] T. Tuomi, K. Naukkarinen, and P. Rabe, phys. stat. sol. (a) **25**, 93 (1974).
[9] M. Hart, J. appl. Cryst. **8**, 436 (1975).
[10] B. K. Tanner, M. Safa, D. Midgley, and J. Bordas, J. Magnetism and magnetic Mater. **1**, 337 (1976).

(Received June 6, 1977)

phys. stat. sol. (a) 43, 479 (1977)

Subject classification: 14.3.1 and 14.3.4; 22.1.2

Forschungslaboratorien der Siemens AG, München

An ESFI SOS Magnetodiode

By

P. LILIENKAMP[1]) and H. PFLEIDERER

Double injection of holes and electrons into an ESFI® SOS film is observed. As is common with magnetodiodes the application of transverse magnetic inductions influences the current–voltage characteristics due to the lifetime profile of the charge carriers across the film. From the dependence of the current on the magnetic induction it is possible to deduce the average bulk lifetime and the surface recombination velocity at the silicon–substrate interface as functions of the voltage. The lifetime is found to increase with voltage, while the surface recombination velocity decreases. These deductions reflect the additional increase in double injection over a certain voltage range and the appearance of magnetoconductance versus voltage extrema. Thus by means of the magnetodiode effect the recombination properties of an ESFI® SOS film may be checked.

Doppelinjektion von Löchern und Elektronen in eine ESFI®-SOS-Schicht wird beobachtet. Wie bei Magnetdioden üblich beeinflußt das Anlegen einer transversalen magnetischen Induktion die Strom–Spannungs-Kennlinien infolge des Lebensdauerprofils der Ladungsträger quer zur Schicht. Aus der Abhängigkeit des Stroms von der magnetischen Induktion lassen sich die mittlere Volumenlebensdauer und die Oberflächenrekombinationsgeschwindigkeit an der Silizium–Substrat-Grenzfläche als Funktionen der Spannung entnehmen. Wie sich zeigt nimmt die Lebensdauer mit der Spannung zu, die Oberflächenrekombinationsgeschwindigkeit dagegen ab. Dieser Befund gibt den zusätzlichen Anstieg der Doppelinjektion in einem gewissen Spannungsbereich und das Auftreten von Extrema in den Magnetokonduktanz–Spannungs-Kurven wieder. Demnach kann der Magnetdiodeneffekt als Mittel zur Kontrolle der Rekombinationseigenschaften einer ESFI®–SOS-Schicht dienen.

1. Introduction

Crossed electric and magnetic fields in the plane of a semiconducting layer produce a Lorentz force normal to the layer. A plasma of holes and electrons follows this force while preserving electric neutrality. The carriers accumulate on one side of the layer and deplete on the other. The carrier redistribution leads to a change in average carrier density if the lifetime of excess carriers varies more or less monotonically across the layer. A Lorentz force which is unidirectional with a lifetime gradient causes an increase in the average carrier density because the layer sections with the longer lifetimes need higher excess carrier densities than the lower lifetime sections in order to maintain the stationary recombination generation balance. Conversely, a Lorentz force in opposition to the lifetime gradient will decrease the average carrier density.

A suitable lifetime function τ of cooridnate y crossing the layer is

$$\frac{1}{\tau(y)} = s_0\,\delta(y) + \frac{1}{\tau_v} + s_b\,\delta(b - y)\,. \tag{1}$$

The layer surfaces are located at $y = 0$ and $y = b$; δ represents the delta function; s_0 and s_b are surface recombination velocities; τ_v designates a *constant* bulk lifetime. An inequality $s_0 < s_b$ provides some sort of (negative) lifetime gradient. The function (1) allows an analytical solution of the diffusion equation for the carrier distribution.

[1]) Present address: Institut für Angewandte Physik der Universität Hamburg, BRD.

The condition (1) is obtainable by treating the two Hall sides of a semiconducting platelet differently. The control of recombination, average carrier density, and current via magnetic induction may be quite efficient if the sample thickness b is comparable with the diffusion length [1]. Intrinsic devices of this kind having ohmic current contacts [2] are called magnetic rectifiers [3], while double injection diodes of pin-structure represent magnetodiodes [4 to 6]. Also p^+pn^+ diodes relying exclusively upon double injection [7] to provide the necessary hole–electron plasma are conceivable.

Thin semiconducting films seem to be promising for the preparation of magnetodiodes [8]. The present paper contributes an investigation of a p^+pn^+ double injection diode fabricated in ESFI® (epitaxial silicon films on insulators) or SOS (silicon on sapphire) technology [9, 10] used to produce CMOS transistors.

An ESFI® film contains many crystal defects (dislocations, stacking faults) which decrease in density with increasing distance from the sapphire substrate [11]. Accordingly the generation lifetime of minority carriers within depleted space charge regions rises exponentially with this distance as measured from the time response of a pulsed MOS capacitor [12, 13] or the reverse current of a gate-controlled diode [14].

2. Experimental Characteristics

Fig. 1 shows the I–U (current–voltage) characteristics of an ESFI diode measured at room temperature for different magnetic inductions B. The voltage was applied in the form of pulses to avoid thermal runaway. The inset diagram shows the p^+pn^+ structure. The heavily doped p^+ and n^+ current contacts inject, when forward biased, holes and electrons into the weakly p-doped base layer. The base dimensions (length a, thickness b, width c) are indicated. A silicon dioxide layer protects the base surface. The current through the base is transported by diffusion only near the contacts. In a middle base region holes and electrons drift. The length of the drift region is termed the effective length a_{eff}.

The double injection current is porportional, at least in the case of weak magnetic induction, to the average lifetime, termed effective lifetime

$$\tau_{eff} = \int_0^b p \, dy \left[\int_0^b \frac{p}{\tau} \, dy \right]^{-1} . \tag{2}$$

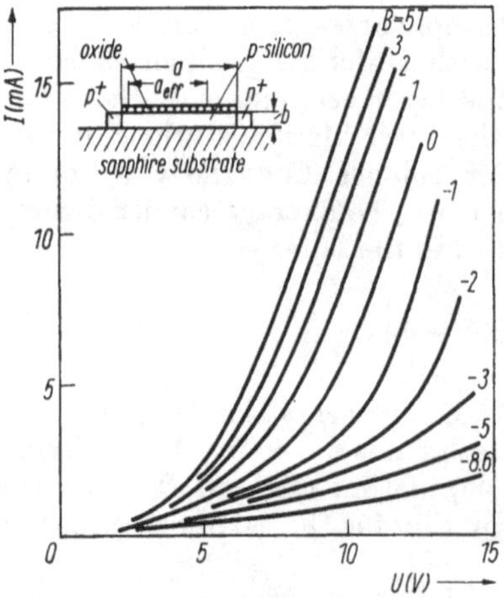

Fig. 1. An ESFI (epitaxial silicon films on insulators) or SOS (silicon on sapphire) diode; I current, U voltage, B external transverse magnetic induction. The inset diagram shows the p^+pn^+ structure; $a = 14$, $b = 0.6$, $c = 30$ μm

The function $p(y)$ represents the excess hole density (assumed to be equal to the excess electron density); the function $\tau(y)$ designates a commonlifetime for holes and electrons. In the

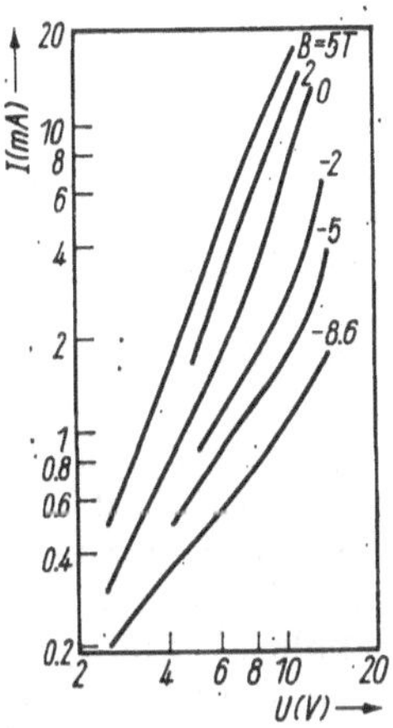

Fig. 2. Characteristics of Fig. 1 in logarithmic scales

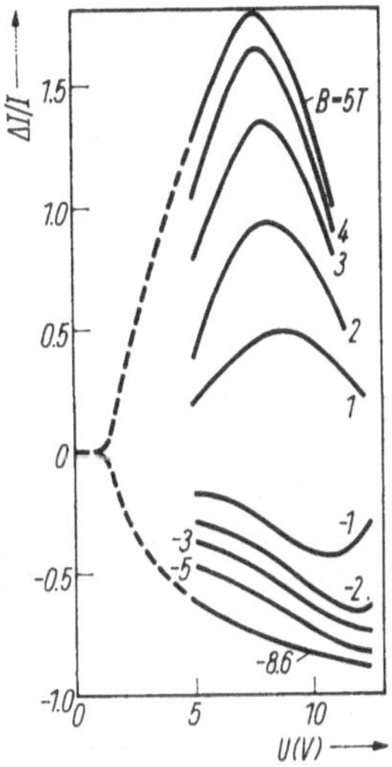

Fig. 3. Another representation of the characteristics in Fig. 1. Relative current change $\Delta I/I$ due to magnetic inductions B vs. voltage U

special case (1) the equation (2) becomes

$$\frac{1}{\tau_{\text{eff}}} = \frac{1}{\tau_{\text{v}}} + \frac{s_0 p(0) + s_b p(b)}{b\langle p\rangle}, \qquad \langle p\rangle = \frac{1}{b}\int_0^b p \, dy\,, \tag{3}$$

where s_0 and s_b denote the recombination velocities at the silicon–silicon dioxide and silicon–substrate interfaces, and $\langle p\rangle$ represents the average excess hole density.

We define the magnetic induction B as positive when the carriers are swept towards the oxide, in the vicinity of which the lifetime is relatively high, i.e. both τ_{eff} and the diode current I increase with $B > 0$, but decrease with $B < 0$. A part of Fig. 1 is redrawn as Fig. 2 using logarithmic scales. Let us look at the characteristic for $B = 0$, which is a straight line with slope 2 up to a voltage $U \approx 7$ V, indicating a relation $I \sim U^2$ known from double injection into a doped base, whereas for $U > 7$ V the slope steepens.

The influence of an inductance B on the current I is stressed by Fig. 3, showing the magnetoconductance defined as the relative current change $\Delta I/I = [I(U, B) - I(U, 0)]/I(U, 0)$ versus voltage U (again using linear scales). While the magnetoconductance cannot, by definition, become smaller than -1, it may well increase beyond $+1$. For $U \to 0$ we have $I(U, B) \to I(U, 0) \to 0$. With respect to this situation our I–U measurements were not sufficiently exact to allow the extraction of the magnetoconductance for the lower voltages. The dashed lines represent more or less arbitrary extrapolations. Most conspicuous in Fig. 3 are the extrema of the magnetoconductance curves. These extrema represent, besides the additional increase in injection current, another peculiarity of ESFI or SOS magnetodiodes.

3. Theoretical Characteristics

The diode characteristics will be described by a simple current–voltage function. Instead of the current I we express the average current density $\langle J\rangle = I/bc$ in the form

$$\langle J\rangle = J_0 \exp\left(\frac{U_0}{nU_{\text{T}}}\right) = J_1 + J_2 + J_3\,. \tag{4}$$

The portion U_0 of the applied voltage U drops across the pn^+ junction. For $U < 1$ V we expect $U \approx U_0$. The related exponential regime of the I–U characteristics does not appear in Fig. 2. According to [15] J_0 is a constant, U_T the temperature voltage, and $n \approx 2$ for SOS diodes. An exponential regime is also known from Si magnetodiodes [16].

For the higher voltages it is the base conductance that determines the shape of the characteristics. The assumption of a homogeneous base enables us to exploit established conductance regimes. First we use the lifetime function (1) and second also the mobilities $\mu_p > 0$ and $\mu_n < 0$ of holes and electrons as well as an acceptor concentration A are treated as constants. None of these assumptions is of course correct. In reality the mobilities across the Si layer vary exponentially [12, 17], like the lifetime, and even an initially homogeneous doping concentration will develop a profile during the oxidation process [18]. At best the parameters μ_p, μ_n, A, and τ_v, considered as constants, may be taken as suitable average values. The bulk lifetime τ_v and surface recombination velocities s_0 and s_b are substituted for the real lifetime function $\tau(y)$. It seems not unreasonable to terminate a realistic lifetime profile with an additional recombination or generation velocity on the substrate surface itself [12, 19]. Note that the velocity s_b of the equivalent function (1) should exceed the "true" velocity associated with a lifetime profile.

We will use a simplified version of the homogeneous base model [20] and distinguish three double injection regimes [7], viz. the ohmic regime

$$J_1 = e\mu_p p_0 E^1 \,, \tag{5}$$

the semiconductor regime

$$J_2 = \frac{(9/8)\, e\, |p_0 - n_0|\, \mu^2 \tau_s E^2}{a_{\text{eff}}} \,, \tag{6}$$

and the insulator regime

$$J_3 = \frac{(125/18)\, \varepsilon \mu^2 \tau_s E^3}{a_{\text{eff}}^2} \tag{7}$$

with e elementary charge, p_0 and n_0 equilibrium densities of holes and electrons (being constant), ε permittivity, $E = (U - U_0)/a_{\text{eff}}$ electric field in the drift region of the base,

$$\mu^2 = \frac{-\mu_p \mu_n}{h^2} > 0 \tag{8}$$

mobility factor, $h = 1 - \mu_p \mu_n B^2 > 1$ magnetoresistance factor, and B magnetic induction. The special lifetime

$$\tau_s = (1 + g_{ks} + f_{ks})\, \tau_v \tag{9}$$

depends on the excess hole density $p(y)$ via the coefficients

$$g_{ks} = \frac{\langle p \rangle}{n_i}, \qquad f_{ks} = -\frac{DB}{Eb} \frac{p(b) - p(0)}{n_i} \tag{10}$$

with n_i intrinsic density, $D = 2D_p D_n/(D_p + D_n)$ ambipolar diffusion constant, $D_p = \mu_p U_T$ and $D_n = -\mu_n U_T$ diffusion constants of holes and electrons, b layer thickness. The transverse drift velocity due to the Lorentz force is $v_L = D^* k$ with

$$D^* = \frac{D}{h} \tag{11}$$

and

$$k = -\frac{(\mu_p - \mu_n)\, EB}{2U_T}. \tag{12}$$

With respect to a cartesian coordinate system the vectors occurring in (12) are represented by their components E_x, k_y, B_z. The coefficients g_{ks} and f_{ks} are functions of k, s_0, and s_b, indicated by the subscripts, and also of D^*, τ_v, b. These functions are explicitly given in [20] (using $k_i = -k$ instead of k). The effective lifetime (3) becomes

$$\tau_{eff} = (1 + g_{ks})\, \tau_v\,. \tag{13}$$

The obvious relation $0 < \tau_{eff} < \tau_v$ corresponds to $-1 < g_{ks} < 0$. In the case of ESFI diodes the coefficient f_{ks} remains insignificant, at least for small B, i.e. $\tau_s \approx \tau_{eff}$. A simple expression for τ_{eff} was proposed in [5]. The formula derived in [21] for τ_{eff} differs somewhat from (13). Since any dependence of μ_p and μ_n on B will be neglected, B affects only the injection currents (6) and (7). In the special case $B = 0$ we have $\tau_{eff} = \tau_s$ exactly [22]. If, additionally, $s_0 = s_b = 0$, then $\tau_{eff} = \tau_v$. For this latter case (6) and (7) were derived originally [23].

In general, one of the current components (5), (6), (7) of (4) dominates over a certain voltage range. The transitional current between two of the regimes is determined by both their components. A more exact and far more complicated descritpion of the entire double injection characteristics was given in [23] and also used in [20]. The diode investigated here does not exhibit an ohmic regime. Instead, an exponential regime merges directly with a semiconductor regime J_2. The inclusion of an ohmic constituent in (4) is nevertheless meaningful because it assumes significance in the case of strong negative magnetic inductions. An inspection of our quantitative analysis revealed that the component J_3 always remains insignificant. Thus the additional rise in injection current noted from Fig. 2 must not be interpreted as a transition to the insulator regime.

Our analysis covered the voltage range $U = 7$ to 11 V. Carrier heating can be expected because electric fields higher than $E = 10$ kV/cm develop in the drift region. We will take this into account with the following rough estimates. Carrier heating is thought to diminish the mobilities in the x-direction of current flow but not in the y-direction of the Lorentz force. We will accordingly introduce a mobility tensor for the holes with unequal even components $\mu_{pxx} \neq \mu_{pyy}$ and vanishing odd components $\mu_{pxy} = \mu_{pyx} = 0$, and simplify the notation by $\mu_{pxx} \to \mu_{px}$ and $\mu_{pyy} \to \mu_{py}$. Analogous definitions apply to the electron mobility. The mobilitiy function $\mu_{px}(E)$ shall depend on a single parameter, the saturation velocity v_s. We assume

$$\mu_{px}(E) = \frac{\mu_{px}(0)}{1 + (\mu_{px}(0)\, E/2v_s)^2} \tag{14}$$

if $E < 2v_s/\mu_{px}(0)$ and otherwise $\mu_{px}(E) = v_s/E$. The same value v_s will be used in the analogous expression for $\mu_{nx}(E)$. We consider $\mu_{py} = \mu_{px}(0)$, $D_{px} = \mu_{px} U_T$, $D_x = 2D_{px}D_{nx}/(D_{px} + D_{nx})$ and so on. As a consequence of the above assumptions (12) becomes

$$k = -\frac{(\mu_{px} - \mu_{nx})\, EB}{2U_T} \tag{15}$$

and (11) splits in

$$D_x^* = \frac{D_x}{h_x}, \qquad D_y^* = \frac{D_y}{h_y} \tag{16}$$

with the factors $h_x = 1 + (\mu_{py} - \mu_{ny})\, D_x B^2/2U_T$, $h_y = 1 + (\mu_{px} - \mu_{nx})\, D_y B^2/2U_T$.

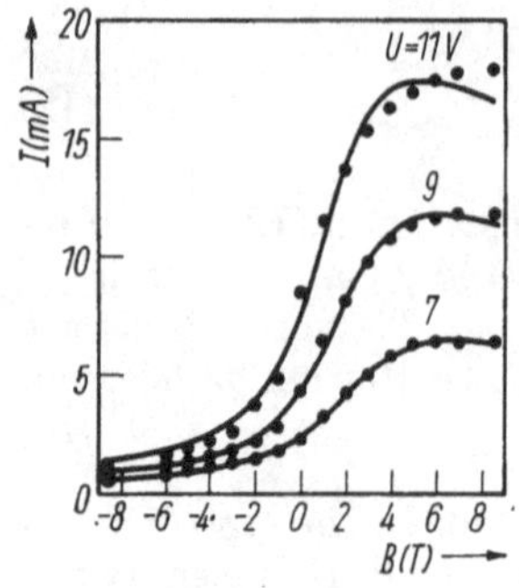

Fig. 4. Theoretical curves and experimental points corresponding to Fig. 1: Current I vs. magnetic induction B; U applied voltage. Theoretical model evaluated with parameter values given in the text and the parameter functions of Fig. 5. ● experiment, —— theory

The coefficients (10) are now functions of D_y^* and k of (15), while (8) becomes

$$\mu^2 = -\frac{\mu_{px}\mu_{nx}}{h_x h_y}. \tag{17}$$

The following analysis was performed with (14) to (16), but (17) multiplied by D_y^*/D_x^*. In any case the straightforward manner to estimate carrier heating only modifies but does not determine the peculiarities of interest here.

4. Evaluation of Characteristics

Fig. 4 shows experimental and theoretical I–B characteristics (current versus magnetic induction). The experimental points were drawn from the I–U measurements in Fig. 1. The additional points marked for strong positive B are not quite reliable because the respective I–U curves lie close together. The above model was evaluated using the following assumptions: pn^+ junction voltage $U_0 = 1$ V, low field mobilities $\mu_p = 300$, and $\mu_n = -600$ cm²/V s as suggested by [10], carrier saturation velocity $v_s = 10^7$ cm/s as given in [24], acceptor concentration $A = 4 \times 10^{15}/\text{cm}^3$ fitted to the characteristic for $U = 7$ V and compatible with the film doping by ion implantation, recombination velocity $s_0 = 0$ at the Si–SiO_2 interface (see comment below). The remaining parameters were chosen to fit the experimental characteristics, viz. the effective length a_{eff}, the bulk lifetime τ_v, and the surface recombination velocity s_b at the Si–sapphire interface, whose values can be taken from Fig. 5.

The I–B characteristics of conventional magnetodiodes are bell-shaped, the maximum being located at $B > 0$, while $I \to 0$ for $|B| \to \infty$ [6, 25]. This picture reflects the inequality $0 < s_0 < s_b$. In the degenerate case $s_0 \to 0$ the characteristics will become S-shaped, as the experimental characteristics of Fig. 4. Our model explains the current saturation $I \to 0$ for $B \to -\infty$ by $\tau_{eff} \to 0$ on account of $s_b \neq 0$, i.e. the suppression of any injection current. On the other side, $B \to \infty$ yields $\tau_{eff} \to \tau_v$ on account of $s_0 = 0$, i.e. saturation of the current at its highest level compatible with a given

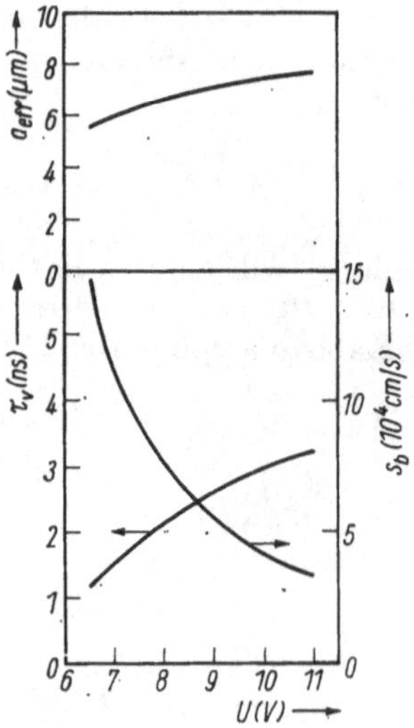

Fig. 5. Effective length a_{eff}, bulk lifetime τ_v, and silicon–substrate recombination velocity s_b vs. voltage U

voltage. The slight maxima of the theoretical curves in Fig. 4 on the right side are caused by the denominator in (17) expressing two-carrier magneto-resistance which is weakened to some extent by the coefficient $f_{ks} > 0$ in (9). Additional direct registrations of I–B characteristics yielded slight maxima of a similar kind also experimentally. In any case the closest agreement between theory and experiment is obtainable by putting $s_0 = 0$. A remark: For vanishing double injection in the case $B \to -\infty$ our magnetodiode model becomes inexact in so far as it neglects the influence of doping on the ambipolar diffusion constant D and the Lorentz force constant k.

In general the lifetime of excess carriers depends on their density. The average carrier density in the base depends on the voltage U and the induction B. Furthermore a transverse carrier distribution $p(y)$ should correspond to a lifetime function $\tau(y)$ [2]. Notwithstanding, our model only admits a constant $\tau \equiv \tau_v$. A function $\tau_v(U, B)$ however is allowed. We are here content to assume a function $\tau_v(U)$ only, being constant along any I–B characteristic. Besides $\tau_v(U)$ we assume functions $s_b(U)$ and $a_{eff}(U)$ and determine all by fitting several I–B curves. The resulting functions, which are plotted in Fig. 5, represent averages over the whole B-range. Due to Fig. 4 the agreement between theory and experiment is good for the lowest voltage $U = 7$ V but deteriorates with increasing U, i.e. the elimination of any dependence on B becomes less successful with increasing injection level. The auxiliary function $a_{eff}(U)$ selects the "best" electric fields and offsets the inaccuracy of putting $U_0 = 1$ V.

The function $\tau_v(U)$ rises while $s_b(U)$ falls as shown in Fig. 5. This behaviour reflects the peculiarities of an ESFI diode. The semiconductor regime means $I \sim U^2$ for $B = 0$ only if τ_v and s_b are constants. Their functional dependence on U allows an additional increase in injection current as seen in Fig. 2, although the carriers begin to be heated up, without an interference of the insulator regime.

The I–U characteristics of conventional magnetodiodes belonging to different B cross each other [20] as do the magnetoconductance versus voltage curves [26]. In Fig. 1 to 3 however no crossing is visible indicating $s_0 \approx 0$. From this it could be concluded, if s_b and τ_v would not depend on U, that with increasing U the magnetoconductance monotonically approaches the saturation values corresponding to $\tau_{eff} \to 0$ for $B < 0$ and $\tau_{eff} \to \tau_v$ for $B > 0$. This conclusion is inconsistent with the observed extrema in Fig. 3. It is the decrease of the function s_b (U)which serves us as an explanation for the anomalous magnetoconductance extrema of our ESFI diode.

Fig. 5 represents the essential result of our analysis. The exact values and shapes of $\tau_v(U)$ and $s_b(U)$ rest on the assumptions made for the mobilities, but the general

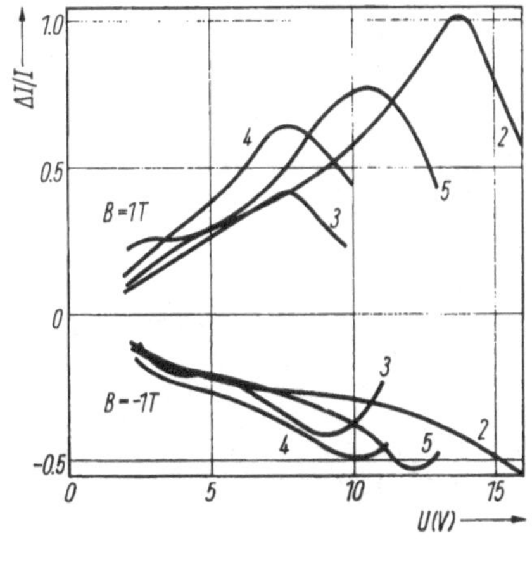

Fig. 6. Four ESFI SOS diodes. Relative current change $\Delta I/I$ due to magnetic inductions $B = \pm 1$ T vs. voltage U. The characteristics of the different diodes are numbered 2 to 5

picture should be correct and could be typical for ESFI magnetodiodes. We found that with increasing voltage equivalent to increasing (average) carrier density the carrier lifetime rises, while the lifetime profile levels off. This behaviour may be understood in terms of deep energy levels as exist in SOS layers [27].

Hitherto a single diode was considered. Fig. 6 taken from [28] now adds the magnetoconductance of four diodes for $B = \pm 1$ T. All the diodes were fabricated simultaneously on different chips of the same wafer. The considerable differences between the curves indicate a remarkable sensitivity of ESFI magnetodiodes. The inspection or evaluation of diode characteristics lends itself to assess the recombination properties of an ESFI film, especially their distribution over a wafer.

Acknowledgements

The authors wish to thank Dr. E. Preuß and Dr. H. Schlötterer for their interest and aid in the present work. They are most indebted to Dr. F. Schwidefsky (†) for his pioneer experiments with ESFI magnetodiodes and also thank Miss Spielberger for further investigations.

This work has been supported under the technological program of the Federal Department of Research and Technology of the FRG.

References

[1] H. Welker, Z. Naturf. **6a**, 184 (1951).
[2] A. Chovet, phys. stat. sol. (a) **28**, 633 (1975).
[3] K. Lehovec, A. Marcus, and K. Schoeni, Phys. Rev. **98**, 229 (1955).
[4] T. Yamada, Proc. Internat. Conf. Physics of Semiconductors, Moscow 1968 (p. 672).
[5] I. Munteanu, Rev. Roumaine Phys. **18**, 323 (1973).
[6] I. Munteanu and C. Constantinescu, phys. stat. sol. (a) **25**, 241 (1974).
[7] R. Baron and J. Mayer, in: Semiconductors and Semimetals, Vol. 6, Chap. 4, Ed. R. K. Willardson and A. C. Beer, Academic Press, New York/London 1970 (p. 201).
[8] D. Lile, J. appl. Phys. **41**, 3480 (1970).
[9] J. F. Allison, D. J. Dumin, F. P. Heiman, C. W. Mueller, and P. H. Robinson, Proc. IEEE **57**, 1490 (1969).
[10] E. Preuss, M. Pomper, Ch. Raetzel, and H. Splittgerber, Siemens Forsch.- u. Entwickl.-Ber. **5**, 338 (1976).
[11] M. S. Abrahams and C. J. Buiocchi, Appl. Phys. Letters **27**, 325 (1975).
[12] J. Tihanyi, European Solid State Device Research Conf., Munich 1973.
[13] D. J. McGreivy and C. R. Viswanathan, Appl. Phys. Letters **25**, 505 (1974).
[14] D. Kranzer, Appl. Phys. Letters **25**, 103 (1974).
[15] D. J. Dumin and R. S. Silver, Solid State Electronics **11**, 353 (1968).
[16] G. A. Egiazaryan, G. A. Mnatsakanyan, V. I. Murygin, and V. I. Stafeev, Soviet Phys. – Semicond. **9**, 829 (1976).
[17] S. T. Hsu and J. H. Scott, RCA Rev. **36**, 240 (1975).
[18] C. D. Maldonado, G. Mandel, and G. Kinoshita, J. appl. Phys. **47**, 2269 (1976).
[19] K. Lehovec and R. Miller, Internat. Electron. Devices Meeting, Washington 1976 (p. 283).
[20] H. Pfleiderer, Solid State Electronics **15**, 335 (1972).
[21] I. Munteanu, phys. stat. sol. (a) **39**, 291 (1977).
[22] R. Hirota, S. Tosima, and M. A. Lampert, J. Phys. Soc. Japan **18**, 535 (1963).
[23] M. A. Lampert and A. Rose, Phys. Rev. **121**, 26 (1961).
[24] N. Kawamura, Japan. J. appl. Phys. **7**, 94 (1968).
[25] H. Pfleiderer and G. Stransky, Nachrichtentech. Z. **30**, 330 (1977).
[26] I. Eisele, H. Pfleiderer, and E. Preuss, Proc. V. Conf. Solid State Devices, Tokyo 1973. Suppl. Japan Soc. appl. Phys. **43**, 418 (1974).
[27] D. J. Dumin, Solid State Electronics **13**, 415 (1970).
[28] H. Pfleiderer and F. Schwidefsky, Symp. Solid State Device Technology, Freudenstadt (FRG) 1976.

(Received June 6, 1977)

phys. stat. sol. (a) **43**, 487 (1977)

Subject classification: 1.1; 22.1.2

Institut für Physik der Universität Dortmund[1] *(a)*
and Atominstitut der Österreichischen Universitäten[2] *(b)*

Oscillatory Structure of Laue Case Rocking Curves

By

U. Bonse (a), W. Graeff (a), R. Teworte (a), and H. Rauch (b)

Perfect crystal Laue case $(n, -n)$ rocking curves (i.e. *convoluted* intrinsic reflection curves) display pronounced oscillations provided both crystals have equal thickness and are rotated exactly through their parallel orientation. It is shown that the oscillations are directly related to the oscillatory structure of the intrinsic curves known from the dynamical diffraction theory, a fact which has been overlooked so far. Experimental oscillatory rocking curves agree very well with calculated curves and can neatly be used to extract the intrinsic distributions. A special feature of the "equal thickness" rocking curves is the extremely narrow central peak. For instance, with the silicon 440 reflection and AgK_α X-rays, the central peak is less than 0.05″ of arc wide. A possible application is in very sensitive small angle measurements with hard X-rays or thermal neutrons.

Laue-Fall-Rockingkurven (d. h. *gefaltete* sog. „intrinsic" Idealkristall-Reflexionskurven) zeigen ausgeprägte Oszillationen für den Fall, daß beide Kristalle exakt gleiche Dicke haben und genau durch ihre Parallelstellung hindurch gedreht werden. Es wird gezeigt, daß die Oszillationen direkt von der Oszillationsstruktur der Idealkristall-Reflexionskurven der dynamischen Theorie herrühren, ein Umstand, der bisher übersehen wurde. Oszillierende experimentelle Rockingkurven stimmen sehr gut mit berechneten überein und gestatten, aus ihnen die „intrinsic" Kurven zu ermitteln. Eine einzigartige Eigenschaft der mit zwei gleichdicken Kristallen gewonnenen Rockingkurven ist die äußerst scharfe Spitze in der Mitte der Kurve, welche z. B. beim Silizium 440-Reflex mit AgK_α-Strahlung weniger als 0,05″ breit ist. Eine mögliche Anwendung ist die Messung der Kleinwinkelstreuung von Röntgenstrahlen und Neutronen mit extremer Auflösung.

1. Introduction

As is well-known from the dynamical diffraction theory for perfect crystals [1], the single crystal reflection curves $I_R(y)$ and $I_T(y)$ for an incident plane wave ('intrinsic' curves) display a pronounced oscillatory structure as long as absorption, characterized by the value of μt (μ absorption coefficient, t specimen thickness), is not too high. The oscillations are due to the interference of the two wavefields with tie points on branches 1 and 2 of the dispersion surface with anomalously high and anomalously low absorption, respectively.

As usual the parameter y measures the deviation from the centre of the reflexion range including refraction corrections. In the symmetrical Laue case y simplifies to

$$y = -\Delta\theta \, \Delta_e^\alpha d^{-1} \, ; \qquad \alpha = \sigma, \pi \, , \tag{1}$$

where $\Delta\theta \equiv \theta - \theta_B$ measures the deviation from the exact Bragg angle θ_B calculated from Bragg's equation. Δ_e^α is the extinction length and d is the spacing of the Bragg planes used. Note that with X-rays Δ_e^α is different for the σ- and π-polarization state by a factor $\cos 2\theta_B$.

Intrinsic curves have been experimentally verified for the Bragg case by Batterman and Hildebrandt [2] and for the Laue case by Kohra and Kikuta [3, 4] and Lefeld-Sosnowska and Malgrange [5].

---

[1]) P.O. Box 50 05 00, D-4600 Dortmund 50, BRD.

[2]) Schüttelstrasse 115, A-1020 Wien, Österreich.

In their experiments, the researchers aimed at producing an incident beam as planar as possible in order to angular scan with it a thin sample. As a function of angle the plane wave Pendellösung oscillations then become clearly visible. Lefeld-Sosnowska and Malgrange [5] were able to observe very distinct oscillations not only in the reflected R-beam but also in the forward diffracted T-beam.

2. Experiment

As we like to point out in this note, in the Laue case the intrinsic curves can also be verified from carefully measured *rocking* curves, i.e. *convoluted* intrinsic curves. With the ray geometry shown in Fig. 1 we have measured $(n, -n)$ rocking curves of a pair of perfect silicon crystals of equal thickness $t = (398.5 \pm 0.5)\ \mu m$ using the 440 reflection and AgK_α X-rays. The results are shown in Fig. 2. Clearly both the rocking curve $R_{RM}(\Delta\theta)$ of the reflected beam and the rocking curve $R_{TM}(\Delta\theta)$ of the transmitted beam display a number of oscillations. The central peak is less than 0.05″ wide. During the measurement great care was taken to shield the apparatus from vibrational and thermal noise and to rotate the crystals through their exact parallel position at $\Delta\theta = 0$, which requires proper alignment also about the ϱ-axis which is defined parallel to the Bragg planes and normal to the θ-axis, i.e. parallel to $\langle 1\bar{1}1\rangle$ in Fig. 1. We find experimentally that the misalignment in ϱ must be smaller than about

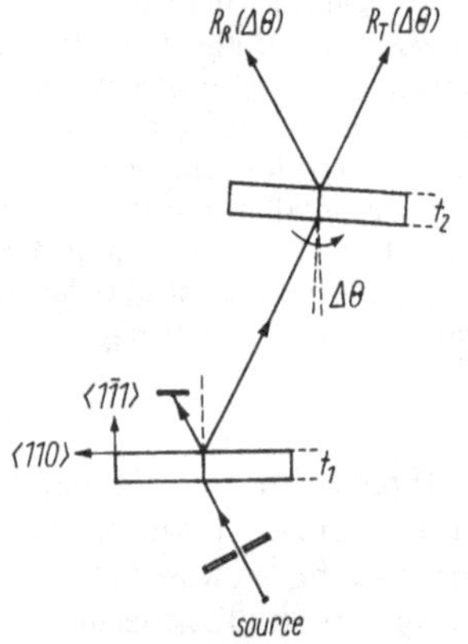

Fig. 1

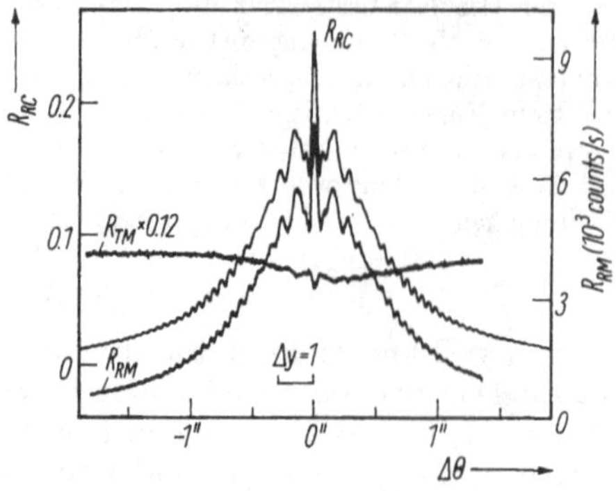

Fig. 2

Fig. 1. Ray geometry used in the measurement of oscillatory Laue case rocking curves $R_R(\Delta\theta)$ and $R_T(\Delta\theta)$

Fig. 2. Measured X-ray rocking curves $R_{RM}(\Delta\theta)$ and $R_{TM}(\Delta\theta)$ of two equally thick perfect crystals of silicon, $t/\Delta_e^\delta = 5.66$, $\Delta_e^\delta = 70.45\ \mu m$, $\mu t = 0.29$, $t_1 = t_2 = (398.5 \pm 0.5)\ \mu m$; Laue case, 440 reflection, AgK_α radiation; ray geometry and crystallographic orientation as shown in Fig. 1. Note the sharp central peak of $R_{RM}(\Delta\theta)$ and the secondary maxima on the wings of the rocking curves. The beat effect in the secondary oscillations is caused by the interplay of σ- and π-polarization. The height of the centre peak is reduced by about 8% due to alinearity of the counting system. Also shown is the corresponding theoretical rocking curve $R_{RC}(\Delta\theta)$ calculated by the convolution of the intrinsic curve in the upper right of Fig. 3. Note the very good correspondence between experimental R_{RM} and calculated R_{RC} what proves the generation of the oscillatory behaviour of the rocking curves by that same behaviour of the intrinsic curves. For this to be true, equal or nearly equal thickness of the two crystals is essential, see text. R_{RC} is shifted vertically with respect to R_{RM} as indicated by the ordinate scales. R_{RM} and R_{RC} have been normalized to equal height of the third "valley" to the right of the centre peak

1″ in order to see the high frequency oscillations on the wings and less than about 5″ in order not to smear the larger oscillations near the centre.

We suspect that there are mainly two reasons why oscillatory rocking curves have not drawn considerable attention so far. The first is that seldom Laue rocking curves of crystals with exactly equal thickness were measured. The second reason is the reported extreme sensitivity of such curves to ϱ-misalignment and general noise. Consequently the two only groups which — to our knowledge — have observed oscillatory rocking curves before worked with two crystal X-ray interferometers [6 to 8], a technique where precise ϱ-alignment and uttermost noise protection is a necessary prerequisite anyhow. At the time of their observations, however, the oscillations were tentatively thought to be due to the action of a six beam diffraction which, because of the highly symmetric crystallographic orientation in the interferometry experiments, might occur at the exact Bragg setting [7].

3. Theory

The main cause for the occurrence of oscillatory Laue case rocking curves is the oscillatory nature of the intrinsic curves, what we will prove by calculating rocking curves from the dynamical theory. Whether there is also an influence by multiple beam diffraction will be left to a forthcoming study.

In the calculations we use the following formulations for $I_R(y)$ and $I_T(y)$ [9], simplified for the symmetrical Laue case:

$$I_R(y) = \exp\left(-\mu t/\cos\theta_B\right) \left|\frac{\sin(AY)}{Y}\right|^2, \tag{2}$$

$$I_T(y) = \exp\left(-\mu t/\cos\theta_B\right) \left|\cos(AY) + iy\frac{\sin(AY)}{Y}\right|^2 \tag{3}$$

with

$$A \equiv \frac{\pi t}{\Delta_e^\alpha} \qquad \alpha = \sigma, \pi \tag{4}$$

and

$$Y \equiv \sqrt{y^2 + \left(\frac{\chi_h \chi_{\bar h}}{|\chi_h \chi_{\bar h}|}\right)^2}\,. \tag{5}$$

χ_h, $\chi_{\bar h}$ are the complex Fourier coefficients of order h, $-h$, respectively, of the dielectric polarizability $\chi(\boldsymbol{r})$ (X-rays) or of the Fermi pseudopotential $V_F(\boldsymbol{r})$ normalized by the free particle energy $\hbar^2k^2/2m_n$ (neutrons). Note that χ_h, $\chi_{\bar h}$ include absorption.

Intrinsic curves calculated for different t/Δ_e and μt values are shown in Fig. 3. As is seen, besides μt the value of t/Δ_e is essential for the structure of the curves. For $|y| > 2$ the number of oscillations within the interval $\Delta y = 1$ is approximately equal to t/Δ_e. Furthermore, the values of $I_R(0)$ and $I_T(0)$ oscillate with varying t with a period Δ_e.

Convoluted curves $R_{RC}(\Delta\theta)$ have been calculated according to

$$R_{RC}(\Delta\theta) = C^\sigma R_R^\sigma(\Delta\theta) + C^\pi R_R^\pi(\Delta\theta)\,; \qquad C^\sigma = 1\,, \quad C^\pi = \cos 2\theta_B\,, \tag{6}$$

$$R_R^\alpha(\Delta\theta) = \int_{-\infty}^{+\infty} I_R(y)\, I_R\left(y + \frac{\Delta_e^\alpha}{d}\,\Delta\theta\right) \mathrm{d}y\,; \qquad \alpha = \sigma, \pi\,. \tag{7}$$

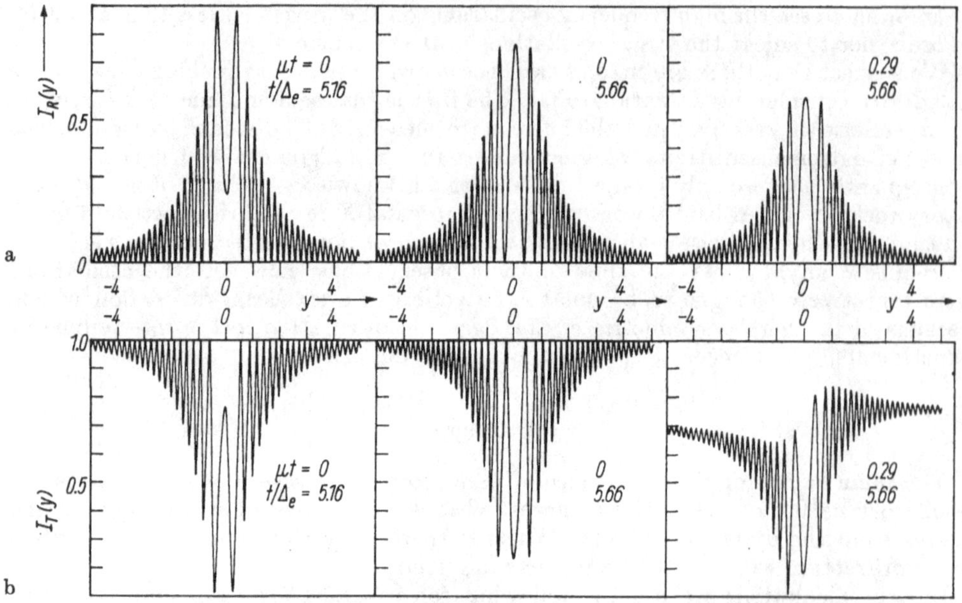

Fig. 3. Calculated intrinsic reflection curves. Laue case; silicon 440 reflection, AgK_α X-rays, σ-state only; a) reflected beam $I_R(y)$, b) forward diffracted beam $I_T(y)$. Note that the value at $y = 0$ depends sensitively on t/Δ_e

An example calculated with the parameter values of the experiment is shown as curve R_{RC} in Fig. 2. As is seen it displays neatly the experimentally observed structure: a sharp centre peak and a number of oscillations on the wings, including the σ–π beat effect, which proves that the oscillatory behaviour of the rocking curves is caused by the oscillatory structure of the intrinsic curves.

The height R_{RC} ($\Delta\theta = 0$) itself of the centre peak oscillates with A as is seen in Fig. 4. The overall structure of the rocking curves depends sensitively on t as is seen from Fig. 5 where two curves are shown with a difference in t of 21 μm. Thus by calculating rocking curves with varying t the height as well as the overall structure of Laue case rocking curves may be used to determine precisely the thickness t of the two crystals, provided t is equal for both.

For this the number of maxima between the centre peak and the σ–π beat can be used to determine the order of t/Δ_e. The exact value of t/Δ_e can be found from the ratio of the height of the central peak alone divided by the overall height of the rocking curve.

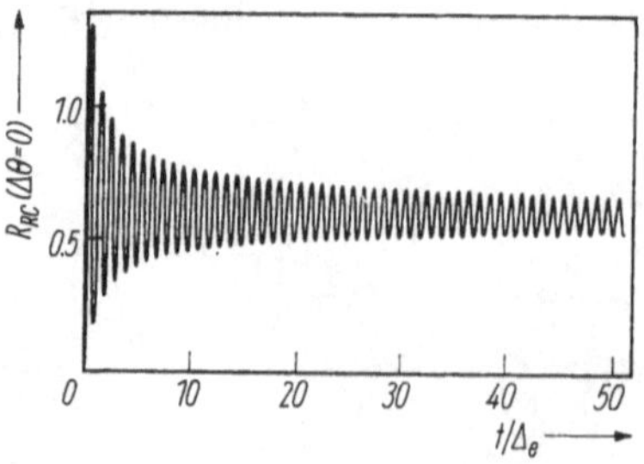

Fig. 4. Height of the central peak of the rocking curves as function of t/Δ_e. Thus from the height the thickness may be determined ($\mu t = 0$)

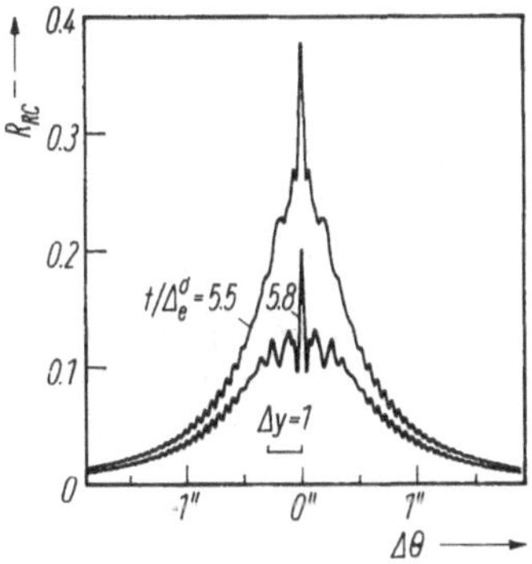

Fig. 5

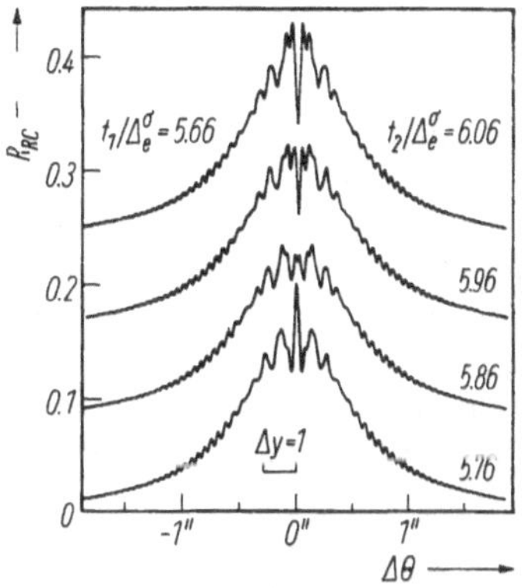

Fig. 6

Fig. 5. Calculated rocking curves for two pairs of crystals with equal thickness $t_1 = t_2 = 5.50\Delta_e^\delta$ and $t_1 = t_2 = 5.80\Delta_e^\delta$, respectively. ($t_1 = t_2 = t$). Note the considerable change of the curve structure when compared to the calculated rocking curve R_{RC} in Fig. 2 with the small thickness change. The other parameter values are those of Fig. 2

Fig. 6. Calculated rocking curves of two crystals with different thickness t_1, t_2 as indicated. All other parameter values are those of Fig. 2. Note that for particular Δt values the centre peak occurring with $t_1 = t_2$ may be converted into a sharp dip. The curves have been shifted vertically against each other by $\Delta R_{RC} = 0.08$

Similarly the curve structure depends very sensitively on thickness difference Δt between the two crystals as is seen in Fig. 6. For particular Δt values the centre peak is converted into a sharp dip.

In the experiment, rather than sharp values of λ and t, the presence of distributions of λ and t is often encountered. The λ- and t-distributions may be combined to a distribution of A. One may wonder how the curve structure is affected if A no longer has a sharp value. For a preliminary answer we have assumed A to be distributed around a value A_0 and with a width ΔA according to a Gaussian:

$$f(A) = \frac{2\sqrt{\ln 2}}{\sqrt{\pi}\,\Delta A} \exp\left\{-\left[(A - A_0)\,\frac{2\sqrt{\ln 2}}{\Delta A}\right]^2\right\}. \tag{8}$$

The influence of averaging over A is shown in Fig. 7 for $t/\Delta_e = 9.82$ and $\Delta A/A_0 = 0.01$ and 0.1, respectively. Obviously the centre peak remains rather unaffected whereas the oscillations on the wings are smeared considerably. Note that the unsmeared curve is *lower* than the smeared curves, what is in general true when the t/Δ_e value is such that the height of the centre peak of the unsmeared curve has a *minimum* what is the case for $t/\Delta_e = 9.82$. The opposite holds, when a t/Δ_e value is chosen where the centre peak of the unsmeared curve has a *maximum*.

Fig. 7. Calculated rocking curves in the presence of a distribution of A values of width ΔA around a value A_0 ($t/\Delta_e = 9.82$, $\mu t = 0$). Note that the sharp centre remains practically unaffected by the diffuseness of A which corresponds to the occurrence of a wavelength distribution $\Delta\lambda$ or a thickness distribution Δt, which is quite common in the actual experiment; σ-state only

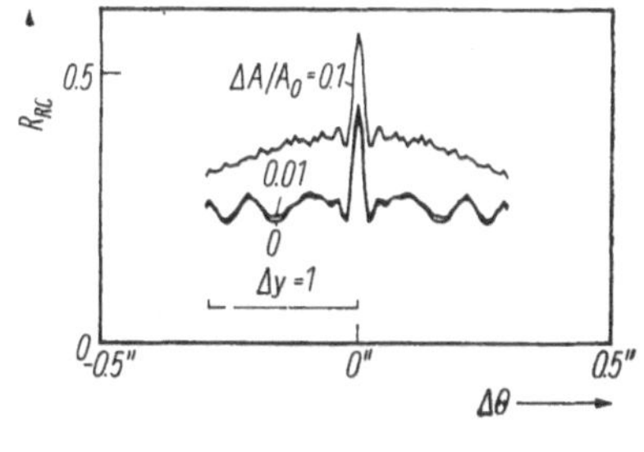

4. Conclusions

The oscillatory structure of rocking curves of equally thick crystals may bear on the beam splitting and beam combining mechanism with Laue case crystals in Angstrom range interferometry. The proper choice of thickness of the wafers of a LLL interferometer may essentially improve the contrast and imaging qualities of such an interferometer.

Furthermore, because of the extreme angular sensitivity of the order of 0.001″, the technique of measuring rocking curves of two equally thick Laue crystals may be used as a high resolution small angle camera with the sample between the two crystals in a similar manner as the double groove-crystal camera [10].

Acknowledgements

The authors gratefully acknowledge fruitful discussions with Dr. P. Becker, Dr. R. D. Deslattes, and Dr. R. Lauer on their observations of oscillatory Laue case rocking curves. We thank Mr. E. Seymann for some numerical calculations.

References

[1] M. von Laue, Röntgenstrahlinterferenzen, 3rd Ed., Akademische Verlagsgesellschaft, Frankfurt/Main 1960.
[2] B. W. Batterman and G. Hildebrandt, phys. stat. sol. **23**, K147 (1967).
[3] K. Kohra and S. Kikuta, Acta cryst. **A24**, 200 (1968).
[4] S. Kikuta and K. Kohra, J. Phys. Soc. Japan **25**, 924 (1968).
[5] M. Lefeld-Sosnowska and C. Malgrange, phys. stat. sol. **30**, K23 (1968).
[6] P. Becker and R. Lauer, private communication (1974).
[7] R. D. Deslattes, A. Henins, R. M. Schoonover, C. L. Caroll, and H. A. Bowman, Phys. Rev. Letters **36**, 898 (1976).
[8] R. D. Deslattes, in: Atomic Masses and Fundamental Constants, Vol. 5, Ed. J. H. Sanders and A. H. Wapstra, Plenum Press, New York/London 1976 (p. 552).
[9] U. Bonse and W. Graeff, in: X-Ray Optics, Ed. H.-J. Queisser, Springer Verlag, Berlin/Heidelberg/New York 1977 (in the press).
[10] U. Bonse and M. Hart, in: Small Angle X-Ray Scattering, Ed. H. Brumberger, Gordon and Breach, New York/London 1967 (p. 121).

(Received June 8, 1977)

phys. stat. sol. (a) 43, 493 (1977)

Subject classification: 10.1; 21.1

Institut A für Physik der Technischen Universität Braunschweig[1])

Etch Pit Study of Dislocation Breakaway in Pre-Yield Deformation of Copper Single Crystals

By

H. NEUHÄUSER and G. SCHULZ

Three-point bending tests on Cu single crystals of low dislocation density and double etching are performed to study the stress dependence of breakaway of grown-in dislocations in the pre-yield region. By an appropriate annealing procedure after each experiment the specimen can recover its original condition which permits to compare experiments with changed parameters (temperature 77 and 300 K, loading time 6.7 ms to 1 s, impurity content, dislocation density). The results indicate that the characteristic sigmoidal dependence of breakaway on stress is mainly determined by the distribution of dislocation segment lengths between nodes in high purity specimens and between additional impurity clusters or jogs impeded by impurities in less pure specimens.

Dreipunkt-Biegeversuche und Doppelätzen werden an Cu-Einkristallen niedriger Versetzungsdichte durchgeführt, um die Spannungsabhängigkeit des Losreißens eingewachsener Versetzungen im preyield-Bereich zu untersuchen. Eine geeignete Glühbehandlung bringt die Proben jeweils in ihren ursprünglichen Zustand zurück, so daß Experimente mit verschiedenen Versuchsparametern (Temperatur 77 und 300 K, Belastungszeit 6,7 ms bis 1 s, Verunreinigungsgehalt, Versetzungsdichte) verglichen werden können. Die Ergebnisse zeigen, daß die charakteristische S-förmige Spannungsabhängigkeit des Losreißens hauptsächlich bestimmt wird durch die Verteilung von Versetzungsbogenlängen zwischen Versetzungsknoten in hochreinen Proben und zwischen zusätzlichen Verunreinigungs-Clustern oder durch Verunreinigungen festgehaltenen jogs in weniger reinen Proben.

1. Introduction

Plastic deformation of crystalline solids sets in by breakaway and propagation of dislocation segments from their grown-in positions. The process of breakaway of single dislocations in Cu single crystals of low dislocation density is studied in the following using the method of multiple etching of dislocations.

For Cu this method, originally introduced for LiF by Johnston and Gilman [1], has been applied by many authors [2 to 13] to study dislocation movement at low stresses. For studying dislocation breakaway in this work we apply load by three-point bending which permits to obtain the stress dependence of breakaway by *one* experiment. By application of an appropriate annealing procedure the original conditions of the sample can be reproduced. This permits to carry out comparable experiments on the same specimen. The influence of temperature, loading time, impurity content, and dislocation density on the breakaway behaviour is investigated and discussed in terms of the dislocation segment length distribution.

2. Experimental Procedures

2.1 Preparation of samples

The etching method for observation of dislocation behaviour requires samples with {111} (or {100}) faces, with few or without subgrain boundaries, and of low dislocation density. Specimens of shape and orientation shown in Fig. 1 were prepared

[1]) 3300 Braunschweig, Mendelssohnstr. 1, BRD.

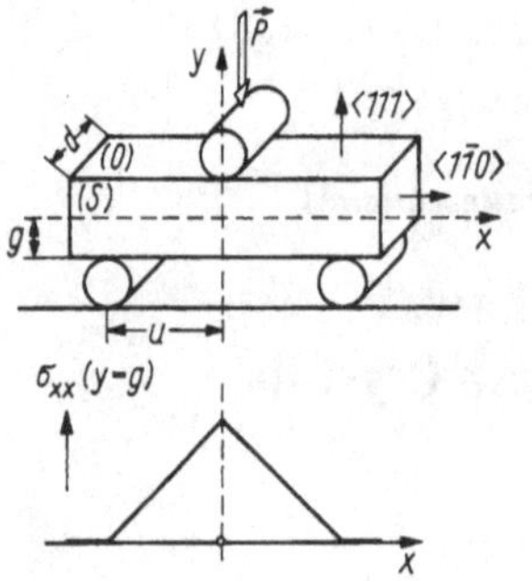

Fig. 1. Specimen shape and orientation; geometry and stress relationship in three-point bending. ($d = 5$ mm, $g = 2.5$ mm, $u = 17$ mm). $\sigma_{xx} = (3P/4dg^3)\, y(u - |x|)$, $\sigma_{yy} = 0$, $\sigma_{xy} = (3P/8dg^3)\,(g^2 - y^2)$

by acid sawing and acid polishing from three types of crystals with increasing purity

i) specimen 1: prepared from crystals grown by the Bridgman method in high purity graphite crucibles in air from electrolytic 99.96% Cu (Kabel-Metall-Werke, Osnabrück);
ii) specimens 2, 3, 4, 5: as i) but prepared from 99.999% Cu (HEK, Lübeck);
iii) specimen 6: prepared from ⟨111⟩-oriented Czochralski crystals of ≈ 25 mm ⌀ grown in vacuum from 99.999% Cu (Johnson-Matthey, London) by Fehmer [14].

A good surface finish has been accomplished by successive polishing with an acid polishing wheel, polishing carefully by hand (sliding the specimen on a cotton cloth soaked with a solution of $FeCl_3$ (density 1.28 g/cm^3)), chemical polishing (20 s at 60 °C in 40 ml HNO_3 (65%) + 127 ml H_3PO_4 (85%) + 31 ml acetic acid (96%)), thorough rinsing with hot water (60 °C), distilled water and methanol, and drying in an air stream.

After preparation the samples were cyclically annealed for 48 h [15] between 830 and 1070 °C in an atmosphere of H_2 and slowly furnace cooled within 24 h. The samples rested on one side face ("S" in Fig. 1) on a boat of high purity graphite. If the initial dislocation density did not exceed 10^6 cm^{-2}, the dislocation density was significantly reduced by this procedure. Annealing resulted in the following dislocation etch pit densities:

specimen 1: $\varrho_0 = 7 \times 10^3$ cm^{-2} (not counting subgrain boundary dislocations),
specimens 2 to 5: $\varrho_0 = 5 \times 10^2$ cm^{-2} (not counting subgrain boundary dislocations),
specimens 6,7: $\varrho_0 = 5 \times 10^2$ cm^{-2} (no subgrains present).

The essential parameter for obtaining low dislocation densities appeared to be the maximum temperature of annealing.

As shown below the annealing treatment described could be used to restore the original specimen condition after a deformation experiment.

2.2 Performance of deformation experiments

Loading of the specimens was carried out in a bending jig by three-point bending (Fig. 1). The force excited magnetically [3] was transmitted to the specimen by cylindrical ceramic rods (⌀ 3 mm) instead of knife edges [2, 3] which would be pressed into the surface. Three-point bending with the {111} faces as the top and bottom surfaces (observation surface "0" in Fig. 1) was used to obtain the entire stress dependence with one single experiment. In the plane of observation ($y = g$) the stress σ_{xx} acting parallel to the sample surface decreases as shown in Fig. 1 from the central rod ($x = 0$) to the support rods ($x = \pm u$). The variation of σ_{xx} with y can be neglected ($\sigma_{xx} \approx$

$\approx \sigma_{xx}(y = g))$ as 100 µm below the surface the stress is only about 4% lower than at the surface for our specimens of 5 mm height.[2])

The stress $\sigma_{xx}(y = g)$ at the crystal surface calculated according to Fig. 1 was resolved to the slip systems by Schmid's law (e.g. for [110]-crystal axis: $\sigma = 0.408\ \sigma_{xx}$ for the $(1\bar{1}1)$, $[0\bar{1}\bar{1}]$, and $[10\bar{1}]$, and for the $(1\bar{1}\bar{1})$, $[0\bar{1}1]$ and, $[101]$ slip systems).

The rectangular load pulse of $t_p = 6.7$ ms to 1 s was generated electromagnetically by a current pulse through a coil in a potshaped magnet. Using elastic dummy samples equipped with strain gauges the shape of the stress pulses generated in the specimens was checked. It was found that the strain (or stress) in the elastic dummy specimens was well proportional to the current through the coil for pulse length $t_p \geqq 5$ ms. For shorter current pulses, however, the shape of the strain pulses differed from the rectangular shape; the specimen carried out a characteristic oscillation with the maximum stress reached only after 1.4 ms (cf. [3]), even if the duration of the rectangular current pulse was shorter than 1.4 ms. In this case the maximum of stress did not only depend on coil current but also on pulse length t_p. The rise time of 1.4 ms primarily depended on the inertia and effective spring constant of the system (essentially the crystal). These results show that this loading method is principally limited to pulse lengths in the range of $\gtrapprox 1.4$ ms. In our experiments only pulses with $t_p \geqq 6.7$ ms were applied to operate certainly with rectangular stress pulses of reproducible shape and height.

The maximum height of loading pulses was 13 N, corresponding to maximum stresses of 1.7 N/mm² at the central loading rod. For the low temperature experiments great care was taken to avoid thermal stresses during cooling and subsequent warming up, extending these procedures over several hours. This was checked by a double etching experiment including the cooling and warming procedures, but without an external loading pulse. This test with a high purity specimen showed a total of less than 1% moved dislocations.

The specimens could be immersed into the cooling bath as well as into the etchant while resting in the bending jig. Etching with a solution of 244 ml H_2O, 100 ml HCl (32%), 54 g $FeCl_3 \cdot 6\ H_2O$, 5 ml HBr (48%) [2, 3] was carried out *in* the loading device immediately before and after loading. In this way uncontrolled mechanical stresses due to handling could be avoided.

The probability of breakaway is $W_l = N_1/N_0$ with N_1 number of depinned dislocations, N_0 total number of grown-in dislocations. The dependence of W_l on stress for each specimen was obtained by counting the etch pits in the microscope (100×) within areas $d \cdot \Delta x$ ($\Delta x = 0.21$ mm, d crystal width (Fig. 1)) shifted along the specimen axis.

3. Results on Breakaway of Grown-in Dislocations

Most experiments were carried out to study the breakaway behaviour of *grown-in* dislocations as a function of applied stress, temperature, loading pulse time, impurity content, and initial dislocation density.

3.1 Stress dependence of Breakaway

Each experiment resulted in an S-shaped stress dependence of the probability of breakaway W_l, as shown for an example in Fig. 2. Similar dependences have been found for various other materials [3, 16 to 18]. The shape of these "breakaway curves"

[2]) Close to the loading rods local stresses arise which may reach up to 1.5 mm into the crystal; this region has to be excluded from evaluation. In the interior an additional shear stress σ_{xy} exists (Fig. 1) which rapidly decreases towards the sample surface. σ_{xy} can be neglected as for $x = \pm u$ ($\sigma_{xx} = 0$) in our experiments no depinned dislocations were observed.

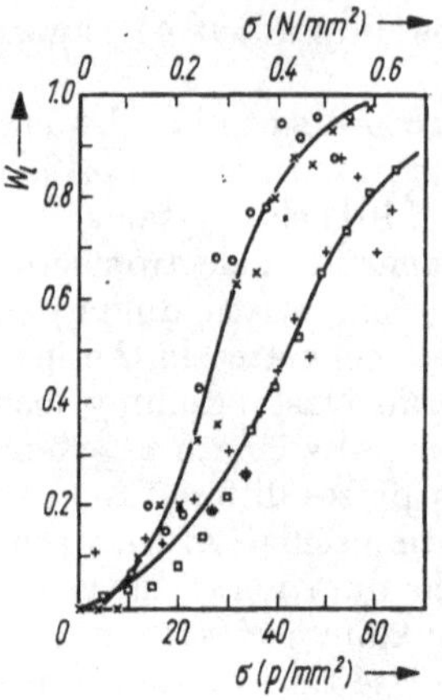

Fig. 2

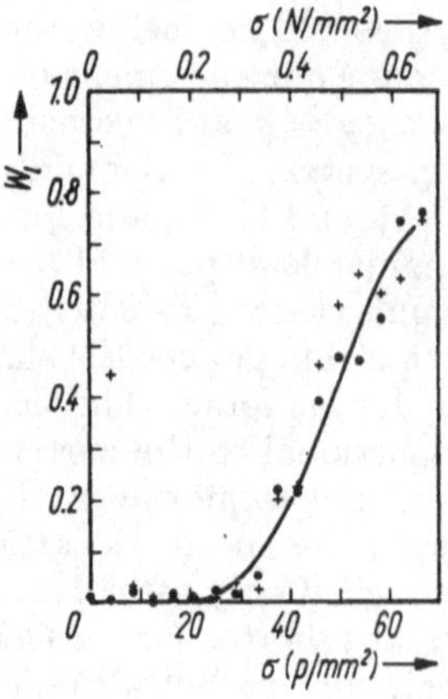

Fig. 3

Fig. 2. Fraction W_l of grown-in dislocations broken away by three-point bending in four experiments with two specimens of type 2 and 3. Before each experiment the specimens were polished and annealed according to Section 2.1. σ is the resolved shear stress, Schmid factor 0.408. Specimen 2: $t_p = 90$ ms, $\varrho_0 = 1.7 \times 10^4$ cm^{-2} (+); $t_p = 1$ s, $\varrho_0 = 0.94 \times 10^4$ cm^{-2} (□). Specimen 3: $t_p = 1$ s, $\varrho_0 = 1.0 \times 10^4$ cm^{-2}; (×); $t_p = 1$ s, $\varrho_0 = 0.5 \times 10^4$ cm^{-2} (○)

Fig. 3. Breakaway dependence evaluated from dilatation surface (+) and compression surface (●) in three-point bending. Specimen 1: $\varrho_0 = 9.5 \times 10^3$ cm^{-2}, $T = 300$ K, $t_p = 1$ s

appeared to be characteristic of the conditions of specimen and loading as shown below, and can be explained by a distribution of dislocation segment lengths (Section 4).

3.2 Comparison of dilatation and compression surfaces

Comparison of the stress dependence of dislocation breakaway as observed on the dilatation and compression surfaces of the same specimen showed essentially the same curves $W_l(\sigma)$: Fig. 3. In some cases on the *compression* side W_l approaches unity presumably due to local stresses arising from the central loading rod (Fig. 1) activating dislocations of all glide systems. On the *dilatation* side W_l approaches a value less than unity as some grown-in dislocations belong to slip systems not activated by the external stress. The increased breakaway probability for very small stresses on the dilatation side arises from local stresses near the support loading rods (cf. Section 2.2).

3.3 Reproducibility of sample conditions after cyclic annealing

The annealing procedure described in Section 2.1 could be used to reproduce the original condition of the specimens after slight deformation. This result is very important, as it allows to carry out series of comparable experiments on various conditions. This reproducibility is shown in Fig. 2 for two different specimens prepared from the same material and with similar dislocation densities: For each of these two specimens we carried out two breakaway experiments with the polishing and annealing procedures 2.1. between the experiments. The curves for different samples clearly differ, but they agree reasonably for the *two* experiments on the *same* sample after cyclic annealing. This result was verified with various samples and formed the basis for the further measurements on the influence of various conditions on the breakaway behaviour.

3.4 Influence of loading pulse time

At room temperature no dependence of breakaway probability was observed on the pulse times used (Fig. 2 and 4a). Only at 77 K we noticed a decreasing probability of breakaway with pulse time. A distinct example is shown in Fig. 4b. This suggests one component of the breakaway process to be thermally activated and to be fairly completed in a time less than about 10 ms at 77 K. At room temperature this breakaway time is by far shorter. Another part of the breakaway proceeds as a practically athermal process for our loading conditions ($t_p \leqq 1$ s).

3.5 Influence of temperature

Contrary to the observation of Marukawa [3] our experiments showed for all specimens a distinct dependence of breakaway on temperature: Fig. 5.[3]) This temperature dependence decreases with increasing purity of the samples. As the pulse time depend-

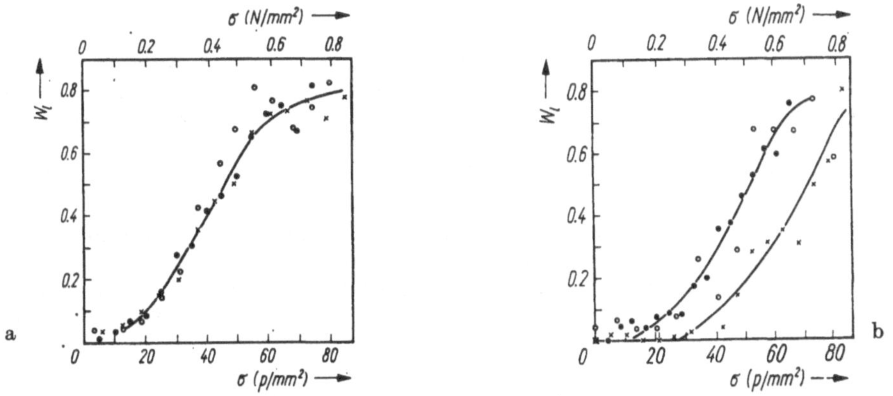

Fig. 4. Pulse time dependence of breakaway a) at room temperature. Specimen 4: $t_p = 1$ s: $\varrho_0 = 1.2 \times 10^4$ cm^{-2} (●), $\varrho_0 = 9.6 \times 10^3$ cm^{-2} (○); $t_p = 6.7$ ms: $\varrho_0 = 1.2 \times 10^4$ cm^{-2} (×); b) at 77 K. Specimen 3: $t_p = 1$ s: $\varrho_0 = 7.4 \times 10^3$ cm^{-2} (●); $t_p = 90$ ms: $\varrho_0 = 4.4 \times 10^3$ cm^{-2} (○); $t_p = 6.7$ ms: $\varrho_0 = 5.8 \times 10^3$ cm^{-2} (×)

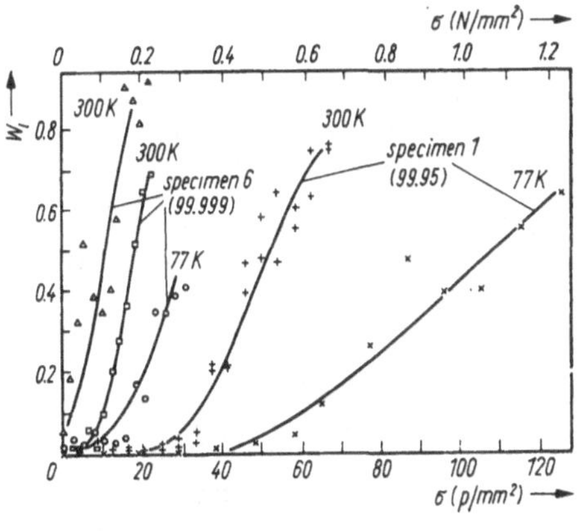

Fig. 5. Temperature dependence of breakaway at pulse times $\geqq 90$ ms. Specimen 1: $T = 77$ K: grown-in dislocation density $\varrho_0 = 6.8 \times 10^3$ cm^{-2} (×), $T = 300$ K: $\varrho_0 = 9.5 \times 10^3$ cm^{-2} (+); specimen 6: $T = 77$ K: $\varrho_0 = 1.3 \times 10^4$ cm^{-2} (○), $T = 300$ K: $\varrho_0 = 1.3 \times 10^4$ cm^{-2} (□), $\varrho_0 = 3 \times 10^3$ cm^{-2} (△)

[3]) Marukawa [3] used different samples at 77 and 300 K. Our result in Section 3.3 (Fig. 2) suggests that the breakaway behaviour of different specimens cannot be compared in general.

ent thermally activated part (Section 3.4) of breakaway has been completed in experiments with $t_p \gtrapprox 90$ ms (see Fig. 4), the temperature effect here seems to arise from obstacles which are practically impenetrable at 77 K but can be oversome at 300 K. The dependence of the breakaway probability on temperature (Fig. 5) may be used as a sensitive indicator for the relative purity of the material studied.

3.6 Influence of grown-in dislocation density and structure

Specimen 6 of highest purity showed different dislocation densities and structures on the two halves of the dilatation surface on the right and on the left of the loading rod. These two parts are evaluated separately in Fig. 5 ($T = 300$ K). The dislocations break away more easily in the area with lower dislocation density. On the other hand, Fig. 2 shows similar breakaway behaviour for the same specimen in spite of different etch pit densities. This suggests that in particular the dislocation structure, not only the overall density, seems to be the important parameter.

3.7 Influence of impurity content

Fig. 5 clearly shows that impurities impede dislocation breakaway, in particular effective at low temperatures. Comparing the experiments with specimens of different purity we find that only for very pure samples the influence of dislocation density and structure (Section 3.6) on breakaway behaviour is noticeable, while with increasing impurity content the influence of the latter dominates over that of the dislocation density.

This is also observed in Fig. 2 (less pure specimens 2 and 3) where no difference of breakaway behaviour is noticed for the *same* specimen in spite of decreasing dislocation density after each annealing procedure. The difference of breakaway curves for specimens 2 and 3 may be due to a difference either in dislocation structure or more likely in the total impurity content.

3.8 Breakaway of fresh dislocations

In specimen 7 of high purity fresh dislocations were produced by a stress pulse of long duration (90 or 175 ms) and sufficient height in a four point bending jig. The breakaway of these fresh dislocations was studied applying a succeeding short stress pulse (6.7 or 13.7 ms) of the same height and etching in the jig before the first, after the first, and after the second pulse. As shown in Table 1, a greater fraction of fresh

Table 1

Breakaway of grown-in and fresh dislocations (specimen 6). σ resolved shear stress for both loading pulses; t_{p1} duration of the first pulse to produce fresh dislocations; t_{p2} second pulse; ϱ_0 grown-in dislocation (etch pit) density, ϱ_{f1} density of fresh dislocations produced by the first pulse, ϱ_{f2} by the second pulse; $S_g(t_{p1})$ percent of grown-in dislocation broken away by the first pulse (t_{p1}), $S_g(t_{p2})$ by the second pulse (t_{p2}), S_f percentage of fresh dislocations produced by the first pulse (t_{p1}) and moved by the second pulse (t_{p2}). $T = 300$ K

σ (p/mm^2)	t_{p1} (ms)	t_{p2} (ms)	ϱ_0 (cm^{-2})	ϱ_{f1} (cm^{-2})	ϱ_{f2} (cm^{-2})	$S_g(t_{p1})$ (%)	$S_g(t_{p2})$ (%)	S_f (%)
6.4	93	6.7	224	612	1600$^+$)	27	43$^+$)	86$^+$)
10.5	93	6.7	82	678	952	57	14	76
10.5	175	13.7	123	749	749	51	19	76

$^+$) Probably the specimen suffered a stronger load pulse inadvertantly between the two pulses of 6.4 p/mm^2.

dislocations (76%) was depinned by a second *short* pulse of the same height than the fraction of grown-in dislocations depinned by the former *long* pulse (55%). Therefore, as expected, fresh dislocations are unpinned considerably more easily than grown-in dislocations at the same stress level. This is confirmed by comparing Table 1 with Fig. 5.

4. Discussion

According to the results in Section 3.4 and 3.5 (Fig. 2 to 5) we attempt to describe the probability $W_l(t, \sigma)$ of breakaway of dislocations at a given temperature T by the simultaneous action of a thermally activated process $U(t, \sigma)$ and a time independent (practically thermally nonactivated) process $S(\sigma)$

$$W_l(t, \sigma) = U(t, \sigma)\, S(\sigma)\,. \tag{1}$$

The time independent part $S(\sigma)$ of W_l is treated as an "athermal" process though a temperature dependence is observed in long pulse time (1 s) experiments. This temperature dependence will be interpreted as arising from different classes of obstacles (impurities) a part of which is practically impenetrable at low temperatures. According to Section 3.4 at room temperature we find $U = 1$ for all pulse times used. Only at 77 K we obtain $U \leqq 1$ which leads to a time dependence of W_l (Fig. 4b).

This tendency was indicated in all short pulse experiments; unfortunately the error limit was not sufficient to allow a safe thermal activation analysis. We therefore restrict the following discussion to the practically athermal part $S(\sigma)$.

As the thermally activated process is certainly accomplished for pulse times $t > > 90$ ms (Fig. 4) we get

$$U(t, \sigma) = 1\,, \qquad W_l(t, \sigma) = S(\sigma) \quad \text{for} \quad t > 90\ \text{ms}\,. \tag{2}$$

The characteristic sigmoidal dependence of the breakaway probability on resolved stress σ is observed in all experiments and is supposed to result from a distribution of segment lengths l. A segment (l) breaks away if according to an Orowan relation

$$\sigma \geqq \alpha G b / l \tag{3}$$

(G shear modulus, $\boldsymbol{b}$ Burgers vector). If the pinning centres of the segments are impenetrable obstacles, e.g. dislocation nodes, α is of the order of 1.

4.1 Distribution of segment lengths between dislocation nodes

Firstly we assume this distribution to arise from a three-dimensional randomly grown-in dislocation network. Models for a network with nodes consisting of three straight segments have been suggested by Johnson and Ashby [19] as well as by Yoshida and Yomogita [20]. The nodes are considered as impenetrable obstacles and the segments between them can break away only if a segment of sufficient length according to (3) extends on a slip plane. For most grown-in dislocations this condition is not fulfilled in general [6, 21]. Only by the motion of jogs longer parts of dislocations can be extended on a slip plane.

The case of completely *immobile jogs* is treated differently in [19] and [20]. Each of these treatments, however, yields results hardly comparable with our experiments. Moreover, if immobile (nonconservative) jogs were solely responsible for breakaway then no temperature dependence of W_l should be observed contrary to our results.

For the case of *easily moving jogs* Yoshida and Yomogita [20] find the number of glissile segments at stress σ

$$M(\sigma) = \delta \frac{m}{6} \left\{1 + \frac{2K}{3} \left(\frac{G}{\sigma}\right)^3 + \frac{K^2}{6} \left(\frac{G}{\sigma}\right)^6\right\} \exp\left(-K \left(\frac{G}{\sigma}\right)^3\right) \tag{4}$$

with m number of segments per unit volume, $K = 8\pi m b^3/9$. The factor δ accounts for an allowed mean deviation of the dislocation segments from the exact slip plane.

Calculating from (4) the relative number of glissile segments and introducing an exponent 2/3 as the node distribution is three-dimensional while we are observing the dislocation movement on the surface, we get for the athermal part of the breakaway probability

$$S(\sigma) = \left\{\frac{fM(\sigma)}{M(\infty)}\right\}^{2/3} = \left\{f\left[1 + \frac{2K}{3}\left(\frac{G}{\sigma}\right)^3 + \frac{K^2}{6}\left(\frac{G}{\sigma}\right)^6\right]\exp\left(-K\left(\frac{G}{\sigma}\right)^3\right)\right\}^{2/3} \tag{5}$$

which is plotted in Fig. 6 for various values of the parameter K using the shear modulus $G = 4.12 \times 10^4$ N/mm² and $f = 2/3$. The product $fM(\sigma)$ accounts for the fact that only 8 of 12 possible Burgers vectors of randomly distributed dislocation characters feel a force from the external stress. (Due to easy movement of jogs, the ratio of the Burgers vectors and not the ratio of slip systems (= 4/9 [3]) is relevant in our case). K is directly connected to the dislocation density (= total dislocation length/volume): $\tilde{\varrho}$ (cm/cm³) $= 6.3 \times 10^{14} K^{2/3}$. From the volume dislocation density $\tilde{\varrho}$ the etch pit density ϱ at the surface can be estimated [22] ϱ (cm^{-2}) $\approx (1/2)\,\tilde{\varrho}$.

Agreement between calculated and observed breakaway probabilities is satisfactory for stresses exceeding the point of inflection; for smaller stresses the measured probabilities exceed that of equation (5). This is assumed to be mainly due to the effect of fluctuations of an internal stress field arising from grown-in dislocations (e.g. [23]) which assist or impede locally dislocation breakaway. Its influence is found to cause a considerable increase of breakaway probability only for low stresses. The magnitude of the effect of internal stress fluctuations is estimated by setting

$$\tilde{S}(\sigma) = \left\{\int_0^\infty \frac{fM(\sigma')}{M(\infty)}\, h(\sigma' - \sigma)\, d\sigma'\right\}^{2/3} \tag{6}$$

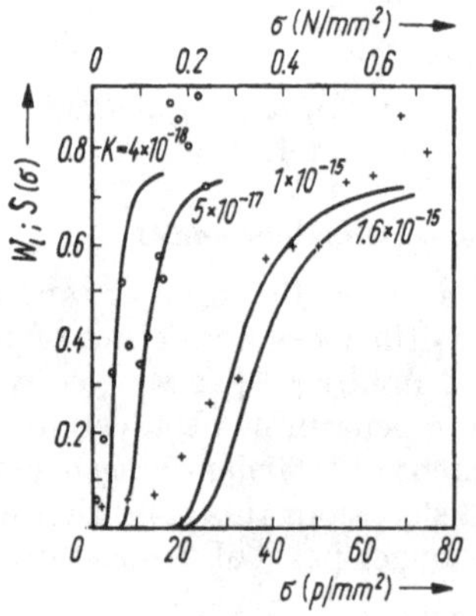

Fig. 6

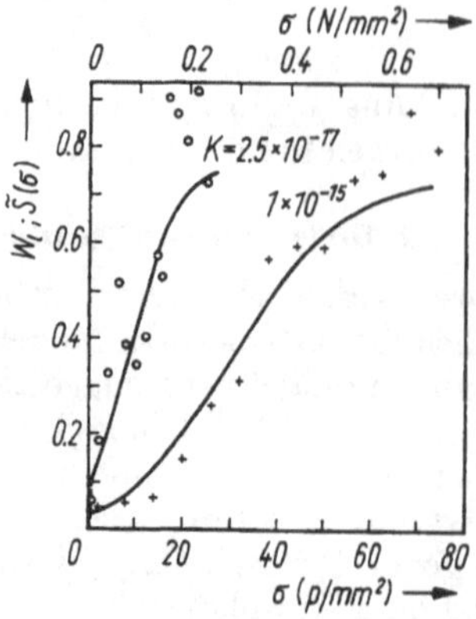

Fig. 7

Fig. 6. Breakaway probability $S(\sigma)$ according to equation (5) plotted against stress σ (full line) and compared with experimental points (specimen 6 (○): $T = 300$ K, $t_p = 1$ s; specimen (4) (+): $T = 300$ K, $t_p = 6.7$ ms); parameter $K = 4 \times 10^{-18}$, 5×10^{-17}, 1×10^{-15}, 1.6×10^{-15}

Fig. 7. Expected breakaway probability $\tilde{S}(\sigma)$ according to equation (6) including a fluctuating internal stress (full line) compared with experimental points (specimen 6 (○): $T = 300$ K, $t_p = 1$ s, $\varrho_0 = 3 \times 10^3$ cm^{-2}, $K = 2.5 \times 10^{17}$; specimen (4) (+): $T = 300$ K, $t_p = 6.7$ ms, $\varrho_0 = 1.5 \times 10^4$ cm^{-2}, $K = 1 \times 10^{-15}$). (see note [4]) with $M(\sigma)$ from equation (4) and a normal distribution $h(\sigma' - \sigma)$ of internal stresses, with a mean square stress $\langle\sigma^2\rangle = (G^2b^2\varrho_0/4\pi) \ln \left(1/(2b\sqrt{\pi\varrho_0})\right)$ (ϱ_0 = grown-in dislocation density) taken according to [24, 25] ("restricted random distribution"). Calculated curves $\tilde{S}(\sigma)$ are shown in Fig. 7 for two examples together with the experimental points.[4] In comparison with Fig. 6 a considerable improvement is noticed. The breakaway curves measured at the compression surface sometimes exceed the asymptotic value of $\tilde{S}(\infty) = (2/3)^{2/3}$ at high stresses due to shear stresses in *all* slip systems near the loading rods (cf. Section 3.2 and note [2]) in Section 2.2).

Table 2 gives the values of dislocation densities ϱ fitted to the experimental curves from equation (6) and the ratios to the grown-in density ϱ_0. Reasonable agreement is found for the specimen of high purity while with increasing amount of impurities, decreasing temperature, and decreasing loading time (at low temperatures) ϱ exceeds ϱ_0. This means that there are strong pinning centers in addition to the threefold nodes, which also explains the increasing ratio ϱ/ϱ_0 with decreasing ϱ_0 (Table 2). These additional strong obstacles may consist of impurity clusters formed at the dislocations during annealing and/or jogs in the dislocation segments impeded by impurities. The influence of impurities is also suggested by the observation that breakaway behaviour is less sensitive to variations in dislocation density for the less pure specimen (Fig. 2) than expected from equation (5).

Table 2

Comparison of grown-in dislocation density ϱ_0 with parameters K and dislocation density $\varrho = 3.15 \times 10^{14} K^{2/3}$ fitted to breakaway experiments (Fig. 2 to 5 and others), according to equation (6) (cf. text). Last column gives the ratio ϱ/ϱ_0 of fitted to grown-in dislocation (etch pit) density. $t_p = 1$ s

Spec. No	T (K)	ϱ_0 (cm^{-2})	K	ϱ (cm^{-2})	ϱ/ϱ_0
6	300	3.0×10^3	2.5×10^{-17}	2.69×10^3	0.90
6	300	1.3×10^4	2.0×10^{-16}	1.08×10^4	0.83
5	300	6.7×10^3	5.0×10^{-16}	1.98×10^4	2.9
4	300	1.3×10^4	1.2×10^{-15}	3.56×10^4	2.8
4	300	1.1×10^4	9.0×10^{-16}	2.94×10^4	2.7
3	300	1.0×10^4	6.0×10^{-16}	2.24×10^4	2.2
3	300	5.0×10^3	3.5×10^{-16}	1.56×10^4	3.1
2	300	9.4×10^3	1.7×10^{-15}	4.49×10^4	4.8
1	300	9.5×10^3	3.5×10^{-15}	7.26×10^4	7.6
6	77	2.5×10^3	1.0×10^{-16}	6.79×10^3	1.9
6	77	1.3×10^4	9.0×10^{-16}	2.94×10^4	2.3
5	77	4.5×10^3	1.6×10^{-15}	4.31×10^4	9.6
5	77	3.6×10^3	1.6×10^{-15}	4.31×10^4	12.0
4	77	1.5×10^4	5.0×10^{-15}	9.21×10^4	6.1
4	77	1.2×10^4	8.5×10^{-15}	1.31×10^5	10.9
3	77	7.8×10^3	3.0×10^{-15}	6.55×10^4	8.4
3	77	7.4×10^3	3.5×10^{-15}	7.26×10^4	9.8
1	77	6.8×10^3	2.2×10^{-14}	2.47×10^5	36.4

[4]) At zero external stress ($\sigma = 0$), $\tilde{S}(\sigma)$ has to be set zero, as one cannot notice by etching those dislocations that have been moved due to internal stress at $\sigma = 0$. The discontinuity arises from treating breakaway as an athermal process.

4.2 Distribution of segment lengths between impurities

The experimental curves W_l are not compatible with the breakaway probability

$$S_K(\sigma) = (1 - c)^{G/\sigma} \tag{7}$$

which results from a segment distribution $N(l) \sim c(1 - c)^{l/b}$ [26, 27] between single impurities distributed randomly with concentration c along the dislocations. A similar behaviour of breakaway probability is determined with Li's [28] segment distribution $N(l) \sim l \exp(-l/\bar{l})$ ($\bar{l}$ mean segment length).

More realistically, the distributions (4) and $N(l)$ have to be combined properly to account for an unzipping mechanism of breakaway. In this case we have to consider two critical lengths of dislocation segments; $\tilde{l}_c = (Gb/\sigma) \cos(\psi/2)$ between penetrable impurities (unzipping angle ψ) and $l_c = \alpha Gb/\sigma$ between dislocation nodes as before. A dislocation breaks away permanently, whenever it contains a segment between impurities of length $\tilde{l} \geqq \tilde{l}_c$ *and* simultaneously has a length $l \geqq l_c$ between nodes. The observed breakaway probabilities significantly exceed the predicted ones for $\sigma \gtrsim 0.3$ N/mm². This may be explained by the assumption that during annealing the impurities form *clusters*, presumably located perferentially at jogs, with greater distances than single impurities would have if distributed randomly along the dislocations.

Thus we conclude that the observed stress dependence of breakaway behaviour of grown-in dislocations essentially results from the distribution of lengths of dislocation segments between nodes and between impurity clusters of jogs fixed by impurities.

5. Conclusions

Experiments on the breakaway behaviour of grown-in dislocations in low dislocation density Cu crystals in the pre-yield region have shown that the original configuration of jogs and impurities on grown-in dislocations can be approximately restored by an appropriate annealing procedure. This finding permits to study the influence of various parameters on breakaway on the same sample. Using three-point bending the entire stress dependence of the breakaway probability $W_l(\sigma)$ could be measured with a single experiment.

For all specimens a sigmoidal shape of $W_l(\sigma)$ is found which can be interpreted as due to the dislocation length distribution [20] between nodes for very pure specimens. With increasing impurity content additional pinning centres divide the segments into shorter parts which need higher stress for breakaway. In particular at low temperatures with decreasing purity the influence of impurities tends to cover that of the node density.

These additional pinning centres may consist of impurity clusters formed at the dislocations during annealing. Possibly, the impeded movement of built-in jogs [21, 19, 6] along the grown-in dislocations during bowing out and extending over the slip plane [21, 29, 30] is the controlling step in breakaway of grown-in dislocations. This is supported by the observation that fresh dislocations produced by a preliminary stress pulse are moved by a succeeding pulse much more easily than grown-in ones, as fresh dislocations are free from the impurity "atmosphere" and from retarding jogs, being extended already in the active slip plane.

The observation of a dependence of breakaway probability on pulse time for low temperatures indicates a thermally activated component of the breakaway process. The mean activation time for breakaway at 77 K is less than 10 ms and much shorter at 300 K. This process seems to control breakaway from relatively weak obstacles (presumably point defects). After its completion further breakaway is governed by the above dislocation segment distribution.

Acknowledgements

For his continuous support and helpful advice we would like to thank Prof. Ch. Schwink. Our sincere thanks are due to Dr. H. Fehmer in the Institute of Physics at the University of Münster, who provided some Cu specimens with exceedingly low dislocation density. Just as many thanks are due to Dipl.-Phys. P. Warnecke for growing several other Cu crystals used in this investigation.

References

[1] W. G. Johnston and J. J. Gilman, J. appl. Phys. **30**, 129 (1959).
[2] F. W. Young, J. appl. Phys. **32**, 1815 (1961); **33**, 963, 3553 (1962).
[3] K. Marukawa, J. Phys. Soc. Japan **22**, 499 (1967).
[4] T. Suzuki, in: Dislocation Dynamics, Ed. A. R. Rosenfield et al., McGraw-Hill Publ. Co., New York 1967 (p. 551).
[5] A. S. Argon and W. T. Brydges, Phil. Mag. **18**, 817 (1968).
[6] P. Petroff and J. Washburn, J. appl. Phys. **37**, 4986 (1966).
[7] Y. Miura and J. Washburn, Crystal Lattice Defects **2**, 197 (1971).
[8] S. Kitajima, H. Tonda, and H. Kaieda, Trans. JIM **9**, Suppl., 740 (1968).
[9] S. Kitajima, H. Tanaka, and H. Kaieda, Trans. Japan. Inst. Metals **10**, 12 (1969).
[10] T. Vreeland and K. M. Jassby, Mater. Sci. Engng. **7**, 95 (1971); Crystal. Lattice Defects **4**, 1 (1973).
[11] J. Dönch, P. Haasen, and R. Labusch, Nachr. Akad. Wiss. Göttingen, math.-phys. Kl. **2**, 15 (1974).
[12] M. Kleintges, Diploma Work Göttingen, Verh. DPG **11**, 670 (1976).
[13] H. Ney, Diploma Work Göttingen, Verh. DPG **12**, 282 (1977).
[14] H. Fehmer and W. Uehlhoff, J. Crystal Growth **13/14**, 257 (1972).
[15] T. Hasegawa, T. Ishii, and S. Karashima, Trans. JIM **11**, 328 (1970).
[16] A. S. Argon and G. E. Padawer, Phil. Mag. **25**, 1073 (1972).
[17] A. A. Predvoditelev, N. K. Rakova, and Nang-Hung-Pin, Soviet Phys. – Solid State **9**, 224 (1967).
[18] N. N. Novikov and M. V. Khimenko, Soviet Phys. – J. exper. theor. Phys. **17**, 892 (1974).
[19] L. Johnson and M. F. Ashby, Acta metall. **16**, 219 (1968).
[20] S. Yoshida and K. Yomogita, Crystal Lattice Defects **4**, 37 (1973).
[21] F. W. Young and F. A. Sherrill, J. appl. Phys. **42**, 230 (1971); **43**, 2949 (1972); F. W. Young, in: Crystal Growth, Ed. H. S. Peiser, Pergamon Press, New York 1966 (p. 789).
[22] G. Schoeck, J. appl. Phys. **33**, 1745 (1962).
[23] I. V. Stratan and A. A. Predvotitelev, Soviet Phys. – Solid State **12**, 1367, 1702 (1970).
[24] M. Wilkens, Acta metall. **17**, 1155 (1969).
[25] B. M. Strunin, Soviet Phys. – Solid State **9**, 630 (1967); **13**, 773 (1971).
[26] J. S. Koehler, in: Imperfections in Nearly Perfect Crystals, Ed. W. Shockley et al., Wiley, New York 1952 (p. 204).
[27] R. Labusch, Z. Phys. **167**, 452 (1962).
[28] J. C. M. Li, in: Physics of Strength and Plasticity, Ed. A. S. Argon, MIT Press, Cambridge 1969 (p. 245).
[29] F. Minari, B. Pichaud, and C. Capella, Phil. Mag. **31**, 275 (1975); **34**, 1121 (1976).
[30] O. Lohne, phys. stat. sol. (a) **25**, 709 (1974); in: Surface Effects in Crystal Plasticity, Ed. R. M. Latanisiou and J. T. Fourie, Noordhoff, Leyden 1977 (p. 487).

(Received May 27, 1977)

phys. stat. sol. (a) 43, 505 (1977)

Subject classification: 11 and 14.3.2; 22.1.2

Zentralinstitut für Kernforschung Rossendorf, Akademie der Wissenschaften der DDR, Dresden[1])

Characterization of Ion Implanted Silicon by the Electrolytic Reverse Current

By

J. HÜLLER and M. T. PHAM

The current–voltage behaviour of ion implanted silicon electrodes in HF electrolyte is investigated. The electrolytic reverse current, i.e. the reaction rate of the minority carrier limited reactions is found to increase. The current increase depends on the implanted dose and layer stripping. Reason for the increased reverse current can be referred to radiation damage acting as generation centres for minority carriers. Measurement of the electrolytic reverse current can be used for determination of damage profiles. Layer stripping is carried out by anodic dissolution in the same electrolyte. The sensitivity of this new method for characterizing ion implanted silicon layers lies at 10^{11} to 10^{12} atoms cm^{-2}.

Das Stromdichte–Potential-Verhalten ionenimplantierter Siliziumelektroden im HF-Elektrolyten wird untersucht. Es wird eine Erhöhung des elektrolytischen Sperrstromes festgestellt, d. h. die Geschwindigkeit der durch die Minoritätsladungsträgerkonzentration bestimmten Reaktionen ist vergrößert. Die Stromerhöhung ist abhängig von der implantierten Dosis und der abgetragenen Schichtdicke. Als Ursache der Stromerhöhung werden Strahlenschäden festgestellt, die als Generationszentren für Minoritätsladungsträger wirken. Die Messung des elektrolytischen Sperrstromes kann zur Bestimmung von Strahlenschädenprofilen verwendet werden. Die Schichtabtragung erfolgt dabei durch anodische Auflösung im gleichen Elektrolyten. Die Empfindlichkeit dieser neuen Methode zur Charakterisierung ionenimplantierten Siliziums liegt bei 10^{11} bis 10^{12} cm^{-2}.

1. Introduction

For the electrochemical properties of silicon it is characteristic that p-Si and n-Si have a different behaviour [1, 2]. At silicon electrodes in hydrofluoric acid solutions the anodic dissolution (1) proceeds over the valence band mechanism and the cathodic hydrogen evolution (2) over the conduction band mechanism [3]

$$\mathrm{Si} + 2\mathrm{p} \rightleftharpoons \mathrm{Si}^{2+} \qquad \text{p hole}\,, \tag{1}$$

$$\mathrm{H}^+ + \mathrm{e} \rightleftharpoons \tfrac{1}{2}\,\mathrm{H}_2 \qquad \text{e electron}, \tag{2}$$

i.e. the anodic reaction consumes holes whereas in the cathodic reaction electrons take part.

If reaction (1) is carried out with n-type silicon electrodes and reaction (2) with p-type Si electrodes, the minority carriers thus in both cases determine the reaction rate and an electrolytic reverse current is observed. This limiting current I_{lim} caused by the supply of minority carrier concentration may be described by relation (3)

$$I_{\mathrm{lim,n}} = e_0 D_\mathrm{n} \frac{n_0}{L_\mathrm{n}} + e_0 S n_0\,; \tag{3}$$

e_0 absolute value of electron charge, D_n diffusion coefficient of electrons, n_0 electron

[1]) 8051 Dresden, Postfach 19, DDR.

concentration in the bulk, L_n diffusion length of electrons, S surface recombination velocity.

An analogous formula as for the shown electron limited saturation current $I_{lim, n}$ is valid for the hole limited saturation current $I_{lim, p}$ with data for holes. Equation (3) shows that the saturation current depends on the diffusion of minority carriers from the semiconductor bulk to the surface and on surface generation.

Limiting currents at Si electrodes, calculated from the diffusion term, are in the order of 10^{-9} A/cm^2, whereas experimentally measured values lie in the region of 10^{-6} A/cm^2, i.e. the saturation current is determined by the surface generation current $I_{gen, n} = e_0 S n_0$. Adsorbed ions or lattice defects in the silicon monocrystals may act as recombination/generation centres.

With ion implanted silicon electrodes we found a strong increase of the electrolytic reverse current [4, 5]. It is the subject of this paper to investigate the current–voltage behaviour of ion implanted silicon electrodes in a HF electrolyte and to show applicative possibilities for the determination of implantation profiles. In this paper only the minority carrier limited reactions are considered, i.e. reaction (1) for n-Si and reaction (2) for p-Si. For majority carrier controlled reactions about which we have already reported [6], a decrease of the reaction rate after ion implantation is typical.

2. Experimental Procedure

2.1 Electronic equipment and measuring cell

The measurement of current–voltage curves (I-U curves) of ion implanted Si must be made fast to minimize the anodic dissolution of the implanted layer due to anodic polarization. For this reason the measurements are carried out potentiodynamically. By means of a potentiostat, driven by a triangular voltage generator, the electrode potential of the silicon working electrode is swept linear through a potential range. The resulting cell current passes a current voltage converter, an analog digital converter, and is stored in a 256 channel memory, operating as multiscaler. The channel number represents the electrode potential and the current is given by the channel content. In comparison with oscillographic measurements this digigal equipment has some advantages: higher accuracy and easy data output. The principle of the assembly is shown in Fig. 1. A detailed description has already been given in [7].

The measuring cell consists of teflon, the counter electrode is made of platinum, and a saturated calomel electrode (SCE) is used as reference electrode. All electrode potentials in this paper are related to the SCE.

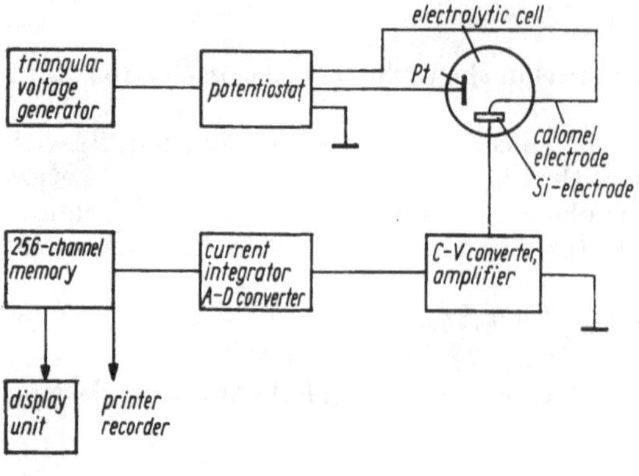

Fig. 1. Principle of the electrochemical arrangement

2.2 Sample preparation

All experiments were performed with silicon wafers oriented in the ⟨111⟩ direction. The sample thickness was 0.3 to 0.6 mm, the resistivity 10 to 400 Ω cm. Before implantation the samples were etch-polished in $HF–HNO_3$ mixtures.

The back side of the wafers was contacted with aluminium by vacuum evaporation. Before each evaporation the back side was implanted with a p^+ (boron for p-Si) or n^+ layer (phosphorus for n-Si) in order to get an ohmic contact. In every case the implantation of the front side was performed after contacting the back side to avoid contamination and thermal treatment.

For experiments p/p^+ (^{11}B in p-Si) and n/n^+ layers (^{31}P, ^{75}As in n-Si) were used to eliminate p–n junction formation. Electrically non-active ions such as ^{84}Kr, ^{12}C, and ^{30}Si were implanted, too. The implantation was performed in the Rossendorf mass-separators at room temperature. Ion doses from 10^{11} to 10^{16} cm^{-2} were used at energies in the range of 30 to 50 keV.

2.3 Measuring conditions

The measurements were carried out in 0.1 n to 2 n HF in the dark at 25 °C. The electrolyte was kept free from oxygen by passing purified nitrogen through it. The potential sweeping speed was 1 V/s as a rule. The polarization was carried out in reverse direction, starting at an electrode potential near to the rest potential. After attaining the predetermined terminal value, the electrode potential jumps back to the rest potential and is kept at this value between the single polarizations. The effective silicon area contacting the electrolyte, is fixed by a gasket and equals about 0.25 mm^2.

The layer removal needed for determination of implantation profiles is carried out galvanostatically or potentiostatically by anodic dissolution in the same electrolyte at current densities smaller than 50 μA/cm^2.

Coloured layers arising during anodic polarization disturb the measurements only insignificantly. These layers consist of amorphous silicon, whose formation can be explained by the mechanism of anodic dissolution of Si in HF, proposed by Memming and Schwandt [3]. The amorphous silicon film is deposited by disproportionation of the primarily formed Si^{2+} ions. The colour of these amorphous silicon layers depends on their thickness, they are porous and not wipeproof.

The layer thickness dissolved is determined coulometrically by means of a current integrator. Also for ion implanted Si electrodes the electrochemical equivalent can be calculated using divalent Si as we found in anodic dissolution experiments with neutron activated silicon samples [8].

3. Results and Discussion

The electrolytic reverse current after ion implantation is largely increased. Shown in Fig. 2 is the increase in the anodic dissolution current for n-Si after implantation with 10^{12} As cm^{-2}, and in Fig. 3 the increase in cathodic hydrogen reduction for p-Si after implantation with 10^{12} B cm^{-2}. For comparison the very low saturation current for the non-implanted substrate and its increase due to generation of minority carriers by illumination is also included in Fig. 3.

In the range of low implantation doses from 10^{11} to 10^{13} cm^{-2} the increased current is still saturated. At higher implantation doses the current in the reverse direction raises exponentially with the electrode potential in the same way as the forward current, and as found by Meek [9] for n^+ Si at reverse bias. The increase in reverse current depends on the dose as shown in Fig. 4 for p-Si in the dose range from 10^{12} to 10^{15} B cm^{-2}. The current–voltage curves as a function of the removed thickness are presented in Fig. 5 for n-Si implanted with 5×10^{14} P cm^{-2}. At current densities

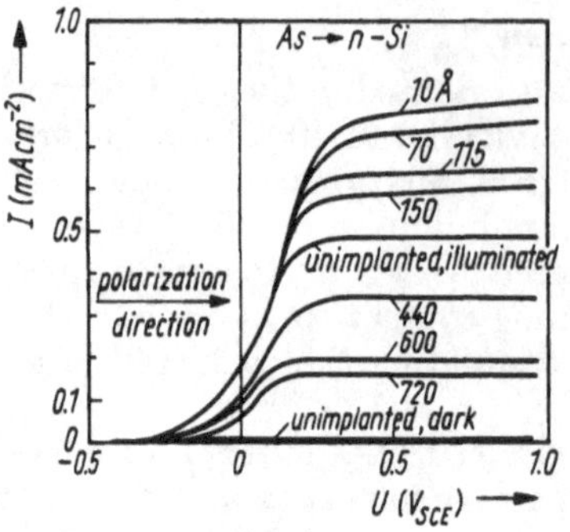

Fig. 2

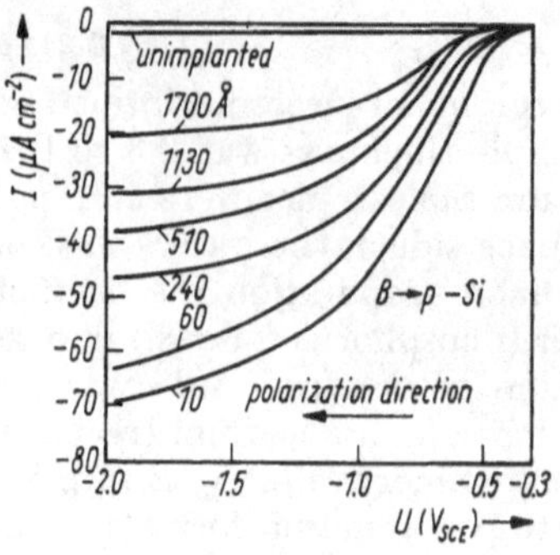

Fig. 3

Fig. 2. Current density I versus electrode potential U for implantation of 10^{12} As cm^{-2} (50 keV, ⟨111⟩ 0.5 nHF) in n-Si. Parameter: layer stripping. Comparison with non-implanted sample with and without illumination

Fig. 3. Current–voltage curves after implantation of 10^{12} B cm^{-2} in p-Si (30 keV, ⟨111⟩, 1 nHF). Parameter: layer stripping. Comparison with non-implanted sample

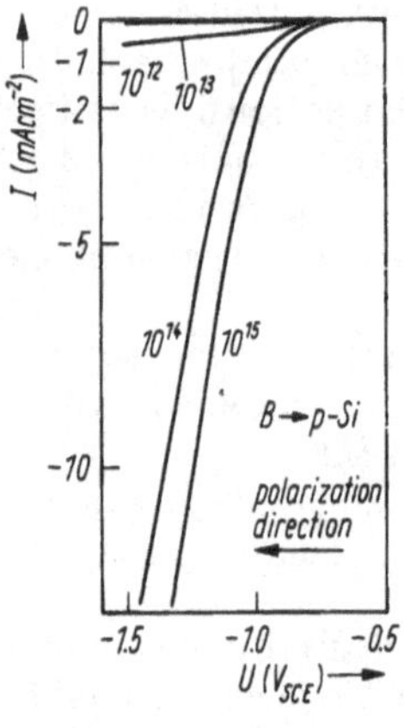

Fig. 4

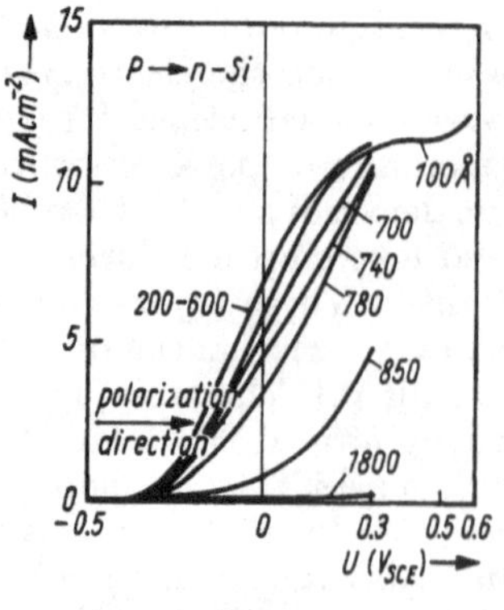

Fig. 5

Fig. 4. Current density I versus electrode potential U after implantation of 10^{12} to 10^{15} B cm^{-2} in p-Si (30 keV, ⟨111⟩, 1 nHF). All curves were measured in the profile maximum

Fig. 5. Current density versus electrode potential after implantation of 5×10^{14} P cm^{-2} in n-Si (30 keV, ⟨111⟩, 2nHF + 1 nNa_2SO_4). Parameter: layer stripping

above some mA/cm^2 there is no linear correlation between the logarithm of the current density and the electrode potential in the form of a Tafel line. The bending off in the I–U curves is due to a depletion of the fluoride ion concentration at the Si surface, which clearly follows from Fig. 5, where they are presented in linear form, especially for polarization up to 1.6 V_{SCE}. A resistance polarization due to surface coverage is in this case superimposed on the charge transfer polarization as a current determining step characteristic of electrochemical reactions at semiconductor electrodes. It follows from the results presented in Fig. 4 and 5 that the increase in reverse current is proportional to the implantation profile, as already reported [10].

The implantation profiles are obtained plotting the current density taken at a fixed electrode potential versus the thickness removed. Such profiles are given in Fig. 6 for doses of 10^{14} and 5×10^{14} P cm^{-2}. For the dose of 5×10^{14} P cm^{-2}, the results obtained from one potentiostatic measurement at a constant electrode potential of $U = \pm 0$ V_{SCE} are compared with those taken at the same electrode potential from a series of consecutive potentiodynamic I–U measurements. It follows from the good agreement that the values for the current from potentiodynamic measurements at a potential sweep rate of $\Delta U = 1$ V/s are assumed to be still stationary.

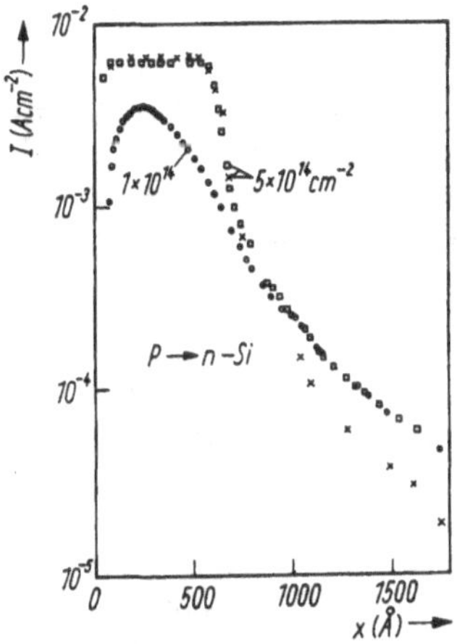

Fig. 6

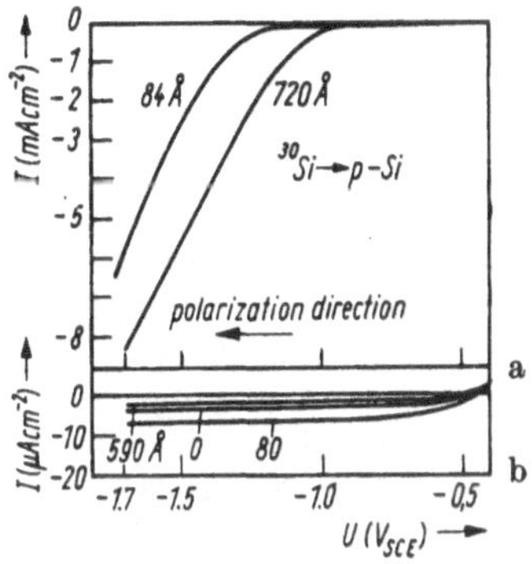

Fig. 7

Fig. 6. Implantation profiles of 1×10^{14} and 5×10^{14} P cm^{-2}, (30 keV, ⟨111⟩ +8°, 2 nHF + + 2 nNa_2SO_4) depicted as function of the current density I at a constant electrode potential $U = \pm 0$ V_{SCE} versus layer removed x. × potentiodynamic from c/v curves, □, ○ potentiostatic

Fig. 7. Comparison of current–voltage curves after implantation of 10^{15} Si cm^{-2} in p-Si a) without and b) with thermal annealing ($T = 500$ °C). Parameter: layer stripping (50 keV, 1 nHF)

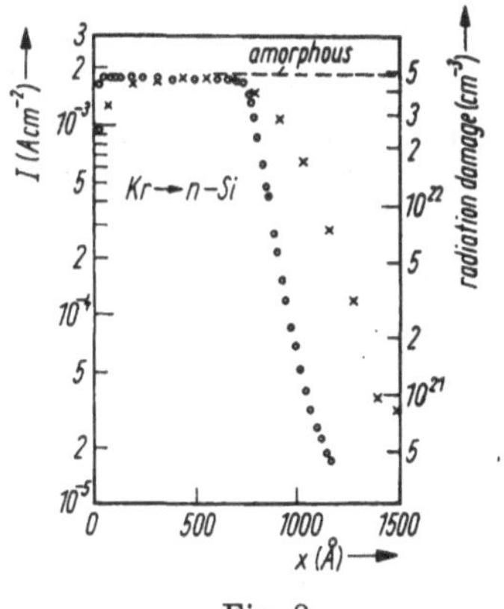

Fig. 8

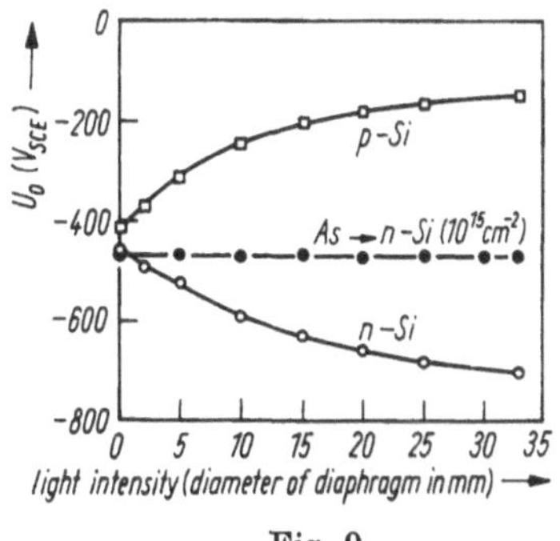

Fig. 9

Fig. 8. Comparison of an electrochemically measured profile (left scale) with a radiation damage profile obtained by Rutherford backscattering measurement (right scale) after implantation of 10^{15} Kr cm^{-2} in n-Si (50 keV, ⟨111⟩ +8°, 1 nHF). × α-backscattering, ○ electrochemical

Fig. 9. Rest potential U_0 of n-Si, p-Si and after implantation of 10^{15} As cm^{-2} in n-Si versus light intensity. Illumination through an UV-interference filter with the wavelength $\lambda = 365$ nm. (1 nHF)

The electrical charge $Q = \int I \, dt$ during a single polarization of the anodic current for n-Si and of the cathodic current for p-Si can be used also as a measure for the ordinate of the profile, if the current density in the investigated potential region varies exponentially with the electrode potential. This representation is given in Fig. 12.

The increase of the reverse current is caused not only by electrically active ions, but also by electrically non-active ions, as we could found with implantation of carbon, silicon, and noble gase ions in Si. The I–U curves are shown in Fig. 7a for 10^{15} Si cm^{-2} in p-Si. The dependence of the increase in reverse current on the removed thickness is seen here, too. In Fig. 8 a comparison is given of a profile obtained from the electrochemical measurement with the damage profile following from Rutherford backscattering measurements for implantation of 10^{15} Kr cm^{-2} in n-Si. The measurements were carried out on the same sample. The agreement is good, i.e. the electrochemical profile reflects also the damage distribution. Both display an amorphization up to the depth of about 700 Å. In addition, the electrochemically measured depth scale is confirmed by the α-backscattering experiments. With this result a further discussion can be added to the profile presented in Fig. 6. The constant profile shape up to about 550 Å at the dose of 5×10^{14} cm^{-2} is interpreted to be correlated with the amorphization, because the amorphization dose for P in Si at room temperature is about $5 \times \times 10^{14}$ cm^{-2} [11]. At the smaller dosis of 10^{14} cm^{-2} a more Gaussian shape of the damage profile is observed.

Furthermore, it is characteristic for ion implanted silicon electrodes that the illumination does not affect the rest potential and the reverse current. Already at an implantation dose of 10^{12} cm^{-2}, the rest potential remains constant on illuminating, as shown in Fig. 9. Only after stripping away the damaged layer or thermal annealing the photovoltaic effect is recoverable.

For non-implanted Si, independent of the conduction type, we found the rest potential at the beginning of the measurement to be $U_0 = (-350 \pm 50)$ mV_{SCE}. It shifts towards the negative direction after implantation. Depending on the implanted dose and the ion kind, the rest potential varies between -400 and -500 mV_{SCE}. After

Fig. 10

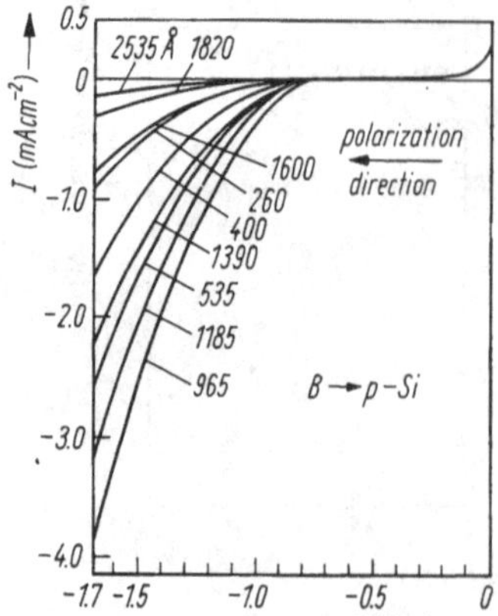

Fig. 11

Fig. 10. Current density versus electrode potential after implantation of 10^{13} B cm^{-2} in p-Si (30 keV, ⟨111⟩, 0.1 nHF) without and with thermal annealing at 500 °C, 30 min

Fig. 11. Current density versus electrode potential after implantation of 10^{14} B cm^{-2} in p-Si. (30 keV, ⟨111⟩, 0.1 nHF). Thermal treatment at 500 °C, 30 min. Parameter: layer stripping

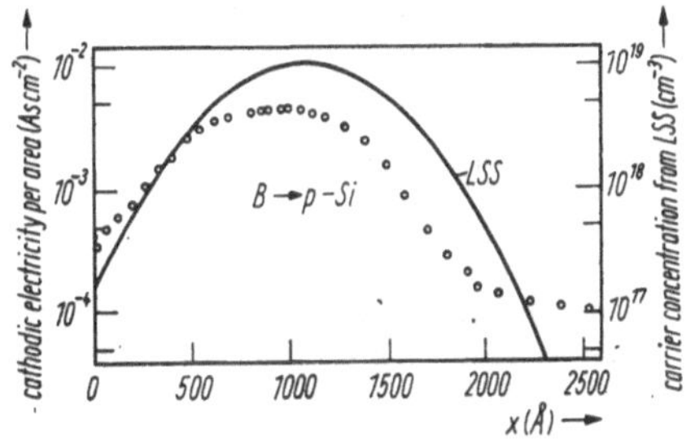

Fig. 12. Implantation profile of 10^{14} B cm^{-2} in p-Si (30 keV, ⟨111⟩, 0.1 nHF) after thermal treatment at 500 °C, 30 min. Comparison between electrochemically measured profile (left scale) and calculated carrier profile (right scale) obtained from the LSS theory

anodically removing thicknesses greater than 50 to 100 Å it adjusts to (−500 + ± 50) mV_{SCE} between the potentiodynamic polarizations.

As we already derived from the impedance measurements and the current–potential behaviour in reaction with majority carriers for the system Si/HF electrolyte [12], the increase in the electrolytic reverse current is interpreted to be caused by implantation damage acting as generation/recombination centres. The surface concentration of minority carriers needed for electrochemical reaction is increased proportionally to the centre density and thus allows an increased reaction rate.

The current increase is indeed to be attributed to radiation damage what is proved by the following facts:

(i) The effect of current increase occurs independently of the kind of ions used for implantation.

(ii) The electrochemical profiles agree well with those from Rutherford backscattering measurements.

(iii) The current increase disappears after annealing or at least reduces largely by incomplete annealing as seen in Fig. 7b for 10^{15} Si cm^{-2} in p-Si and in Fig. 10 for 10^{13} B cm^{-2} in p-Si.

The annealing behaviour at temperatures above 500 °C was not systematically studied so far in order to prevent a degradation of the Al contact on the back side of the sample due to the alloying temperature of 577 °C for Al/Si.

If the implantation dose for electrically active ions is so high that after annealing the bulk carrier density becomes greater than 10^{18} cm^{-3}, the reverse current is high. In this case, the surface generation due to the tunneling mechanism is dominant as found by Meek for n^+ Si [13].

We observed, indeed, for implantation doses above 10^{14} cm^{-2} (this corresponds at 30 keV to about 5×10^{18} B cm^{-3} at the profile maximum) still high currents as seen in Fig. 11 for 10^{14} B cm^{-2} in p-Si after annealing at 500 °C for 30 min. The electrochemical profile in Fig. 12 derived from these *I–U* measurements represents thus a superposition of the remaining damage profile with a carrier concentration profile. For comparison a carrier concentration profile calculated after the LSS theory [14] is also shown in Fig. 12.

4. Conclusions

The increase in electrolytic reverse current represents a very sensitive and new method for analysing ion implanted surfaces. The sample needs not be thermally activated, because the measuring effect is caused by implantation induced radiation damage. The sensitivity is of the order of 10^{11} to 10^{12} cm^{-2}, whereas it is 10^{14} to 10^{15} cm^{-2} for Rutherford backscattering for example. It is advantageous that the layer stripping can be done in the same cell with the same electrolyte by anodic dissolution in HF. The coulometric determination of the dissolved Si amount allows a high depth resolution of about 10 to 20 Å.

Acknowledgements

We wish to thank Dr. Klabes and his co-workers for conducting the Rutherford backscattering measurements.

References

[1] H. GERISCHER, in: Advances in Electrochemistry and Electrochemical Engineering, Vol. I, Interscience Publ., New York 1961 (p. 139).
[2] V. A. MYAMLIN and YU. V. PLESKOV, Elektrokhimiya Poluprovodnikov, Izd. Nauka, Moskva 1965.
[3] R. MEMMING and G. SCHWANDT, Surface Sci. **4**, 109 (1966).
[4] J. HÜLLER, ZfK-Rep. **294**, 202 (1975).
[5] H. MICHEL, ZfK-Rep. **294**, 186 (1975).
[6] PHAM MINH TAN, Z. phys. Chem. (1977), to be published.
[7] J. HÜLLER, ZfK-Rep. **294**, 297 (1975).
[8] J. HÜLLER and H. MICHEL, ZfK-Rep. **294**, 207 (1975).
[9] R. L. MEEK, J. Electrochem. Soc. **118**, 437 (1971).
[10] J. HÜLLER, Internat. Conf. Ion Implantation, Budapest 1975 (p. 333).
[11] J. W. MAYER, L. ERIKSSON, and J. A. DAVIES, Ion Implantation in Semiconductors, Academic Press, New York 1970.
[12] M. T. PHAM and J. HÜLLER, J. appl. Electrochem. (1977), to be published.
[13] R. L. MEEK, Surface Sci. **25**, 526 (1971).
[14] J. LINDHARD, M. SCHARFF, and H. E. SCHIOTT, Kong. Danske Vid. Selsk., mat.-Fys. Medd. **33**, No. 14 (1963).

(Received August 5, 1977)

phys. stat. sol. (a) **43**, 513 (1977)

Subject classification: 13 and 14.3.4; 6

Tbilisi State University

Hot Electron Current–Voltage Curves in a Transverse Magnetic Field with Account of Phonon Heating

By

T. O. Gegechkori and Z. S. Kachlishvili

It is shown that the phonon heating does not lead to the disappearance of the negative differential conductivity part of the I–U curve in the fixed current regime in contrast to the fixed field regime. The I–U curve has no singularities in a strong magnetic field, while in a weak magnetic field both N- and S-type I–U curves may be observed. The critical values of the field strength corresponding to the extremal points of the I–U curves are determined.

Es wird gezeigt, daß Phononenaufheizung nicht zu einem Verschwinden des negativen Teils der differentiellen Leitfähigkeit der I–U-Charakteristik im festgehaltenen Stromregime führt, im Gegensatz zum festgehaltenen Feldregime. Die I–U-Charakteristik hat keine Singularitäten in einem starken Magnetfeld, während im schwachen Magnetfeld sowohl I–U-Kurven vom N- als auch S-Typ beobachtet werden können. Die kritischen Werte der Feldstärke, die den Extrempunkten der I–U-Kurve entsprechen, werden bestimmt.

1. Introduction

The low-temperature phonon distribution may suffer a substantial departure from equilibrium due to the interaction of hot electrons with the long-wave length phonons (LP). This in its turn may change the qualitative behaviour of a semiconductor in a strong electric field. This problem was studied in a number of papers [1 to 5] both in the absence of a magnetic field [1] and in the presence of strong non-quantizing [2] and quantizing magnetic fields [3, 5]. The effect of the phonon heating on the path of electrons was also studied [4]. In all of these papers the I–U curves were obtained mostly in the fixed field regime and the effect of a weak magnetic field was neglected. The fixed current regime in the presence of a quantizing magnetic field was considered only in [5].

We consider a non-degenerate semiconductor in the fixed current regime. The conductivity may be represented in the form

$$\sigma = en\mu_{1,i}(1 + K) , \tag{1}$$

where $K = \mu_{2,i}^2/\mu_{1,i}^2$, $\mu_{1,i}$ and $\mu_{2,i}$ are the mobility coefficients corresponding to drift and Hall current, respectively, e is the electron charge, and n the hot electron concentration.

The state of the non-equilibrium LP may be described by a Planck distribution function with the temperature either coinciding with or differing from the electron temperature T_e. In the first case the momentum mean free path may be represented as

$$l_i(y) = l_i^0 y^{(t+1)/2} \theta^r . \tag{2}$$

Then one obtains

$$\frac{\mu_{1,i}}{\mu_0^i} = \frac{\theta^{(2r-1)/2}}{\Gamma\left(\frac{t+5}{2}\right)} \int\limits_0^\infty \frac{e^{-y}\, y^{(t+3)/2}\, dy}{1 + h_i y^t \theta^{2r-1}} , \tag{3}$$

$$\frac{\mu_{2,i}}{\mu_0^i} = \frac{\theta^{2r-1}}{\Gamma\left(\frac{t+5}{2}\right)} \sqrt{h_i} \int\limits_0^\infty \frac{e^{-y}\, y^{(2t+3)/2}\, dy}{1 + h_i y^t \theta^{2r-1}}. \tag{4}$$

Here $y \equiv \varepsilon/T_e$, $\theta \equiv T_e/T$, $h_i = (eHl_i^0)^2/2mc^2k_0T$, μ_0^i is the zero-field mobility in absence of a magnetic field corresponding to the i-th momentum scattering mechanism. The values of t and r for all the well-known scattering mechanisms are given in Table 1. Other notations are conventional.

Table 1
Values of the parameters t and r

momentum scattering mechanisms		r	t
I	charged impurities	2	3
N	neutral impurities	1/2	0
p	dipole centres	1	1
ac. def.	non-equilibrium acoustic phonons (deformation)	−1	−1
ac. piez.	non-equilibrium acoustic phonons (piezoelectric)	0	1

In distinction from [5, 6] the strong magnetic field is to be defined here by the condition

$$h_i \theta^{2r-1} \gg 1 \qquad (\bar{y} \approx 1)\,. \tag{5}$$

If the inequality sign is reversed the magnetic field is weak.

The other condition $\sqrt{K} \gg 1$ used in [5, 6] appears to be rather stringent. In fact, at $\sqrt{K} \ll 1$ and neglecting K the conductivity σ given by (1) coincides with that in the fixed field regime. Hence nothing new is to be expected in the fixed current regime at a weak magnetic field. However, as shown in what follows there are some momentum scattering mechanisms for which K becomes of the order of one at $h_i\theta^{2r-1} \ll 1$. This leads to some essentially new results: a) not only S-type, but N-type I–U curves exist in the fixed current regime; b) the phonon heating does not lead to the disappearance of the negative differential conductivity part of the I–U curves (in contrast to the resuls of [4] which apply to the fixed field regime); c) the possibility is pointed out to determine the critical values of the field strength corresponding to the extremal points of the I–U curves.

It is to be noted that the role of the correction terms in the expression for the conductivity, which lead to the qualitatively new results, was first discussed in [7]. In this work the investigation was carried out using the Boltzman equation and assuming the equilibrium phonon distribution, the hot-electron distribution function was assumed to be normalizable.

The LP heating is considered employing the method developed in [1, 2]. The LP get the energy and crystal momentum from the electrons. However, the crystal momentum is lost at the boundaries of the sample with a characteristic time $\tau^{(p)} \approx L/S$, where L and S are the size of the sample and the velocity of sound, respectively. In the case considered $\tau^{(p)} < \mathrm{Min}\,(\tau_{ff}, \tau_{fe})$. Here τ_{ff} and τ_{fe} are the times required respectively to transfer the crystal momentum and energy from LP to the thermal phonons (heat reservoir) and to receive the same by LP from the hot electrons. When the conditions just stated are fulfilled, the LP distribution is isotropic and the eventual development of acoustic instability is ruled out. As shown in [2], the conditions are well satisfied for n-Ge and n-GaAs at low lattice temperatures ($T \leq 20$ K), and for the samples of $L \leq 1$ cm. As stated above we consider the case of LP having a Planck distribution with the electron temperature. According to [1, 2] this is possible if a) $1 < \theta < < k_0T/8mS^2$ and $\tau_{ff} \gg \tau_{fe}$ or b) $\theta > k_0T/8mS^2$; $\tau_{ff} \gg \tau_{fe}$. The first case corresponds to the soft electron heating, charge carriers interact only with the LP, the other phonons acting as a heat reservoir for the latter. The second case is that of a strong electron heating, the carriers emit superheated phonons for which there is no heat reservoir. The stationary state is established by the energy transfer from the LP to the warm phonons (the first case) or by the energy transfer to the medium surrounding the sample (the second case), the time required is $\tau^{(\varepsilon)} \gg \tau^{(p)}$.

2. S- and N-Type *I–U* Curves

Using (1) the condition for the S- and N-type *I–U* curves to appear may be represented in the form

$$\frac{\mathrm{d}\ln n(\theta)}{\mathrm{d}\theta} + \frac{\mathrm{d}\ln \mu_{1,i}(\theta)}{\mathrm{d}\theta} + \frac{\mathrm{d}\ln[1 + K(\theta)]}{\mathrm{d}\theta} \pm \frac{\mathrm{d}\ln p(\theta)}{\mathrm{d}\theta} = 0\,. \tag{6}$$

Here $p(\theta) = p_{\mathrm{fl}}(T)\,(\theta - 1)\,\theta^{\alpha}$ is the power transferred either to the thermal phonons or to the surrounding of the sample.

Let the energy scattering be due to the interaction of the electron with the hot acoustic deformation or piezoelectric phonons. Then [1, 2] irrespective of the magnetic field strength $\alpha = 2$ or $\alpha = \frac{3}{2}$ in presence and absence of a "heat reservoir", respectively, $e = f$ and $e = b$. The upper and lower signs in (6) correspond to the N- and S-type *I–U* curves, respectively.

In a strong magnetic field (5) one obtains

$$\frac{\mu_{1,i}}{\mu_i^0} = \frac{\Gamma\left(\frac{5-t}{2}\right)}{\Gamma\left(\frac{5+t}{2}\right)} \frac{\theta^{(1-2r)/2}}{h_i}\,, \qquad \frac{\mu_{2,i}}{\mu_0^i} = \frac{\Gamma\left(\frac{5}{2}\right)}{\Gamma\left(\frac{5+t}{2}\right)} \frac{1}{\sqrt{h_i}}\,, \tag{7}$$

$$K = \frac{\Gamma^2\left(\frac{5}{2}\right)}{\Gamma^2\left(\frac{5-t}{2}\right)} h_i \theta^{2r-1}\,, \tag{8}$$

while in a weak field

$$\frac{\mu_{1,i}}{\mu_0^i} = \theta^{(2r-1)/2}\,, \qquad \frac{\mu_{2,i}}{\mu_0^i} = \frac{\Gamma\left(\frac{2t+5}{2}\right)}{\Gamma\left(\frac{t+5}{2}\right)} \sqrt{h_i}\,\theta^{2r-1}\,, \tag{9}$$

$$K = \frac{\Gamma^2\left(\frac{2t+5}{2}\right)}{\Gamma^2\left(\frac{t+5}{2}\right)} h_i\, \theta^{2r-1}\,. \tag{10}$$

Now we assume the concentration to be independent of the applied field. Considering the strong heating case ($\theta \gg 1$) we get with the help of (7) and (9)

$$\frac{2r-1}{2}\frac{K-1}{K+1} \pm (\alpha+1) = 0 \tag{11}$$

and

$$\frac{2r-1}{2}\frac{3K+1}{K+1} \pm (\alpha+1) = 0 \tag{12}$$

in a strong and weak magnetic field, respectively.

K is given by (8) and (10) in the first and second case, respectively. From these expressions it is seen that in both cases K is proportional to the parameter $h_t\theta^{2r-1}$. However, the coefficients vary in a wide range depending upon the value of t, the difference between K and the parameter might be considerable especially in the case of (10). It is easily seen that (11) has no "physical" solutions. Thus the I–U curve has no singularities in a strong magnetic field in this case. Each of the equations (12) has one "physical" solution. The solution exists at $r = -1$ $(t = -1)$ for the upper sign and at $r = 2$ $(t = 3)$ for the lower sign.

Consequently in case of the energy scattering mechanisms considered both N- and S-type I–U curves might be observed in a weak magnetic field. An N-type curve should appear if the momentum is scattered by the deformation acoustic phonons, while an S-type curve is to be expected if the ionized impurity scattering is dominant.

The corresponding critical temperatures are given by

$$\theta_{\text{kp}}^{\text{N}} = \left(\frac{9\pi}{4(2\alpha-1)}\,h_\alpha\right)^{1/3}, \tag{13}$$

$$\theta_{\text{kp}}^{\text{S}} = \left(\frac{2\alpha-1}{7-2\alpha}\,\frac{1}{24\pi h_I}\right)^{1/3}. \tag{14}$$

Solving the energy balance equation with respect to E_x one can find values of the critical field strengths (corresponding to the critical electron temperatures (13) and (14)) at which the N- and S-type regions appear in the I–U curves:

$$\left(\frac{E_x}{E_0^\alpha}\right)_{\text{kp}}^{\text{N}} = \sqrt{\frac{9}{8\alpha+5}}\left(\frac{9\pi}{4(2\alpha-1)}\,h_\alpha\right)^{(2\alpha+5)/12}, \tag{15}$$

$$\left(\frac{E_x}{E_0^I}\right)_{\text{kp}}^{\text{S}} = \sqrt{\frac{7-2\alpha}{6}}\left(\frac{2\alpha-1}{7-2\alpha}\,\frac{1}{24\pi h_I}\right)^{(2\alpha-1)/12}. \tag{16}$$

Here $E_0^i = (p_{\text{ef}}(T)/en\mu_0^i)^{1/2}$.

It is to be noted that the ranges within which the electric and magnetic fields may be varied are determined by the conditions of a strong heating $\theta \gg 1$ and a weak magnetic field.

The qualitative shapes of the I–U curves are shown in Fig. 1a and b for the cases considered.

Particularly interesting is the case when both S- and N-type I–U curves may be obtained in the same sample. In fact it is seen from (2) that the charged impurity scattering dominates at $1 \ll \theta \ll \theta_0$ where $\theta_0 = (l_0^\alpha/l_0^I)^{1/3}$. Then an S-type I–U curve is obtained if h_I is chosen accordingly. Scattering by non-equilibrium acoustic phonons becomes important on further increasing the electric field. It is easily seen that an N-type I–U curve is obtained at a value of the magnetic field which is considerably greater than that for the S-type curve. Thus, besides an increase in the electric field, one has to increase the magnetic field as well, to realize the N-type I–U curve.

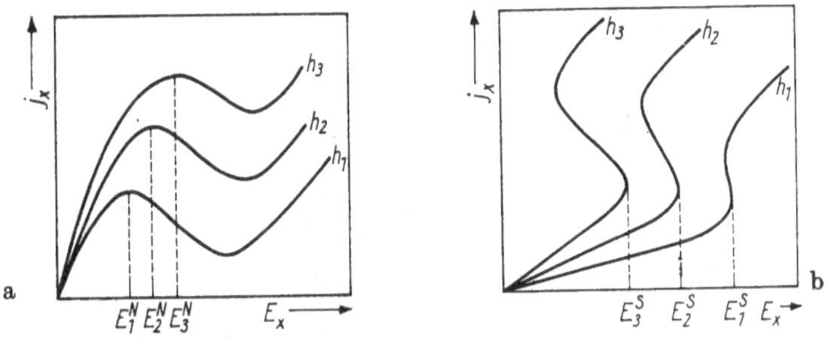

Fig. 1. Dependence of the a) N- and b) S-type I–U curves upon the magnetic field. $h_3 > h_2 > h$

3. Dependence of the Electron Temperature and Current Density upon the Electric Field Strength

On solving the energy balance equation with respect to θ one obtains
a) in a strong magnetic field at $K \gg 1$ ($\theta \gg 1$)

$$\theta = \left[\frac{\Gamma^2\left(\frac{5}{2}\right)}{\Gamma\left(\frac{5+t}{2}\right)\Gamma\left(\frac{5-t}{2}\right)}\right]^{2/(2\alpha-2r+3)} \left(\frac{E_x}{E_0^i}\right)^{4/(2\alpha-2r+3)}. \tag{17}$$

b) and in a weak magnetic field at $K \ll 1$ ($\theta \gg 1$)

$$\theta = \left(\frac{E_x}{E_0^i}\right)^{4/(2\alpha-2r+3)}. \tag{18}$$

For the currents we obtain correspondingly

$$\frac{j}{j_0^i} = \left[\frac{\Gamma^2\left(\frac{5}{2}\right)}{\Gamma\left(\frac{5+t}{2}\right)\Gamma\left(\frac{5-t}{2}\right)}\right]^{\frac{2(\alpha+1)}{2\alpha-2r+3}} \left(\frac{E_x}{E_0^i}\right)^{\frac{2\alpha+2r+1}{2\alpha-2r+3}} \tag{17a}$$

and

$$\frac{j}{j_0^i} = \left(\frac{E_x}{E_0^i}\right)^{\frac{2\alpha+2r+1}{2\alpha-2r+3}}, \tag{18a}$$

where $j_0^i = en\mu_0^i E_0^i$.

For all mechanisms of the momentum scattering the electron temperature and the current density increase on raising the electric field strength. This is in agreement with the results of [1, 2].

When the momentum is scattered by ionized impurities the approximate solution (18) is valid for very weak magnetic fields only ($K \ll 1$). If the magnetic field is not too weak then $K \approx 1$ and the energy balance equation for this scattering mechanism ($r = 2$, ($t = 3$)) is solved exactly at $\alpha = 2$.

The solution is given by

$$\theta_{12} = \left(\frac{1}{48\pi h_I}\right)^{2/3} \left(\frac{E_0^I}{E_x}\right)^{4/3} \left[1 \pm \sqrt{1 - 96\pi h_I \left(\frac{E_x}{E_0^I}\right)^4}\right]^{2/3}. \tag{19}$$

At $96\pi h_I(E_x/E_0^I)^4 \ll 1$ expanding the r.h.s. of (19) in terms of the small parameters one obtains

$$\theta_1 \approx \left(\frac{1}{24\pi h_I}\right)^{2/3} \left(\frac{E_0^I}{E_x}\right)^{4/3}, \qquad \theta_2 \approx \left(\frac{E_x}{E_0^I}\right)^{4/3}. \tag{20}$$

In this case the temperature and current density are double-valued functions of the electric field strength possessing with rising and falling branches. At $\alpha = 7$ the condition $96\pi h_I(E_x/E_0^I)^4 = 1$ (at which the temperature becomes a single-valued function of the field strength) determines the critical field for strength for the S-type $I-U$ curve. It is physically clear that the energy balance equation does not have any real solution at $96\pi h_I(E_x/E_0^J)^4 > 1$.

The following is to be noted when comparing these results with those of [7 to 9]: neglecting the correction term in the expression for the conductivity in [8, 9] the S-type $I-U$ curve is obtained in a strong magnetic field, if momentum and energy are scattered by ionized impurities and by equilibrium (deformation or piezoelectric) acoustic phonons, respectively. However, taking account of the correction term it has been shown in [7] that the N-type $I-U$ curve is obtained for these scattering mechanisms. Heating of the acoustic (deformation and piezoelectric) phonons in a strong magnetic field leads to the disappearance of both N- and S-type regions. They appear instead in a weak magnetic field.

Acknowledgement

The authors would like to thank Prof. V. L. Bonch-Bruevich for helpful discussions.

References

[1] L. E. GUREVICH and T. M. GASIMOV, Fiz. tverd. Tela 9, 106 (1967).
[2] L. E. GUREVICH and T. M. GASIMOV, Fiz. Tekh. Poluprov. 1, 774 (1967).
[3] L. E. GUREVICH and T. M. GASIMOV, Fiz. tverd. Tela 10, 3258 (1968).
[4] T. M. GASIMOV and L. E. GUREVICH, Fiz. tverd. Tela 11, 2946 (1969).
[5] T. M. GASIMOV, Fiz. Tekh. Poluprov. 4, 733 (1970).
[6] V. L. GUREVICH and U. L. FIRSOV, Zh. eksper. teor. Fiz. 40, 199 (1961).
[7] K. KH. ASRATYAN and Z. S. KACHLISHVILI, Fiz. Tekh. Poluprov. 9, 394 (1975).
[8] F. G. BASS, Zh. eksper. teor. Fiz. 48, 275 (1965).
[9] I. B. LEVINSON, Fiz. tverd. Tela 6, 2113 (1964); 7, 1362, 2417 (1965).

(Received June 16, 1977)

phys. stat. sol. (a) 43, 519 (1977)

Subject classification: 17; 22.5.2

I. Physikalisches Institut der Universität Gießen

Thermally Stimulated Exoelectron Emission of NaF

Electron Bombardment and Exoelectron Energy Investigations

By

U. BRUNSMANN[1]) and A. SCHARMANN

Clean NaF single crystal surfaces are prepared by baking in ultra high vacuum and monitored by Auger electron spectroscopy. They show an intense thermally stimulated exoelectron emission after electron bombardment. The energy of excitation yields information about bulk effects and effects of the outermost surface layers. Thus it is used as an adjustable parameter for a detailed discussion of current exoelectron emission concepts based on energy distributions. The energy measurements are performed with a plane retarding field arrangement as well as with a cylindrical mirror analyzer. An overthermal spread of all the distributions is observed, which may be explained by a patch field model.

Saubere NaF-Einkristalloberflächen werden durch Ausheizen im Ultrahochvakuum präpariert und durch Auger-Spektroskopie kontrolliert. Sie zeigen eine intensitätsstarke thermisch stimulierte Exoelektronenemission nach Elektronenbeschuß. Die Energie der Anregung liefert Information über Effekte des Volumens und der äußersten Oberflächenschichten. Sie wird daher als veränderlicher Parameter zu einer detaillierten Diskussion gängiger Modellvorstellungen der Exoelektronenemission benutzt, die auf Energieverteilungen basieren. Die Energiemessungen selbst werden mit einer ebenen Gegenfeldanordnung sowie mit einem Zylinderspiegelanalysator durchgeführt. Für alle Energieverteilungen wird eine überthermische Breite beobachtet, die sich am einfachsten in einem Fleckenfeld-Modell erklärt.

1. Introduction

Thermally stimulated exoelectron emission (TSEE) of a solid surface, excited by ionizing radiation, can be defined as that part of electron emission that additionally occurs to ordinary thermionic emission. Both, surface (a, b) and volume (c) concepts basing on different electron trapping during excitation, have been discussed in order to explain the phenomena observed on ionic crystals (reviewed in [1 to 5]):

a) adsorbates are involved in the trapping or emission process,

b) electrons are directly emitted from intrinsic surface states,

c) exoelectron emission (EE) mainly reflects properties of bulk centers, depending on electron delocalisation, diffusion towards the surface, and overcoming of the electron affinity.

This paper shall be devoted to a common problem of all emission models: the overthermal width of the exoelectron energy distributions.

Since basic experiments of Lepper [6] and Bohun [7] and their interpretation given by Nassenstein [8] a thermionic escape mechanism has frequently been suggested for the volume concept of EE, though it seems not to be supported by any energy measurement [4, 5]. Recent considerations show that there may be a space charge layer, which accelerates the electrons towards the surface at least after excitation by electron bombardment [9, 10]. Direct simulation of the field assisted electron transport by a Monte Carlo method taking electron–phonon interactions into account yields

[1]) Part of Thesis, Gießen (D 26).

a width of the energy distribution, which is in the experimental order of magnitude [11]. The experimental tests were performed on BeO [11, 12]. Later on, however, strong arguments have been given for an interpretation of TSEE in terms of a surface concept rather than by a volume emission model [13].

At present little is known about the nature and position of surface energy levels which are responsible for TSEE. Thus, it can only be speculated that an energy transfer, for instance due to electron detachment [14] or surface recombination processes [15], may contribute to the exoelectron energy distribution.

Looking for a volume explanation of exoelectron energies, which neglects effects of surface contamination and grain boundaries, it is indispensable to prepare clean single crystal surfaces that have not been used so far for energy measurements. Cleaving [16] or evaporating [17] in ultra high vacuum (UHV) has been applied to ensure that LiF exhibits exoelectron emission independent of any surface coverage. TSEE and thermal bleaching of the optical absorption have revealed that bulk centers are partly involved in the emission process [16, 18]. By recent measurements, using Auger electron spectroscopy (AES) as a highly sensitive tool to monitor the surface composition, we could confirm these results for NaF, too [19]. Substantiated by the appearance of selective optically stimulated exoelectron emission, which shows the same wavelength dependence as the optical absorption of bulk centers [20, 21], it may generally be accepted now, that these alkali halides of lowest electron affinity [22] can be used as model substances for the volume concept of EE.

Since the lowest exoelectron mean energies after excitation by electron bombardment have been obtained on NaF [23], such clean single crystal surfaces are favoured for a detailed discussion of possible volume escape mechanisms based on exoelectron energy variations, which will be given in the following, using TSEE intensity and temperature as well as the exciting beam energy as adjustable parameters.

2. Experimental

The investigations were carried out in a bakeable UHV apparatus (Varian) equipped with sorption pumps, a titanium sublimation pump and a triode sputter ion pump. During the measurements the pressure ranged from 10^{-9} to 10^{-10} Torr.

Platelets of about $0.7 \times 5 \times 5$ mm^3 in size were cleaved in air from vacuum grown NaF single crystals (Korth, NaF-suprapur). No OH^- signals in their infrared spectra could be detected by the manufacturer. The examined surfaces (100-planes) were cleaned by resistive heating to more than 600 °C in UHV [19, 24]. The surface composition, monitored by a standard cylindrical mirror Auger system (Varian, grazing incidence electron excitation) before and after each set of TSEE measurements, did not show any Auger signals except those of the host lattice. Thermal stimulation was performed by heating with approximately 100 K/min in a constant current mode. The temperature of the samples could be measured by a NiCr–Ni thermocouple. The AES electron gun served as exoelectron excitation source in the energy range between 200 eV and 3 keV. Below 200 eV a simple cathode system of a television tube was applied. The exoelectrons were detected by channeltron-, pulse-, and multichannel analyzer-techniques.

Three different methods were applied to get the exoelectron energy distributions:

1. A plane retarding field analyzer as already described in [23]. This equipment is only sensitive to normal velocity components. Therefore it is favourable for thermionic approximations.

2. A cylindrical mirror analyzer (CMA) to measure the total energy spectrum [25]. For that purpose, resolution and signal to noise ratio of the CMA is superior to that of the retarding field device whenever the emitter can be considered as a point source

(focused excitation by the AES gun). To ensure constant transmission characteristics, the CMA could not be used without preacceleration of the exoelectrons to 10 eV. Thereby the resolution of the analyzer was reduced to $\Delta E/E = 1\%$.

3. A retarding potential difference (RPD) technique, given in [23]. It was applied in order to get differential energy distributions if the CMA could not be used because of large emitting surface areas.

The energy measurements were restricted to a few seconds of nearly constant emission intensity and temperature. During these periods the heating current was switched off to avoid disturbances due to magnetic fields.

3. Evaluation of Exponential Energy Distributions

Whenever the exoelectron energy spectra can be approximated by a Maxwellian velocity distribution, a plane retarding field analyzer yields results according to the well known starting current formula of thermionic emission:

$$I = I_S \exp\left(\frac{e_0(U - U_S)}{e_0 U_M}\right), \quad \text{if} \quad U < U_S\,,$$

$$I = I_S\,, \qquad \text{if} \quad U \geqq U_S\,,$$

where I is the detected current, I_S the saturation current, U the retarding grid voltage, U_S the onset of saturation current, e_0 elementary charge, and $e_0 U_M$ the mean energy. Due to charging effects U_S is not in any case identical with the contact potential difference. For thermionic emission

$$e_0 U_M = kT$$

holds, where k is Boltzmann's constant and T the absolute temperature. In order to achieve an evaluation of the measured values as objective as possible, we applied a numerical curve fitting procedure based on a linearization by taking the logarithm and on the principle of least squares. Its quality can be checked by the correlation coefficient r^2, which is defined by

$$r^2 = S_{UI}^2/S_U^2 \cdot S_I^2\,,$$

where S_U, S_I denote the standard deviations of U, I, respectively, and S_{UI} the co-variance. If all measured values obey a linear equation, $r^2 = 1$ holds, otherwise $0 < r^2 < 1$. The fitting procedure was started in the range of high retarding voltages only for values of I that did not deviate statistically from the exponential decay by more than 25%. First there was a rise of r^2 due to the increasing number of pairs I, U. Since the exoelectrons of lowest energy do not follow the exponential shape of curve [26], the evaluation was stopped as soon as r^2 reached its maximum.

4. Results

4.1 TSEE intensity

Fig. 1 shows a glow curve of negatively charged particles that were emitted from an UHV-cleaned NaF surface after bombardment with 10^{14} electrons of 30 eV. The emission current density was in the order of 10^{-14} A/cm². Each of the three typical ranges of strong emission intensity had already been detected by Bohun by means of an open air point counter [18]. The hatched areas indicate the temperature positions and the width of the temperature intervals which we used for energy measurements. On account of the narrow TSEE energy distributions above 200 °C, which will be given in the following section, we were able to prove by crossed electric and magnetic fields that the emitted particles solely were electrons. Below the measuring point c

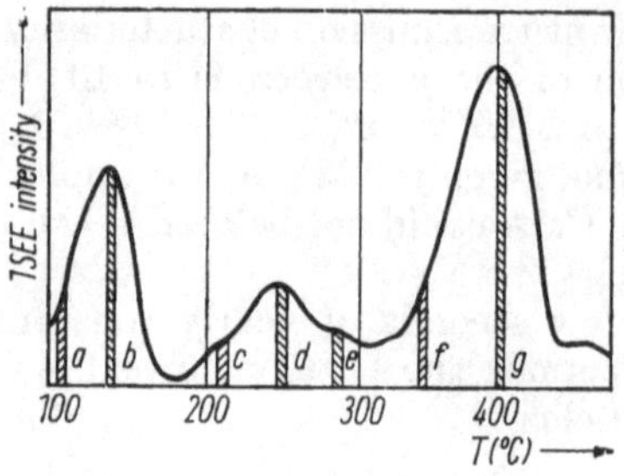

Fig. 1. TSEE of a cleaned NaF surface (pressure < $< 10^{-9}$ Torr), excitation: bombardment with 10^{14} electrons of 30 eV, hatched areas: temperature ranges of energy measurements

an ion emission could not be excluded. Even there, however, the main part of the emission current was due to electrons.

The TSEE intensity of different temperature ranges strongly depends on the energy of excitation. If it exceeded 200 eV the TSEE above 200 °C dominated, whereas excitation below 20 eV caused an intense emission at 130 °C. After bombardment with electrons of lowest energies (< 15 eV) the 400 °C maximum could not be detected at all. Three typical glow curves are represented in Fig. 2.

We want to point at a stepwise increase of TSEE intensity which was observed with increasing energy of excitation between 16 and 19 eV, though we cannot give an explanation at present. It is independent of TSEE temperature and could not simply be referred to a surface charging dependent on the first cross-over voltage V_p^I at which the total secondary emission yield is equal to unity. Following Gibbons [27] and Wei [28], V_p^I of NaF should occur between 20 and 60 V and should vary with the temperature of the sample whereas between 20 and 100 °C the stepwise behaviour did not. Reliable energy measurements could not be obtained on its low energy side because of the weak TSEE intensity.

In order to calibrate the scale of that low energy range, a disappearance potential spectroscopy (DAPS) arrangement [29], modified for low current densities, was applied. During bombardment the accelerating voltage between electron gun and sample continuously increased from 5 to 35 V. Simultaneously those electrons, which were elastically reflected from the sample, could be selected by the retarding field analyzer. As soon as the threshold energy of a new inelastic channel is obtained, electrons

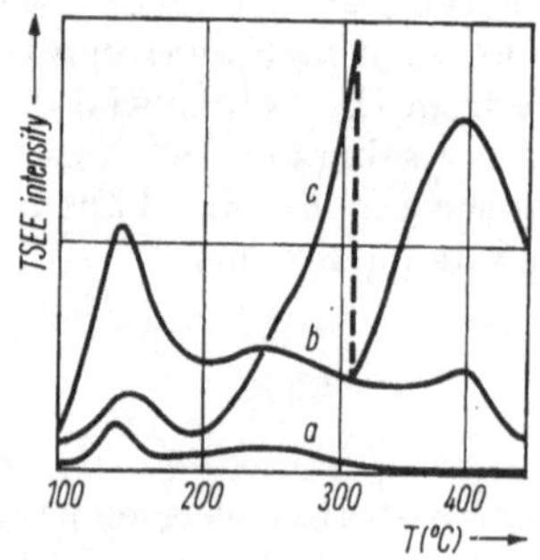

Fig. 2

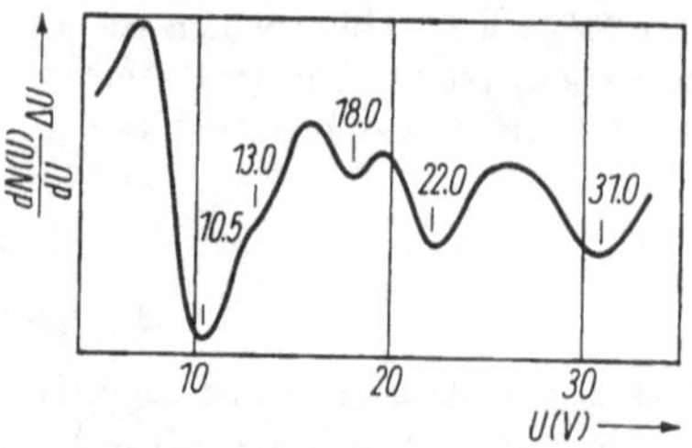

Fig. 3

Fig. 2. TSEE of NaF after bombardment with 10^{14} electrons, energy of excitation: (a) 10 eV, (b) 17 eV, (c) 3 keV

Fig. 3. Differential DAPS spectrum of a clean NaF surface. The scale of U has been adjusted with espect to a correspondence of the first minimum with exciton absorption. (On the abscissa U is the electron accelerating voltage.)

disappear from the elastic reflection intensity $N(U)$ and cause a variation which can be enhanced by differentiation. Using the RPD technique [23] the current density was kept below 10^5 e_0/cm^2s and thus sample charging could be neglected. The result is shown in Fig. 3. The accelerating voltage scale has been shifted so that the first intensity loss coincides with the exciton absorption edge of NaF at 10.5 eV according to the measurements of Hilsch [30]. Due to this adjustment there is a satisfactory agreement of the remaining structure with energy loss measurements [31, 32] if the low resolution of the plane analyzer is taken into account. It may be concluded from this measurement that the reported intensity step occurs about 5 eV above that threshold energy, which would be expected if defects like anion vacancies are occupied during excitation by interband transitions. Only an interband transition between 17 and 18 eV correlates with the observed effect [31, 32].

4.2 TSEE energy

We have already pointed out that the strong TSEE about 400 °C can be understood in terms of a thermionic emission process [26]. Equivalent low mean energies could not be obtained at lower temperatures. Moreover, it is impossible to use an exponential approximation within the whole temperature range, similar to results

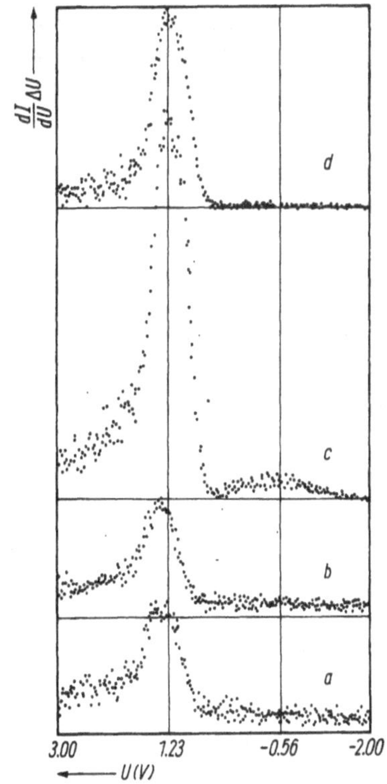

Fig. 4

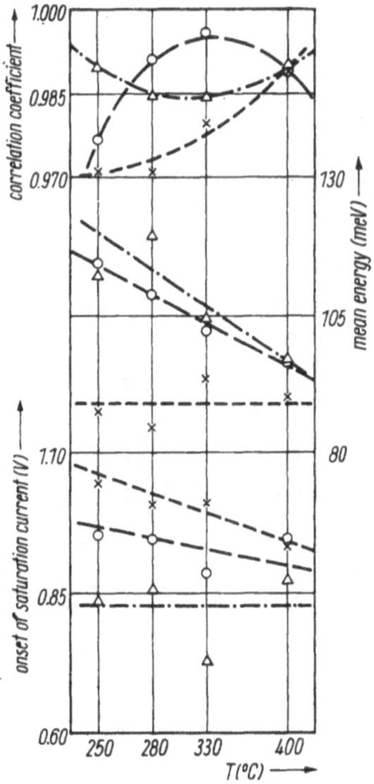

Fig. 5

Fig. 4. RPD energy distributions measured during the TSEE of NaF after bombardment with electrons of 30 eV. The temperature ranges a, b, c, d refer to Fig. 1, $\Delta U = 0.6$ V. (On the abscissa U is the retarding voltage)

Fig. 5. Parameters of the Maxwellian energy approximation dependent on the emission temperature and on the energy of excitation: × 20, ○ 25, △ 30 eV obtained on LiF surfaces [33]. The low temperature measurements a, b, c yielded a high energy tail which clearly deviates from a Maxwellian distribution and which is more intense for lower excitation energies. This is shown in Fig. 4 by the RPD technique. Note that because of the low intensity a RPD of 0.6 V has become necessary, which broadens the full width at half maximum in the low energy part of the distribution by a factor of two. Additionally, the maximum position is shifted about 0.3 V towards positive retarding voltages.

The high-energy tail can only be observed up to measuring point d. An exponential fitting above 250 °C yields correlation coefficients $r^2 > 0.9$. Fig. 5 summarizes the three parameters of the Maxwellian approximations r^2, e_0U_M, and U_S dependent on TSEE temperature and energy of excitation. For low temperature and low exciting beam energy the TSEE intensity is weak. Thus the lowest correlation coefficient is obtained for these values. The mean energy decreases at lower energies of excitation or at higher temperatures. However, it does not fall below 90 meV. The onset of saturation current is shifted in a similar way. All variations occurred in the lowest range of exciting beam energies that could be used, whereas between 30 eV and 3 keV no remarkable change was obtained.

In the latter range a single peaked shape of the total energy spectrum is typical for most of the measurements performed with the CMA device as described in [26]. The mean full width at half maximum is 0.8 eV within a standard deviation of 10%. Only for 200 eV, 2 keV, and 3 keV excitation an intermediate additional broadening of about 0.3 eV could be observed due to a temperature dependent shift of the maximum position (Table 1). Such a shift can also be seen after low energy excitation in the RPD measurements of Fig. 4.

Table 1

CMA full widths at half maximum (eV)

energy of excitation (eV)	TSEE temperature						
	110 °C (a)	130 °C (b)	210 °C (c)	250 °C (d)	280 °C (e)	330 °C (f)	400 °C (g)
200	0.99	1.03	0.86	1.27	0.83	0.95	0.91
500	0.85	0.69	0.73	0.80	0.78	0.84	0.69
1000	0.83	0.69	0.76	0.73	0.80	0.73	0.83
2000	0.83	0.83	0.90	0.97	1.09	0.90	0.76
3000	0.80	0.90	0.97	1.21	1.11	0.85	0.88

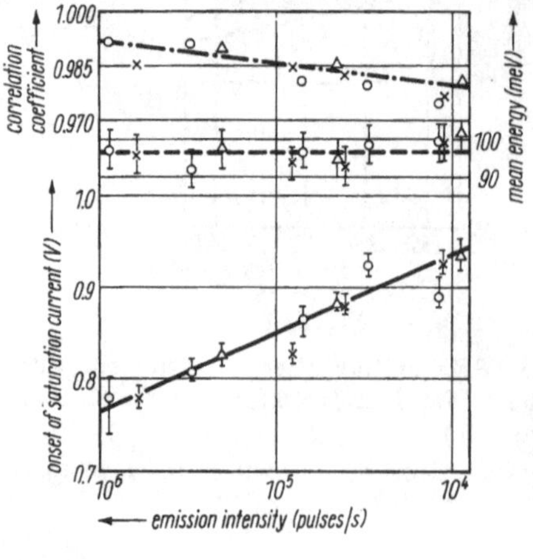

Fig. 6. Changes of correlation coefficient, mean energy, and onset of saturation current with the time dependent decay of TSEE intensity. For each of the measurements ×, ○, △ one excitation was performed

It has been demonstrated that the exoelectron energy distribution of LiF does not change if the TSEE intensity is varied by excitation dose [33]. The strong emission yield of NaF at about 400 °C can be used to check, if this holds for the time dependent decay of the emission rate, too. Fig. 6 shows the results of three series of measurements, which were performed on the same cleavage plane. Their limits of error refer to the numerical approximation. Again the quality of the approximation decreases with decreasing emission intensity. The mean energy does not change, whereas the onset of saturation current is shifted to positive values after prolonged exoelectron emission. This clearly indicates that a quick energy measuring system is necessary to fit the apparatus on the nonstationary emission process. Otherwise, disturbances due to sample charging cannot be excluded.

5. Discussion

5.1 TSEE intensity

The strong TSEE intensity of clean sample surfaces restricts an explanation of the emission mechanism to defects of the host lattice [19], however, it does not allow to decide wether any kind of transport process is involved or not. The exciting beam energy can be used as a guide to handle this problem. Following Fischer [34], who showed that Young's formula for the penetration depth of electrons [35] can be applied to estimate the thickness d of the defect layer created by electron bombardment of NaF, it seems to be reasonable that for energies of excitation above some hundreds of eV, where d is greater than 100 Å, most of the thermally liberated electrons will undergo electron–phonon collisions prior to reaching the surface. Since for that excitation strongest emission intensity occurs in the elevated temperature range, both, high TSEE temperature and high energy of the exciting electron beam may be favourable to discuss exoelectron energy distributions in terms of a diffusion towards the surface.

Low energy electron impact has already been used by Kirihata and Akutsu [36] in order to restrict the exoelectron excitation of an alkali halide to the outermost surface layers. However, their explanation of the TSEE as an electron detachment from negative ions weakly chemisorbed at the crystal surface, which has been derived from GM gas flow counter measurements, can hardly be accepted with respect to the results of UHV investigations of clean crystal surfaces [16].

The glow peak at 130 °C, which is preferably excited by lowest electron energies could not be correlated with a thermal bleaching of the optical absorption of F, M, and R centers, contrary to the glow maxima of higher temperatures [26]. Heating results in an increase of the optical absorption between 300 and 900 nm. Together with the results obtained on LiF in UHV [16] and the investigations of Bohun [18], who got an intense low temperature TSEE of NaF after cleaving of the crystals without any excitation by ionizing radiation, these experiments point to intrinsic surface states to be involved in the emission process at 130 °C.

Regarding alkali halides, it has to be noted that changes of the TSEE intensity, which depend on the exciting beam energy, can also be caused by radiolytic defect formation [37] that possibly provides an explanation for the observed intensity step. Optical absorption [38], plasma loss [39], and Auger electron spectroscopy [38, 40] as well as mass spectroscopy of lattice ions and neutrals emitted during and after excitation [38] yield information about the surface of alkali halides damaged by ionizing radiation. Simultaneous measurements of exoelectron emission should follow in order to reveal correlations especially in the low energy range of excitation.

5.2 TSEE energy

Since we are dealing with clean surfaces we need not consider here sophisticated models in which the overthermal spread of exoelectron energy distributions is referred to short-lived compound states of adsorbates [41]. Different explanations that do not require an adsorbate will be discussed in the following.

Kortov suggested two independent components for BeO [12] contributing to the measured energy distribution: a temperature independent emission intensity of high energetic electrons due to local electric fields, e.g. caused by microcracks, and a thermionic-like emission intensity that varies with temperature. This model demands an inverse relationship between TSEE intensity and exoelectron mean energy which, however, could not be found on NaF as can be seen from Fig. 6.

Various kinds of Auger processes [42], which sometimes have been used to explain TSEE, are not very probable for the whole temperature range of the emission [43]. Moreover, energy distributions on alkali halides did not prove such a process up to now. Again following the reasoning given in Section 5.1, one should expect a broadening of the energy distributions with increasing energy of excitation due to an increasing diffusion length and the electron-phonon interactions thereon [44]. Such an effect can hardly be recognized (Table 1).

Since a dc-yield enhancement of secondary electron emission is known for NaF in the keV range [45], it may be concluded that the TSEE energy width as well as the frequently assumed negative electron affinity [18, 20] can be related to a field assisted electron emission process. Only recently promising results have been given by Fitting et al. [10] which show for SiO_2 that fields up to 10^5 V/cm are built up due to electron bombardment. A first order estimation has pointed to a satisfactory agreement between exoelectron energy calculations for BeO assuming such fields as well as electron–phonon interactions and our measurements on NaF, regarding the shape of the distribution curve and the full width at half maximum [46]. The model predicts alterations of the distribution width if the electron accelerating field is varied by excitation parameters. That goal could not be achieved by our measurements as can be seen from Table 1. The variations with temperature, too, can at present not be explained in terms of this model. Besides, two serious objections can be made against a simple transfer of the theoretical results cited above to alkali halides. First, concerning the field calculations, an ion emission during electron bombardment must additionally be taken into consideration. Secondly, as to the simulation of exoelectron energies, it should be noted that the maxima of most of the measured distribution functions fall below 0.5 eV, if the distribution is assumed to start at zero energy. Within this electron energy range the mean free path becomes comparable to the lattice constant. The applicability of an electron–phonon interaction Hamiltonian based on a macroscopic polarization field and on plane-wave electron states seems quite questionable under such circumstances [44].

Moreover, a field assisted emission process is questionable in the case of TSEE excitation by low energy electron impact. Fig. 5 demonstrates that considerable changes of exoelectron energy parameters just occur in that range of excitation. The different onsets of saturation current can be related to Fermi level shifts [47, 48]. Its temperature dependence has already been explained by electronic and ionic conductivity [49].

Concerning the overthermal spread of the exoelectron energy distributions we like to take up a patch field explanation, because we did not find convincing evidence for any of the emission processes discussed above, although dramatic changes within the investigated temperature range, as demonstrated in Fig. 4, seem to point to different escape mechanisms. Patch field broadening does not afford an electron transport process and thus may also occur at lowest emission temperature as well as at lowest energy of excitation. Even exoelectron energies up to some keV could be related to the lateral charge distribution of freshly cleaved NaF surfaces by Linke and Wollbrandt [50]. On the other hand, EE itself is regarded to cause observable changes of the surface potential relief [51].

The results of our investigations suggest that the lowest measurable mean energies at highest emission temperature are determined by the surface charge relief, too. An inverse relationship between mean energy and temperature may be due to a homogenizing of the surface potential. It seems to be reasonable that such an effect increases with the energy of excitation, as can be inferred from Fig. 5. A simple numerical simulation of patch field influences on Maxwellian energy distributions has been given in a recent paper [52].

Acknowledgements

We have to thank Dr. W. Kriegseis and Dipl.-Phys. U. Wiessler for valuable discussions.

References

[1] A. Bohun, phys. stat. sol. **3**, 779 (1963).
[2] A. Bohun, Proc. III. Internat. Symp. Exoelectrons, PTB Mitteil. **80**, 323 (1970).
[3] A. Scharmann, Proc. IV. Internat. Symp. Exoelectron Emission and Dosimetry, Ed. A. Bohun, Liblice 1973 (p. 12).
[4] H. Käämbre, ibid. (p. 57).
[5] H. Käämbre, Proc. V. Internat. Symp. Exoelectron Emission and Dosimetry, Ed. A. Bohun and A. Scharmann, Zvikov 1976 (p. 89).
[6] J. Lepper, Z. Naturf. **10a**, 47 (1955).
[7] A. Bohun, Czech. J. Phys. **5**, 224 (1955).
[8] H. Nassenstein, Z. Naturf. **10a**, 944 (1955).
[9] J. Drenckhan, H. Gross, and H. Glaefeke, phys. stat. sol. (a) **2**, K51 (1970).
[10] H. J. Fitting, H. Glaefeke, and W. Wild, Proc. V. Internat. Symp. Exoelectron Emission and Dosimetry, Ed. A. Bohun and A. Scharmann, Zvikov 1976 (p. 292).
[11] V. S. Kortov and P. P. Zolnikov, phys. stat. sol. (a) **31**, 331 (1975).
[12] V. S. Kortov, phys. stat. sol. (a) **19**, 59 (1973).
[13] M. Euler and A. Scharmann, phys. stat. sol. (a) **34**, 297 (1976).
[14] M. Euler and A. Scharmann, Z. Phys. **B25**, 313 (1976).
[15] I. V. Krylova, Proc. V. Internat. Symp. Exoelectron Emission and Dosimetry, Ed. A. Bohun and A. Scharmann, Zvikov 1976 (p. 40).
[16] P. Gordan and A. Scharmann, Z. Phys. **217**, 309 (1968).
[17] L. I. Samuelsson, phys. stat. sol. (a) **32**, K155 (1975).
[18] A. Bohun, phys. stat. sol. **2**, 992 (1962).
[19] U. Brunsmann and A. Scharmann, Vakuum-Technik **26**, 78 (1977).
[20] P. Petrescu, phys. stat. sol. (a) **4**, K3 (1971).
[21] A. A. Alybakov, V. A. Gubanova, and S. Halmursaev, Proc. V. Internat. Symp. Exoelectron Emission and Dosimetry, Ed. A. Bohun and A. Scharmann, Zvikov 1976 (p. 153).
[22] Y. Iguchi, H. Sugawara, and T. Sasaki, III. Internat. Conf. Vacuum Ultraviolet Radiation Physics, Tokyo 1971.
[23] U. Brunsmann and A. Scharmann, phys. stat. sol. (a) **26**, K123 (1974).
[24] J. Estel, H. Hoinkes, H. Kaarmann, H. Nahr, and H. Wilsch, Surface Sci. **54**, 393 (1976).
[25] H. Z. Sar-El, Rev. sci. Instrum. **38**, 1210 (1967).
[26] U. Brunsmann and A. Scharmann, Proc. V. Internat. Symp. Exoelectron Emission and Dosimetry, Ed. A. Bohun and A. Scharmann, Zvikov 1976 (p. 278).
[27] D. J. Gibbons, in: Handbook of Vacuum Physics, Vol. 2, Part. 3, Ed. A. H. Beck, Pergamon Press, Braunschweig 1966 (p. 336).
[28] P. S. P. Wei, Surface Sci. **24**, 219 (1971).
[29] J. Kirschner and P. Staib, Phys. Letters **42A**, 335 (1973).
[30] R. Hilsch, Z. Phys. **77**, 427 (1932).
[31] P. E. Best, Proc. Phys. Soc. **79**, 133 (1962).

[32] M. CREUZBURG, Z. Phys. 196, 433 (1966).
[33] U. BRUNSMANN, R. HUZIMURA, and A. SCHARMANN, phys. stat. sol. (a) 26, K149 (1974).
[34] F. FISCHER, Z. Phys. 154, 534 (1959).
[35] R. YOUNG, J. appl. Phys. 27, 1 (1956).
[36] H. KIRIHATA and J. AKUTSU, phys. stat. sol. (a) 35, K73 (1976).
[37] P. D. TOWNSEND, J. Phys. C 9, 1871 (1976).
[38] T. E. GALLON, I. G. HIGGINBOTHAM, M. PRUTTON, and H. TOKUTAKA, Surface Sci. 21, 224 (1970).
[39] L. S. COTA ARAIZA and B. D. POWELL, Surface Sci. 51, 504 (1975).
[40] D. G. LORD and T. E. GALLON, Surface Sci. 36, 606 (1973).
[41] D. FINTELMANN, Proc. V. Internat. Symp. Exoelectron Emission and Dosimetry, Ed. A. BOHUN and A. SCHARMANN, Zvikov 1976 (p. 288).
[42] P. T. LANDSBERG, phys. stat. sol. 41, 457 (1970).
[43] V. BICHEVIN and H. KÄÄMBRE, phys. stat. sol. (a) 4, K235 (1971).
[44] J. LLACER and E. L. GARWIN, J. appl. Phys. 40, 2766 (1969).
[45] J. LLACER and E. L. GARWIN, J. appl. Phys. 40, 2776 (1969).
[46] V. S. KORTOV and V. G. ISAKOV, Proc. V. Internat. Symp. Exoelectron Emission and Dosimetry, Ed. A. BOHUN and A. SCHARMANN, Zvikov 1976 (p. 212).
[47] A. KASK, L. GONCHAROVA, and H. KÄÄMBRE, phys. stat. sol. (a) 11, K127 (1972).
[48] U. BRUNSMANN and A. SCHARMANN, Proc. IV. Internat. Symp. Exoelectron Emission and Dosimetry, Ed. A. BOHUN, Liblice 1973 (p. 137).
[49] Y. HAYAKAWA and N. ODA, phys. stat. sol. (a) 29, K117 (1975).
[50] E. LINKE and J. WOLLBRANDT, Proc. IV. Internat. Symp. Exoelectron Emission and Dosimetry, Ed. A. BOHUN, Liblice 1973 (p. 75).
[51] V. S. POSTNIKOV, V. M. EVLEV, and V. A. AMMER, Soviet Phys. – Solid State 17, 608 (1975); Fiz. tverd. Tela 17, 946 (1975).
[52] U. BRUNSMANN and A. SCHARMANN, phys. stat. sol. (a) 42, K79 (1977).

(Received June 24, 1977)

phys. stat. sol. (a) 43, 529 (1977)

Subject classification: 13.4 and 20.3; 22.2.2

Sektion Physik der Karl-Marx-Universität Leipzig,
Arbeitsgemeinschaft A^{III}–B^{V}-Halbleiter

Determination of the Concentration of Substitutional O in GaP by Means of Donor–Acceptor Pair Luminescence

By

R. BINDEMANN, H. FISCHER, and K. KREHER

The concentration of substitutional oxygen in GaP is determined from a comparison of the donor–acceptor pair luminescence of shallow S–C-pairs, excited via a two-step excitation and the same luminescence excited by band-to-band transitions. The measured concentrations in the order of 10^{16} cm^{-3} are in good agreeement with values reported in the literature.

Die Konzentration von substitutionell in GaP eingebautem Sauerstoff wird aus dem Vergleich der über einen Zweistufenprozeß angeregten Lumineszenz flacher S–C-Paare mit der über Band–Band-Übergänge angeregten Paarlumineszenz verglichen. Die gemessenen Konzentrationen in der Größenordnung von 10^{16} cm^{-3} sind in guter Übereinstimmung mit Literaturwerten.

1. Introduction

Oxygen is a residual impurity in GaP. It is incorporated as a substitutional impurity at P sites forming a deep donor (binding energy $E_D = 896$ meV) or it forms complexes with more or less complicated structure thus influencing the electrical and luminescence properties of the material [1 to 4].

The total oxygen concentration can be determined by means of ^{3}He activation analysis [5, 6] or by proton activation analysis of ^{18}O doped material [7].

The substitutional O concentration can be found from high-temperature Hall effect [8, 9], spectroscopic [10], or photocapacitance [11, 12] measurements.

In this paper we present a new method for measuring the substitutional O concentration based on the following principle: The donor–acceptor pair (DAP) luminescence can be excited via the deep donor level of oxygen [13], the intensity depending on the O concentration. As an absolute measurement is difficult, this luminescence is compared with the luminescence excited via band-to-band transitions so that only relative intensities are needed.

2. Theory

The intensity of the DAP luminescence I_L at low temperature for a pair distance R is given by [14]

$$I_L = I_0(R)\, f(R) \tag{1}$$

self-absorption and internal reflexions being neglected. $I_0(R) = BR^4 \exp\left(-(4\pi/3)\, NR^3\right) \times P(R)$ is the intensity for very strong excitation, $f(R)$ is the probability that the pair with distance R is occupied.

In the case of homogeneous excitation Hagston [15] found

$$f(R) = \left(1 + \frac{P(R)}{A\sigma_e(R)\, n}\right)^{-1}. \tag{2}$$

$P(R)$ is the probability for radiative transitions, $\sigma_e(R)$ the capture cross section, N the sum of the concentrations of shallow impurities, A and B are constants. Formula (2) implies that the capture of electrons is the rate-determining process [16].

2.1 Band-to-band excitation

In the case of inhomogeneous excitation the electron concentration depends on the distance x from the illuminated surface. As long as the diffusion length L is larger than the pair distance R, the intensity of the pair luminescence is determined by the local electron density $n(x)$ and we find

$$I_{L1} = I_0(R) \frac{1}{d} \int_0^d \left(1 + \frac{P(R)}{A\sigma_e(R)\, n(x)}\right)^{-1} dx\,. \tag{1a}$$

d is the thickness of the sample.

If the absorption is not too strong ($\alpha_1 d \approx 1$, $\alpha_1 L \ll 1$) the influence of diffusion and surface recombination can be neglected and we get

$$n(x) = I_{exc\,1}\, \alpha_1 \tau \exp(-\alpha_1 x) = n_1 \exp(-\alpha_1 x)\,, \tag{3}$$

τ is the lifetime of the free electrons, α_1 the absorption constant, $I_{exc\,1}$ the excitation intensity.

From (1a) and (3) we get

$$I_{L1} = \frac{I_0}{\alpha_1 d} \ln\left(\frac{1 + A\sigma_e(R)\, n_1/P(R)}{1 + A\sigma_e(R)\, n_1 \exp(-\alpha_1 d)/P(R)}\right). \tag{1b}$$

2.2 Two-step excitation

For the two-step excitation with intensity $I_{exc\,2}$ via the donor O the electron density n_2 and the absorption constant α_2 can be found from a simple recombination model according to Fig. 1 [13]

$$n_2 = \frac{N_0 \sigma_{PI} \sigma_{PN} \tau^* \tau I_{exc\,2}^2}{1 + (\sigma_{PI} + \sigma_{PN})\, \tau^* I_{exc\,2}}, \tag{4}$$

$$\alpha_2 = \frac{N_0 \sigma_{PN} (1 + 2\sigma_{PI} \tau^* I_{exc\,2})}{1 + (\sigma_{PI} + \sigma_{PN})\, \tau^* I_{exc\,2}}. \tag{5}$$

N_0 is the concentration of substitutional oxygen. σ_{PN} and σ_{PI} are the cross sections for photoneutralization and photoionization of O.

As the absorption constant is low the sample is excited homogeneously and we find

$$I_{L2} = I_0(R) \left(1 + \frac{P(R)}{A\sigma_e(R)\, n_2}\right)^{-1}. \tag{1c}$$

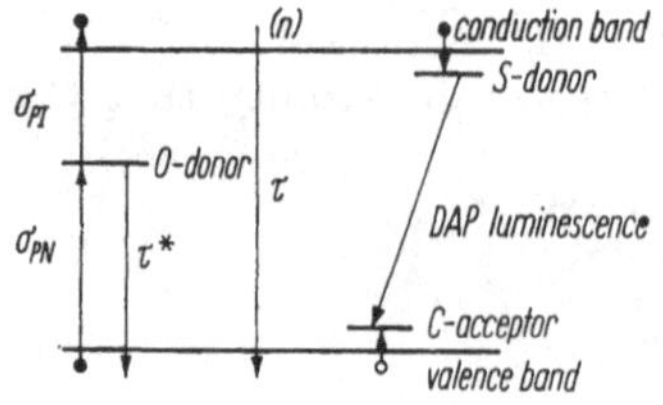

Fig. 1. Recombination model for two-step excitation. σ_{PN} and σ_{PI} are the optical cross-sections for photoneutralization and photoionization, τ and τ^* are the lifetimes of electrons in the conduction band and the oxygen level. Recapture into the O-level is neglected

2.3 Comparison of two-step and band-to-band excitation

The DAP spectrum has a peak at an energy

$$h\nu_{max} = E_g - (E_D + E_A) + E_c^{max},$$

the Coulomb energy $E_c^{max} = e^2/4\pi\varepsilon\varepsilon_0 R_{max}$ corresponds to pairs with distance R_{max}.

The peak position E_c^{max} depends on the concentration N of shallow impurities and on the excitation intensity I_{exc}.

First we consider the case of homogeneous band-to-band excitation ($\alpha_1 d \ll 1$). If we find a pair of excitation intensities $I_{exc\,1}$ and $I_{exc\,2}$ which give the same peak position E_c^{max}, we have obviously $n_1 = n_2$ and $I_{L1}^{max} = I_{L2}^{max}$. Using (3) and (4) we get

$$\frac{I_{exc\,2}}{I_{exc\,1}} = \frac{\alpha_1}{N_0}\left(\frac{1}{\sigma_{PI}} + \frac{1}{\sigma_{PN}}\right)\left(1 + \frac{1}{(\sigma_{PI} + \sigma_{PN})\,\tau^* I_{exc\,2}}\right). \tag{6}$$

Extrapolation to high excitation intensity gives

$$N_0 = \alpha_1\left(\frac{1}{\sigma_{PI}} + \frac{1}{\sigma_{PN}}\right)\frac{I_{exc\,1}}{I_{exc\,2}}\bigg|_{I_{exc\,2}\to\infty}. \tag{7}$$

In this case nothing must be known about the shape of the DAP spectrum and about a possible dependence of τ on the carrier concentration.

In the case $\alpha_1 d \gtrsim 1$ higher luminescence intensities can be obtained, the determination of the O concentration is based on the same principle but is more involved.

In a first step the concentration N of shallow impurities must be determined. This can be done from the DAP spectrum at extremely low excitation, for in this case the peak position is given [14, 17] by

$$E_c^{max}|_{I_{exc}\to 0} = \frac{e^2}{4\pi\varepsilon\varepsilon_0}\sqrt[3]{\frac{2\pi N}{3}}. \tag{8}$$

Knowing N the peak position can be calculated as a function of An from (1b) and (1d). The results for $N = 5 \times 10^{17}\ \mathrm{cm^{-3}}$ are shown in Fig. 2.

If we find a pair of excitation intensities $I_{exc\,1}$ and $I_{exc\,2}$ belonging to the same peak position we take the corresponding values An_1 and An_2 from the theoretical curves of Fig. 2. Using (3) and (4) we get

$$\frac{An_1}{An_2} = \frac{\alpha_1}{N_0}\frac{I_{exc\,1}}{I_{exc\,2}}\left(\frac{1}{\sigma_{PI}} + \frac{1}{\sigma_{PN}}\right)\left(1 + \frac{1}{(\sigma_{PI} + \sigma_{PN})\,\tau^* I_{exc\,2}}\right) \tag{6a}$$

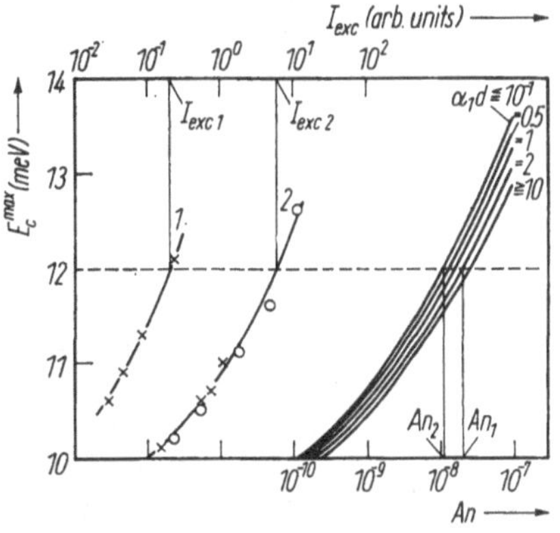

Fig. 2. Right part (lower scale): Peak position E_c^{max} versus free carrier concentration n according to (1b). $\sigma_c \sim R^2$ and $P(R) \sim \exp(-R/a)$ with $a = 9.1$ Å [18] have been assumed, $N = 5 \times 10^{17}\ \mathrm{cm^{-3}}$. Left part (upper scale): Measured peak position versus excitation intensity for sample SL 19/20. (1) Band-to-band-excitation, (2) two-step excitation, + tungsten iodine lamp, ○ He–Ne laser

and the extrapolation to high excitation intensities gives

$$N_0 = \alpha_1 \left(\frac{1}{\sigma_{PI}} + \frac{1}{\sigma_{PN}}\right) \left(\frac{I_{exc\,1}}{I_{exc\,2}} \frac{An_1}{An_2}\right)_{I_{exc\,2} \to \infty} . \tag{7a}$$

In the case of very strong absorption ($\alpha_1 L \gg 1$) carrier diffusion and surface recombination must be taken into consideration. This will be the subject of a forthcoming paper.

3. Experimental

We present results from four GaP crystals. The samples 143 and 154 were solution-grown from liquid Ga at $T = 1000$ K, the samples SL 19/20 and SL 14/15 are not intentionally doped and highly compensated grown by the synthesis solute diffusion (SSD) technique. Band-to-band excitation and two-step excitation ($h\nu = 1.959$ eV) at low intensity were done using a 250 W tungsten iodine lamp which was controlled by an electronic equipment [19] and a SPM 2 monochromator.

Higher intensities for two-step excitation were accomplished by a He–Ne-laser HNA 188. In order to have equal geometrical conditions the light of the laser also passed through the monochromator.

The light of the S–C-pair luminescence was analyzed with a grating monochromator GDM 1000 and recorded using the lock-in technique [25].

All measurements were done at liquid He temperature. The samples were carefully shielded from thermal radiation of warm parts of the cryostat so that no carrier redistribution due to infrared radiation can influence the peak position [20]. The values $\sigma_{PN}(1.959 \text{ eV}) = 1.1 \times 10^{-16}$ cm^2 and $\sigma_{PI}(1.959 \text{ eV}) = 2.5 \times 10^{-16}$ cm^2 were taken from the literature [21, 22].

4. Results and Discussion

Band-to-band excitation was done at 520 nm ($h\nu = 2.384$ eV) corresponding to $\alpha_1 = 25$ cm^{-1}.

Fig. 2 shows the peak position versus the excitation intensity for band-to-band excitation (curve 1) and two-step excitation (curve 2) for sample SL 19/20.

The parameters of the samples and the measured substitutional oxygen concentrations are summarized in Table 1.

Peters and Vink [9] find from high-temperature Hall effect data for undoped solution-grown samples $N_0 < 3.3 \times 10^{15}$ cm^{-3}, for solution-grown samples doped with Ga_2O_3 N_0 varies between 8×10^{15} cm^{-3} and 7.4×10^{16} cm^{-3}. All these values are in good agreement with ours. The increase of the O solubility by Zn doping found in [9] could not be seen in our sample 143/3.

In principle the measurement of the O concentration can be influenced by other impurities. Two cases must be distinguished.

Table 1

sample	doping	type of conductivity	d (10^{-2} cm)	N_0 (cm^{-3})
SL 19/20	without	n	7.6	6.3×10^{15}
SL 14/15	without	n	4.5	3.2×10^{15}
154/1	Ga_2O_3	n	4.3	1.2×10^{16}
143/3	Zn, Ga_2O_3	p	6.8	1.1×10^{16}

The known deep-level impurities Cu, Co, and Fe could be responsible for two-step excitation with threshold energies of 1.68 eV (Cu), 1.84 eV (Fe), and 1.93 eV (Co) [23]. No such thresholds were seen in the excitation spectrum of the two-step luminescence [24].

Shallow impurities can influence the carrier concentrations n_1 and n_2 differently if the photoionization cross sections at $h\nu_1$ and $h\nu_2$ are different. This influence was measured using infrared radiation with energies between 0.2 and 1.4 eV additionally to the two-step excitation. Indeed this radiation caused an increase of the luminescence which could be used for measuring the spectral dependence of the photoionization cross section of sulfur donors, but this increase can be detected only using very sensitive modulation techniques [24]. Thus we conclude that our measurements of the O concentration are not influenced by other impurities.

5. Conclusions

The new measuring method for the substitutional oxygen concentration gives results in agreement with other methods. It has a high sensitivity and is not influenced by the presence of other impurities. Therefore it is especially well suited for the measurement of low oxygen concentrations.

References

[1] P. J. Dean, Appl. Solid State Sci., Vol. 1, Ed. R. Wolfe, Academic Press, New York 1969.
[2] R. N. Bhargava, Philips tech. Rev. **32**, 261 (1971).
[3] R. Z. Bachrach, O. G. Lorimor, L. R. Dawson, and K. B. Wolfstirn, J. appl. Phys. **43**, 5098 (1972).
[4] J. S. Jayson, R. Z. Bachrach, P. D. Dapkus, and N. E. Schumaker, Phys. Rev. B **6**, 2357 (1972).
[5] C. K. Kim, Radiochem. Radioanal. Letters **2**, 53 (1969).
[6] C. K. Kim, Analyt. Chim. Acta **54**, 407 (1971).
[7] E. C. Lightowlers, J. C. North, A. S. Jordan, L. Derrick, and J. L. Merz, J. appl. Phys. **44**, 4758 (1973).
[8] L. M. Forster and J. Scardefield, J. Electrochem. Soc. **116**, 494 (1969).
[9] R. C. Peters and A. T. Vink, Inst. Phys. Conf. **24**, 254 (1975).
[10] J. M. Dishman, M. DiDomeniko, and R. Caruso, Phys. Rev. B **2**, 1988 (1970).
[11] H. Kukimoto, D. H. Henry, and G. L. Miller, Appl. Phys. Letters **21**, 251 (1972).
[12] H. Kukimoto, C. H. Henry, and F. R. Merritt, Phys. Rev. B7, 2486 (1973).
[13] A. Schindler, R. Bindemann, and K. Kreher, phys. stat. sol. (b) **59**, 439 (1973).
[14] R. Bindemann and K. Unger, phys. stat. sol. (b) **66**, 133 (1974).
[15] W. E. Hagston, J. Lum. **3**, 253 (1971).
[16] M. J. Dishman, Phys. Rev. B **5**, 2258 (1972).
[17] T. N. Morgan, T. S. Plaskett, and G. D. Pettit, Phys. Rev. **180**, 845 (1969).
[18] A. T. Vink, J. Lum. **9**, 159 (1974).
[19] H. Fischer, Exper. Tech. Phys. **25**, 227 (1977).
[20] R. Bindemann, H. Fischer, and H. Kreher, phys. stat. sol. (a) **34**, K63 (1976).
[21] R. Olsson, J. Phys. Chem. Solids **33**, 549 (1972).
[22] G. Björklund and H. G. Grimmeis, Solid State Electronics **14**, 589 (1971).
[23] A. E. Yunovich, Fortschr. Phys. **23**, 317 (1975).
[24] R. Bindemann, H. Fischer, and K. Kreher, to be published.
[25] H. Fischer and R. Bindemann, Jenaer Rundschau **22**, 239 (1977).

(Received August 8, 1977)

phys. stat. sol. (a) 43, 535 (1977)

Subject classification: 1.1

Institute of Materials Engineering, Warsaw Technical University[1])

Determination of the Crystal Orientation from Intersections of Kikuchi Lines

By

J. A. Kozubowski

A method for the calculation of the exact direction of the electron beam in the crystal is described. Directions represented by the intersection points of three pairs of Kikuchi lines are determined at first. Then the direction of the electron beam is calculated from the position of the center of the diffraction pattern relative to these intersection points. An example given shows that the inherent accuracy of the method approaches 1 min of arc. Comparison with other methods and with results obtained using different sets of initial data indicates that at present the deformation of the foil in the area selected for diffraction and the distortion of the microdiffraction pattern are the main factors limiting the accuracy of the orientation determination.

Es wird eine Berechnungsmethode der exakten Richtung des Elektronenstrahls im Kristall beschrieben. Zuerst werden Richtungen bestimmt, die durch die Schnittpunkte dreier Paare von Kikuchilinien bestimmt sind. Dann wird die Richtung des Elektronenstrahls aus der Lage des Zentrums des Beugungsmusters relativ zu diesen Schnittpunkten berechnet. Ein Beispiel zeigt, daß die innere Genauigkeit der Methode 1 Bogenminute erreicht. Ein Vergleich mit anderen Methoden und den Ergebnissen, die mit verschiedenen Sätzen von Anfangswerten erhalten werden, zeigt, daß gegenwärtig die Deformation der Schicht in der für die Beugung ausgewählten Fläche und die Verzerrung des Mikrobeugungsmusters die Hauptfaktoren sind, die die Genauigkeit der Orientierungsbestimmung begrenzen.

1. Introduction

Kikuchi lines [1] are well known to electron microscopists as the best means for an exact determination of the crystal orientation [2]. Several methods estimating the direction of the electron beam in a crystal have been proposed [3 to 8]. The methods of Ryder and Pitsch [4] and of Pumphrey and Bowkett [8], considered to be most accurate, are based on the measurement of the relative position of Kikuchi lines and diffraction spots. In the RP method use is made of three Kikuchi lines connected with three identified reflections not belonging to the same zone. The great advantage of the RP method consists in the fact that there is no need to know the camera length, the electron wavelength, and the lattice parameter. The accuracy of the RP method is close to 3′ if well defined reflections are used. The PB method is based on measurements of the distances between Kikuchi lines and the corresponding spots; since a pair of reflections is used, the 180° uncertainty has to be resolved in advance. The electron wavelength and the lattice parameter must be used in the calculations. The accuracy of the PB method has been estimated as better than 2′ [9].

In practice it is often difficult to achieve the maximum accuracy of the above-mentioned procedures, especially in thick foils where the reflections are diffuse, and bright Kikuchi lines are broad when situated close to reflections.

The principle of the method proposed in this paper is based on the observation that the centre of the electron diffraction pattern is normally surrounded by several dark Kikuchi lines, and that their intersection points are well defined even in relatively thick

[1]) Narbutta 85, 02-524 Warsaw, Poland.

foils. Provided that the crystal directions related to these points have been calculated, one can determine the beam direction from the positions of the intersection points relative to the centre of the diffraction pattern.

2. Description of the Method

The present method is based on a similar principle as that of Hälbig et al. [10], used by these authors for the orientation determination from intersections of Kossel lines [11]. Differences are due to the characteristic geometry of Kikuchi lines.

Suppose a selection is made of three identified Kikuchi lines which form a triangle enclosing a trace of the primary beam[2]) (Fig. 1). Point p_{ij} is the trace of the direction $\bar{\boldsymbol{p}}_{ij}$ of intersection of the two cones of electron beams diffracted by the planes with normals $\bar{\boldsymbol{n}}_i$ and $\bar{\boldsymbol{n}}_j$. Intersections of these cones with the photographic plate form the corresponding Kikuchi lines (i) and (j). The directions $\bar{\boldsymbol{p}}_{ij}$ can be found from the following set of equations:

$$\bar{\boldsymbol{n}}_i \cdot \bar{\boldsymbol{p}}_{ij} = |\bar{\boldsymbol{g}}_i| \frac{\lambda}{2}, \qquad \bar{\boldsymbol{n}}_j \cdot \bar{\boldsymbol{p}}_{ij} = |\bar{\boldsymbol{g}}_j| \frac{\lambda}{2}, \qquad |\bar{\boldsymbol{p}}_{ij}| = 1, \tag{1}$$

where $\boldsymbol{n}_i$, $\boldsymbol{n}_j$ are unit vectors in the directions of the reciprocal lattice vectors $\boldsymbol{g}_i$, $\boldsymbol{g}_j$, respectively (Fig. 2a, b), and λ is the electron wavelength.

To identify a chosen Kikuchi line consistently with others, the direction of its $\bar{\boldsymbol{g}}_i$ vector must be taken in such a way that it points from the trace of the diffracting plane to this Kikuchi line. The electron wavelength can be determined for the whole series of diffraction patterns by the method of Pumphrey [12]. Equations (1) have in general two solutions for $\bar{\boldsymbol{p}}_{ij}$; the proper one is selected by comparison with the approximate beam direction determined by the direction opposite to the zone axis of $\bar{\boldsymbol{g}}_i$, $\bar{\boldsymbol{g}}_j$.

Solutions of (1) for three pairs of Kikuchi lines ($i, j = 1, 2, 3$) give the directions $\bar{\boldsymbol{p}}_{12}$, $\bar{\boldsymbol{p}}_{23}$, $\bar{\boldsymbol{p}}_{31}$ corresponding to the intersection points p_{12}, p_{23}, p_{31} of Kikuchi lines (1), (2), and (3). The actual angular scale of the diffraction pattern is determined by comparing — for each pair of points p_{ij} — the angle calculated from the cosine formula with the distance between the points measured on the print (or enlarger screen). For better accuracy, the mean value calculated from the three pairs is subsequently used. With the use of the angular scale and measured distances of points p_{ij} from the centre, the angles α_{ij} between beam direction and $\bar{\boldsymbol{p}}_{ij}$ directions can easily be determined. Knowing the values of these angles and using the $\bar{\boldsymbol{p}}_{ij}$ directions in pairs, one can find

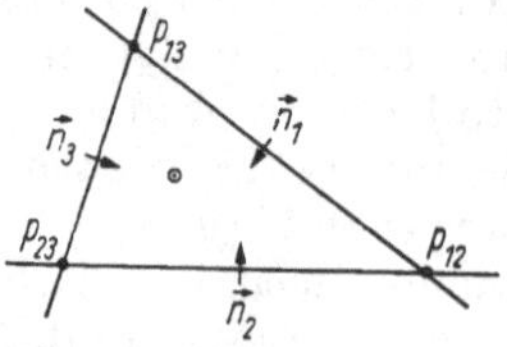

Fig. 1. Points of intersection p_{12}, p_{13}, p_{23} of three Kikuchi lines resulting from diffraction on the planes with normals $\bar{\boldsymbol{n}}_1$, $\bar{\boldsymbol{n}}_2$, $\bar{\boldsymbol{n}}_3$

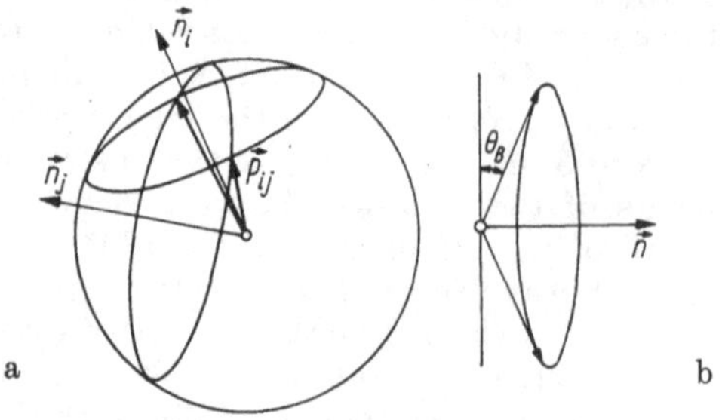

Fig. 2. a) Direction of intersection $\bar{\boldsymbol{p}}_{ij}$ of two Kikuchi cones; b) with normals $\bar{\boldsymbol{n}}_i$, $\bar{\boldsymbol{n}}_j$ to the diffracting planes

---

[2]) This requirement is not necessary, since the centre of the diffraction pattern can equally well be situated outside the chosen triangle of Kikuchi lines.

three estimates of the true beam direction $\boldsymbol{r}$. The third direction $\bar{\boldsymbol{p}}_{ij}$ serves each time for the selection of the proper solution of the set of equations

$$\bar{\boldsymbol{r}} \cdot \bar{\boldsymbol{p}}_{ik} = \cos \alpha_{ik}\,, \qquad \bar{\boldsymbol{r}} \cdot \bar{\boldsymbol{p}}_{kj} = \cos \alpha_{kj}\,, \qquad |\bar{\boldsymbol{r}}| = 1; \qquad i, j, k = 1, 2, 3\,. \tag{2}$$

The mean value of the three estimates represents the best approximation to the true beam direction, and its standard deviation is a measure of the inherent accuracy of the method.

Because in this method the same type of a set of three equations is solved six times, this can best be done by using a computer program.

3. Discussion

An example of the application of this method is shown in Fig. 3a, b. Four triangles of Kikuchi lines ABC, ABD, GEH, ACF were used for the calculation of the beam direction. Measurements of the distances were made on a 24 × 30 cm print. The

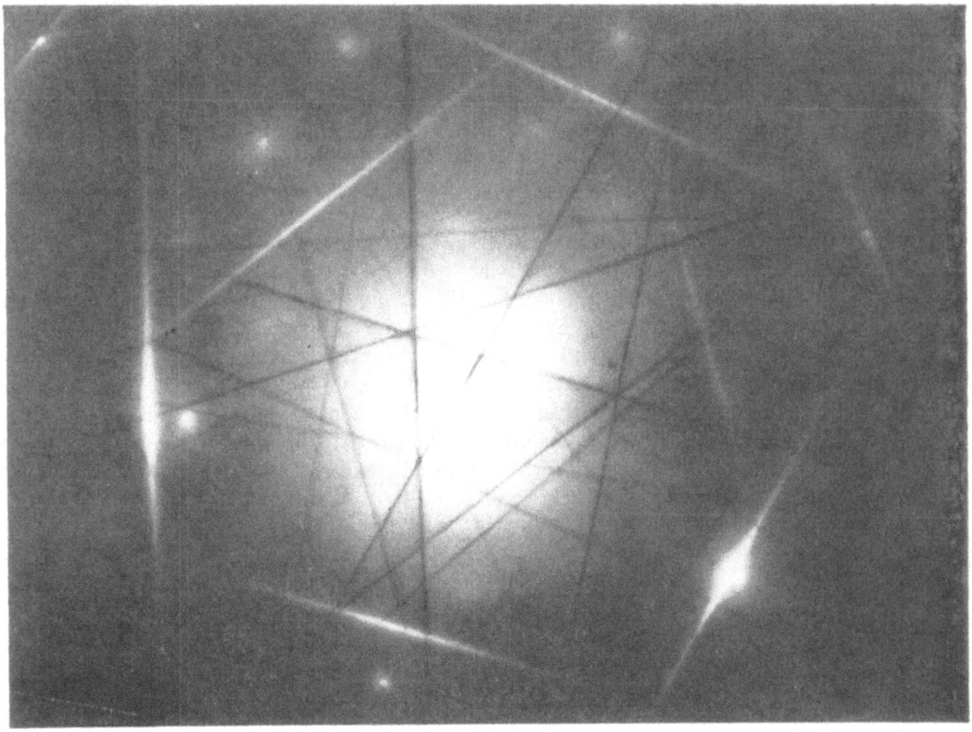

a

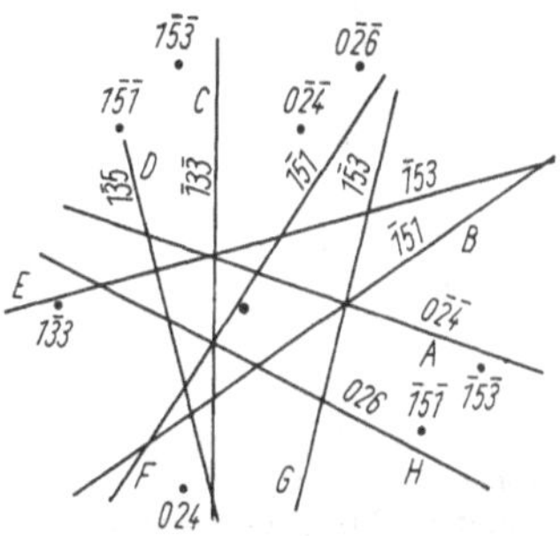

Fig. 3. a) Diffraction pattern of a thin foil of aluminium.
b) Indices of reflections and dark Kikuchi lines

b

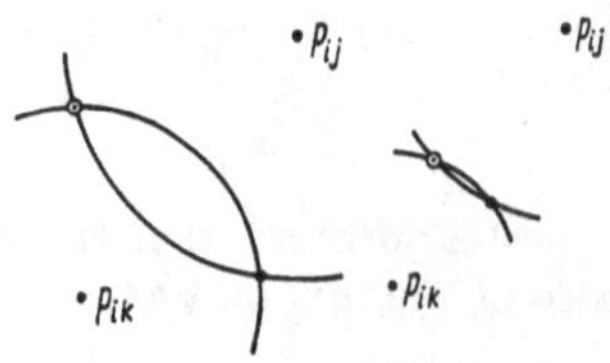

a b

Fig. 4. a) Proper choice of intersections of Kikuchi lines. b) Improper choice of intersections; small errors in distance measurements lead to a relatively great error in location of the beam direction

results are presented in Table 1. For the sake of comparison, the beam directions determined by the Ryder and Pitsch (RP) and Pumphrey and Bowkett (PB) methods are also included in Table 1. These results indicate that the inherent accuracy of the present method is better than 1′ (values of Δ_1 remain within the range of 0.4′ to 1.0′). When very small triangles of Kikuchi lines are used for calculation, the greater relative errors of distance measurements reduce the final accuracy (estimate No. 3). However, the use of excessively large triangles also increases the error (estimate No. 4), owing to distortions of different kind being always present in the SAD pattern (even before printing).

Relatively large differences (Δ_2) between the results obtained for various sets of initial data are most probably related to small deformation of the foil and the distortion of the diffraction pattern introduced by TEM and printing. The unexpectedly high value of $\Delta_2 = 9'$ in estimate No. 5 probably originates from small errors of measurements of the position of diffuse spots $\bar{1}5\bar{1}$ and $1\bar{3}3$, and of the position of the corresponding bright Kikuchi lines. Consequently, the small value of $\Delta_2 = 2'$ in estimate No. 6 seems at least partly to be accidental.

For the best accuracy in this method, pairs of points of intersection situated on a line running nearly through the centre (Fig. 4) should be avoided, because in this case the effect of the errors in distance measurements is greatest.

These considerations lead to the more general conclusion that if the orientations determined by this method or another one of similar accuracy (RP or PB) are further used (e.g., for calculation of the disorientation of grains), then the final error is due to the residual deformation of the foil and to the distortions of diffraction pattern rather than to the inherent accuracy of the method actually used for orientation determination.

Table 1

No.	Kikuchi lines used for calculations of [*hkl*]	*h*	*k*	*l*	Δ_1	Δ_2
1	ABC	−0.975599	−0.201301	0.087664	0.007° (0.4′)	0.060° (4′)
2	ABD	−0.975788	−0.200593	0.087184	0.007° (0.4′)	0.000° (0′)
3	ACF	−0.975955	−0.199350	0.088158	0.014° (0.8′)	0.095° (6′)
4	GEI	−0.975893	−0.200873	0.085346	0.016° (1′)	0.094° (6′)
5	mean value of estimates 1 to 4	−0.975809	−0.200529	0.087088	—	—
6	ACF (RP)	−0.975632	−0.201525	0.086769	—	0.148° (9′)
7	AF (PB)	−0.975900	−0.200169	0.086902	—	0.035° (2′)

Δ_1 inherent accuracy of single estimate of the beam direction (standard deviation of a mean).
Δ_2 deviation from the mean direction (estimate No. 5).

4. Summary

The present method for finding the exact direction of an electron beam in crystals provides an accuracy of an order of 1′ if the area selected for diffraction is not deformed. The accuracy does not depend on the tilting of the foil. This method, though in its second phase similar to that described by von Heimendahl [5, 6], is more accurate than the latter. In the present method no geometrical constructions are necessary for finding the position of the points used in the calculations; all what is needed is to pin-point the well defined intersections (and to find the exact position of the centre of the diffraction pattern — a common difficulty in all actually used methods). The method could be useful in studies of grain boundaries between crystals with small or special (coincidence) disorientations, in which the distribution of linear grain boundary defects (intrinsic) is extremely dependent on the mutual orientation of the grains [13, 14].

Acknowledgements

The author is grateful to Prof. M. W. Grabski and to Dr. W. Łojkowski for stimulating discussions.

References

[1] S. Kikuchi, Proc. Imp. Acad. Japan, **4**, 271 (1928).
[2] P. B. Hirsch, A. Howie, R. B. Nicholson, D. W. Pashley, and M. J. Whelan, Electron Microscopy of Thin Crystals, Butterworths, London 1965.
[3] H. M. Otte, J. Dash, and H. F. Schaake, phys. stat. sol. **5**, 527 (1964).
[4] P. L. Ryder and W. Pitsch, Phil. Mag. **18**, 807 (1968).
[5] M. von Heimendahl, W. Bell, and G. Thomas, J. appl. Phys. **35**, 3614 (1964).
[6] M. von Heimendahl, phys. stat. sol. (a) **5**, 137 (1971).
[7] S. S. Sheinin and C. D. Cann, phys. stat. sol. **11**, K1 (1965).
[8] P. H. Pumphrey and K. M. Bowkett, phys. stat. sol. (a) **2**, 339 (1970).
[9] P. H. Pumphrey and K. M. Bowkett, phys. stat. sol. (a) **3**, 375 (1970).
[10] H. Hälbig, P. L. Ryder, and W. Pitsch, Mikrochim. Acta (Suppl. **3**), 201 (1968).
[11] H. Hälbig, P. L. Ryder, and W. Pitsch, Proc. V. Internat. Congr. X-Ray Optics and Microanalysis, Tübingen 1968, Ed. G. Mollenstedt and K. H. Gaukler, Springer-Verlag, Berlin 1969 (p. 388).
[12] P. H. Pumphrey, Proc. V. Europ. Conf. Electron Microscopy, Manchester 1972, Inst. of Physics, London 1973 (p. 466).
[13] S. Amelinckx and W. Dekeyser, Adv. Solid State Phys. **8**, 327 (1959).
[14] W. Bollmann, Crystal Defects and Crystalline Interfaces, Springer-Verlag, Berlin 1970.

(Received June 21, 1977)

phys. stat. sol. (a) **43**, 541 (1977)

Subject classification: 1.4; 10.2; 22

Rijksuniversitair Centrum Antwerpen[1])

A Cluster Model for the Transition State and Its Study by Means of Electron Diffraction

III. Generalisations of the Theory and Relation to the SRO Parameters[2])

By

D. VAN DYCK, R. DE RIDDER, G. VAN TENDELOO, and S. AMELINCKX [3])

In Part I of this paper a cluster model is proposed for the explanation of the geometrical features of diffuse intensity contours observed in electron diffraction patterns from binary substitutionally disordered systems. Part II is mainly devoted to the application of the model to some particular systems. In this paper the cluster model is extended to multicomponent systems and to non-primitive lattices. It is furthermore shown that for binary systems the goodness of any particular cluster relation can be evaluated with the aid of the SRO parameters and that the predominant cluster relation can be derived from an eigenvalue problem. Applications are given.

Im ersten Teil dieser Arbeit wurde ein Clustermodell für die Erklärung der geometrischen Züge der diffusen Intensitätskonturen vorgeschlagen, die in den Elektronenbeugungsbildern binärer substitutionsfehlgeordneter Systeme beobachtet werden. Der zweite Teil behandelte die Anwendung des Modells auf einige spezielle Systeme. In dieser Arbeit wird das Clustermodell auf Vielkomponentensysteme und auf nichtprimitive Gitter ausgedehnt. Darüberhinaus wird gezeigt, daß für binäre Systeme die Qualität irgendeiner speziellen Clusterbeziehung mit Hilfe der SRO-Parameter berechnet werden kann, und daß die vorherrschende Clusterbeziehung aus einem Eigenwertproblem abgeleitet werden kann. Anwendungen werden angegeben.

1. Introduction

The transition state was previously [1] defined as the state in the ordering process of substitutionally disordered systems for which the diffuse intensity in the electron and X-ray diffraction patterns is confined to the vicinity of a periodically repeated geometrical locus in reciprocal space. As described in Part I of this paper [2] for the case of binary compounds, this transition state is considered as being composed of predominant clusters of a specific polyhedral type with the common feature that the occupation operators of their sites obey the same linear relation. The polyhedron and the cluster relation can be derived from the Fourier representation of the geometric locus.

Part II [3] was mainly devoted to the application of the method to some particular systems for which also ordering models could be proposed. In this paper, the theory is in part generalised to n-component systems and extended to non-primitive lattices. The importance of any arbitrary cluster relation can be calculated from the measured SRO parameters and the deduction of the predominant cluster relation can be formulated as an eigenvalue problem.

---

[1]) B 2020 Antwerpen, Middelheimlaan 1, Belgium.

[2]) Part I and II see phys. stat. sol. (a) **38**, 663 (1976) and **40**, 669 (1977), respectively.

[3]) Also at S.C.K., Mol, Belgium.

2. Theory

According to the kinematical theory the amplitude of electrons or X-rays diffracted by a perfect lattice occupied by atoms of different types is given by

$$a(\boldsymbol{g}) = \sum_i f_i(\boldsymbol{g})\, e^{2\pi i \boldsymbol{g}\cdot\boldsymbol{r}_i}\,, \tag{1}$$

where the summation extends over the lattice, with $f_i(\boldsymbol{g})$ representing the scattering amplitude in the direction $\boldsymbol{g}$ of the atom at position $\boldsymbol{r}_i$. In this expression, the physical information about the occupation (order) can be separated from the lattice geometry in the following manner: calling n the number of different atom species involved, and denoting their fractions and scattering amplitudes by m_α and $f^\alpha(\boldsymbol{g})$, respectively, ($\alpha = 1, \dots, n$) with

$$\sum_{\alpha=1}^{n} m_\alpha = 1 \tag{2}$$

we can define the mean atomic scattering amplitude by the weighted average:

$$\hat{f}(\boldsymbol{g}) = \sum_{\alpha=1}^{n} m_\alpha f^\alpha(\boldsymbol{g}) \tag{3}$$

so that

$$f^\alpha(\boldsymbol{g}) = \hat{f}(\boldsymbol{g}) + \Delta f^\alpha(\boldsymbol{g}) \tag{4}$$

with

$$\Delta f^\alpha(\boldsymbol{g}) = f^\alpha(\boldsymbol{g}) - \sum_{\beta=1}^{n} m_\beta f^\beta(\boldsymbol{g}) \tag{5}$$

representing the deviation of the scattering power of the α atom from the average; one then has

$$\sum_i \Delta f_i(\boldsymbol{g}) = N \sum_{\beta=1}^{n} m_\beta\, \Delta f^\beta(\boldsymbol{g}) = N\left[\sum_{\beta=1}^{n} m_\beta f^\beta(\boldsymbol{g}) - \sum_{\beta=1}^{n} m_\beta \sum_{\alpha=1}^{n} m_\alpha f^\alpha(\boldsymbol{g})\right] = 0 \tag{6}$$

(N is the total number of atoms in the crystal).

Substitution of (4) into (1) yields for the amplitude of the diffracted wave:

$$a(\boldsymbol{g}) = a_\mathrm{B}(\boldsymbol{g}) + a_\mathrm{D}(\boldsymbol{g}) \tag{7}$$

with

$$a_\mathrm{B}(\boldsymbol{g}) = \hat{f}(\boldsymbol{g}) \sum_i e^{2\pi i \boldsymbol{g}\cdot\boldsymbol{r}_i} \tag{8}$$

and

$$a_\mathrm{D}(\boldsymbol{g}) = \sum_i \Delta f_i(\boldsymbol{g})\, e^{2\pi i \boldsymbol{g}\cdot\boldsymbol{r}_i} \tag{9}$$

$a_\mathrm{B}(\boldsymbol{g})$ contains all the information about the perfect *lattice*, occupied by average atoms, in the form of delta functions at the Bragg positions. $a_\mathrm{D}(\boldsymbol{g})$ is the amplitude of the diffuse scattering, containing the information about the *ordering*. According to (6) this amplitude is zero at the Bragg positions so that for all $\boldsymbol{g}$:

$$a_\mathrm{B}(\boldsymbol{g})\, a_\mathrm{D}(\boldsymbol{g}) \equiv 0 \tag{10}$$

and hence

$$I(\boldsymbol{g}) = a(\boldsymbol{g})\, a^*(\boldsymbol{g}) = a_\mathrm{B}(\boldsymbol{g})\, a_\mathrm{B}^*(\boldsymbol{g}) + a_\mathrm{D}(\boldsymbol{g})\, a_\mathrm{D}^*(\boldsymbol{g}) = I_\mathrm{B}(\boldsymbol{g}) + I_\mathrm{D}(\boldsymbol{g})\,. \tag{11}$$

As a result of the uncoupling of the intensities, the study of the ordering process can be restricted to the investigation of the experimentally available diffuse intensity $I_\mathrm{D}(\boldsymbol{g})$.

When information is contained in the geometry of the distribution of $a_\mathrm{D}(\boldsymbol{g})$ as it is the case for the transition state [1 to 3] it is convenient to separate the $\boldsymbol{g}$ dependence of $\Delta f_i(\boldsymbol{g})$ from the sum in (9). This can be achieved by rewriting (5) by means of a vector dot product as

$$\Delta f^{\alpha}(\boldsymbol{g}) = \boldsymbol{F}(\boldsymbol{g}) \cdot \sigma^{\alpha} \tag{12}$$

with the row vector

$$\boldsymbol{F}(\boldsymbol{g}) = [f^1(\boldsymbol{g}), f^2(\boldsymbol{g}), \dots, f^n(\boldsymbol{g})] \tag{13}$$

and the occupation operator written as a column vector

$$\sigma^{\alpha} = \begin{pmatrix} 0 \\ 0 \\ \vdots \\ 0 \\ 1 \\ 0 \\ \vdots \\ 0 \end{pmatrix} - \begin{pmatrix} m_1 \\ m_2 \\ \\ \vdots \\ m_\alpha \\ \vdots \\ \\ m_n \end{pmatrix} = \begin{pmatrix} -m_1 \\ -m_2 \\ \vdots \\ -m_{\alpha-1} \\ 1 - m_\alpha \\ -m_{\alpha+1} \\ \vdots \\ -m_n \end{pmatrix}, \tag{14}$$

where

$$\sum_{j=1}^{n} (\sigma^{\alpha})_j = 0 \,. \tag{15}$$

Substitution in (9) yields for the diffuse amplitude

$$a_{\mathrm{D}}(\boldsymbol{g}) = \boldsymbol{F}(\boldsymbol{g}) \cdot \sum_i \sigma_i \, \mathrm{e}^{2\pi i \boldsymbol{g} \cdot \boldsymbol{r}_i} \,. \tag{16}$$

The expression for the diffuse amplitude is finally separated into a factor containing only the information on the nature of the atoms and a Fourier series, containing the information about the occupation of the lattice sites. Notice also the duality between $a_{\mathrm{B}}(\boldsymbol{g})$ and $a_{\mathrm{D}}(\boldsymbol{g})$: $a_{\mathrm{B}}(\boldsymbol{g})$ contains the information of a periodic cell in direct space, in the form of delta functions in reciprocal space; $a_{\mathrm{D}}(\boldsymbol{g})$ contains the information of the occupation operators (delta functions) in direct space, in the form of a periodic function in reciprocal space. The occurrence of periodicity in the reduced diffuse scattering is thus a direct consequence of the substitutional character of the disorder.

3. Geometrical Locus of Diffuse Intensity

3.1 General expression

Like in binary compounds [2] the equation for the geometrical locus of diffuse intensity, which has the periodicity of the reciprocal unit cell, takes the form

$$f(\boldsymbol{g}) = \sum_k \omega_k \, \mathrm{e}^{-2\pi i \boldsymbol{g} \cdot \boldsymbol{r}_k} = 0 \,. \tag{17}$$

The assumption that the diffuse intensity to be located exactly on this surface this can be expressed mathematically as

$$a_{\mathrm{D}}(\boldsymbol{g}) \, f(\boldsymbol{g}) \equiv 0 \tag{18}$$

or using (16) and (17)

$$\boldsymbol{F}(\boldsymbol{g}) \cdot \sum_i \sigma_i \, \mathrm{e}^{2\pi i \boldsymbol{g} \cdot \boldsymbol{r}_i} \sum_k \omega_k \, \mathrm{e}^{-2\pi i \boldsymbol{g} \cdot \boldsymbol{r}_k} \equiv 0 \tag{19}$$

so that

$$\boldsymbol{F}(\boldsymbol{g}) \cdot \sum_{i,k} \sigma_i \omega_k \, \mathrm{e}^{2\pi i \boldsymbol{g} \cdot (\boldsymbol{r}_i - \boldsymbol{r}_k)} \equiv 0 \,. \tag{20}$$

Three cases have to be distinguished, depending on whether or not the unit cell contains internal lattice points.

a) *Primitive lattice*: If the lattice can be described as primitive (eg. f.c.c., b.c.c., ..) $\boldsymbol{r}_i - \boldsymbol{r}_k = \boldsymbol{r}_j$ is again a lattice vector, and hence

$$F(\boldsymbol{g}) \cdot \sum_j \left(\sum_k \sigma_{j+k}\omega_k\right) e^{2\pi i \boldsymbol{g}\cdot\boldsymbol{r}_j} \equiv 0 . \tag{21}$$

For each coefficient in the Fourier series we thus have

$$\sum_k \omega_k \sigma_{j+k} = 0 \qquad \text{(for all } j\text{)} \tag{22}$$

or in matrix notation

$$[\sigma]_j\, \omega = 0 \tag{23}$$

with $[\sigma]_j$ the matrix with column vectors σ_{j+k} and ω the column vector with components ω_k.

Like in the case of binary compounds, the geometrical locus of diffuse intensity allows to deduce a relation of the type (22) for the occupation operators σ; this is the so called "cluster-relation" which has to be satisfied in each lattice point j.

According to (14) and (15) this cluster relation consists of a system of n linear homogeneous equations of order $n - 1$. Notice also the bijective correspondence between the linear relations for the occupation operators in direct space and the periodic surfaces in reciprocal space.

b) *Primitive lattice with omissions*: If a primitive lattice can be generated by adding a limited number of missing points per unit cell to the original lattice (e.g. h.c.p.) then: $\boldsymbol{r}_i - \boldsymbol{r}_k = \boldsymbol{r}_j$ is again a lattice vector.

If the occupation operators σ_j are taken zero in these new positions, relation (15) and the cluster-relation (22) still holds.

c) *General lattice*: In case the non-primitive unit cell contains more points in general positions $\boldsymbol{R}_1, \boldsymbol{R}_2, \dots, \boldsymbol{R}_m$, the crystal can be considered as consisting of m identical sublattices, related by translations over vectors $\boldsymbol{R}_j$ and the amplitude of the diffuse scattering can be written from (16),

$$a_\mathrm{D}(\boldsymbol{g}) = F(\boldsymbol{g}) \cdot \sum_i \left[\sum_{j=1}^{m} \sigma_{\boldsymbol{r}_i+\boldsymbol{R}_j}\, e^{2\pi i \boldsymbol{g}\cdot\boldsymbol{R}_j}\right] e^{2\pi i \boldsymbol{g}\cdot\boldsymbol{r}_i}$$

and hence, the cluster relation (22) becomes

$$\sum_{j=1}^{m} e^{2\pi i \boldsymbol{g}\cdot\boldsymbol{R}_j} \left[\sum_k \omega_{\boldsymbol{r}_k} \sigma_{\boldsymbol{r}_i+\boldsymbol{r}_k+\boldsymbol{R}_j}\right] = 0 \tag{24}$$

for all $\boldsymbol{r}_i$.

A sufficient condition for (24) to be satisfied is that the cluster relation (22) is obeyed in each of the m sublattices. The application of this result to the description of the ordering of interstitial atom strings in a complex oxide is published elsewhere [4].

3.2 Information from planar sections in reciprocal space

In 100 keV electron diffraction experiments, where the Ewald sphere is nearly planar, diffraction patterns represent planar sections through the origin of reciprocal space. Although three-dimensional reconstruction of the diffuse locus is possible by tilting experiments, information can already be obtained from a single basic section. Consider as an example, the intersection of the periodic intensity locus with the (001) plane, which is a periodic curve represented by the two-dimensional Fourier series:

$$f(h, k) = \sum_l \omega_l\, e^{-2\pi i(hx_l + ky_l)} = 0 . \tag{25}$$

From (16) one has for the amplitude of the diffracted wave

$$a(h, k, 0) = F(\boldsymbol{g}) \cdot \sum_i \sigma_i \, e^{2\pi i(hx_i + ky_i)} \tag{26}$$

and since

$$a(h, k, 0) \, f(h, k) \equiv 0 \, .$$

One obtains similarly (for all j)

$$\sum_l \omega_l \sigma^{\mathrm{P}}_{j+l} = 0 \, , \tag{27}$$

where σ^{P} is the projected occupation operator, given by

$$\sigma^{\mathrm{P}}_{x_i, y_i} = \sum_{z_i} \sigma_{x_i, y_i, z_i} \, . \tag{28}$$

From the intensity curve in a basic planar section of reciprocal space one can thus unambiguously deduce the cluster relation for the *projected occupation operators.*

3.3 Intensity distribution along the geometrical locus

If all possible clusters obeying the same relation occur at random in the system, it is believed that the intensity should be distributed uniformly along the intensity surface. However, it may happen that not all of these clusters are present; this could give rise to a concentration of the diffuse intensity on parts of the locus. This argument was used to explain the honeycomb-networks of diffuse intensity occurring in some vacancy disordered two-dimensional hexagonal lattices (see [3] Section 6).

If during the ordering process, a second independent cluster relation becomes important, the intensity becomes concentrated near the intersection of the two corresponding loci. The occurrence of a third dominant independent cluster relation causes the intensity to concentrate into points of reciprocal space; i.e. in the LRO superstructure reflections. It should be noted that even for very similar systems with the same LRO structure, the sequence of appearance for these dominant clusters in the ordering process can be reversed, as clearly demonstrated in [1] for the alloys Ni_4Mo and Au_4V.

A necessary and sufficient condition for the LRO superstructure reflections to be situated on the diffuse intensity locus, is that the corresponding cluster relation is satisfied for the LRO structure. When this structure is known, the number of possible cluster relations is limited.

On the other hand, all the cluster relations corresponding to surfaces through the LRO superlattice reflections have to be fulfilled in the LRO structure.

Finally it should be mentioned that the theory, up to this point, is valid beyond the kinematical approximation since the periodic form of a cluster locus is not influenced by multiple Bragg scattering.

4. Relation between the Cluster Description and the SRO Parameters

4.1 General expressions for a binary system

Only in the case of a system containing two types of atoms A and B with respective fractions m_{A} and m_{B}, the vector expression (16) for the diffuse amplitude can be reduced to a scalar expression. With the aid of (15) one can write

$$a_{\mathrm{D}}(\boldsymbol{g}) = [f_{\mathrm{A}}(\boldsymbol{g}) - f_{\mathrm{B}}(\boldsymbol{g})] \sum_i \sigma_i \, e^{2\pi i \boldsymbol{g} \cdot \boldsymbol{r}_i} \, , \tag{29}$$

where the σ_i are the well known Flinn operators, defined by

$$\sigma_i = \quad m_{\mathrm{B}} \text{ for A at position } \boldsymbol{r}_i\,,$$
$$\sigma_i = -m_{\mathrm{A}} \text{ for B at position } \boldsymbol{r}_i$$

and hence

$$I_{\mathrm{D}}(\boldsymbol{g}) = [f_{\mathrm{A}}(\boldsymbol{g}) - f_{\mathrm{B}}(\boldsymbol{g})]^2 \sum_{i,k} \sigma_i\sigma_k\, \mathrm{e}^{2\pi i \boldsymbol{g}(\boldsymbol{r}_k - \boldsymbol{r}_i)} \tag{30}$$

$$= N m_{\mathrm{A}} m_{\mathrm{B}} [f_{\mathrm{A}}(\boldsymbol{g}) - f_{\mathrm{B}}(\boldsymbol{g})]^2 \sum_j \alpha_j\, \mathrm{e}^{2\pi i \boldsymbol{g}\cdot\boldsymbol{r}_j}\,, \tag{31}$$

where α_j are the Warren-Cowley SRO parameters given by

$$\alpha_j = \frac{1}{m_{\mathrm{A}} m_{\mathrm{B}}} \frac{1}{N} \sum_i \sigma_i \sigma_{i+j} = \frac{\langle \sigma_i \sigma_{i+j} \rangle}{m_{\mathrm{A}} m_{\mathrm{B}}} \tag{32}$$

with

$$\alpha_0 = 1\,. \tag{33}$$

These real symmetric SRO parameters can be determined from intensity measurements by numerical Fourier inversion of the reduced diffuse intensity (31)

$$\alpha_j = \frac{1}{N m_{\mathrm{A}} m_{\mathrm{B}} V^*} \int\limits_{V^*} \frac{I_{\mathrm{D}}(\boldsymbol{g})\, \mathrm{e}^{-2\pi i \boldsymbol{g}\cdot\boldsymbol{r}_j}}{[f_{\mathrm{A}}(\boldsymbol{g}) - f_{\mathrm{B}}(\boldsymbol{g})]^2}\, \mathrm{d}\boldsymbol{g}\,, \tag{34}$$

where V^* is the volume of the reciprocal unit cell.

4.2 Relative goodness of an arbitrary cluster relation: The normalised variance

Let us consider an arbitrary cluster relation with corresponding locus

$$f(\boldsymbol{g}) = \sum_k \omega_k\, \mathrm{e}^{-2\pi i \boldsymbol{g}\cdot\boldsymbol{r}_k} = 0 \tag{35}$$

so that using (29) we can write

$$a_{\mathrm{D}}(\boldsymbol{g})\, f(\boldsymbol{g}) = [f_{\mathrm{A}}(\boldsymbol{g}) - f_{\mathrm{B}}(\boldsymbol{g})] \sum_{i,k} \omega_k \sigma_i\, \mathrm{e}^{2\pi i \boldsymbol{g}(\boldsymbol{r}_i - \boldsymbol{r}_k)} = [f_{\mathrm{A}}(\boldsymbol{g}) - f_{\mathrm{B}}(\boldsymbol{g})] \sum_j \varepsilon_j\, \mathrm{e}^{2\pi i \boldsymbol{g}\cdot\boldsymbol{r}_j} \tag{36}$$

where ε_j is the cluster residue at the site $\boldsymbol{r}_j$

$$\varepsilon_j = \sum_k \omega_k \sigma_{j+k}\,. \tag{37}$$

In case the intensity is exactly confined to the locus, the cluster relation is satisfied everywhere so that all ε_j are equal to zero; if on the other hand, the cluster relation is violated, a measure for the deviation is given by the variance

$$\langle|\varepsilon|^2\rangle = \frac{1}{N} \sum_i |\varepsilon_i|^2 = \frac{1}{N} \sum_{i,j} \varepsilon_j \varepsilon_i^* \delta_{i,j} =$$

$$= \frac{1}{N} \sum_{i,j} \varepsilon_j \varepsilon_i^* \frac{1}{V^*} \int\limits_{V^*} \mathrm{e}^{2\pi i \boldsymbol{g}\cdot(\boldsymbol{r}_j - \boldsymbol{r}_i)}\, \mathrm{d}\boldsymbol{g} =$$

$$= \frac{1}{N V^*} \int\limits_{V^*} \frac{I_{\mathrm{D}}(\boldsymbol{g})\, |f(\boldsymbol{g})|^2}{[f_{\mathrm{A}}(\boldsymbol{g}) - f_{\mathrm{B}}(\boldsymbol{g})]^2}\, \mathrm{d}\boldsymbol{g}\,. \tag{38}$$

Equation (38) expresses the principle that the better a certain cluster relation is satisfied in the ordering system, the more the diffuse intensity is confined to regions where $|f(\boldsymbol{g})|^2$ is small.

A more convenient parameter however is the normalised variance

$$S^2 = \frac{\langle|\varepsilon|^2\rangle}{\langle|\varepsilon_0|^2\rangle} \tag{39}$$

$\langle|\varepsilon_0|^2\rangle$ being the variance of the residue distribution for the same relation in the completely disordered system.

4.3 Relation of the normalised variance to the SRO parameters

Substitution of (31) and (35) in (38) gives

$$\langle|\varepsilon|^2\rangle = \frac{1}{NV^*} N m_A m_B \sum_{i,j,k} \alpha_j \omega_i \omega_k^* \int_{V^*} e^{2\pi i \boldsymbol{g}\cdot(\boldsymbol{r}_j - \boldsymbol{r}_i + \boldsymbol{r}_k)}\, d\boldsymbol{g} = m_A m_B \sum_{i,k} \omega_i \alpha_{i-k} \omega_k^* \tag{40}$$

or in matrix notation

$$\langle|\varepsilon|^2\rangle = m_A m_B \tilde{\omega}^* \alpha \omega = m_A m_B \langle\omega|\ \alpha\ |\omega\rangle \tag{41}$$

with

$$\omega = \begin{pmatrix} \omega_0 \\ \omega_1 \\ \vdots \end{pmatrix}, \tag{42}$$

hereafter called the "cluster" vector and

$$\alpha_{i,j} = \alpha_{(i-j)}\,. \tag{43}$$

In the case of complete disorder, one has

$$\alpha_{i,j} = \delta_{i,j} \tag{44}$$

so that α is the unit matrix and hence

$$\langle|\varepsilon_0|^2\rangle = m_A m_B \tilde{\omega}^* \omega = m_A m_B \langle\omega\mid\omega\rangle \tag{45}$$

which expresses Parseval's theorem. For the normalised variance (39), one finally deduces from (41) and (45)

$$S^2 = \frac{\langle\omega|\ \alpha\ |\omega\rangle}{\langle\omega\mid\omega\rangle}. \tag{46}$$

The knowledge of the α parameters, obtained from diffuse intensity measurements allows one to determine the "goodness" of any particular cluster relation in any state of the ordering process by means of the normalised variance S^2.

It follows from (38) and (39) that the normalised variance is a positive quantity with $S^2 = 0$ corresponding to a relation being exactly satisfied in each lattice point, $S^2 = 1$ corresponding to the case of complete disorder. Notice the existence of cluster relations for which $S^2 > 1$, corresponding to clusters whose frequency of occurrence in the system is less than the statistical one.

4.4 Derivation of the predominant cluster relation

The predominant cluster relation can be defined as the one yielding the smallest normalised variance for a given system. Its derivation can be reduced to finding the minimum for

$$S^2 = \frac{\langle\omega|\ \alpha\ |\omega\rangle}{\langle\omega\mid\omega\rangle}. \tag{47}$$

This problem is very similar to the well-known problem in quantum mechanics of finding the ground state energy level and the corresponding wave function for a given Hamiltonian. It can be solved by means of a variational principle resulting in an eigenvalue/eigenvector equation similar to the Schrödinger equation:

$$\alpha\,|\omega\rangle = S^2\,|\omega\rangle\,. \tag{48}$$

The predominant cluster vector $|\omega\rangle$ is thus the eigenvector corresponding to the lowest eigenvalue; this eigenvalue is the normalised variance S^2 of the cluster relation.

4.5 Properties of the α-operator

a) Since α is a real symmetric matrix its eigenvalues and eigenvectors are real. As a result only real cluster relations have to be considered, consistent with their physical meaning.

b) From (38) it follows that $\langle\omega|\,\alpha\,|\omega\rangle$ is a non-negative quadratic form; all the eigenvalues are therefore non-negative. This property can be used as a test for the consistency of the measured SRO parameters.

c) Experimentally obtained values for α_0 differ in general from unity. Using "corrected" parameters $\alpha_i' = \alpha_i/\alpha_0$ would give the same eigenvector but with a new normalised variance $S'^2 = S^2/\alpha_0$. A substitution of α_0 by 1 on the other hand would also leave the eigenvectors unchanged but would replace S^2 by $S^2 + 1 - \alpha_0$.

d) The sum of the eigenvalues equals the trace of the matrix so that

$$\sum_i^n S_i^2 = \sum_i^n \alpha_{ii} = n \qquad \text{the dimension of } \alpha$$

and hence

$$\frac{1}{n}\sum_i^n S_i^2 = 1 \tag{49}$$

the average of the eigenvalues is one (the normalised variance for a cluster relation in a totally disordered system).

e) The product of the eigenvalues is equal to the determinant of the matrix

$$\prod_{i=1}^n S_i^2 = \det \alpha\,. \tag{50}$$

A necessary and sufficient condition for a cluster relation to be satisfied is $S^2 = 0$ and hence $\det \alpha = 0$.

f) In principle the dimensions of the α matrix and of the cluster vectors are infinite. However, if only a finite number r of cluster points is considered, the cluster vector can be written as

$$|\omega\rangle = \begin{pmatrix} \omega \\ \hline 0 \end{pmatrix} \tag{51}$$

and (46) becomes

$$S^2 = \frac{\widetilde{(\omega \,\vdots\, 0)} \left(\begin{array}{c|c} \alpha_r & \\ \hline & \end{array}\right) \begin{pmatrix} \omega \\ \hline 0 \end{pmatrix}}{\widetilde{(\omega \,\vdots\, 0)} \begin{pmatrix} \omega \\ \hline 0 \end{pmatrix}} \frac{\tilde{\omega}\alpha_r\omega}{\tilde{\omega}\omega}. \tag{52}$$

The matrix α may thus be replaced by its leading principal submatrix of order r without changing the eigenvalues and eigenvectors of the form (51). Since all leading principal minors of a non-negative definite quadratic form are non-negative definite the foregoing properties remain valid.

g) A particular r-point cluster can be characterised by the vector consisting of its occupation operators

$$\begin{pmatrix} \sigma_0 \\ \sigma_1 \\ \vdots \\ \sigma_r \end{pmatrix} \tag{53}$$

calling $|a_i\rangle$ ($i = 1, \ldots, 2^r$) all such possible r point clusters and n_i their respective fractions, the residue of the cluster $|a_i\rangle$ for an arbitrary relation $|\omega\rangle$ is given from (37) by

$$\varepsilon_i = \langle \omega \mid a_i \rangle \tag{54}$$

and hence

$$N m_A m_B = \langle \omega | \alpha | \omega \rangle \sum_i |\varepsilon_i|^2 = \sum_i n_i \langle \omega \mid a_i \rangle^2 = \sum_i n_i \langle \omega \mid a_i \rangle \langle a_i \mid \omega \rangle$$

and thus

$$N m_A m_B \alpha = \sum_i n_i |a_i\rangle \langle a_i| \tag{55}$$

which is consistent with (32).

h) It should be noted that arbitrarily small values for S^2 can be obtained by increasing the number of points in the eigenvalue calculation. However, the eigenvectors thus obtained correspond to increasingly complicated loci that contain the diffuse intensity surface but do not allow any simple physical interpretation. For this reason, preference should be given to simple cluster relations with relatively low variance.

4.6 Degeneracy

If due to the translation symmetry of the lattice and the symmetry of the diffuse intensity distribution, the lattice points used in the calculation contain m equivalent predominant clusters, the degeneracy will be m-fold and the eigenvectors will form an m-dimensional vector space of predominant cluster vectors. The difficulty can be avoided by starting the eigenvalue calculation with a limited number of points that can be gradually increased.

In some symmetric cases the symmetry of the diffuse intensity can be exploited. Suppose the points chosen in the calculations exhibit a point symmetry operation R of the diffuse intensity such that

$$R\alpha = \alpha R \tag{56}$$

and from (48)

$$\alpha |\omega\rangle = S^2 |\omega\rangle \,. \tag{57}$$

In case the eigenvector $|\omega\rangle$ is non-degenerate, it follows from the commutation of α and R that $|\omega\rangle$ is also an eigenvector of R, i.e.

$$R |\omega\rangle = m |\omega\rangle \,. \tag{58}$$

If R is a permutation of the n-th order i.e. if

$$R^n = 1$$

one has

$$m^n = 1$$

and hence

$$m = \pm 1$$

leading to

$$R |\omega\rangle = \pm |\omega\rangle \,. \tag{59}$$

The cluster vector is thus symmetric or antisymmetric with respect to the permutation R of its elements, which reduces the number of different coefficients ω.

5. Application to Some Particular Systems

5.1 Discussion of the experimental methods

Only very few reliable experimental SRO parameter values for systems clearly exhibiting a transition state are available from the literature. SRO parameters can in principle be obtained by measuring the reduced diffuse intensity in a large number of points, closely sampling the reciprocal unit cell, and numerically calculating the Fourier transform according to (34). Since the kinematical theory has to be applicable only X-ray or neutron diffraction measurements can be used.

In electron diffraction the more crude assumption is made that the diffuse intensity has constant values at parts of a geometric locus with a given width, and is zero elsewhere. Values for the SRO parameters can then likewise be obtained from numerical Fourier integration. The reliability of this method can be questioned since electron diffraction has to be described by means of the dynamical theory and moreover visual estimation of intensities from photographic plates leads to approximate values only.

5.2 Directly measured SRO parameters

5.2.1 Vacancy ordering in VC_x $(0.70 \leqq x \leqq 0.89)$

As described in Part II of this paper, VC_x has a defective sodium chloride structure, one of the f.c.c. sublattices being occupied by metal atoms and the other by carbon atoms and vacancies.

From careful measurements of neutron diffraction intensities, Sauvage et al. [5] were able to obtain reliable results for the SRO parameters. Using these values an eigenvalue calculation with 14 points, yielded for the predominant cluster relation

$$\sum_{\text{octahedron}} \sigma_i - 0.9 \sum_{\text{cube}} \sigma_i = 0 \tag{60}$$

in agreement with the electron diffraction results, described in Part II.

For the normalised variance we obtained a slightly negative value, but still equal to zero within the experimental error,

$$S^2 = -0.015 \, .$$

As a test the method was also applied to the LRO parameters of V_6C_5, yielding the expected cluster relation

$$\sum_{\text{octahedron}} \sigma_i = 0 \tag{61}$$

with $S^2 = 0$.

5.2.2 Substitutional ordering in Au_4V

The Au_4V structure consists of an f.c.c. lattice occupied by Au and V atoms; it shows transition state behaviour in the annealing stage [1] Fürnrohr [6] measured a number of SRO parameters from X-ray diffraction intensities, for different annealing times. We used these values in a 14-point eigenvalue calculation. For all annealing times we find the same predominant cluster relation

$$\sigma_{000} + \sigma_{220} + \sigma_{202} + \sigma_{022} = \sigma_{200} + \sigma_{020} + \sigma_{002} + \sigma_{222} \, , \tag{62}$$

which expresses the fact that the composition of the tetrahedral cluster consisting of the cube points (000, 220, 202, 022) is equal to that of the tetrahedral cluster (200, 020, 002, 222). Note that this relation is independent of composition.

For the normalised variance as a function of annealing time, we obtained

annealing time at 400 °C	S^2
0	0.504
10 min	0.245
20 min	0.034
∞	−0.001

In accordance with expectation the value of S^2 decreases with increasing annealing time, showing that the cluster relation becomes increasingly better obeyed.

5.2.3 *Ionic ordering in lithium ferrite: $LiFeO_2$*

This compound has the NaCl structure. As described in Part II, it consists of two f.c.c. sublattices, one is occupied by O^{2-} ions and the other one by a disordered mixture of Li^+ and Fe^{3+} ions. Only few SRO parameters were obtained from X-ray diffraction measurements by Brunel et al. [7]. As a result our eigenvalue calculation had to be restricted to six points, revealing the cluster relation

$$\sum_{\text{octahedron}} \sigma_i = 0 \tag{63}$$

with

$$S^2 = -0.17 \text{ (quenched from 742 °C)},$$
$$S^2 = 0.13 \text{ (quenched from 742 °C) for another specimen},$$
$$S^2 = 0.21 \text{ (quenched from 857 °C)},$$
$$S^2 = 0.01 \text{ LRO}.$$

Taking into account that an extension of the cluster relation was impossible due to the limited set of SRO parameters, relation (63) can be considered as in agreement with previous results [3]. Note that S^2 is larger for the higher temperature, expressing the fact that at the higher temperature the cluster relation is less well obeyed as one would expect.

5.3 Geometrically estimated SRO parameters

5.3.1 *Vacancy ordering in VC_x* $(0.70 \leqq x \leqq 0.89)$

In an earlier paper, describing the ordering in VC_x, Sauvage et al. [8] calculated the SRO parameters using the geometrical technique described in Section 5.1.

SRO parameters were obtained under two different assumptions.

a) The geometrical locus has a constant width. For the SRO parameters thus obtained, a 14-point eigenvalue calculation yielded the cluster relation

$$\sum_{\text{octahedron}} \sigma_i + 0.13 \sum_{\text{cube}} \sigma_i = 0 \tag{64}$$

with $S^2 = 0.027$.

b) The geometrical locus has zero width. In this case the cluster relation takes the form

$$\sum_{\text{octahedron}} \sigma_i = 0$$

with $S^2 = -0.001$.

5.3.2 *Vacancy ordering in illite*

Illite is a layer-silicate with partial substitution of the Si^{4+} ions by Al^{3+} ions, Ba^{2+} ions between adjacent silicate sandwiches compensate the electrical charge deficiency. These Ba^{2+} ions are assumed to occupy interstices in the vicinity of the missing charges giving rise to a two-dimensional hexagonal metal-vacancy ordering system. SRO parameter calculations using the geometrical technique were performed by Besson et al. [9]. With these values a 19-point eigenvalue calculation could be carried out, yielding the predominant cluster relation

$$\sigma_{21} + \sigma_{\bar{1}\bar{2}} + \sigma_{\bar{1}1} = \sigma_{\bar{2}\bar{1}} + \sigma_{1\bar{2}} + \sigma_{1\bar{1}}$$

with $S^2 = 0.026$ which expresses the fact that the composition of the cluster (2, 1) ($\bar{1}$, $\bar{2}$) ($\bar{1}$, 1) is equal to that of ($\bar{2}$, $\bar{1}$), (1, $\bar{2}$) (1, $\bar{1}$) in agreement with the results of Part II.

6. Conclusion

The cluster model for the transition state in substitutionally disordered binary systems, revealed by the occurrence of a periodic diffuse intensity locus in reciprocal space and described in terms of relations between the occupation parameters of cluster sites, can be extended to multicomponent systems and to non-primitive lattices. In the case of a binary system a method could be developed for the evaluation of the "goodness" of any particular cluster relation with the aid of the SRO parameters. The predominant cluster relation can be obtained as the eigenvector corresponding to the smallest eigenvalue of the matrix consisting of the SRO parameters. This eigenvalue is the normalised variance of the cluster relation measuring the predominance of the corresponding cluster in the system. The technique is applied to some particular ordering systems for which SRO parameters are available in the literature, such as VC_x ($0.70 \leq x \leq 0.89$), $LiFeO_2$, Au_4V and illite where the derived cluster relations are in agreement with previous results.

It is found that S^2, which measures how well the cluster relation is satisfied, increases with increasing temperature and decreases with increasing annealing time at a given temperature. This behaviour is in accordance with expectations if S^2 is to describe a state of partial order.

References

[1] R. de Ridder, G. van Tendeloo, and S. Amelinckx, Acta cryst. **A32**, 216 (1976).
[2] R. de Ridder, G. van Tendeloo, D. van Dyck, and S. Amelinckx, phys. stat. sol. (a) **38**, 663 (1977).
[3] R. de Ridder, D. van Dyck, G. van Tendeloo, and S. Amelinckx, phys. stat. sol. (a) **40**, 669 (1977).
[4] R. de Ridder, G. van Tendeloo, D. van Dyck, and S. Amelinckx, phys. stat. sol. (a) **41**, 555 (1977).
[5] M. Sauvage, E. Parthé, and W. B. Yellon, Acta cryst. **A30**, 597 (1974).
[6] Fürnrohr, Ph. D. Thesis, Stuttgart 1976.
[7] M. Brunel and F. De Bergevin, J. Phys. Chem. Solids **30**, 2011 (1969).
[8] M. Sauvage and E. Parthé, Acta cryst. **A28**, 607 (1972).
[9] G. Besson, C. Thoubar, and J. Méring, J. appl. Cryst. **7**, 345 (1974).

(Received August 2, 1977)

phys. stat. sol. (a) **43**, 553 (1977)

Subject classification: 1.3 and 1.4; 21.1; 21.6

Rijksuniversitair Centrum Antwerpen[1])

Long-Period Antiphase Boundary Structures in the Ordered Alloy Au_4Zn [2])

I. One-Dimensional Long-Period Antiphase Boundary Structure

By

G. van Tendeloo and S. Amelinckx[3])

The one-dimensional long-period superstructure in Au_4Zn is studied using both electron microscopy and electron diffraction. Direct lattice imaging of the periodic or pseudo periodic antiphase boundaries gives a straightforward interpretation for deviations from the generally accepted $M = 2$ periodicity of the antiphase boundaries. Spacing and orientation anomalies in the diffraction pattern are explained by changes in the periodicity of the fault planes and jogging of the antiphase boundaries. This model is confirmed by optical simulation.

Mit Elektronenmikroskopie und Elektronenbeugung wird die eindimensionale langperiodische Superstruktur in Au_4Zn untersucht. Direkte Gitterabbildung der periodischen oder pseudoperiodischen Antiphasengrenzen ergibt eine direkte Interpretation für Abweichungen von der allgemein angenommenen $M = 2$ Periodizität der Antiphasengrenzen. Die Abstands- und Orientierungsanomalien im Beugungsbild werden mit Änderungen in der Periodizität der Fehlerebene und Sprüngen der Antiphasengrenzen erklärt. Dieses Modell wird durch optische Simulation bestätigt.

1. Introduction

Long-period ordered alloys have been the subject of intense study for several years. The use of present day high resolution electron microscopy opens the possibility for a more detailed interpretation of the diffraction patterns than has been possible hitherto. In particular it has now become feasible to correlate unambiguously sequences of antiphase boundaries observed in the image with pecularities of the electron diffraction pattern. In a series of papers on the long-period antiphase boundary (LPAPB) structures in Au_4Zn we will try to combine the structural knowledge derived from X-ray diffraction [1 to 5] with the electron microscopic observations and selected area diffraction patterns of bulk Au_4Zn.

Since in particular cases the correct interpretation of diffraction patterns with non-integer values of M is still somewhat in doubt, we shall show in this first paper that in Au_4Zn it has been possible to interpret in detail the diffraction pattern in terms of a model, based on the direct observation of sequences of antiphase boundaries.

In a second paper the growth mechanism of the long-period superstructures will be treated. It will be shown that the two-dimensional APB structure is generated by the propagation of "hairpins" of antiphase boundaries across a one-dimensional sequence of APB's.

A subsequent paper will deal with the particular behaviour of the alloy under irradiation with high-energy electrons of the HVEM. Two-dimensional areas do not

---

[1]) Middelheimlaan 1, B-2020 Antwerpen, Belgium.

[2]) Work performed under the auspices of the association R.U.C.A.–S.C.K.

[3]) Also at S.C.K./C.E.N., Boeretang, B-2400 Mol, Belgium.

suffer any visible radiation damage, whilst one-dimensional areas are damaged at electron energies exceeding 450 kV.

2. Theoretical Considerations

2.1 *Diffraction by a perfectly periodic array of antiphase boundaries or shear planes*

The kinematical theory for the diffraction of electrons by a crystal containing a perfectly periodic sequence of parallel antiphase boundaries or shear planes has been presented in [6]. It was shown that the diffraction pattern exhibits sequences of equidistant superlattice spots, with an interspot distance corresponding with the period $2M$ (in f.c.c. unit cells) of the antiphase boundary structure. The position, with respect to the spot $\boldsymbol{g}$, of the sequence of superlattice spots associated with $\boldsymbol{g}$ is determined by the value of $\boldsymbol{g} \cdot \boldsymbol{R}$, where $\boldsymbol{R}$ is the displacement vector of the APB's or shear planes. The basic spot $\boldsymbol{g}$ divides the distance between two successive superlattice spots in the ratio $\boldsymbol{g} \cdot \boldsymbol{R}/(1 - \boldsymbol{g} \cdot \boldsymbol{R})$.

This feature has been used to determine $\boldsymbol{R}$ from the diffraction pattern of shear structures [6] or alloys containing periodic, non-conservative, antiphase boundaries [7]. In the present study the method is used to determine the displacement vector of the APB's giving rise to the long-period superstructure.

2.2 *Diffraction by a uniformely mixed periodic arrangement*

The kinematical theory for the diffraction by a crystal containing uniformely mixed sequences of antiphase boundaries has been developed in the one-dimensional case by Fujiwara [8]. The results which are of application here, can be summarized as follows. Let M be the average width of a domain strip expressed in a number of f.c.c. unit cells. The number M need not to be an integer; however, if it is not the sequence must contain domain strips of different width of which M is the average. Fujiwara studied the case of uniform mixing, which he defines by means of a periodic step function, with period $2M$ (Fig. 1). The unit cells are numbered and represented by points; depending on the interval of the step function in which the point representing a given unit cell comes, this unit cell will be displaced or not. This model ensures that in each arbitrary chosen interval the average is approximated in an optimal manner. Fig. 2 shows an actual case with $M = 2.25$ which is of interest here; the sequence can be described by means of the symbol $|2\,\bar{2}\,2\,\bar{3}\,|\,2\,\bar{2}\,2\,\bar{3}|$ ... The real period contains 9 unit cells, the average period, however, is $2M = 4.5$.

It was shown by Fujiwara that for such a one-dimensional sequence along one of the cube directions, e.g. $\boldsymbol{a}_3$, the diffraction pattern exhibits maxima at $l = n \pm$ $\pm (2m + 1)/2M$ (n and m integers). Large maxima are formed at $l = n \pm 1/2M$ and smaller secondary maxima at $l = n \pm 3/2M$; as calculated by Fujiwara [8] the intensity of successive maxima decreases rapidly. The fractional displacement of the

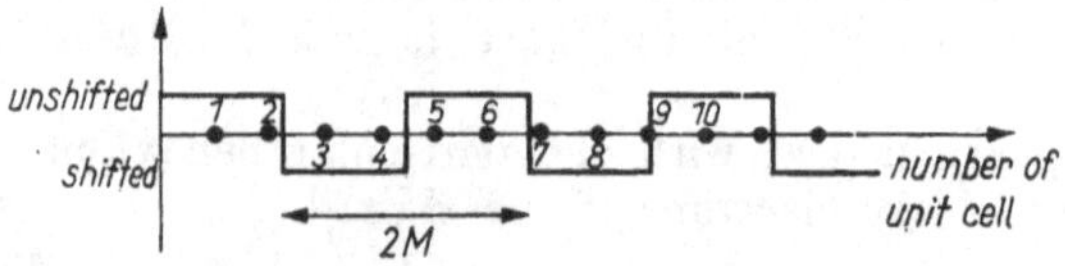

Fig. 1. Schematic representation of a periodic step function with period $2M$ along the [001] row. Successive points along this row represent f.c.c. unit cells: depending on the interval the corresponding unit cell is shifted or not (after Fujiwara [8])

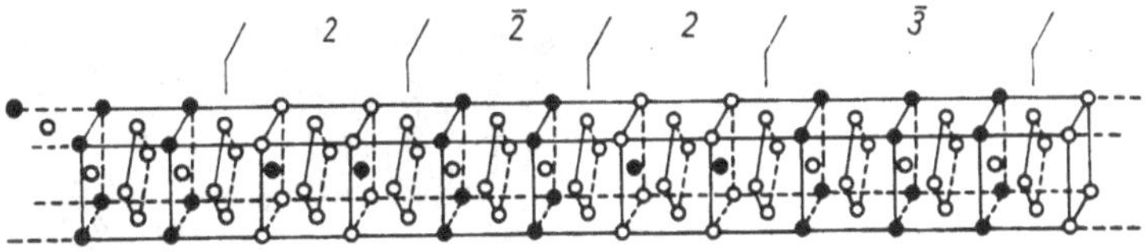

Fig. 2. Actual model of a long-period antiphase boundary structure with $M = 2.25$. The model can be described as $2\,\bar{2}\,2\,\bar{3}$. The block diagram of this structure is shown in Fig. 1

sequence of long-period superlattice spots with respect to the basic spots is still given by the ratio $\boldsymbol{g} \cdot \boldsymbol{R}/(1 - \boldsymbol{g} \cdot \boldsymbol{R})$.

The displacement of the type $\boldsymbol{R} = 1/2[110]$ gives rise to a phase factor $e^{i\alpha}$ with $\alpha = 2\pi \boldsymbol{g} \cdot \boldsymbol{R} = \pi(h + k)$ in the structure amplitude. If $h + k =$ even no phase shift will occur; such reflections will remain unaffected, but for reflections such that $h + k =$ odd, $\boldsymbol{g} \cdot \boldsymbol{R}$ is an integer + one half; the basic spots will apparently be split into symmetrically situated satellites separated by a distance $1/2M$ left and right of the position of the basic reflections. This is shown schematically in Fig. 3.

One should note that as a result the strong spots such as 020 also acquire apparently satellites, at a very small distance, corresponding in reciprocal space to

$$\frac{1}{a} - \frac{2}{Ma} = \frac{M-2}{M}\,\frac{1}{a}$$

i.e. $1/9a$ in the present case. In the case shown the corresponding period in direct space is about $9a = 36$ Å; this is equal to the real period $2\,\bar{2}\,2\,\bar{3}$ (Fig. 2), i.e. nine unit cells. This result is general; if the sequence is $2\,\bar{2}\,2 \ldots \bar{2}\,3$ (n terms) the satellite spacing is $[(M - 2)/M]\,(1/a)$ with $M = (2n + 3)/(n + 1)$ i.e. $[1/(2n + 3)]\,(1/a)$ and the corresponding fringe spacing is $(2n + 3)\,a$.

It has been shown in [9] that in the case of nonintegral structures images formed by the interference between beams corresponding to satellites associated with different basic spots exhibit Moiré fringes with a period corresponding to the reciprocal of the spacing of the satellites. In the present case such fringes reveal the repeat distance of the sequences, i.e. 36 Å. Such Moiré fringes were shown to represent a map of the orientation and spacing anomalies [9].

In [10] it was shown that orientation anomalies in diffraction patterns of shear structures are associated with jogging of shear planes. The same of course also applies to periodic antiphase boundaries. The change in average orientation of the antiphase boundaries can also in the present case be achieved by jogging. Below we shall discuss a model based on these considerations.

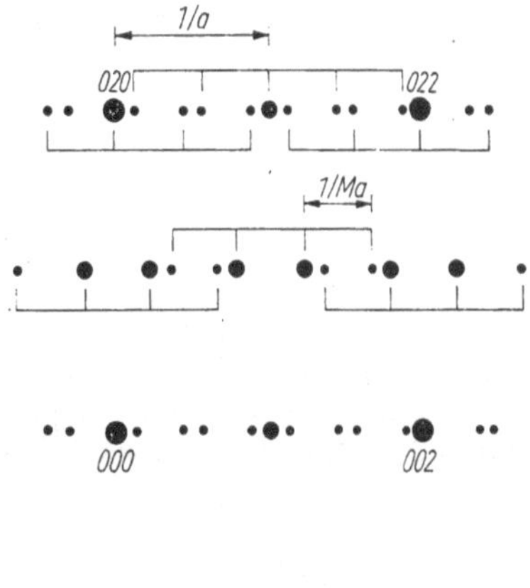

Fig. 3. Schematic representation of a [100]* diffraction pattern with non-integral M value. Satellites belonging to the same basic reflection are connected by brackets

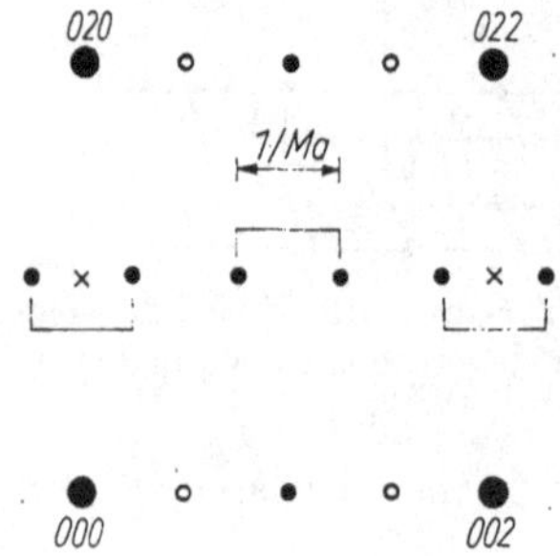

a

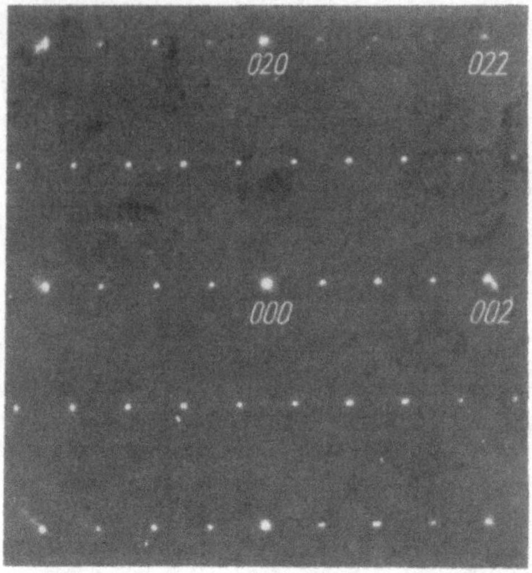

b

Fig. 4. a) Schematic representation of a [100]* diffraction pattern with integral M value ($2M = 4$). Open circles represent forbidden reflections. b) Actually observed pattern of Au_4Zn with $2M = 4$

3. Experimental Analysis

3.1 Specimen preparation

Au_4Zn material was prepared by melting together both elements in an evacuated quartz capsule. No significant weight loss was measured after melting. The ingots were rolled to a 100 μm thickness and disks were punched out of such foil. All annealing treatments were performed with 3 mm discs of 100 μm thickness in evacuated quartz tubes. Energy dispersive analysis of the produced X-rays in the electron microscope showed a composition between Au_4Zn and Au_3Zn.

Thinning was performed by electropolishing using an electrolyte containing 500 cm^3 of H_2O, 50 cm^3 of HCl, and 50 cm^3 of H_2SO_4 [11]. The voltage used was 30 V and a current density of 100 A/cm^2 was obtained.

Most observations were performed with a 100 kV electron microscope equiped with a top entry goniometer stage and a special high magnification projector lens. For the present investigation all specimens have been first heat treated at 480 °C for one day, quenched in water and annealed at 250 °C for periods varying between 1 and 10 d. In such specimens both the one-dimensional long-period superstructure termed Au_3Zn (H) by Iwasaki et al. [4, 5] and the two-dimensional long-period superstructure termed $Au_{3+}Zn$ by Wilkens et al. [3] have been observed. Occasionally grains having the $Au_3Zn(R)$ structure were also found. In the present paper we will *only* deal with the $Au_3Zn(H)$ structure, the $Au_{3+}Zn$ phase will be treated in Part II whilst an investigation of the $Au_3Zn(R_1)$ and $Au_3Zn(R_2)$ phases is in progress.

3.2 Diffraction evidence

3.2.1 Integral superstructure

The simplest type of diffraction pattern which was observed in Au_4Zn exhibits a fourfold long superstructure spacing, i.e. $2M = 4$. The pattern is shown schematically in Fig. 4a and an observed pattern is reproduced in Fig. 4b. In this case the average spacing between antiphase boundaries is equal to two times the size of the unit cell. We also occasionally observed integral superstructures with $2M = 6$.

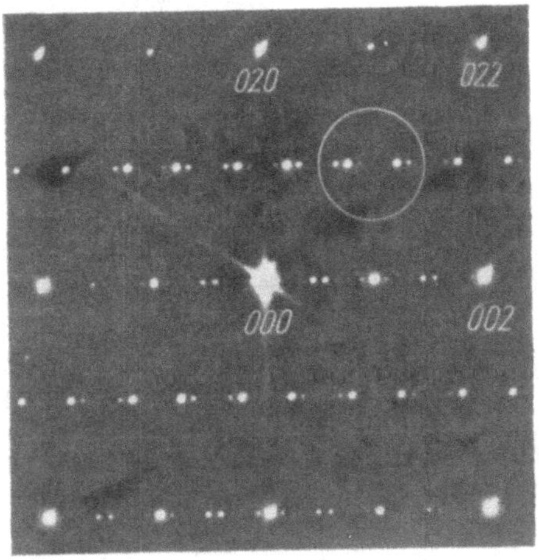

Fig. 5. Observed [100]* electron diffraction pattern corresponding to $M = 2.25$. Reflections selected for direct lattice imaging as in Fig. 10 have been indicated

3.2.2 *Non-integral superstructures*

The diffraction patterns mostly exhibit a "spacing anomaly" characteristic of a superstructure with a non-integral M-value. Fig. 3 shows schematically such a pattern, whilst an observed pattern is reproduced in Fig. 5. The M-value is somewhat variable: $2 < M \leq 2.25$; in the case of Fig. 5 it is 2.25. Each spot has satellites on each side of the position of the basic spot; they decrease rapidly in intensity with their distance to these basic spots. Usually only one spot is visible on each side. Such groups of spots are joined by brackets in Fig. 3; where they meet they clearly fail to match, giving rise apparently to satellites very close to the main spot.

3.2.3 *Non-integral superstructures with orientation anomaly*

In some cases patterns of the type reproduced in Fig. 6b and exhibiting an orientation anomaly are observed. Such a pattern is shown schematically in Fig. 6a where the orientation anomaly has been exaggerated somewhat. Patterns of this type are

Fig. 6. a) Schematic representation of a diffraction pattern exhibiting as well a spacing as an orientation anomaly; satellites belonging to the same group are indicated. b) [100]* electron diffraction pattern of Au_4Zn clearly showing an orientation anomaly between satellites of different groups

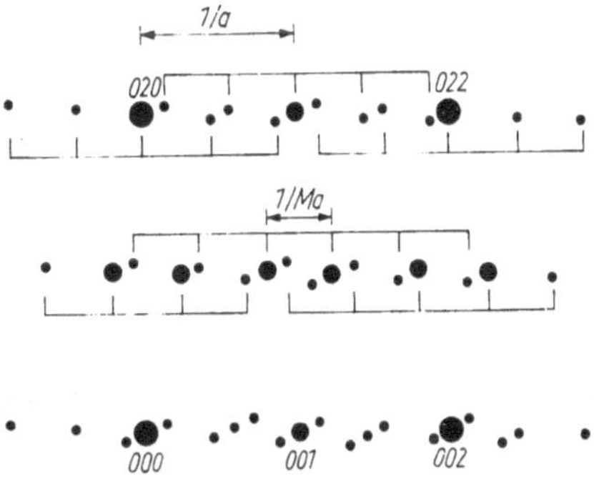

a

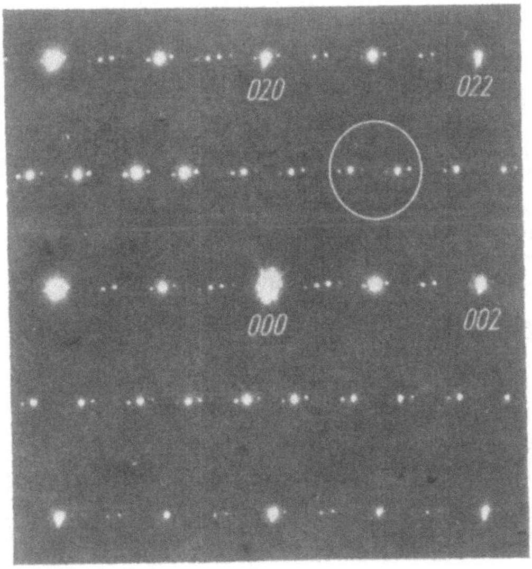

b

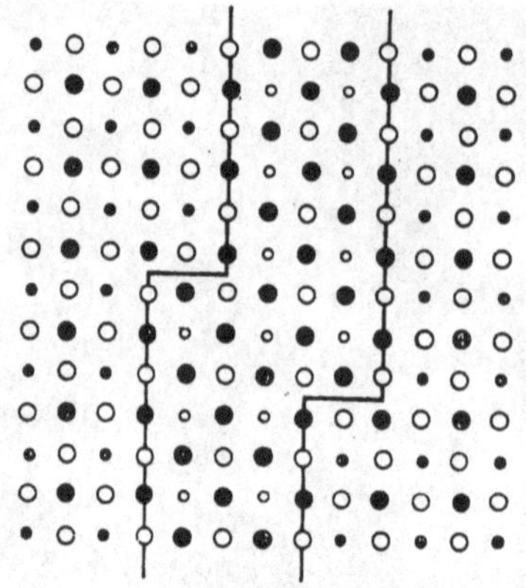

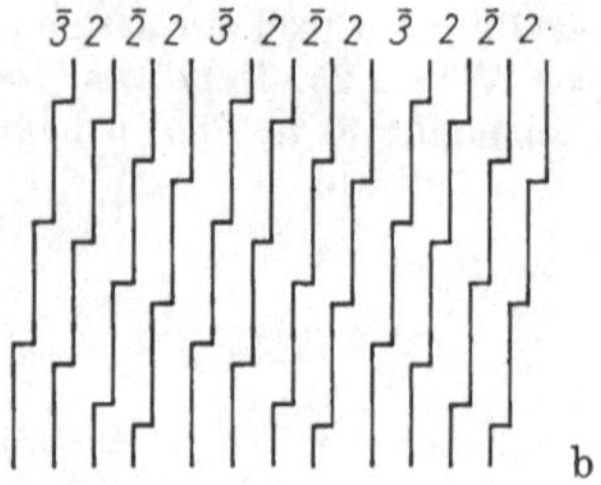

Fig. 7. a) Atomic scale representation of jogging of the APB's. Au: ● at $x = 0$, ○ at $x = 0.5$; Zn: ● at $x = 0$, ○ at $x = 0.5$. b) Schematic representation of the APB jogging giving rise to inclined average APB planes and to the related Moiré-like contrast. The "concentration" of ledges has been exaggerated in this scheme

related to deviations of the average orientation of the antiphase boundaries from the simple cube orientation. As will be shown below by optical simulations, such deviations result from systematic jogging of the antiphase boundary plane in the manner shown schematically in Fig. 7a and 7b, giving rise to ledges or strips in the plane of the APB. Such a geometry allows to achieve a more uniform distribution of APB's. The strip which is three unit cells wide is in fact fragmented into uniformely distributed segments with this width and located between two ledges.

3.2.4 Diffraction patterns due to several variants

It is clear that the long period can develop with equal probability along any one of the three cube directions. Diffraction patterns which are clearly due to the superposition of two or three variants are often observed. An example due to two variants is shown in Fig. 8. The analysis in its component patterns is straightforward. One should note that the one-dimensional structure becomes slightly tetragonal, the c-axis being perpendicular to the APB's. This explains the splitting of the fundamental reflections (e.g. $(220)_{\text{f.c.c.}}$ and the asymmetry around superstructure reflections, e.g. $(110)_{\text{f.c.c.}}$ (Fig. 8). This type of pattern should be distinguished from the pattern due to a two-dimensional LPAPB structure.

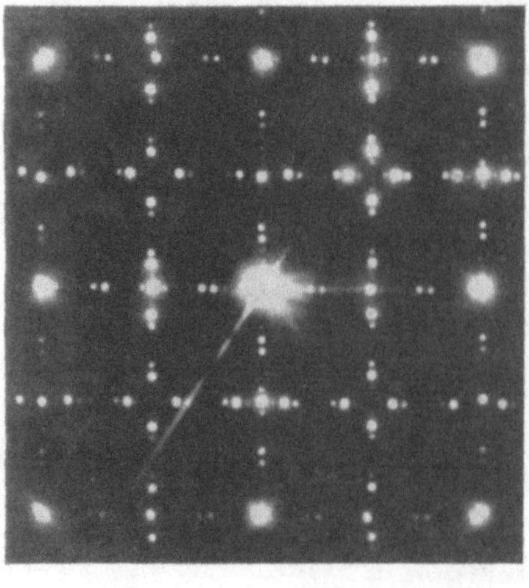

Fig. 8. Electron diffraction pattern due to a superposition of two orientation variants each with non-integral M-value

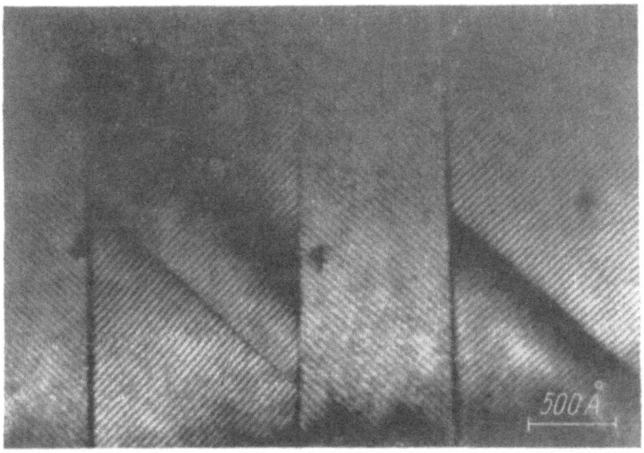

Fig. 9. Image of the 36 Å Moiré-like fringes at a rather low magnification. Different orientation variants are separated by $(101)_{f.c.c.}$ planes

3.3 Lattice images

Lattice images were formed by collecting satellites belonging to two different basic reflections. The circle in Fig. 5 indicates the reflections collected in the aperture.

Two different microscopes were used. In one of these the resolving power was not sufficient to reveal the 8 Å spacing corresponding to the width of two f.c.c. unit cells; only the "Moiré" type fringe with a 36 Å spacing is revealed. In the nomenclature of Fujiwara $2\nu M = 36$ Å [8]. A typical image is produced in Fig. 9; it exhibits clearly two orientation variants, as well as translation variants.

Lattice images made with a microscope allowing to achieve in sufficiently thin Au_4Zn foils a resolving power better than 4 Å, exhibit the individual antiphase boundaries; and also the individual unit cells if two basic spots are collected in the aperture. Usually they also show broader fringes; these are the Moiré-like fringes with a 36 Å spacing referred to above. They reveal the true repeat pattern in the ideally periodic sequence.

These "Moiré" fringes are often variable in spacing and orientation as is the true period. Where they are parallel to the antiphase boundaries the diffraction pattern is of the type shown in Fig. 3. The images of antiphase boundaries reveal two different spacings corresponding to the width of either two or three unit cells (Fig. 10); the most frequent sequence is $2\,\bar{2}\,2\,\bar{3}$ but also $2\,\bar{2}\,2\,\bar{2}\,3$ occurs locally. Where they form an angle with the antiphase boundaries (Fig. 11) the diffraction pattern also shows an orientation anomaly and is of the type of Fig. 6b. The images of the antiphase boundaries then show jogs of the type shown schematically in Fig. 7. The Moiré fringes follow closely the geometrical locus of these jogs. These jogs may propagate along the APB, displacing the latter laterally [12].

Fig. 10. Direct imaging of the "Moiré" fringes together with the individual antiphase boundaries by collecting the reflections encircled in Fig. 5. Moiré fringes are parallel to the APB, therefore no orientation anomaly is observed in the diffraction pattern of Fig. 5

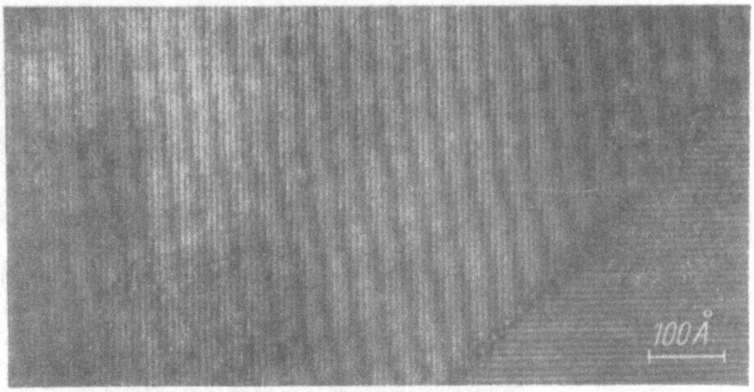

Fig. 11. Lattice imaging when collecting all reflections encircled in Fig. 6 b. Moiré fringes are making an angle with respect to the individual APB's. Individual jogs can be observed in the inset

When APB's are eliminated or change spacing, hairpin configurations are formed. In Fig. 12 where the one-dimensional LPAPB structure emerges into the Cu_3AuI-type structure, the APB's are joining together by pairs and form the well-known hairpin arrangements. The inset of Fig. 12 shows the diffraction pattern taken across the boundary region; it is clearly the superposition of the diffraction patterns due to both structures. When the APB spacing changes abruptly as in Fig. 13 a, b where the spacing changes from 8 Å ($M = 2$) to 12 Å ($M = 3$) antiphase boundaries are always generated (or eliminated) by pairs.

This configuration is presumably related to the propagation mechanism of the APB's and will be treated in more detail in Part II of this paper. It is consistent with the fact that the displacement vector is half a lattice vector of the ordered structure.

Occasionally the "Moiré" fringes are absent; the diffraction pattern is integral in these cases: $M = 2$ or $M = 3$. This confirms our explanation of the origin of the Moiré-like fringes.

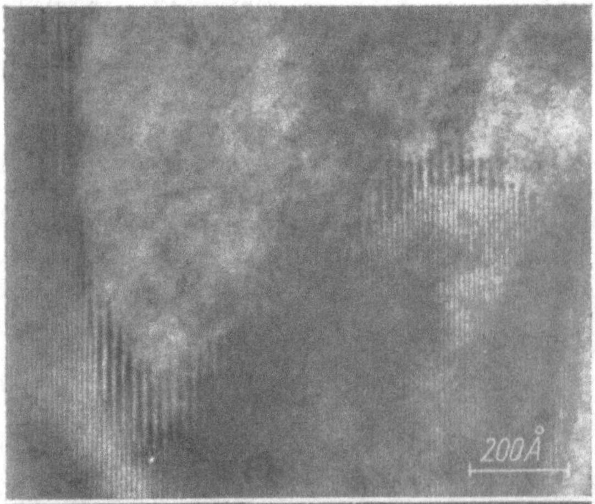

Fig. 12. Antiphase boundaries in Au_4Zn are always generated or eliminated by pairs. The inset electron diffraction pattern is taken over the boundary region. Note the tetragonal deformation revealed in the diffraction pattern

a

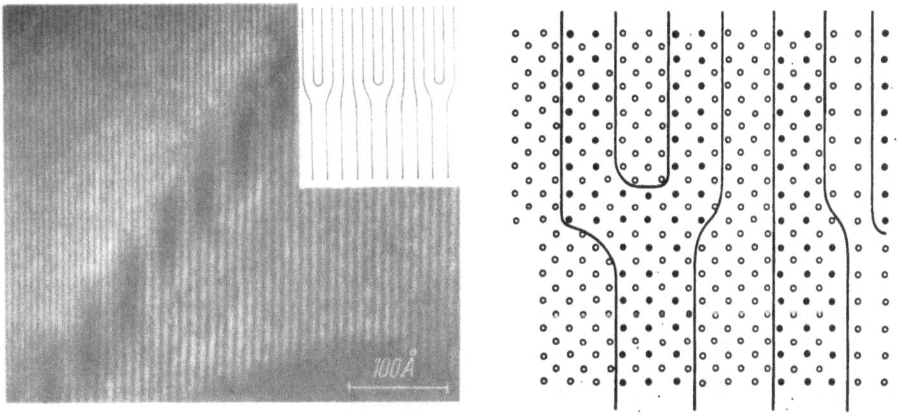

b

Fig. 13. a) APB's changing spacing from 8Å ($M = 2$) to 12 Å ($M = 3$). The supplementary fault planes are generated (or eliminated) by pairs; this is shown more clearly in the inset. b) Atomic scale representation of the change in periodicity from $M = 2$ to $M = 3$. Only one cube plane has been drawn for simplicity

4. Model for the Long-Period Superstructure

The diffraction pattern (Fig. 4) shows that $\boldsymbol{g} \cdot \boldsymbol{R} = 0$ (mod integer) if $h + k =$ $=$ even and $\boldsymbol{g} \cdot \boldsymbol{R} = 1/2$ (mod 1) if $h + k =$ odd. From this it can be concluded that the displacement vector should be of the form $a/2[110]$, where a is the lattice parameter of the f.c.c. unit cell. This vector is consistent with the observations of hairpin arrangement since $a/2[110] + a/2[110] = a[110]$, i.e. a lattice vector. Depending on the fault plane (100) or (001) the APB may be conservative or non-conservative. For the one-dimensional LPAPB structure all faults are apparently of the conservative type. In the two-dimensional LPAPB structure one family is conservative whereas the other is non-conservative (Part II).

A model for the antiphase boundaries in the Cu_3AuI structure is well known already for a long time and the model is represented in Fig. 2. It consists of conservative APB's in the (001) plane all with a displacement vector $\boldsymbol{R} = 1/2[110]$. The non-integral M values being attributed to mixing of APB spacings in the manner proposed by Fujiwara [8]. While these interpretations were mostly based on diffraction evidence, which unavoidably averages the structure over large crystal areas, the present observations give direct microscopic and local evidence for the proposed model.

The lattice fringes directly reveal the width of the successive domain strips, as well as the jogging or "ledging" of the APB's. The interpretation in terms of a struc-

Fig. 14. Direct lattice imaging of mixed sequences of antiphase boundaries. A combination of 2 2 2 3 and 2 2 2 2 3 is observed

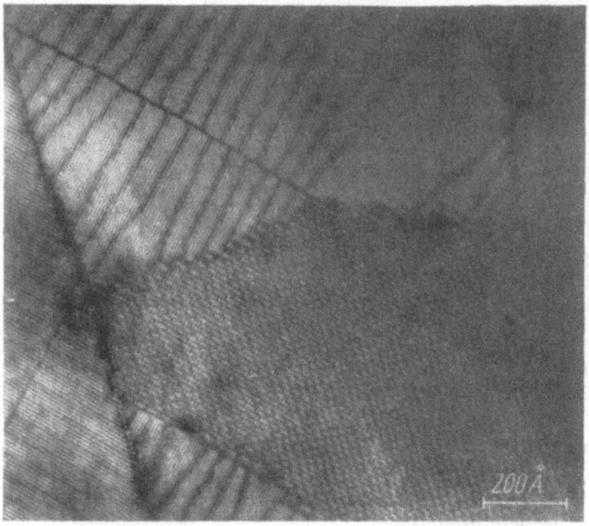

Fig. 15. The Moiré-like fringes may be very irregular and can even stop abruptly in the vicinity of a domain boundary, or a hairpin. Note the presence of region with stacking sequence $2\bar{2}2\bar{3}$ in the left bottom part

tural model is therefore straightforward. The value of M as deduced from the diffraction pattern is somewhat variable; from the corresponding lattice images one finds that the sequences are also different as well as the spacing of the "Moiré" fringes. The most frequently observed sequence is $2\,\bar{2}\,2\,\bar{3}$ ($M = 2.25$), but sometimes other sequences such as $2\,\bar{2}\,2\,\bar{2}\,3\,\bar{2}\,2\,\bar{2}\,2\,\bar{3}$ giving rise to $M = 2.2$ can also be observed (Fig. 14).

In the vicinity of hairpins or other structural defects the Moiré fringes may become quite irregular (Fig. 15). If this is only local, no diffraction effect will be observed. All these effects can be understood from the model depicted in Fig. 7.

It pictures how jogs in APB's may look like on an atomic scale. Fig. 16 is a schematic representation of an actually observed situation where Moiré fringes are changing orientation around a hairpin APB.

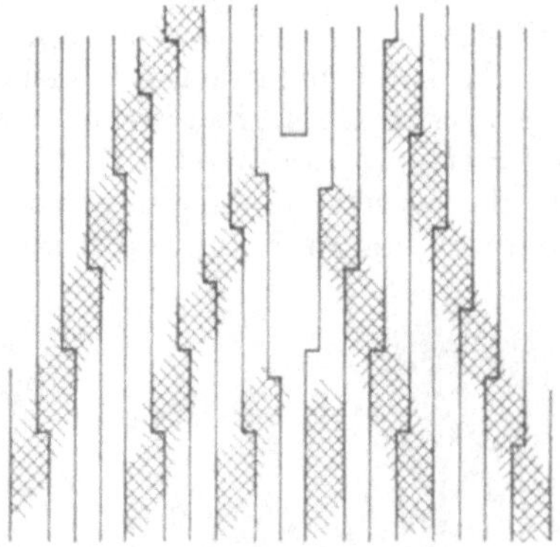

Fig. 16

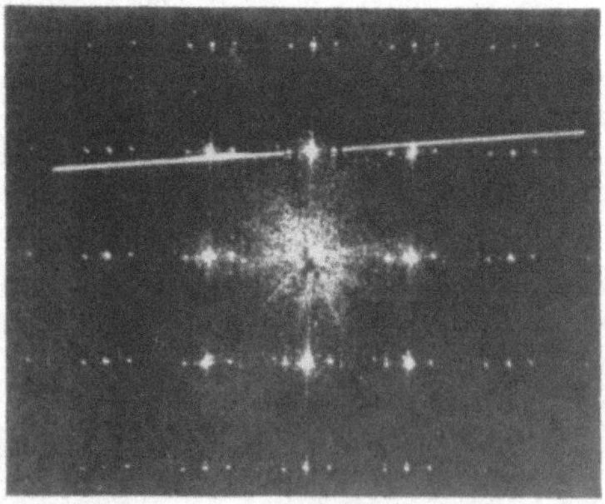

Fig. 17

Fig. 16. Schematic representation of the "Moiré" configuration in the vicinity of a hairpin

Fig. 17. Optical simulation of a grating constructed by introducing jogs causing a change in orientation of the average APB plane with respect to the cube plane. For simplicity only one (100) layer was drawn

Fig. 18. Direct image showing different orientation variants in the initial stages of ordering. Where two variants meet a "checker-bord" pattern is produced

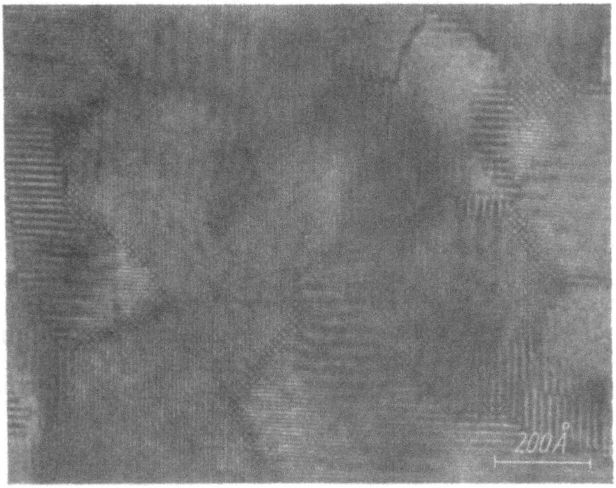

As an additional means of testing the validity of the proposed model we performed optical simulation experiments with laser light. Only one plane of the structure is represented and used as an optical grating.

The average direction of the APB's encloses a small angle with the cube directions; this change in orientation being brought about by regular "jogging" of the APB's (Fig. 7). In the microscope image the Moiré fringes would enclose the same angle with the cube direction.

The resulting optical diffraction pattern is shown in Fig. 17. It is clear that the sequences of satellite spots around each basic spot enclose an angle with the rows of basic spots which corresponds to the misorientation of the APB's.

Also the spacing of the satellite spots is equal to the reciprocal of the "average" APB spacing; as follows from Fujiwara's theory. Since the model is only two-dimensional and contains one layer only, superlattice spots cannot be differentiated from basic spots.

When differences in M value (i.e. differences in average APB spacing) have to be accommodated this can be done by introducing jogs or ledges as described before. However, if the differences are too large, e.g. when M is abruptly changing from 2 to 3, the accommodation takes place by the hairpin mechanism (Fig. 13a). A schematic representation on an atomic scale is shown in Fig. 13b.

Where two one-dimensional areas meet, regions exhibiting a two-dimensional checker-board patterns are often produced (Fig. 18). In most cases this pattern is not

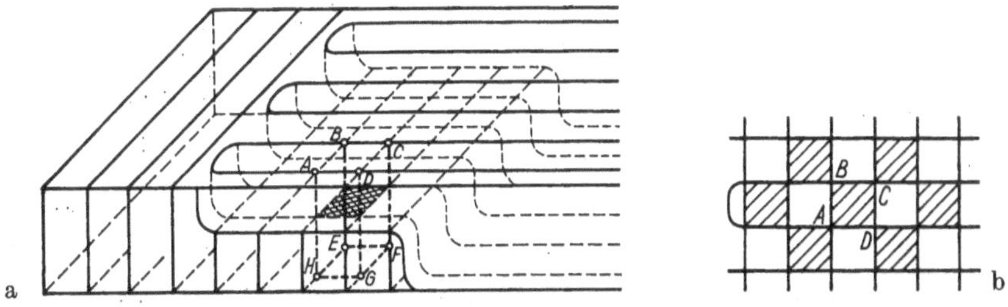

Fig. 19. a) Spatial representation of overlapping orientation variants, each variant having a one dimensional APB structure. b) View along normal to the foil plane

the image of a two-dimensional antiphase boundary structure but is produced by overlap. The phase change due to the interface between two overlapping domains such as that shown by cross hatching in Fig. 19 is either π or 0. In successive squares along a given domain strip the phase difference will be either π or 0, producing the checker-board pattern of contrast which produces the pattern of apparent lines along the (110) direction.

5. Conclusion

In this paper lattice imaging using high resolution electron microscopy has proved to be important for the study of long-period superstructures. For various reasons the Au–Zn system has been chosen as an example. Most of the X-ray results have been directly confirmed and a model explaining all the features of the diffraction patterns and of the images has been proposed. Jogging of the antiphase boundaries turned out to be an important feature as well in understanding details of the diffraction pattern, such as orientation anomalies, as in understanding the propagation of APB's. It is also worthwhile to notice that the present observations do not seem to confirm the dea that the spacing of one-dimensional APB's in Au_3Zn(H) is fixed at $M = 2$ [2, 4].

Acknowledgement

Thanks are due to Prof. Dr. J. van Landuyt for numerous discussions.

References

[1] K. SCHUBERT, B. KIEFER, and M. WILKENS, Z. Naturf. **9a**, 987 (1954).
[2] K. SCHUBERT, B. KIEFER, M. WILKENS, and R. HAUFLER, Z. Metallk. **46**, 692 (1955).
[3] M. WILKENS and K. SCHUBERT, Z. Metallk. **48**, 550 (1957); **49**, 633 (1958).
[4] H. IWASAKI, M. HIRABAYASHI, K. FUJIWARA, D. WATANABE, and S. OGAWA, J. Phys. Soc. Japan **15**, 1771 (1960).
[5] H. IWASAKI, J. Phys. Soc. Japan **17**, 1620 (1962).
[6] J. VAN LANDUYT, R. DE RIDDER, R. GEVERS, and S. AMELINCKX, Mater. Res. Bull. **5**, 353 (1970).
[7] G. VAN TENDELOO, J. VAN LANDUYT, P. DELAVIGNETTE, and S. AMELINCKX, phys. stat. sol. (a) **25**, 697 (1974).
[8] K. FUJIWARA, J. Phys. Soc. Japan **12**, 7 (1957).
[9] R. DE RIDDER, G. VAN TENDELOO, and S. AMELINCKX, phys. stat. sol. (a) **33**, 383 (1976).
[10] R. DE RIDDER, J. VAN LANDUYT, and S. AMELINCKX, phys. stat. sol. (a) **9**, 551 (1972).
[11] R. WOLF, to be published.
[12] J. VAN LANDUYT and S. AMELINCKX, J. Solid State Chem. **6**, 222 (1973).

(Received June 20, 1977)

phys. stat. sol. (a) 43, 565 (1977)

Subject classification: 14.3.1 and 14.3.4; 22.9

Department of Physics, University of Malaya, Kuala Lumpur[1])

Space-Charge-Limited Currents in Evaporated Films of Pentacene

By

K. O. LEE and T. T. GAN

Electrical conductivity measurements are made at various fixed temperatures on evaporated films of pentacene deposited at different substrate temperatures. Analysis of the resulting SCL-currents enables to determine both the trapping levels and trap densities. The density of traps is shown to increase exponentially with decreasing substrate temperature. The possible origin of the trapping centres in the pentacene films is discussed.

Des mesures de conductivité électrique ont été effectuées à plusieurs températures sur des couches minces de pentacène deposées par évaporation sur des supports portés à différentes températures. L'analyse des résultats de courants limités par la charge d'espace permet de déterminer à la fois le niveau des pièges et leur densité. On constate que la densité de pièges croit exponentiellement lorsque la température du support sur lequel est effectuée l'évaporation décroit. L'origine possible des centres de piège dans les couches minces de pentacène est discutée.

1. Introduction

Extensive literature exists concerning the optical and electrical properties of smaller members of the aromatic compounds such as naphthalene [1], anthracene [2, 3], and tetracene [4 to 6]. However, very little information is avaible concerning the electronic properties of pentacene. Early work on evaporated films of pentacene shows an exponential temperature dependence of the conductivity [7], and more recently there has been report of charge carrier injection into pentacene thin layers through metallic contacts [8]. In the present paper we report the results of a systematic study on the current–voltage behaviour of pentacene films, prepared by vapour deposition onto substrates maintained at various fixed temperatures. The experimental observations show that the electronic properties of the evaporated films are highly sensitive to the temperature at which the film was deposited.

2. Preparation of Samples

The pentacene film samples were deposited onto fused quartz plates by evaporation from a molybdenum boat in a vacuum system at pressure between 10^{-5} and 10^{-6} Torr. During evaporation, the temperature of the substrates could be controlled and varied from -160 to 100 °C. The deposition rate was maintained constant at approximately 0.21 μm/min. The film thickness, ranging between 0.30 and 1.40 μm, was measured using an interference microscope. A detailed description of the experimental set-up has been reported in a previous paper [9].

X-ray diffraction patterns show that the deposits evaporated at temperatures above -50 °C are polycrystalline and that those evaporated at $T_s = -100$ and -150 °C are amorphous. Optical absorption results indicate a structural transition from polycrystalline to amorphous deposit occurring at around $T_s = -90$ °C [9].

[1]) Kuala Lumpur, Malaysia.

3. DC Conduction

dc conductivity was measured in pentacene films, in sandwich-type structure with both Al–film–Al and Al–film–Au configurations. I–U measurements were made at different fixed temperatures between 230 and 380 K, with all the films initially annealed in vacuum to a temperature of about 390 K for consistency. The currents were measured in a pressure below 10^{-3} Torr using a Keithley 610C electrometer with a Fluke 412B power supply. A large number of films of varied thicknesses was prepared at different substrate temperatures for the dc measurements. The I–U behaviour of the films studied revealed no difference between the two electrode systems.

Typical results observed for the current, as a function of applied voltage, are shown in Fig. 1 and 2. Well-defined regions can be seen from these I–U curves: a low-field Ohm's law region, and a square-law region at higher voltages followed by a section of steep current rise with voltages. These observations conform to the general features predicted by Lampert [10] who treated the problem of charge injection into an insulator with trapping centres, and thus may be interpreted in terms of space-charge-limited (SCL) conduction.

It has been established by Silinsh et al. that the intrinsic dark conductivity in pentacene thin layers is p-type [8]. The same authors observed SCL current in the organic solid as a result of hole injection by metallic contacts using the sandwich-type systems: Al–Pc–Au and Au–Pc–Au. Hole injection into pentacene through a metallic contact is highly possible considering the low ionisation energy of 5.07 eV for pentacene [8]. This is supported by the fact that at a metallic contact the rate of hole injection increases with decreasing ionisation energy of the organic crystal [11], as it has been found that gold forms ohmic contacts for hole injections with tetracene (ionisation energy 5.25 eV) [5] and perylene crystals (ionisation energy 5.33 eV) [11]. With the above consideration one may conclude that, in the conduc-

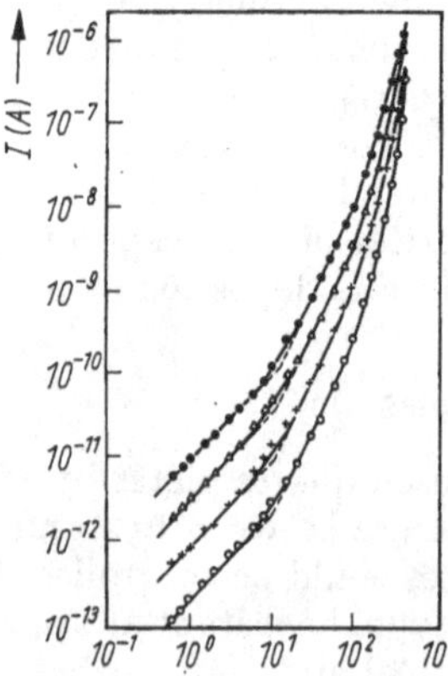

Fig. 1

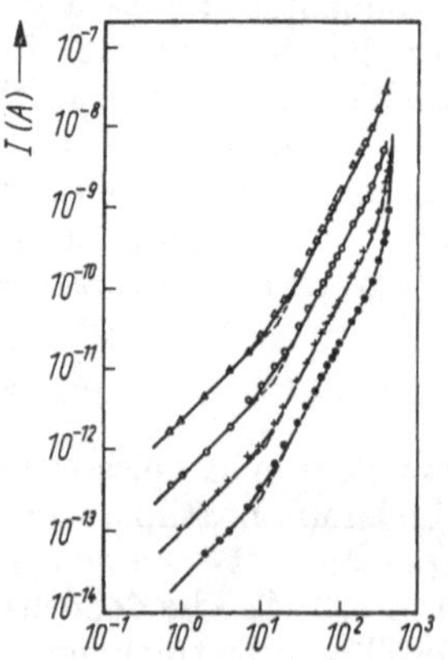

Fig. 2

Fig. 1. Current–voltage curves at several different temperatures for pentacene film evaporated at $T_s = 50$ °C. $L = 0.89$ μm. $A = 0.72$ cm². Al–film–Al system. ● $T = 377$, △ 353, + 328, ○ 299 K

Fig. 2. Current–voltage curves at several different temperatures for pentacene film evaporated at $T_s = -100$ °C. $L = 0.59$ μm. $A = 0.72$ cm². Al^-–film–Au^+ system. △ $T = 373$, ○ 348, + 323, ● 300 K tivity in pentacene films studied here, the majority charge carriers involved are holes both in the ohmic and SCL regions.

At low fields the ohmic current, supported by thermally generated carriers, is given by

$$I_\Omega = n_c e \mu_0 \frac{U}{L} . \tag{1}$$

At higher applied voltages, injected charges being predominant, Lampert's theory on solids containing a single trapping level predicts a space-charge current given by

$$I_{SC} = \frac{9}{8} \theta_n \varepsilon \mu_0 \left(\frac{U^2}{L^3}\right), \tag{2}$$

where μ_0 and ε are the microscopic mobility and the permittivity, respectively, and θ_n is the fraction of total carriers which are free. The presence of trapping centres is responsible for the factor θ_n which greatly reduces the space-charge current.

The transition from ohmic to quadratic behaviour occurs at a voltage $U_{\Omega C}$ which can be defined by equating the two currents

$$U_{\Omega C} = \frac{8}{9} \frac{n_c e L^2}{\varepsilon \theta_n} . \tag{3}$$

The value of the critical applied voltage at which the current increases steeply, corresponds to the filling of traps and can be related to the trap density, N_m, in the solid by the expression

$$U_{TFL} = \frac{N_m e L^2}{2\varepsilon} . \tag{4}$$

In order to check that our experimental data are correctly interpreted in terms of SCL conduction without ambiguity, it is highly desirable that both ohmic and SCL currents exhibit the correct thickness dependence. For this purpose pentacene films were prepared systematically of thicknesses ranging from 0.30 to 1.30 μm, and I–U curves at a fixed temperature of these films were carefully determined. The

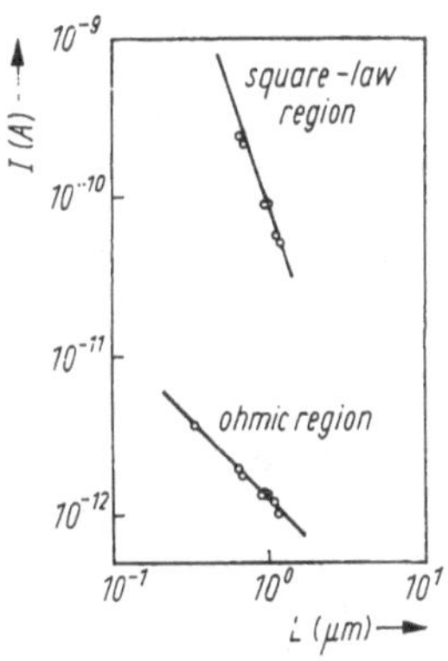

Fig. 3

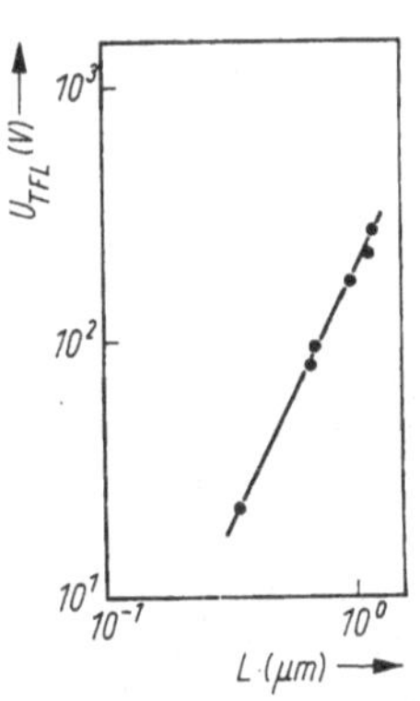

Fig. 4

Fig. 3. Thickness dependence of currents for ohmic region ($U = 1$ V) (slope $= -1$) and SCL region ($U = 30$ V) (slope $= -3$) for pentacene films measured at 74 °C. ($T_s = 27$ °C)

Fig. 4. Thickness dependence of the trap-filled-limit voltage for pentacene films measured at 74 °C. ($T_s = 27$ °C) (slope $= 2$)

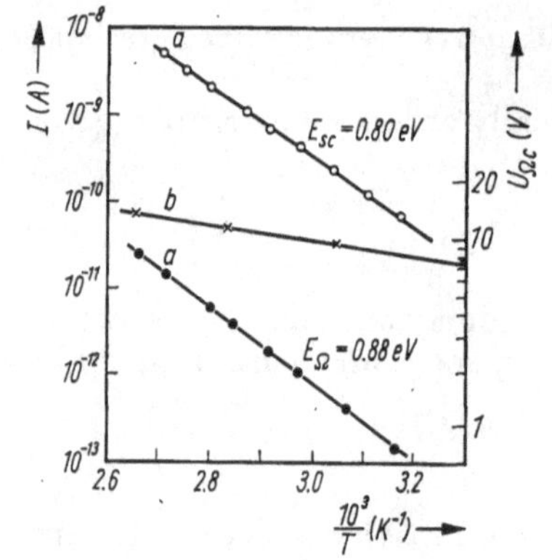

Fig. 5. (a) Variation of current with temperature for $U = 5$ V (ohmic region) and $U = 40$ V (square-law region) for pentacene film described in Fig. 1. (b) Variation of transition voltage $U_{\Omega C}$ with temperature for curves shown in Fig. 1 (slope = 0.08 eV)

results shown in Fig. 3 and 4 provide an excellent evidence that a SCL conduction interpretation is appropriate in the present case.

The trap-filled-limit voltage, $U_{\rm TFL}$, taken at the onset of a nearly vertical section of the I–U curve, furnishes information concerning the density of traps that exist in the solid. The excellent $U_{\rm TFL} \sim L^2$ behaviour (Fig. 4) lends us confidence that reliable quantitative deduction could be obtained for the films.

Roberts and Schmidlin, in their statistical treatment of a wide band-gap material containing localized levels, demonstrated how measurements of thermal activation energies for electrical conduction could be used in the deduction of both depth and concentration of localized levels [12]. Pentacene, being an organic material, can be regarded as a wide band-gap semiconductor. In the following, Roberts' theory is applied to the conductivity measurements thus leading to parameters concerning the trap levels being deduced for pentacene films.

The case in which the ohmic and SCL activation energies are different is termed non-extrinsic, and n_c, the density of conduction carriers, and θ_n, the ratio of free to total carriers, are respectively given by

$$n_c = N_c \left(\frac{N_q}{N_m}\right)^{1/2} \exp - \frac{(E_c - E_m) + \frac{1}{2}(E_m - E_q)}{kT} \tag{5}$$

and

$$\theta_n = \frac{N_c}{N_m} \exp - \frac{E_c - E_m}{kT}, \tag{6}$$

where suffix c refers to conduction band level, m and q refer to 'dominant' electron and hole levels, respectively.

Current measurements in pentacene films at fixed voltages made as function of temperature yielded different activation energies for ohmic and SCL regions. Fig. 5 (a) shows a plot of lg I versus $10^3/T$ for $U = 5$ V, where the current is ohmic, and for $U = 40$ V, where the current is space-charge-limited. Fig. 5 (b) shows that the activation energy for the transition voltage, $U_{\Omega C}$, from ohmic to quadratic behaviour is equal to the difference between the two activation energies.

Using the appropriate notations for the present case of hole injection, and applying the measured activation energy values in (5) and (6), we obtain $(E_q - E_v) = 0.80$ eV for the dominant hole level, and $(E_m - E_v) = 0.96$ eV for the dominant electron level for pentacene film No. 71. Equation (4), for $U_{\rm TFL}$, and a dielectric constant of $\varepsilon = 4$, yields a trap density $N_q = 7.3 \times 10^{16}$ cm^{-3}. There being no reported value for the dielectric constant for pentacene, the only available relevant information is that the mean refractive index $n = 2$ [13].

Applying the conductivity parameters measured for ohmic and SCL regions to the appropriate equations, we deduce the other important quantities for the film: the electron trap density $N_m = 4.6 \times 10^{16}$ cm^{-3}, and the product $\mu_0 N_v = 2.9 \times 10^{16}$ (V cm s)$^{-1}$.

The trap densities for pentacene films, in the range of 10^{16} to 10^{17} cm^{-3}, obtained here are comparable to the corresponding value for tetracene single crystals, 10^{15} cm^{-3}, obtained by Baessler et al. [5], and to that for naphthalene single crystals, of the order of 10^{16} to 10^{18} cm^{-3}, obtained by Campos [1].

Table 1

Electronic parameters deduced from measured SCL currents for pentacene films evaporated at different substrate temperatures

(a) Al–film–Al system

film No.	thickness L (μm)	substrate temperature T_s (°C)	activation energies E_Ω (eV)	E_{sc} (eV)	average trap densities N_q (10^{16} cm^{-3})	N_m (10^{16} cm^{-3})	product $\mu_0 N_v$ 10^{16} (V cm s)$^{-1}$
81	1.29	90	0.88	0.80	2.9	3.4	7.7
71	0.89	50	0.88	0.80	7.3	4.6	6.1
33	0.80	27	0.89	0.80	8.4	9.8	3.0
63	0.68	−50	0.90	0.80	13.0	35.0	2.8
54	0.56	−100	0.91	0.80	35.0	57.7	4.1
42	0.32	−150	0.91	0.80	215.0*)	69.5*)	0.3*)
(b) Al^-–film–Au^+ system							
82	1.05	90	0.88	0.80	4.0	2.8	13.5
72	1.02	50	0.88	0.80	5.9	3.5	4.4
61	0.61	−50	0.90	0.80	24.0	21.8	3.2
51	0.59	−100	0.91	0.80	48.0	34.8	1.5

*) A trap-filled-limit was not reached on the experimental curves for this film. These values were deduced assuming $U_{TFL} = 500$ V.

The procedure of SCL conduction analysis was applied to a number of pentacene films evaporated onto quartz substrates maintained at various different temperatures. The results are summarized in Table 1.

The spread in the trap depth, the $(E_m - E_v)$ value ranging from 0.96 to 1.02 eV as calculated from the activation energy values, is acceptable within the experimental accuracy limit.

Film No. 42, the thinnest of the series, did not furnish a steep current rise region on the I–U plots. The measurements were not taken beyond the applied voltage of 350 V to avoid thermal destruction of the film. If a value of 500 V is taken for U_{TFL}, we obtain for the film trap densities of $N_q = 215 \times 10^{16}$ cm^{-3} and $N_m = 69.5 \times \times 10^{16}$ cm^{-3}.

If the effective density of states in the valence band is taken to be approximately given by the molecular density, i.e., $N_v = 2.8 \times 10^{21}$ cm^{-3}, the mobility value, μ_0, computed from the SCL current measurements is of the order of 10^{-5} cm^2/Vs.

4. Discussion

The above analysis is seen to be effective for locating trapping levels in the pentacene films, but it does not offer a physical explanation of the nature of the trapping

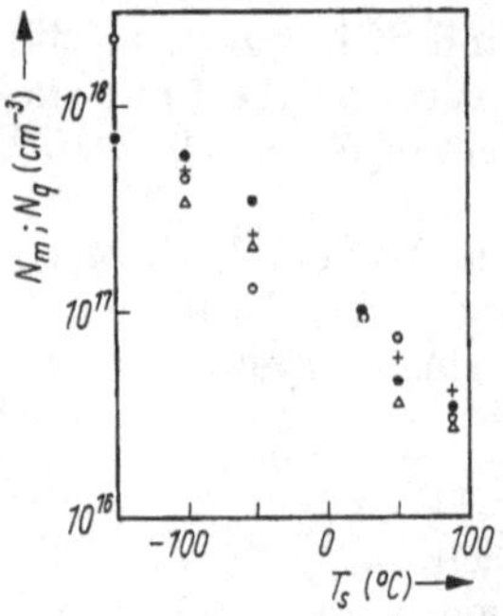

Fig. 6. Variation of trap densities, N_m and N_q, with substrate temperature, T_s; ○ and ● belong to the Al–film–Al system; + and △ belong to the Al–film–Au system. (N_m: ● and △; N_q: ○ and +)

centres. Two major factors can be responsible for the existence of trapping centres in the films, that is, (i) the chemical impurity in the form of adsorbed oxygen or the oxide complex of pentacene, and (ii) the structural defects in the films. There exists experimental evidence in support of the belief that the trapping in organic solids is associated with structural defects rather than with chemical impurities. For examples, Thomas et al. [14] found that trapping centres in crystalline anthracene are mainly due to lattice imperfections and essentially independent of chemical impurity; Baessler et al. [5] deduced that traps in tetracene crystals are of structural rather than chemical origin; experiments performed by Kuroda and Flood [15] revealed no detectable change in the dark conductivity in pentacene films on exposure to an oxygen atmosphere.

In the pentacene films studied in the present work, it would be very plausible to suggest that traps are associated with the degree of molecular disorder in the films. Although the trapping levels are similar for all films, the densities of traps show an exponential dependence on temperature at which the film was produced (Fig. 6). This is in agreement with the optical absorption results [9] which show that films deposited at high substrate temperatures consist of well-oriented large size crystallites, while films produced at low substrate temperatures contain small crystallites and molecules of irregular orientations, the degree of disorder aggravating with decreasing substrate temperature. However, the trapping in the pentacene films may not be completely independent of the presence of oxygen. Pentacene is highly reactive toward oxygen and there are occasions during the course of the present experiments in which oxygen molecules and hence the oxide complex can find their presence in the films. The films were prepared in a vacuum of 10^{-5} to 10^{-6} Torr, an atmosphere not exactly free of oxygen, and were annealed in an atmosphere of pressure 10^{-3} to 10^{-4} Torr at 390 K prior to the dc measurements being taken. Storage of the films in air, which is unavoidable in between measurements, induces the formation of oxide complexes of pentacene, as was revealed by optical observations [9]. Riemer [16] reported a high permeation rate of oxygen through tetracene films when these were exposed to oxygen after deposition. It was believed that oxygen permeation occurred as a result of the high density of grain boundaries. Oxygen permeation of this kind through pentacene films results in large scale oxidation taking place in the bulk of the solid, a process becoming more important in the films containing a large density of imperfections. In the discussion of the origin of the trapping centres in pentacene films, we cannot rule out the role played by the presence of oxygen, which is closely associated with the degree of structural disorder of the films.

Acknowledgements

We would like to thank M. Schott and P. Delannoy for their helpful discussions.

References

[1] M. Campos, Mol. Cryst. liquid Cryst. **18**, 105 (1972).
[2] W. Mehl and B. Funk, Phys. Letters A **25**, 364 (1967).
[3] I. Gränacher, Solid State Commun. **3**, 331 (1965).
[4] T. Sakurai and S. Hayakawa, Japan. J. appl. Phys. **13**, 1733 (1974).
[5] H. Baessler, G. Hermann, N. Riehl, and G. Vaubel, J. Phys. Chem. Solids **30**, 1579 (1969).
[6] W. Hofberger, phys. stat. sol. (a) **30**, 271 (1975).
[7] D. C. Northrop and O. Simpson, Proc. Roy. Soc. **234** A, 124 (1956).
[8] E. A. Silinsh, A. I. Belkind, D. R. Balode, A. J. Biseniece, V. V. Grechov, L. F. Taure, M. V. Kurik, J. I. Vertzymacha, and I. Bok, phys. stat. sol. (a) **25**, 339 (1974).
[9] K. O. Lee and T. T. Gan, to be published.
[10] M. A. Lampert, Phys. Rev. **103**, 1648 (1956).
[11] B. Korsch, F. Willig, H. J. Gaehrs, and B. Tesche, phys. stat. sol. (a) **33**, 461 (1976).
[12] G. G. Roberts and F. W. Schmidlin, Phys. Rev. **180**, 785 (1969).
[13] A. F. Prikhotko, A. F. Skorobogatko, and L. I. Tsikora, Optika i Spektroskopiya **26**, 966 (1969).
[14] J. M. Thomas, J. O. Williams, and L. M. Turton, Trans. Faraday Soc. **64**, 2505 (1968).
[15] H. Kuroda and E. A. Flood, Canad. J. Chem. **39**, 1981 (1961).
[16] W. Riemer, Thin Solid Films **14**, 79 (1972).

(Received June 23, 1977)

phys. stat. sol. (a) 43, 573 (1977)

Subject classification: 18.4; 18.1; 22.8

Institute for Low Temperature and Structure Research, Polish Academy of Sciences, Wroclaw

Magnetic Properties of UGeY Compounds (Y = S, Se, Te)

By

A. ZYGMUNT

As a continuation of systematic magnetic studies of UXY-intermetallics of tetragonal crystal structure, where X = O, P, As, Sb, Bi, Si, Ge, Sn and Y = S, Se, Te, the magnetic properties of UGeY compounds are investigated within the range 4.2 to 300 K. UGeS, UGeSe, and UGeTe are antiferromagnetics with Néel temperatures 88, 40, and 73 K, respectively. The splitting diagram of the $^3H_4$ term of U^{4+} ion in the crystal field of C_{4v} symmetry, the crystal field parameters, and the temperature function of magnetic susceptibility calculated from the point-charge model are reported. Some divergence between calculated and experimental results is observed.

Получены результаты магнитных измерений соединений в области температур 4,2 до 300 K, которые являются продолжением наших систематических исследований интерметаллических тетрагональных соединений UXY (X = O, P, As, Sb, Bi, Si, Ge, Sn; Y = S, Se, Te). Соединения UGeS, UGeSe и UGeTe являются антиферромагнитными с точкой Нееля 88, 40 и 73 K, соответственно. Приведены диаграмма расщепления мультиплета $^3H_4$ иона U^{4+} в кристаллическом поле симметрии C_{4v}, параметры кристаллического поля и температурная зависимость восприимчивости, причем использовалась модель точечных зарядов. Показано, что магнитные свойстяв исследованных образцов не подчиняются этой модели.

1. Introduction

During the last years systematic structural and magnetic studies of UXY (X = O, P, As, Sb, Bi, Si, Ge, Sn and Y = S, Se, Te) ternary intermetallics crystallizing in two competitive tetragonal types of structure: PbFCl (P4/nmm) and UGeTe (I4/mmm), have been carried out in our laboratory [1 to 3]. Among the UXY series there are such compounds as UOY which have non-metallic character. The other are typical intermetallics. Magnetic and electric properties of UOY and UXY compounds [X = P, As, Sb, Bi] have been investigated previously [2 to 5]. Contrary to the latter which are ferromagnetics the former are antiferromagnetics. For all investigated UXY compounds in the paramagnetic region, the temperature dependence of the reciprocal susceptibility does not follow the Curie-Weiss law. The curvilinear temperature dependence of the reciprocal susceptibility seems to manifest the temperature population of crystal field levels. The values of the paramagnetic moment of UXY compounds change from 2.80 to 3.58 μ_B, which is equal to the magnetic moment of the free U^{4+} ion. It could suggest that the total crystal field splitting of the $^3H_4$ term is comparable with the measurement temperatures coming up to 1000 K. UGeY compounds which have well-refined crystal structure, have been chosen as the subject of investigations [6].

2. Experiment

UGeY compounds (Y = S, Se, Te) were obtained by heating a stoichiometric mixture of the powdered components in evacuated and sealed ampoules at 900 °C for two weeks. The samples were examined by the X-ray powder method using a

Guinier camera. The X-ray analysis confirmed the PbFCl-type structure for UGeS and UGeTe-type structure for UGeSe and UGeTe. Magnetic measurements were performed for the powdered material on the PAR parallel field vibrating sample magnetometer Model 150A in the temperature range 4.2 to 300 K.

3. Results and Discussion

The structural and magnetic experimental results of UGeY compounds are presented in Fig. 1 to 5 and Table 1. Irrespective of the type of structure UGeS, UGeSe, and UGeTe are antiferromagnetics with the Néel temperatures 88, 40, and 73 K, respectively. Two of them: UGeS and UGeTe, belonging to different types of structure, show a spin–flop-like dependence of magnetization on magnetic field strength at temperatures below T_N (Fig. 2 and 5). We suppose that such a behaviour of UGeS and UGeTe is connected with their first-order phase transition in magnetic field.

In the paramagnetic region the magnetic susceptibility of UGeY compounds does not obey the Curie-Weiss law. The curvilinear temperature dependence of the reciprocal susceptibility seems to manifest the temperature population of crystal field levels. It is also known that the temperature dependence of the reciprocal susceptibility in the presence of exchange interaction can be expressed by the equation

$$\chi_{\text{exp}}^{-1}(T) = \chi_{\text{vv}}^{-1}(T) - \lambda \,, \tag{1}$$

where $\chi_{\text{vv}}^{-1}(T)$ is given by the van Vleck theory and λ is the molecular field constant. The $\chi_{\text{exp}}^{-1}(T)$ and $\chi_{\text{vv}}^{-1}(T)$ curves are shifted parallel by the constant value λ and have the same slope and effective magnetic moment.

A few assumptions have been made to calculate the magnetic susceptibility $\chi_{\text{vv}}^{-1}(T)$ of UGeS.

Table 1

Lattice constants and magnetic characteristics of UGeY compounds

compound	crystalline structure: space group	crystalline structure: type	lattice constants (Å)	magnetic ordering	critical field for magnetic transition H_c (kOe)	Néel temperature T_N (K)	Weiss constant θ (K)	paramagnetic moments μ_{eff} (μ_B)
UGeS	P4/nmm	PbFCl	$a =$ 3.820 $c =$ 8.323	AF.	25	88	+77	2.60
UGeSe	I4/mmm	UGeTe	$a =$ 3.933 $c =$ 16.966	AF.	—	40	−20	3.03
UGeTe	I4/mmm	UGeTe	$a =$ 4.110 $c =$ 17.599	AF.	17	73	+27	2.84

The U^{4+} ion with $5f^2$ localized electrons exists in the crystal lattice and the formal charge of the sulphure ligand is (−2). Since the U–S distance in the uranium coordination polyhedron (Fig. 6) is approximately the sum of U^{4+} and S^{2-} ionic radii, this assumption seems to be acceptable. The crystal field splitting may be treated as a perturbation. It means that the splitting is much less than the spin–orbit coupling and Coulomb interaction, so J is a good quantum number and $^3H_4$ is the correct term.

The crystal field Hamiltonian of the uranium ion in the $[UGe_4S_5]$ polyhedron of C_{4v} symmetry has the form

$$H = B_2^0\hat{O}_2^0 + B_4^0\hat{O}_4^0 + B_4^4\hat{O}_4^4 + B_6^0\hat{O}_6^0 + B_6^4\hat{O}_6^4 \,, \tag{2}$$

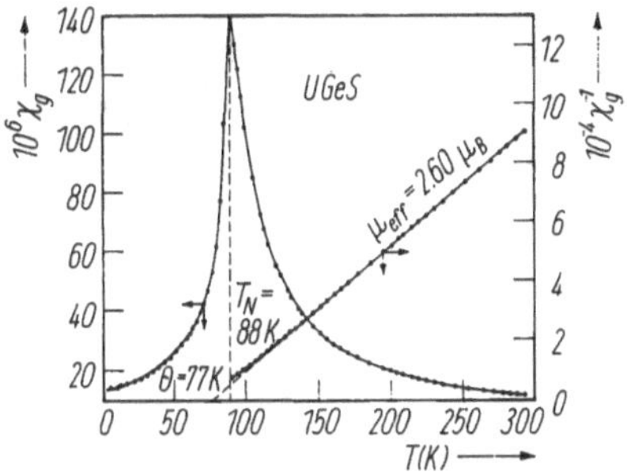

Fig. 1. Magnetic properties of UGeS in the temperature range between 4.2 and 300 K; $H = 4$ kOe

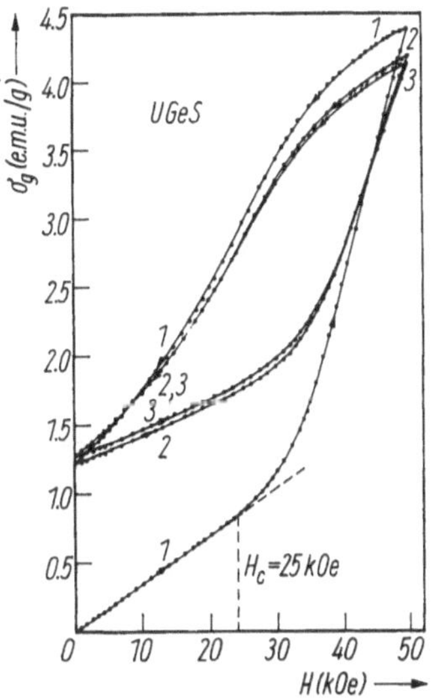

Fig. 2. Dependence of magnetization on magnetic field strength $\sigma_g = f(H)$ at 4.2 K for powder UGeS

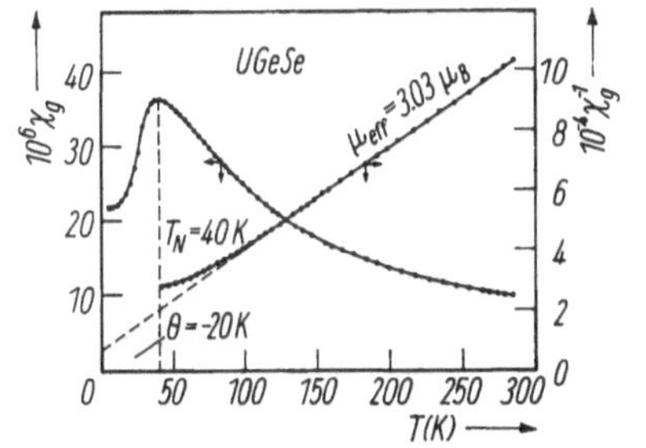

Fig. 3. Magnetic properties of UGeSe in the temperature range between 4.2 and 300 K; $H = 4$ kOe

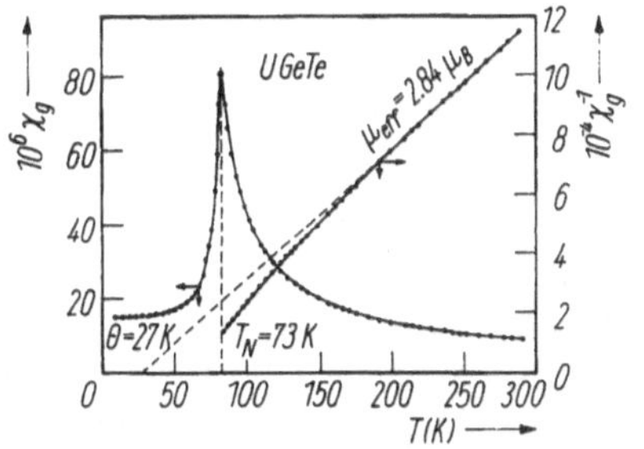

Fig. 4. Magnetic properties of UGeTe in the temperature range between 4.2 and 300 K; $H = 4$ kOe

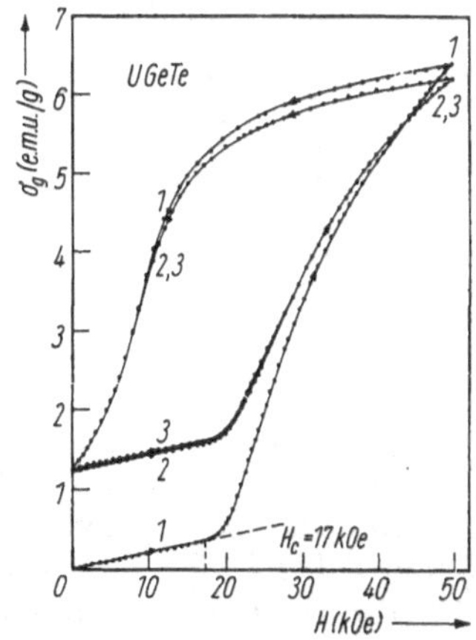

Fig. 5. Dependence of magnetization on magnetic field strength $\sigma_g = f(H)$ at 4.2 K for powder UGeTe

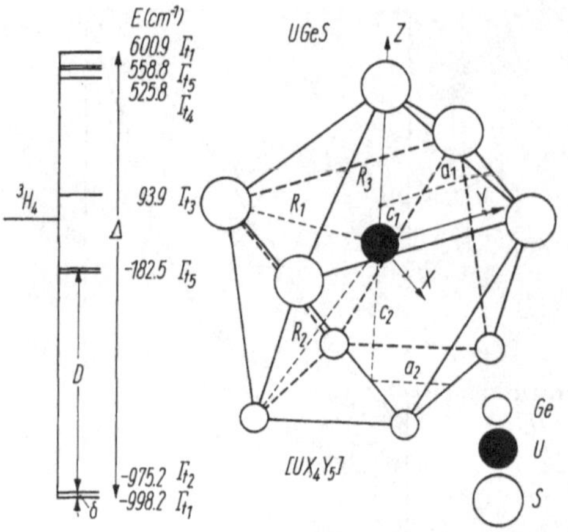

Fig. 6. The coordination polyhedron of U in UGeS and diagram of crystal field levels for $^3H_4$ in tetragonal C_{4v} symmetry

where B_n^m denote crystal field parameters and $\hat{O}_n^m$ are Steven's equivalent operators [7].

The crystal field parameters B_n^m have been calculated basing on the point-charge model:

$$B_2^0 = (4e^2A_2^0/R_1^3 + 4e^2A_2^{0'}/R_2^3 + e^2/R_3^3)\,\alpha\langle r^2\rangle\,, \tag{3}$$

$$B_4^0 = (e^2A_4^0/R_1^5 + e^2A_4^{0'}/R_2^5 + e^2/4R_3^5)\,\beta\langle r^4\rangle\,, \tag{4}$$

$$B_4^4 = (-35e^2A_4^4/2R_1^5 + 35e^2A_4^{4'}/2R_2^5)\,\beta\langle r^4\rangle\,, \tag{5}$$

$$B_6^0 = (e^2A_6^0/4R_1^7 + e^2A_6^{0'}/4R_2^7 + e^2/8R_3^7)\,\gamma\langle r^6\rangle\,, \tag{6}$$

$$B_6^4 = (-\,63e^2A_6^4/4R_1^7 + 63e^2A_6^{4'}/4R_2^7)\,\gamma\langle r^6\rangle\,, \tag{7}$$

where $R_i = (2a_i^2 + c_i^2)^{1/2}$ are distances between the i-th ligands and the uranium ion, expressed by c_i and a_i, the parameters of the crystal cell, α, β, γ are constants, $\langle r^n\rangle$ is the mean n-th power of the radius of 5f electrons (8), A_n^m geometric coefficients obtained by expanding the polyhedron potential of the coordination polyhedron into power series:

$$A_2^0 = [(c_1/a_1)^2 + 1]/[(c_1/a_1)^2 + 2]\,, \tag{8}$$

$$A_2^{0'} = [(c_2/a_2)^2 + 1]/[(c_2/a_2)^2 + 2]\,, \tag{9}$$

$$A_4^0 = [(c_1/a_1)^4 - 6(c_1/a_1)^2 + 3/2]/[(c_1/a_1)^2 + 2]^2\,, \tag{10}$$

$$A_4^{0'} = [(c_2/a_2)^4 - 6(c_2/a_2)^2 + 3/2]/[(c_2/a_2)^2 + 2]^2\,, \tag{11}$$

$$A_4^4 = 1/[(c_1/a_1)^2 + 2]^2\,, \tag{12}$$

$$A_4^{4'} = 1/[(c_2/a_2)^2 + 2]^2\,, \tag{13}$$

$$A_6^0 = [2(c_1/a_1)^6 - 30(c_1/a_1)^4 + 45(c_1/a_1)^2 - 5]/[(c_1/a_1)^2 + 2]^3\,, \tag{14}$$

$$A_6^{0'} = [2(c_2/a_2)^6 - 30(c_2/a_2)^4 + 45(c_2/a_2)^2 - 5]/[(c_2/a_2)^2 + 2]^3\,, \tag{15}$$

$$A_6^4 = [5(c_1/a_1)^2 - 1]/[(c_1/a_1) + 2]^3\,, \tag{16}$$

$$A_6^{4'} = [5(c_2/a_2)^2 - 1]/[(c_2/a_2) + 2]^3\,. \tag{17}$$

The crystal field parameters have the following values (in cm^{-1}): $B_2^0 = -3.273 \times 10$, $B_4^0 = -6.972 \times 10^{-2}$, $B_4^4 = 1.077$, $B_6^0 = 6.709 \times 10^{-5}$, $B_6^4 = 8.766 \times 10^{-3}$. The energy values and the eigenvectors for each crystal field level calculated for UGeS are listed in Table 2. The obtained splitting scheme of the $^3H_4$ term is shown in Fig. 6. It can be observed that the nine-field degenerate $^3H_4$ term splits up into two doublets

Table 2

Energy levels and eigenvectors of the ground $^3H_4$ term in tetragonal C_{4v} symmetry for UGeS as derived from the point-charge model

irreducible representation	degeneracy	energy (cm^{-1})	eigenvectors
Γ_{t_1}	1	+600.9	$0.0847\,\|+4\rangle + 0.0847\,\|-4\rangle + 0.9928\,\|0\rangle$
Γ_{t_5}	2	+558.8	$0.2313\,\|-3\rangle + 0.9729\,\|+1\rangle$ $0.2313\,\|+3\rangle + 0.9729\,\|-1\rangle$
Γ_{t_4}	1	+525.8	$0.7071\,\|+2\rangle + 0.7071\,\|-2\rangle$
Γ_{t_3}	1	+ 93.9	$0.7071\,\|+2\rangle - 0.7071\,\|-2\rangle$
Γ_{t_5}	2	−182.5	$0.9729\,\|-3\rangle - 0.2313\,\|+1\rangle$ $0.9729\,\|+3\rangle - 0.2313\,\|-1\rangle$
Γ_{t_2}	1	−975.2	$0.7071\,\|+4\rangle - 0.7071\,\|-4\rangle$
Γ_{t_1}	1	−998.2	$0.1198\,\|0\rangle - 0.7020\,\|-4\rangle - 0.7020\,\|+4\rangle$

Γ_{t5} and five singlets Γ_{t1}, Γ_{t1}, Γ_{t2}, Γ_{t3}, Γ_{t4}. The total splitting of $^3H_4$ is about 1600 cm^{-1}. The ground state is composed of two close singlets Γ_{t1} and Γ_{t2}. The doublet Γ_{t5} are the first excited levels. Such pattern of levels gives the temperature dependence of magnetic susceptibility:

$$\chi_{vv} = 1/3\,\chi_{\parallel} + 2/3\,\chi_{\perp}\,, \tag{18}$$

$$\chi_{\parallel} = N\beta^2[40.96\,(1 - e^{-\delta/kT})\,\delta + 9.936\,e^{-D/kT}]\,(1 + e^{-\delta/kT} + 2\,e^{-D/kT})^{-1}\,, \tag{19}$$

$$\chi_{\perp} = N\beta^2[2.714\,(1 - e^{-D/kT})/D + {} + 2.423\,(1 - e^{-(D-\delta)/kT})/(D - \delta)]\,(1 + e^{-\delta/kT} + 2\,e^{-D/kT})^{-1}\,, \tag{20}$$

with effective magnetic moment not less than $3.20\mu_B$ at 300 K. The experimental effective magnetic moment for UGeS is equal to $2.60\mu_B$ and cannot be fitted with the diagram of point-charge crystal levels.

We are going to find the actual crystal field parameters for the UGeY series from the temperature dependence of the magnetic susceptibility within a wider temperature range and the experimental value of the magnetic moment in the ordered state.

Acknowledgement

The author is grateful to Prof. B. Stalinski for his stimulating interest in this work.

References

[1] A. Murasik, W. Suski, R. Troć, and J. Leciejewicz, phys. stat. sol. **30**, 61 (1968).
[2] W. Trzebiatowski, J. Niemiec, and A. Sępichowska, Bull. Acad. Polon. Sci., Ser. sci. chim. **9**, 373 (1961).
[3] A. Zygmunt, A. Murasik, S. Ligenza, and J. Leciejewicz, phys. stat. sol. (a) **22**, 75 (1974).
[4] A. Zygmunt and M. Duczmal, phys. stat. sol. (a) **9**, 659 (1972).
[5] A. Zygmunt and A. Czopnik, phys. stat. sol. (a) **18**, 731 (1973).
[6] A. J. Klain Haneveld and F. Jellinek, J. less-common Metals 18, 123 (1969).
[7] K. W. H. Stevens, Proc. Phys. Soc. **65A**, 209 (1952).
[8] A. J. Freeman, J. P. Desclaux, G. H. Lander, and J. Faber, Jr., Phys. Rev. B **13**, 1168 (1976).

(Received June 21, 1977)

phys. stat. sol. (a) **43**, 579 (1977)

Subject classification: 10.1 and 11; 21.1

School of Science, Griffith University, Nathan[1])

Defect Recovery in Zn after Low Temperature Electron Irradiation

By

S. MYHRA

Some new experimental results for Stage I and III recovery in electron-irradiated Zn are presented. Stage I (< 20 K) is found to consist of more sub-stages than previously resolved. There is little evidence for long range migration below 20 K although the possibility cannot be ruled out. Stage III (> 70 K) exhibits several features different from those reported previously. In particular, a sub-stage centered at 140 K, thought to be dose-dependent, exhibits no observable shift. A possible explanation is suggested. Several recovery models are investigated; the totality of available results cannot readily be explained fully by either of the two currently most widely accepted models.

Es werden neue experimentelle Ergebnisse für Erholung der Stufen I und III in elektronenbestrahltem Zn mitgeteilt. Es wird gefunden, daß Stufe I (< 20 K) aus mehr Unterstufen besteht als früher angenommen. Nur geringe Hinweise für weitreichende Verlagerungen unterhalb 20 K werden gefunden, obwohl die Möglichkeit nicht ausgeschlossen werden kann. Stufe III (> 70 K) zeigt einige Merkmale, die von den früher berichteten unterschiedlich sind. Insbesondere zeigt ein bei 140 K gelegener Subzustand, von dem angenommen wird, daß er dosisabhängig ist, keine beobachtbare Verschiebung. Eine mögliche Erklärung wird vorgeschlagen. Einige Erholungsmodelle werden untersucht. Die Gesamtheit der vorhandenen Ergebnisse läßt sich von keinem der beiden zur Zeit am verbreitetsten akzeptierten Modelle vollständig erklären.

1. Introduction

The interpretation of recovery data for irradiated metals has been a controversial subject for some time with regard to the f.c.c. metals in general and Cu versus Au in particular. Sound arguments based on good data have been advanced for both the 1- and the 2-interstitual models. Recently, the dispute has found a new home in some of the h.c.p. metals such as Cd and Zn. This has occurred for two reasons; a great deal of reliable data has become available for these metals and their anisotropic structures leave rather more room for a variety of interpretations.

Seeger and Gösele [1] have discussed the recovery of electron and neutron irradiated Cd in some detail and proposed that anisotropic interstitial migration can account for the concentration dependent Stage III (peak temperature 101 to 143 K). Roggen, Cornelis, Nihoul, and Stals (henceforth abbreviated to RCNS) have recently found evidence for an analogous stage in α-particle irradiated Zn [2 to 4] which seemingly supports the Seeger-Gösele model. Previously, Vandenborre, Nihoul, and Stals [5] (VNS) have reported recovery results for electron-irradiated Zn.

The present paper presents new results for the recovery of electron-irradiated Zn and compares these to previously available data.

---

[1]) Nathan, Queensland, 4111, Australia.

2. Experimental

The general experimental procedure, the cryostat, the monitoring systems and the beam optics have been described previously [6, 7].

The sample material was supplied by Metals Research Ltd. in ribbon form having an initial stated purity of six nines. The ribbon was cold rolled down to a 0.0025 cm foil from which ≈ 0.02 cm wide strips were cut. After etching in dilute HNO_3, the samples were rinsed and annealed in vacuum at 300 °C for sereval hours. The strips were soldered in a zig-zag pattern to copper mounting studs on a sapphire rod sample holder. (Sample holder and cold-finger arrangements have been described elsewhere [7].) The final as-mounted resistance ratios of the samples, $R_{290\,K}/R_{5\,K}$, were typically 10^3. Total sample length exposed to the 1 MeV electron beam was ≈ 12 cm. A dummy sample, shielded from the beam, was included and served essentially as a secondary thermometer.

The experimental resolution of the incremental resistivity measurements was typically $\approx 6 \times 10^{-7}$ μΩcm; a Leeds and Northrup six-dial potentiometer and a measuring current of 100 mA stabilized to better than 5 parts in 10^5 were used. The temperature measurements were made with the aid of a calibrated Rh(Fe) resistance thermometer embedded in the sample holder; the resolution and absolute accuracy of this sensor were better than 0.01 K. Uniform irradiation of the sample aperture was achieved by sweeping the beam along orthogonal axes with a parallel-plate capacitor arrangement driven by ramp-function power supplies. The annealing was done isochronally with 5 min heating pulses; the approach and departure slopes of the pulses were such that heating and cooling times for temperature increments between successive pulses were less than 30 and 5 s at 250 and 10 K, respectively.

3. Annealing Results

A typical full recovery curve is shown in Fig. 1. The irradiations were performed below 7 K for the lowest dose runs and the annealing programme was terminated at 250 K in most instances. Relevant results for the total irradiation dose and the induced resistivities of the five runs are shown in Table 1. Details of the annealing results below and above 20 K will now be presented under two headings.

3.1 $T_A < 20$ K

Direct and differential recovery curves are shown in Fig. 2 and 3. The data for run 4 are somewhat truncated because irradiation annealing encroached. The temperature range 14 to 18 K is of particular interest since this is the region where a dose-dependent analogue to the Cu I_E sub-stage might be found. It is seen, however, that the two minor sub-stages centered just below 15 K and around 17 K are dose independent within the experimental resolution. Other first-order sub-stages can be resolved at 7.5,

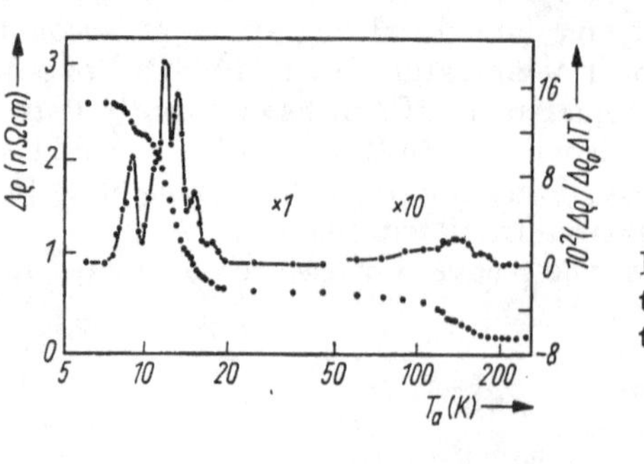

Fig. 1. Direct and differential recovery curves for the temperature range 7 to 250 K for run 3. Note the change of vertical scale for differential recovery at 50 K

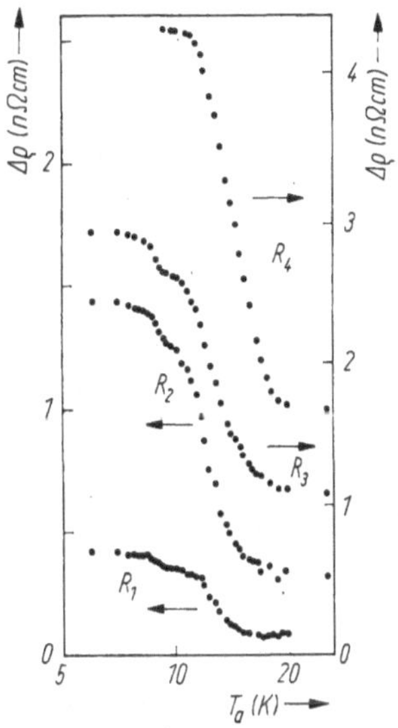

Fig. 2. Direct recovery curves for $T_A < 20$ K from runs 1, 2, 3, and 4. Substantial beam heating has encroached into the lower part of the recovery for run 4

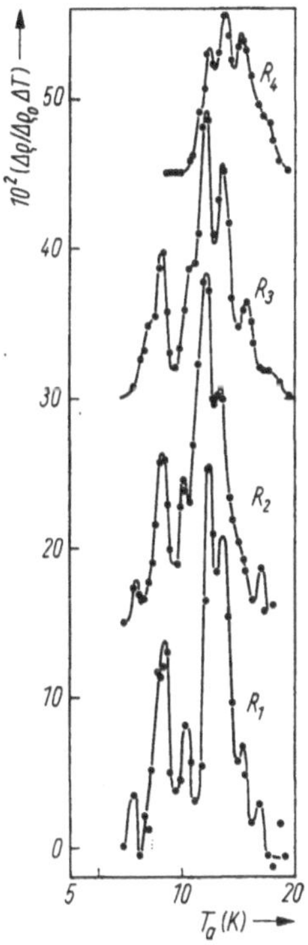

Fig. 3. Differential recovery curves for $T_A < 20$ K from runs 1, 2, 3, and 4

9.0, 10.4, 11.8, and 12.8 K. These results are with one exception identical to those obtained by α-irradiation [3]. Rather less structure is reported by VNS although the overall features are identical. RCNS have reported a sub-stage at 15.4 K but no further Stage I recovery. The sub-stage at ≈ 17.5 K, resolved by the present data, is in evidence in the work by VNS only as a slight high-temperature tail. A cold-worked specimen was investigated during runs 3 and 4. Except for some enhancement of the

Table 1
Characteristics data for annealing runs

run	dose (e/cm²)	$\Delta\varrho_0$ (nΩ cm)	$\Delta\varrho_{I}$ (< 20 K) (nΩ cm)	$\Delta\varrho_{III}$ (> 70 K) (nΩ cm)	$\Delta\varrho_{III}$ (>120K) (nΩ cm)	comments
1	1.89×10^{15}	0.42	0.32	0.10	N.D.	annealing programme terminated at 20 K
2	5.04×10^{16}	1.39	1.05	0.31	N.D.	annealing programme terminated at 20 K
3	1.83×10^{17}	2.94	2.16	0.65	0.38	slight beam heating below 10 K
4	7.98×10^{17}	4.28	2.10	1.69	0.98	most of Stage I lost to beam heating — hence the low value for $\Delta\varrho_I$ (< 20 K)
5	2.44×10^{18}	2.23	0	2.23	1.58	lower part of Stage II lost to beam heating — hence the low value for $\Delta\varrho_{III}$ (> 70 K)

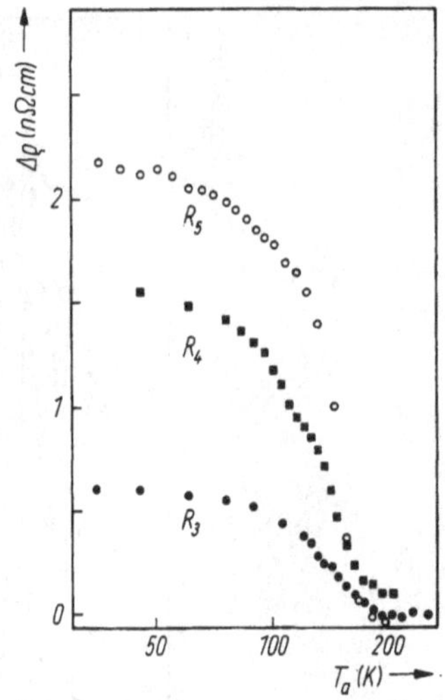

Fig. 4. Direct recovery curves for $T_A > 20$ K from runs 3, 4, and 5. Substantial beam heating has encroached into the lower part of the recovery for run 5

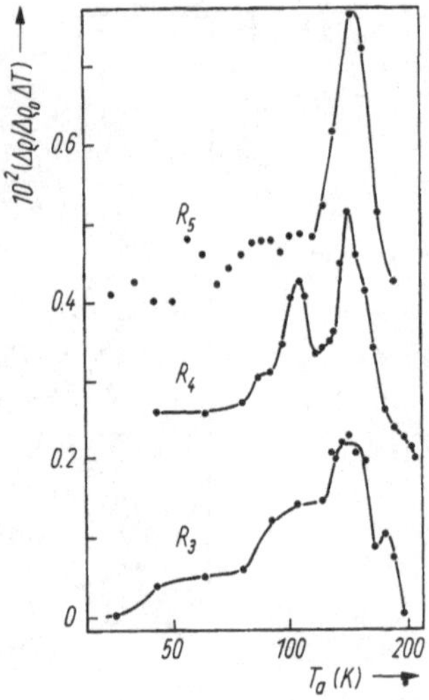

Fig. 5. Differential recovery curves for $T_A >$ > 20 K from runs 3, 4, and 5

sub-stage at 17.5 K no significant differences were observed in the annealing characteristics.

3.2 $T_A > 20$ K

Direct and differential recovery curves are shown in Fig. 4 and 5. The data for run 5 are largely invalid below 100 K because of radiation annealing resulting from beam heating. (High dose runs become unreasonable time-consuming unless the beam current is increased.) Two prominent sub-stages are resolved with centre temperatures of 100 and 145 K. Two further sub-stages appear to centred at ≈ 85 and ≈ 175 K. No clear-cut dose dependence is observed. Furthermore, these results have little in common with those previously published by RCNS [2] who found dose independent sub-stages at ≈ 190 and 163 K as well as a prominent dose dependent sub-stage between 120 and 140 K. However, VNS [5] have reported sub-stages at 85 and 98 K in addition to the dose dependent one at 130 to 154 K; no recovery is found above 160 K.

4. Discussion

4.1 $T_A < 20$ K

Using the present results and those of RCNS and VNS it is now possible to assemble a listing of sub-stages which are resolvable below 20 K: I_1 (5 K), I_2 (5.9 K), I_3 (7.5 K), I_4 (8.8 K), I_5^* (10.5 K), I_6^* (11.8 K), I_7 (12.8 K), I_8 (15.4 K), and I_9^* (17.5 K). Those marked with a star have been resolved for the first time in the present work. The observed lack of dose dependence and the narrow peak widths, $\Delta T_{1/2}/T_c \approx 0.095$, tend to support the supposition that no long range migration occurs for $5 < T_A <$ < 20 K. However, until the characteristics of I_7 and I_8 have been investigated in some detail with isothermal recovery programmes, impurity doping, etc., it is impossible to rule out long-range migration in this temperature range.

Sub-stages I_4, I_5, I_6, and I_7 were fitted to the first order chemical rate equation

$$\ln\left(\frac{n_i}{n_f}\right)_{j\text{-th anneal}} = \nu\tau \exp\left(-\frac{E_A}{kT_j}\right),$$

where n_i and n_f are the initial and final defect concentrations, respectively, for the j-th annealing pulse at T_j, ν is a frequency factor, τ is the pulse time, k is Boltzmann's constant and E_A is the activation energy for recombination. First order fit was confirmed for I_4 and $E_A = (15 \pm 3)$ meV was calculated from an Arrhenius plot. The other well-resolved peaks I_5, I_6, and I_7 can be handled the same way although with less confidence because the stages partially overlap.

A procedure [8, 9], similar to that used for resolutions of individual half-lives in a complex nuclear decay spectrum, was used to sequentially "peel off" the sub-stages I_7, I_6, and I_5. The resultant activation energies were $E_A(I_5) = (32 \pm 5)$ meV, $E_A(I_6) = (32 \pm 4)$ meV, and $E_A(I_7) = (40 \pm 3)$ meV. I_8 and I_9 are not well enough defined for any meaningful analysis to be done. However, if it is assumed that free migration in either three, two or one dimensions does occur, one may choose reasonable values for the parameters relevant to I_8 and I_9 and calculate expected peak shifts for these sub-stages when the specimens are subjected to the actual irradiation doses. The result is that ΔT (shift) should be at least 1 K; this should be resolvable in the present experiment.

4.2 $T_A > 20$ K

The following sub-stages have been reported previously (VNS and RNCS): III_1 at 85 K, III_2 at 100 K, III_3 at 101 K, III_4 at 130 to 155 K, III_5 at 165 K, and III_6 at 190 K. The labelling is arbitrary and should not be thought of as implying a presumed interpretation. Sub-stage III_4 appears to be dose dependent (RCNS and VNS). The present data exhibit clearly resolved peaks at 110 and 140 K which fit III_3 and III_4. There is a shoulder at ≈ 90 K and this may therefore be III_1. The best recovery curve (run 4) has two high temperature shoulders at ≈ 160 and 185 K which can be identified with III_5 and III_6. Hence the gross features of the three independent experiments are similar although the number of high-temperature sub-stages are rather more numerous than for the f.c.c. metals. However, the present annealing data show no peak shift for III_4 even though any reasonable assumptions with respect to the activation energy and dimension of the diffusion process for an alleged free migration process should result in substantial and visible peak shifts. Also, the differences between the present work and that of VNS must be accounted for.

One possible way out of the dose dependence dilemma may be that the apparent peak shift (or lack of such) is indeed not real but rather due to fortuitous circumstances. Consider the following scenario. There are two over-lapping but resolvable sub-stages of roughly similar magnitude. If a dose-dependent fractional recovery per sub-stage is postulated, the relative encroachment of the sub-stages with dose will alter. Hence a spurious peak shift to lower temperature can be generated if the following conditions are satisfied:

(i) Two sub-stages with high relative encroachment are present.

(ii) The fractional recovery must increase with dose for the sub-stage with the lowest center temperature.

The opposite situation, such that a "real" peak shift is masked, may take place when the dose-dependent sub-stage is followed at a higher temperature by a dose-independent encroaching stage such that its fractional recovery contribution is increasing with dose. This case is rather less likely to be a problem as the resolution improves with dose when the shifting sub-stage move away from the encroaching partner.

Similarly, the situation may be reversed such that a diminishing dose independent sub-stage precedes the shifting sub-stage.

The most complicated situation occurs, of course, when three stages are overlapping, the fractional recovery for all stages are changing relatively with dose and dose dependence may (not) be present.

Unfortunately, the third case may be the one which applies for Zn. The sub-stages III_3, III_4, and III_5 are encroaching and the relative peak heights are clearly changing with dose. This effect can also account for the poor correspondence in peak position when the present data is compared to those of RCNS and VNS. The three studies differ primarily in their method of damage introduction; 1 MeV (SM), 2.5 MeV (VNS) electrons and 6.1 MeV (RCNS) α-particles respectively. The maximum energy transfers can readily be calculated from the expressions below:

$$T_M = \frac{560.8}{A} (x + 2) \qquad \text{(electrons)}$$

$$T_M = \frac{4m_\alpha A}{(m_\alpha + A)^2} E \qquad \text{(nonrelativistic } \alpha\text{-particle)} ,$$

where m_α and A are the atomic weights of the α-particle and the target atom respectively, x is electron beam energy divided by the electronic rest mass energy. For Zn we have $T_M(1\ \text{MeV}\beta) \approx 68$ eV, $T_M(2.5\ \text{MeV}\beta) \approx 290$ eV, and $T_M\ (6.1\ \text{MeV}\alpha) = 1.33$ MeV. However the average energy transfers are more relevant. These are given by

$$\langle T\rangle = T_D\left(\ln\left(\frac{T_M}{T_D}\right) - 1 + \frac{Z}{137}\right) \qquad \text{(electrons)}$$

$$\langle T\rangle = T_D \ln\left(\frac{T_M}{T_D}\right) \qquad (\alpha\text{-particles}) ,$$

where T_D is the threshold displacement energy for the target material, 14.5 eV for Zn.

We find that $\langle T\rangle$ is 11 eV (1 MeVβ), 26 eV (2.5 MeVβ), and 166 eV (6.1 MeVα). As a result of the large difference in average energy transfer for β- and α-irradiation, it is likely that the proportions of distinct defect configurations will be different in the present work as compared to that of RCNS. Other studies have exhibited variations in relative magnitudes for the sub-stages for different methods of defect production. Further differences may arise from the presence of impurities, excess vacancy concentrations, dislocations, etc. Two of the studies presently considered (SM and RCNS) used specimens with nearly identical background characteristics; resistance ratios 1000 versus 750 and both well-annealed while the relevant number for the 2.5 MeV β-irradiation case was 8000 to 9000.

5. Some Model Considerations

The following characteristics must be explained by a successful model: (i) A large number of dose-independent sub-stages, which can be described by first order rate equations, are found below 20 K. (ii) There is some evidence for free migration in the range $5 < T < 20$ K. (iii) At least six sub-stages are found in the range $80 < T < 200$ K. (iv) There is considerable evidence for a dose dependent sub-stage between 130 and 145 K; the lack of shift for the present work may be due to a fortuitous combination of several close sub-stages. (v) There is little, if any, recovery between 20 and 70 K. (vi) There is substantial similarity of recovery features for several h.c.p. metals.

It is unlikely that anything but close-pair recovery can account for sub-stages I_1 through I_7; I_8 and/or I_9, on the other hand, cannot be ruled out as being dose dependent. Referring back to Fig. 3, it is, for example, possible to argue that I_9 is gradually disappearing under I_8 with increasing dose. Furthermore, this sub-stage appears only as a high-temperature tail on the I_8 substage in the two other studies [3, 5]. This is not unreasonable as one would expect the higher energy transfer during α-irradiation to create a larger proportion of non-close defect pairs. From the results of RNCS [3] it is moreover possible to detect a trend towards a diminishing tail at higher α-doses. Similarly, the data from VNS are a result of much higher defect concentrations. The plethora of close-pair stages (compared to Cu) is not surprising when the h.c.p. structure of Zn is considered; because of the substantial deviation (1.85 versus $(8/3)^{1/2} = 1.63$) of c/a from the ideal value there should be a large number of non-equivalent close-pair interstitial sites.

It is difficult to envisage an energy hierarchy of close-pair interstitial sites which does not end in sites leading to free migration. Some possibilities which could account for the lack of an I_E-type sub-stage can be thought of: a) The presence of substantial impurity concentation such that free migration is suppressed. b) Instant free interstitial conversion to a more stable defect. c) A very large capture radius such that free migration only occurs at low defect concentrations where the experimental resolution is poor.

Non-linear damage rates, observed even at irradiation temperatures below most of the close-pair stages, have been found for Zn [5] and several other h.c.p. metals. Thus c) can be supported. Even though high purity specimen materials (six 9's) have been used in the present experiment, there is a possibility that a) or a combination of a) and c) may suppress free migration recovery. If hypothesis a) is operative, substantial stage II recovery should occur. Possibly the sub-stages labelled III_1, III_2, and III_3 may fit this picture; the lack of dose dependence in the corresponding temperature range is in accord with such an interpretation. These sub-stages are relatively enhanced for the α-irradiation study; this observation agrees with the postulated higher non-close defect concentration and the slightly poorer resistance ratio for the specimen material used by RCNS. However, specimens used by VNS have a much higher resistance ratio while at the same time an enhanced recovery below 100 K is observed. Alternately, some or all of the sub-stages III_1, III_2, and III_3 may be interpreted, in the Gösele-Seeger model, as close-pair interstitial sub-stages preceding III_4, an alleged interstitial free migration stage; presumably this is a Type-II interstitial or a converted Type-I configuration. Returning to the 1-interstitial picture, sub-stage III_4 could then rather be associated with vacancy migration followed by vacancy release from traps and/or vacancy complex formation in III_5 and III_6. Therefore, both models would seem to fit the available experimental data. There are two serious objections to identifying III_1, III_2, and III_3 as interstitial release from traps. Firstly, they are too well-defined and, secondly, the magnitudes are rather too high unless unreasonably large impurity concentrations and/or anomalous high interstitial specific resistivities are assumed. On the other hand, their relative magnitudes versus III_4 and lack of first-order shape makes a 2-interstitial interpretation unlikely. This leaves us with the vacancy as the most likely candidate but with the need for a dose-independent mechanism such as correlated migration to defect traps or impurity sites. Substage III_4 can then quite naturally be interpreted as being due to vacancy free migration. Subsequently, III_5 and III_5 may be accounted for by invoking release of vacancies from traps and/or vacancy-complex formation.

Previous results for quenching effects and vacancy characteristics in Zn[10] point to substantial vacancy migration in the temperature range 140 to 180 K with slight continuous recovery up to 350 K. This strongly supports the RCNS model and, con-

Table 2

Summary of model considerations

sub-stage	T_Q (K)	observed characteristics	Seeger-Gösele model	I-interstitial model
III_1	85	seen by: VNS, SM increasing percentage with dose. no D-D 1st order kinetics shape	close pair or type I interstitial release from traps	interstitial release from traps
III_2	100	seen by: VNS, RNCS small for 2.5 MeV β-irr. large for 6.1 MeV α-irr. no D-D	type II close pair	interstitial release from traps or correlated vacancy migration
III_3	110	seen by: RNCS, SM not seen for high-purity specimens prominent for α-irr. no D-D	type II close pair	interstitial release from traps or correlated vacancy migration
III_4	130–155	seen by: RNCS, VNS, SM D-D seen by RNCS, VNS and consistent with free migration. $E_A = 0.24$eV	type II-interstitial free migration	vacancy free migration
III_5	165	seen by: RNCS, SM not seen for high-purity specimens	type II-interstitial release from traps	vacancy complex formation
III_6	190	seen by: RNCS, SM not seen for high-purity specimens D-D not ruled out	vacancy migration	vacancy release from traps

versely, speaks against vacancy-migration in III_4. Following the RCNS argument it is then necessary to look for dose dependence in sub-stages III_5 or III_6; this is not present in the α-irradiation study while it cannot be ruled out for the present electron irradiation data. The VNS data on high-purity Zn in fact exhibit no recovery above sub-stage III_4. The shoulders on the high-temperature side of III_4 in Fig. 5 appear to diminish at higher doses.

However, the discrepency between alleged vacancy migration characteristics in irradiated versus quenched metals has a long history [11]. If the mismatch cannot be used as the deciding factor for various recovery models in, say, the f.c.c. metals, it would be unreasonable to place a greater reliance on it for the h.c.p. metals.

The model considerations from above have been summarized in Table 2.

6. Conclusions

1. Several new sub-stages have been resolved for recovery below 20 K. There is some evidence for free migration for $5 < T_A < 20$ K with the proviso that I_8 and I_9 need further study.

2. The general recovery features in the present work for annealing above 20 K are similar to those previously observed. However, there are major and significant discre-

pancies for the peak positions. As well, the role of the dose dependent sub-stage III_4 remains unresolved. An explanation for these variations has been advanced by considering the role of encroachment and relative sub-stage enhancement with dose.

3. The totality of available recovery features in Zn has been considered in terms of the 1- and 2-interstitial models and variations on these. It appears that none of the models can account fully for all the results unless unreasonable assumptions and assignments are made. Further detailed measurements with the present techniques as well as new experimental approaches are clearly required.

Acknowledgements

The present work was supported by the Australian Institute of Nuclear Science and Engineering and made extensive use of experimental facilities at the Australian Atomic Energy Commission's research laboratories. The research project was commenced while the author was employed by the University of Queensland. It is a pleasure to acknowledge experimental help from Dr. R. B. Gardiner and Miss J. E. Charles as well as Mr. K. Fenn.

References

[1] A. Seeger and U. Gösele, Radiat. Eff. **24**, 123 (1975).
[2] S. Roggen, J. Cornelis, J. Nihoul, and L. Stals, Phil. Mag., in the press.
[3] S. Roggen, J. Nihoul, J. Cornelis, and L. Stals, J. nuclear. Mater. (Amsterdam), in the press.
[4] S. Roggen, J. Nihoul, J. Cornelis, and L. Stals, Scripta metall., in the press.
[5] H. Vandenborre, J. Nihoul, and L. Stals, Crystal Lattice Defects **5**, 89 (1974).
[6] S. Myhra and R. B. Gardiner, Radiat. Eff. **18**, 39 (1973).
[7] S. Myhra and R. B. Gardiner, Radiat. Eff. **15**, 209 (1972).
[8] T. N. O'Neal and R. L. Chaplin, Phys. Rev. B **5**, 3810 (1972).
[9] S. Myhra and R. B. Gardiner, Radiat. Eff. **27**, 35 (1975).
[10] F. Maury, A. Lucasson, and P. Lucasson, Crystal Lettice Defects **2**, 47 (1971).
[11] M. W. Thompson, Defects and Radiation Damage in Metals, Cambridge University Press, 1969.

(Received July 4, 1977)

phys. stat. sol. (a) 43, 589 (1977)

Subject classification: 10.2; 22

Institute of Ultrahard Materials[1]) *(a), Institute of Metal Physics*[2]) *(b), Academy of Sciences of the Ukrainian SSR, Kiev, and Institute of Physical Metallurgy*[3]) *(c), Czechoslovak Academy of Sciences, Brno*

Positron Annihilation in Boron Nitride Crystals

By

A. S. VISHNEVSKII (a), I. YA. DEKHTYAR (b), E. G. MADATOVA (b), V. S. MIKHALENKOV (b), and A. ČÍŽEK (c)

The dislocation structure of the zincblende form of BN is investigated. The crystals of 3 to 5 μm in diameter are compressed under 60 kbar at temperatures of 20, 500, and 1500 °C and the angular correlation spectra of annihilation photons are measured. The results are interpreted as follows: The positrons are trapped at the Bdislocations and the fraction of dangling bonds $\alpha_d \approx 0.6$ corresponds to that of GaAs for Gadislocations.

Была исследована дислокационная структура боразона. Порошек 3 до 5 μm) нитрида бора BN со сфалеритной структурой был спрессован (давление 60 kbar) при температурах 20, 500 и 1500 °C и были измерены кривые углового распределения аннигиляционных фотонов. Результаты можно интерпретировать таким образом: Ловушками для позитронов являются ядра дислокаций, состоящие из атомов бора и доля занятых дислокационных уровней $\alpha_d \approx 0{,}6$ согласна с величиной полученной для Ga-дислокаций в GaAs.

1. Introduction

The positron annihilation method (PA) has allowed to discover and to appreciate qualitatively characteristic features of dislocations in semiconducting compounds, i.e. GaAs [1]. On the other hand, it is well known that under standard conditions the diamond-type compounds are dielectrics and acquire semiconducting properties in the presence of lattice imperfections. In particular, the boron nitride BN belongs to these compounds.

The close form of BN with the diamond cubic structure, commonly called the zincblende (ZnS) structure, is characterized by a tetrahedral distribution of atoms: Each atom is surrounded by four equidistant nearest atoms of the other kind, arranged at the vertices of a tetrahedron. The zincblende lattice may be treated as two f.c.c. sublattices interpenetrating one quarter of the way along the main diagonal of the cube. Due to the distribution of valence electrons between the neighbour atoms, the electron-pair covalent bond is formed in the ⟨111⟩ direction.

The 60° dislocations seem to be the most numerous ones in crystals with the diamond structure and are introduced by plastic deformation [2]. The fraction of atoms on the edge of the extra-half plane of dislocation has unsaturated "dangling" bonds that may serve as acceptors for electrons.

The dislocation structure of the zincblende is alike to that of diamond with the exception that there are two types of dislocations in these compounds, i.e. the α- and β-dislocations [3].

The purpose of the present paper is to study the relation between the results obtained by PA and the dislocation structure of BN.

[1]) Avtozavodskaya 2, 252 153 Kiev, UdSSR.

[2]) Prospekt Vernadskogo 36, 252142 Kiev, USSR.

[3]) Žižkova 22, 61662 Brno, Czechoslovakia.

2. Experimental Procedure

The change in the angular correlation distribution of annihilating photons (ADAP) has been investigated in dependence on the compression temperature of BN powders with the zincblende structure. The powder dimensions were 3 to 5 μm in all samples, the compression was 60 kbar and the temperatures of sintering 20, 500, and 1500 °C.

The ADAP spectra were performed by using the long-slit geometry and the angular resolution of 1 mrad. The positron source was ^{22}Na of 5 mCi activity.

3. Results and Their Analysis

The measured ADAP curves were analysed on a computer by two approaches:

a) The resulting curve was supposed to be composed of two Gaussians;

b) the resulting curve was fitted as the sum of an inverted parabola and a Gaussian. The analysis of the second-squares differences of the experimental and calculated points showed that the case b agrees much better. Thus the ADAP curves may be written in the form

$$J(\theta) = h_p\,[1 - (\theta/\theta_F)^2] + h_g \exp\,[-(\theta/\sigma\sqrt{2})^2]\,,$$

where h_p and h_g are the intensities of the parabolic and Gaussian component, respectively, at the angle $\theta = 0$ between the annihilating photons, and σ is the standard deviation of the Gaussian. The angle θ_F corresponds to the Fermi impulse of the electron.

Besides, the fractions of annihilating photon intensities at the angles $\theta = 0$ and $\theta = 8$ mrad, $f = J(0)/J(8)$, were evaluated; this quantity seems to be characteristic of the shape of the ADAP curves. The results are shown in Table 1; the parameters of boron at room temperature are listed as well.

Table 1

Shape characteristics of the ADAP curves in BN and B

material	BN			B (room temperature)
compression temperature (°C)	20	500	1500	
f	3.40	3.50	5.54	6.07
θ_F	10.29 ± 0.16	10.36 ± 0.13	10.02 ± 0.26	9.34 ± 0.06
σ	4.97 ± 0.03	4.68 ± 0.03	4.29 ± 0.09	4.24 ± 0.05

4. Discussion

It is seen from Table 1 that the parameter f does not practically change at low compression temperatures and is growing at high ones. Since the increase of material imperfections generally leads to the rise of f, it may be supposed that a plastic deformation originates in BN crystals at high sintering temperatures under compression of 60 kbar, and is accompanied by the formation of defects trapping the positrons.

The parameter f has a saturation tendency in specimens subjected to deformation at high temperatures and the parameter σ is identical in BN, sintered at 1500 °C with that of pure B (see Table 1). This fact may indicate that the B atoms seem predominantly to be the traps of positrons.

On the other hand, one cannot suppose in this case that the divacancies surrounded by the B atoms appear as trapping centres, because such a distribution is excluded in the surrounding of a divacancy in the two-layer structure. Therefore it can be assumed that the dislocation cores consisting of B atoms are the traps of positrons.

The nature of such centres lies in unpaired electrons on the edge of the extra-half plane in the 60° dislocations in semiconductors with diamond structure. Consequently, such a dislocation represents an array of traps for the thermalized positrons. The negatively charged dislocation line consisting of B atoms repels the conduction electrons so that a cylindrical space charge region forms around the dislocation. According to Read [4], the radius of the cylinder is defined by the electroneutrality condition so that it should contain an amount of positive charge equal to the negative one on the dislocation, and the supply of electrons to broken bonds on the dislocation line is being realized by the centres in the cylinder described. If R is the cylinder radius, a is the spacing between broken dangling bonds on the cylinder axis, N_d and N_a denote the volume concentrations of donors and acceptors, respectively, then the fraction of occupied dislocation levels may be written as

$$\alpha_d = \pi R^2 a (N_d - N_a) . \tag{1}$$

The quantity α_d also determines the electrostatic potential of a positron. It may be regarded as the potential of a continuous charged line lying along the axis of the uniformly and positively charged cylinder and is then expressed by

$$\varphi(r) = \frac{\alpha_d e^2}{\varepsilon a} \left(2 \ln \frac{R}{r} - 1 + \frac{r^2}{R^2} \right), \tag{2}$$

where $a/\alpha_d \leqq r \leqq R$ is the distance from the dislocation and ε denotes the permittivity.

On the other hand, the interaction potential of a positron and a dislocation may be determined from the constancy of the chemical potential of the electronic system, if the external initiation evokes a deformation of the ion distribution. The Fermi energy E_F seems to be the chemical potential and must be constant in the whole crystal. The Fermi level may be generally found somewhere in the middle of the forbidden band of a semiconductor, i.e. between the top of the valence band and the bottom of the conduction band.

If an acceptor level is introduced in a crystal after plastic deformation, then for E_F = const the boundaries of the valence and conduction band are lightly bended in connection with an elastic deformation around the dislocation. For the determination of the electrostatic potential, generated by a dislocation and acting on the positron, Read's model [4] should be a little modified. A redistribution of ions and electrons around a dislocation will be assumed so that the boundaries of the valence and conduction band would not be changed. The Fermi level must then be locally modified in the dislocation surrounding.

In the semi-classical approximation, the chemical potential μ around a dislocation may be represented by a relationship relating the local Fermi impulse $p_F(\boldsymbol{r})$, with the equilibrium Fermi impulse p_F in the form

$$\mu = \frac{p_F^2}{2m} = \frac{\langle p_F^2(\boldsymbol{r}) \rangle}{2m} - e\varphi(\boldsymbol{r}) = \text{const} , \tag{3}$$

where the brackets signify the mean value of the trapped states of positrons. Then, the mean potential energy of the trapped positrons around a dislocation is equal to

$$\langle -e\varphi(\boldsymbol{r}) \rangle = \frac{p_F^2}{2m} - \frac{\langle p_F^2(\boldsymbol{r}) \rangle}{2m} = \frac{mc^2}{2} (\theta_F^2 - \theta_{Fd}^2) , \tag{4}$$

where θ_F and θ_{Fd} are the values of the annihilation angles corresponding to the Fermi impulses in the pure and damaged regions of the crystal, respectively. It is seen in Table 1 that the values at 20 and 1500 °C are: θ_F = 10.29 mrad and θ_{Fd} = 10.02 mrad. The relation (4) then yields $\langle e\varphi(\boldsymbol{r}) \rangle \approx 1.8$ eV.

Further, if $f_0 = J(0)/J(8)$ defines the annihilation parameter for the defectless crystal, f_d denotes it for the crystal with dislocation density ϱ_d, and f_s is this quantity for the state, when all positrons are practically trapped by dislocations and annihilate, then according to the trapping model [5], the expression may be written

$$f_d = f_0 + (f_s - f_0) \frac{\nu c_d}{\lambda_f + \nu c_d}, \tag{5}$$

where λ_f is the rate of positron annihilation in an ideal crystal, νc_d is the rate of positron trapping, and c_d is the relative concentration of trapping centres. The quantity νc_d makes it possible to determine the cross-width w for the trapping of positrons by a dislocation from the relation $\nu c_d = \varrho_d v w$, where v is the mean rate of thermalized positrons.

As mentioned above, the parameter f is growing at high temperatures. However, it may be admitted that it cannot be greater than the parameter f of pure B, since the dislocations are formed by the B atoms as being repelled in the earlier analysis. Therefore, $f_s = 6.07$ is taken. The value of silicon crystal is used for λ_f, since the silicon valency corresponds to the mean one of the BN crystal. Then, $\lambda_f = 4.4 \times 10^9\ s^{-1}$ is obtained [6]. This value nearly coincides with the annihilation rate resulting for the GaAs crystal [1]. For the BN crystals deformed at 1500 °C, $f_d = 5.54$ and $f_0 = 3.40$ (deformation at 20 °C) may be calculated; the value $c_d = 17.6 \times 10^9\ s^{-1}$ was then evaluated from (5).

The upper limit of the dislocation density in crystals with covalent bond may be taken as $\approx 10^9\ cm^{-2}$. Then, if the rate of thermalized positrons is used $v \approx 10^7$ cm/s, the value of $w = 1.8 \times 10^2$ Å may be obtained for B dislocations. This value is of the same order as that one of GaAs crystals [1].

For the determination of the fraction of occupied dislocation levels in B dislocations, the expression may be derived from (1) and (2) in the form

$$\varphi(\boldsymbol{r}) = \frac{\alpha_d e^2}{\varepsilon_a} \left[2 \ln \frac{\alpha_d^{1/2} (N_d - N_a)^{-1/2}}{(\pi a)^{1/2}\, r} + \frac{\pi a (N_d - N_a)\, r^2}{\alpha_d} - 1\right]. \tag{6}$$

Further, it can be supposed [1] that the positron is trapped by a cylindrical region around a dislocation of radius $r = w/2 < R$. Under these conditions, if the positron is trapped, it is accelerated and its kinetic energy rises to the value of $\varphi(\boldsymbol{r}) \approx 1.8$ eV at $r = w/2$. This means that the trapping potential gap equals 1.8 eV at $r = w/2$. For the determination of $\varphi(\boldsymbol{r})$ from (6) the following values are used: $a = 3.61$ Å, $\varepsilon = 4.15$ [7] and $N_d - N_a = 3 \times 10^{16}\ cm^3$ (as in GaAs) [1]. Then $\alpha_d \approx 0.6$.

5. Conclusion

The time spectra of positrons have been used for the study of the dislocation structure of compound semiconductors, i.e. GaAs. In the present paper, the same results have been obtained from the angular correlation distribution of annihilating photon spectra in boron nitride crystals.

References

[1] E. Kuramoto, S. Takeuchi, M. Noguchi, T. Chiba, and C. Tsuda, J. Phys. Soc. Japan. **34**, 103 (1973).

[2] W. C. Dash, J. appl. Phys. **31**, 2275 (1960).

[3] D. B. Holt, J. Phys. Chem. Solids **23**, 1353 (1962).

[4] W. T. Read, Phil. Mag. **45**, 775 (1954).

[5] W. Brandt, in: Positron Annihilation, Ed. A. T. Stewart and L. O. Roellig, Academic Press, New York/London 1967 (p. 180).

[6] H. Weisberg and S. Berko, Phys. Rev. **154**, 249 (1967).

[7] V. G. Samsonov, in: Non-Metallic Nitrides, Metallurgizdat, Moscow 1969 (p. 89) (in Russian).

(Received July 27, 1977)

phys. stat. sol. (a) 43, 593 (1977)

Subject classification: 18.3; 22.8.2

Hirst Research Centre, The General Electric Company Ltd., Wembley[1])

Non-Reciprocal Propagation of Magnetostatic Volume Waves

By

N. D. J. MILLER[2])

The general case of backward travelling magnetostatic volume waves propagating within a tangentially magnetised dielectric layered structure is considered. It is theoretically shown that the structure, which consists of an epitaxial YIG film separated from a ground plane by means of a dielectric layer, allows no-reciprocal propagation of magnetostatic volume waves. Supporting experimental evidence is provided.

Der allgemeine Fall rückwärts laufender magnetostatischer Volumenwellen wird betrachtet, die sich in einer tangential magnetisierten dielektrischen Schichtstruktur ausbreiten. Es wird theoretisch gezeigt, daß die Struktur, die aus einem epitaxialen YIG-Film besteht, der von einer leitenden Ebene durch eine dielektrische Schicht getrennt ist, die Ausbreitung nicht reziproker magnetostatischer Volumenwellen erlaubt. Die Theorie wird durch experimentelle Ergebnisse ergänzt.

1. Introduction

The general case of volume (backward travelling) and surface magnetostatic waves propagating within a tangentially magnetised ferrimagnetic plate was first treated theoretically by Damon and Eshbach [1]. Since then the investigation, both theoretical and experimental, of surface waves propagating within an epitaxial yttrium-iron garnet (YIG) film have been actively pursued [2, 3]. With few exceptions [4], the surface waves considered have been those propagating at right angles to the magnetic field. This emphasis upon surface wave propagation has arisen from their non-reciprocal nature and potential applications in the field of signal processing at microwave frequencies [5].

In this paper the Damon and Eshbach theory is extended to the general case of volume waves propagating within a tangentially magnetised dielectric layered structure. The structure consists of a YIG film, of thickness d and having infinite dimensions in the yz plane, separated from a conducting plane by a dielectric layer of thickness t. As shown in Fig. 1, the internal magnetic field (H) and magnetisation (M) lie within the yz plane and are inclined at an angle θ to the z axis. Propagation effects are predict-

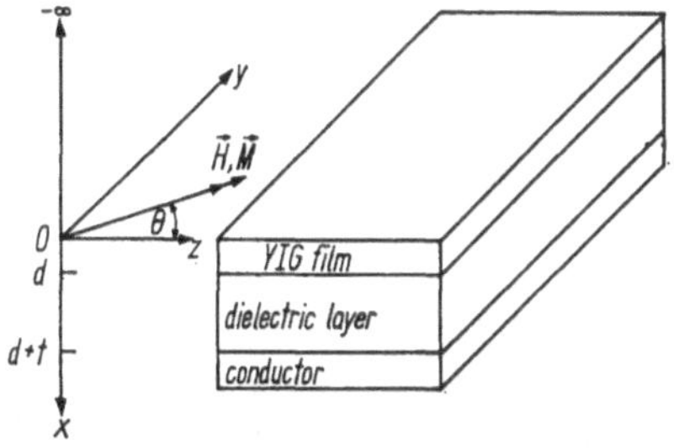

Fig. 1. Schematic of coordinate system

[1]) Wembley, England.

[2]) Now at Microwave & Electronic Systems Ltd. Newbridge, Midlothian, Scotland.

ed by solving Maxwell's equation in the magnetostatic limit ($\nabla \times \boldsymbol{h} = 0$) with exchange effects being neglected. It is shown that the dielectric layered structure allows the propagation of non-reciprocal volume waves.

2. Theory

It is assumed that the YIG film is lossless and that the medium is magnetically saturated. When a harmonic time dependence is assumed the equation of motion for the spin system may be solved and expressed in the usual form

$$\boldsymbol{b} = \boldsymbol{h} + 4\pi \boldsymbol{m} = \mu(\theta)\, \boldsymbol{h}\,. \tag{1}$$

Here $\mu(\theta)$ is the general permeability tensor which for the co-ordinate system under consideration may be written as

$$\mu(\theta) = \begin{vmatrix} \mu_{xx} & \mu_{xy} & \mu_{xz} \\ \mu_{yx} & \mu_{yy} & \mu_{yz} \\ \mu_{zx} & \mu_{zy} & \mu_{zz} \end{vmatrix}\,, \tag{2}$$

where

$$\mu_{xx} = \mu\,, \quad \mu_{xy} = i\mu_1 \cos\theta = -\mu_{yz}\,, \qquad \mu_{xz} = -i\mu_1 \sin\theta = -\mu_{zx}\,,$$
$$\mu_{yy} = \mu \cos^2\theta + \sin^2\theta\,, \qquad \mu_{yz} = (1-\mu)\sin\theta\cos\theta = \mu_{zy}\,,$$
$$\mu_{zz} = \mu \sin^2\theta + \cos^2\theta\,;$$
$$\mu = 1 - 4\pi M H\,[(\omega/\gamma)^2 - H^2]^{-1}\,, \qquad \mu_1 = 4\pi M(\omega/\gamma)\,[(\omega/\gamma)^2 - H^2]^{-1}$$

and ω is the angular frequency of the wave, $\gamma/2\pi$ the gyromagnetic ratio (2.8 MHz/Oe) and $4\pi M$ is the saturation magnetisation (1740 G) of YIG. Because $\nabla \cdot \boldsymbol{b} = 0$ the magnetostatic fields are governed by the two equations

$$\nabla \cdot \mu(\theta)\, \nabla \Phi = \mu_{xx} \frac{\partial^2 \Phi}{\partial x^2} + \mu_{zz} \frac{\partial^2 \Phi}{\partial x^2} = 0 \qquad \text{within the YIG film}\,, \tag{3}$$

$$\nabla \cdot \nabla \Phi = \frac{\partial^2 \Phi}{\partial x^2} + \frac{\partial^2 \Phi}{\partial z^2} = 0 \qquad \text{outside the YIG film}\,. \tag{4}$$

Equations (3) and (4) have assumed that the fields are uniform in the y direction. The magnetostatic approximation which enables the r.f. magnetic field to be written as the gradient of the scalar magnetic potential (Φ) has also been assumed. A trial scalar magnetic potential function for volume wave propagation that satisfies equations (3) and (4) may be written as

$$\left.\begin{aligned} &\Phi = A \exp\{|k|\, x\} \exp\{-ikz\}\,; && -\infty < x < 0\,; \\ &\Phi = [B \cos(\beta\, |k|\, x) + C \sin(\beta\, |k|\, x)] \exp\{-ikz\}\,; && 0 < x < d\,; \\ &\Phi = [D \exp\{|k|\, x\} + E \exp\{-|k|\, x\}] \exp\{-ikz\}\,; && d < x < d + t\,; \end{aligned}\right\} \tag{5}$$

where k is the wavenumber and β, obtained from (3), is given by

$$-\beta^2 = \sin^2\theta + \frac{\cos^2\theta}{\mu}. \tag{6}$$

Applying the boundary conditions that b_x and Φ should be continuous across the layered structure together with the requirement that b_x be zero at the conductor, yields the following dispersion equation:

$$2\mu\beta \cot(\beta\, |k|\, d) = (\mu_1 s)^2 + (\mu\beta)^2 - 1 + [(\mu\beta)^2 + (1 - \mu_1 s)^2] \exp\{-2\, |k|\, t\}\,, \tag{7}$$

where

$$s = (k \sin \theta)/|k| = \pm \sin \theta .$$

This general dispersion equation describes two waves which propagate in the $\pm z$ direction depending on whether s takes on the value $\pm \sin \theta$.

These waves will be referred to as positive or negative depending on the sign of s. The presence of the s term, as with magnetostatic surface waves [2], removes the directional degeneracy and may allow non-reciprocal propagation.

This general dispersion equation for a dielectric layered structure will now be examined for its limiting cases. Firstly when the dielectric layer is absent ($t = 0$) the dispersion equation reduces to

$$\mu\beta \cot (\beta |k| d) = (\mu_1 s)^2 + (\mu\beta)^2 - \mu_1 s \tag{8}$$

which because of the term in s again allows non-reciprocal propagation. In the limit $\theta = 0$ this equation reduces to

$$\beta \cot (\beta |k| d) = -1 . \tag{9}$$

Secondly when the conductor is completely removed ($t \to \infty$) the dispersion equation reduces to

$$2\mu\beta \cot (\beta |k| d) = (\mu_1 s)^2 + (\mu\beta)^2 - 1 \tag{10}$$

which is precisely that obtained by Damon and Eshbach [1]. Because s appears as an even power the wave is reciprocal. In the limit of $\theta = 0$ the equation reduces to

$$2\beta \cot (\beta |k| d) = \beta^2 - 1 . \tag{11}$$

The general and limiting cases of the dispersion equations contained in equations (8) to (11) are all governed by (6) which requires β to be real for volume wave propagation. Propagation is thus contained between an upper (ω_1) and lower (ω_2) frequency limit given by

$$\omega_1 = \gamma[H(H + 4\pi M)]^{1/2} , \qquad \omega_2 = \gamma[H(H + 4\pi M \sin^2 \theta)]^{1/2} . \tag{12}$$

Thus it is seen that, unlike the surface wave, propagation of the volume wave may occur for all angles θ.

The general dispersion equation which may be solved numerically for k as a function of ω allows the theoretical curves contained in Fig. 2 to be plotted. However the non-

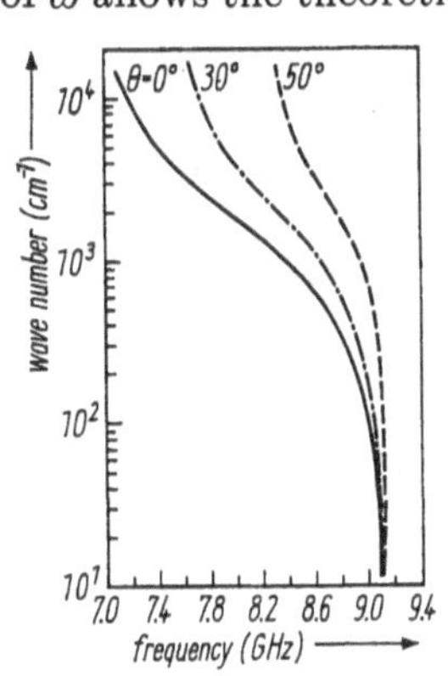

Fig. 2

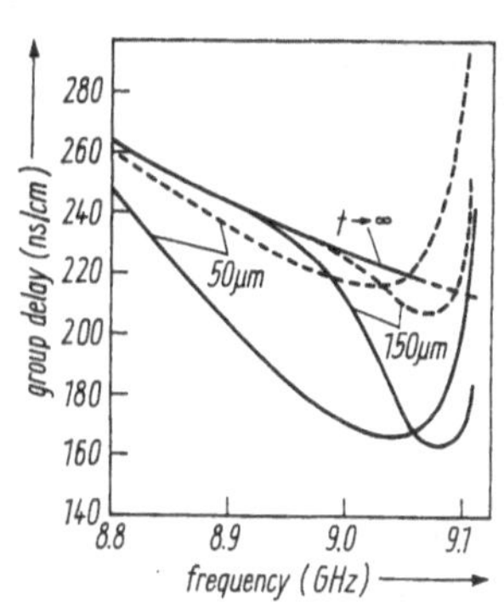

Fig. 3

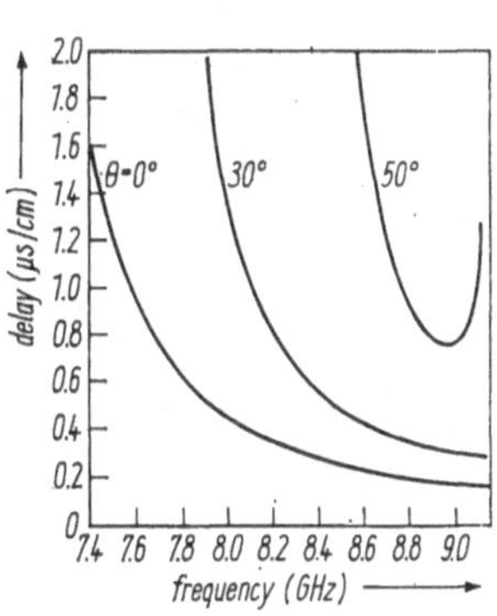

Fig. 4

Fig. 2. Theoretical dispersion curves as a function of propagation angle. $H = 2500$ Oe, $d = 10$ µm, $t \to \infty$

Fig. 3. Theoretical group delay curves as a function of dielectric thickness. $H = 2500$ Oe, $d = 10$ µm; propagation angle 20°; ——— negative wave; – – – positive wave

Fig. 4. Theoretical group delay curves as a function of propagation angle. $d = 10$ µm, $t \to \infty$, $H = 2500$ Oe reciprocal nature of volume wave propagation is best illustrated both theoretically and experimentally by considering the group time delay (τ) which may be calculated from the group velocity ($\partial\omega/\partial k$) using the result

$$\tau = 1/(\partial\omega/\partial k) = \frac{(F + G)\sin^2(\beta\,|k|\,d) + \mu\beta k d\;\partial\beta/\partial\omega}{d\beta^2\mu - t[(\mu\beta)^2 + (1 - \mu_1 s)^2]\sin^2(\beta\,|k|\,d)\exp\{-2\,|k|\,t\}}, \tag{13}$$

where

$$F = [\mu\beta(1 + \exp\{-2\,|k|\,t\}) - \cot(\beta\,|k|\,d)]\,(\mu\,\partial\beta/\partial\omega + \beta\,\partial\mu/\partial\omega)\,,$$

$$G = s[\mu_1 s - (1 - \mu_1 s)\exp\{-2\,|k|\,t\}]\,\partial\mu_1/\partial\omega\,.$$

Because the group velocity is negative the volume waves under consideration are backward travelling. In passing it should be noted that the character (backward travelling) of the volume wave does not, unlike the surface wave, depend on the dimensions of the layered structure. Fig. 3 and 4 contain theoretical predictions for the group time delay which are complementary to the dispersion curves of Fig. 2.

The following points emerge from the group delay curves. Firstly as the dielectric to film thickness ratio approaches zero the influence of the ground plane extends over an increasingly larger frequency and wave number range. Conversely as the ratio becomes large the propagation, except for small wave numbers, is unaffected by the ground plane. Secondly as the angle θ increases it is seen that the group velocity decreases. Finally the group delay curves for all values of dielectric thickness pass through a minimum as the angle θ increases.

Thus far the analysis has shown that the presence of a conducting layer leads to non-reciprocal delay characteristics. A further interesting property of the system under consideration is the manner in which the magnetic scalar potential varies within the YIG film. From (5) the scalar potential within the YIG film may be written as

$$\Phi = C\left[\cos(\beta\,|k|\,x) + \frac{1 - \mu_1 s}{\mu\beta}\sin(\beta\,|k|\,x)\right]\exp\{-ikz\}\,, \tag{14}$$

where C is a constant. It is seen that the potential distribution depends on the direction of propagation and that this dependence does not require the presence of a ground plane. Fig. 5 contains typical potential distributions both within and outside the YIG film from which the following points emerge.

When $\theta = 0$ the potential distributions for the plus and minus wave are degenerate. The magnitudes of the potential at the surfaces of the film are approximately equal and in the limit when the ground plane is removed they become equal.

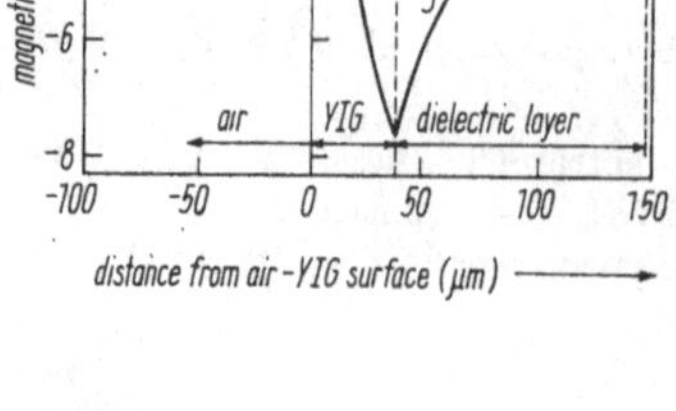

Fig. 5. Magnetic scalar potential within a layered structure. $H = 3000$ Oe; frequency 10.37 GHz; $t = 110.6$ μm, $d = 38$ μm. (1) $\theta = 40°$, positive wave ($k = 142$ cm^{-1}), (2) $\theta = 0°$, degenerate wave, (3) $\theta = 40°$, negative wave ($k = 163$ cm^{-1})

In the more general case when $\theta \neq 0$ the plus and minus wave are not degenerate and further the magnitude of the potential at the film surfaces are not equal. It should be noted that these effects do not depend on the presence of a conducting layer merely that such a layer enhances them. As the potential distributions are different for the plus and minus wave so too are the energy distributions. Although a complete analysis has not been carried out it is expected that the propagation losses for the two waves will also differ. This situation is analogous to the case of surface wave propagation [6] within a dielectric layered structure. The predictions of non-reciprocal propagation for magnetostatic volume waves will be supported by experimental results in the following sections.

3. Experimental Propagation Results

As the non-reciprocal nature of backward travelling volume waves is most clearly illustrated by considering the group time delay, a pulse echo technique was employed. A schematic of the delay line structure is contained within Fig. 6. Excitation and detection of the magnetostatic wave was achieved through the use of high impedance ($\approx 50\,\mu$m wide) open circuit microstrip lines. The experimental techniques for measuring the group time delay have previously been reported [7].

Because the magnetostatic approximation does not enable any dependence upon the electric properties of the YIG or dielectric layer to be inferred, it has been assumed that their dielectric constants are identical. However in practice this assumption is not satisfied and it is necessary to replace the physical thickness (T) of the dielectric layer by its electrical length (t) using the expression

$$t = T(\varepsilon/\varepsilon_1)^{1/2} ,$$

where ε and ε_1 are the dielectric constants of the dielectric layer and YIG, respectively. Thus when computing the theoretical results it is the electrical length that is always used. Further, because of the uncertainty in the value of the small anisotropy field the internal magnetic field has been used as an adjustable parameter when comparing the experimental results with theory.

A comparison of the experimental and theoretical results for various propagation angles are contained within Fig. 7. The results which are in good agreement with theory are for a negative wave propagating within a 38 μm YIG film over a distance of 3 mm. Because the alumina dielectric layer is 635 μm thick the results approach those for an unbacked ($t \to \infty$) film where evidence of non-reciprocal propagation would only be observed for small wave numbers ($k \leqq 40$ cm^{-1}). However such small wave number magnetostatic waves are unable, because of cut-off effects [2], to propagate efficiently within the structure used such that non-reciprocal effects were not observed. The results show that the group delay increases with the propagation angle and that for large angles a region of non-dispersive propagation ($\partial^2 k/\partial\omega^2 \approx 0$) exists. Evidence of non-reciprocal propagation is contained within Fig. 8 where

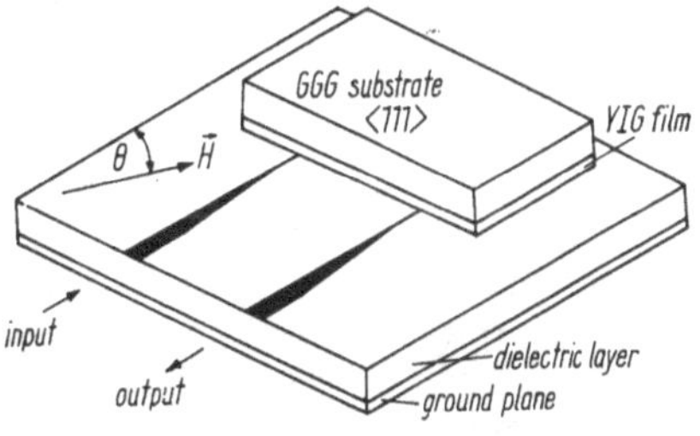

Fig. 6. Schematic of the magnetostatic volume wave delay line

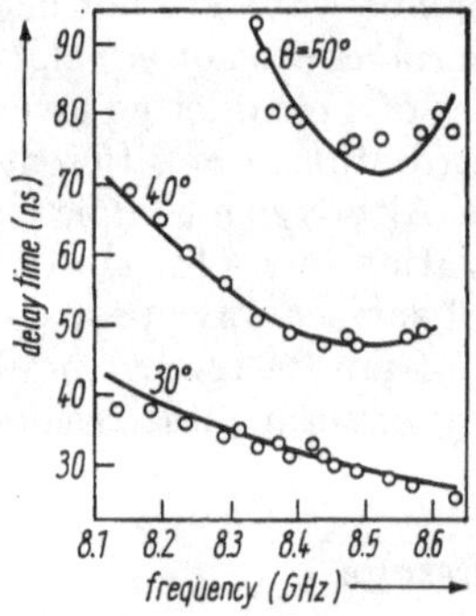

Fig. 7

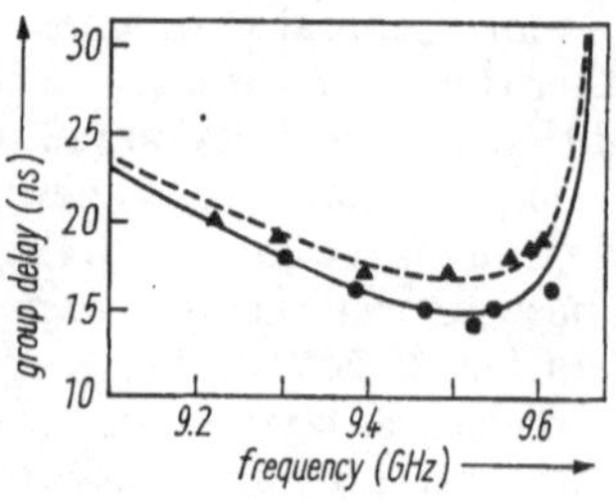

Fig. 8

Fig. 7. Comparison of experimental and theoretical group time delay as a function of propagation angle. $H = 2350$ Oe, $\varepsilon = 9.9$, $T = 635$ μm, $d = 38$ μm, propagation path length 3 mm

Fig. 8. Comparison of experimental and theoretical group time delay as a function of frequency. $H = 2690$ Oe, $\varepsilon = 2.35$, $t = 110$ μm, $d = 38$ μm, $\theta = 10°$, 3 mm propagation path. ▲ positive, ● negative wave

agreement with theory is again seen to be good. The results of Fig. 8 also indicate that the presence of a ground plane not only leads to non-reciprocal propagation (when $\theta \neq 0$) but that regions of non-dispersive delay also exist.

4. Experimental Transmission Results

Typical transmission responses which clearly illustrate the non-reciprocal effects are contained within Fig. 9. The results for a 4 mm propagation path length and a 106 μm thick dielectric layer were obtained by adjusting the sample position so as to yield a smooth low loss transmission response. Positive and negative wave responses were then obtained by interchanging the input and output connections to the delay line. When comparing the results it is seen that those for a 20° propagation angle have a narrower bandwidth, however this is primarily caused by the use of a thinner film and not by the increased propagation angle. An important point that is not evident from the results of Fig. 9 is that as the propagation angle increases from zero the insertion loss decreases, passing through a minimum that is about 3 dB up on the level observed for propagation in the direction of the magnetic field. Indeed the insertion loss for angles less than about 20° is smaller than that obtained when propagation is along the field direction. When the propagation angle is increased beyond 20° the insertion loss rises quite rapidly because of the greater delay time.

Although the calculation has not yet been performed it is expected that because the two waves have different energy distributions they will also have different propagation losses, a situation that is analogous to the surface wave case [6]. However the difference in propagation loss is unlikely to account for the magnitude of the observed effects. Instead the non-reciprocal transmission effects are caused by a field displacement action that leads to non-reciprocal transduction. From Fig. 5 it is seen that the negative wave ($\theta \neq 0$) contains a greater concentration of energy at the film surface nearest the conductor than that for the positive wave. Because the transducers reside on the surface nearest the conductor it is the negative wave that they preferentially excite. Thus the bi-directional nature of the input transducers is destroyed which results in a total theoretical (for a perfect match) transduction loss approaching 3 dB as opposed to 6 dB when the angle of propagation is zero.

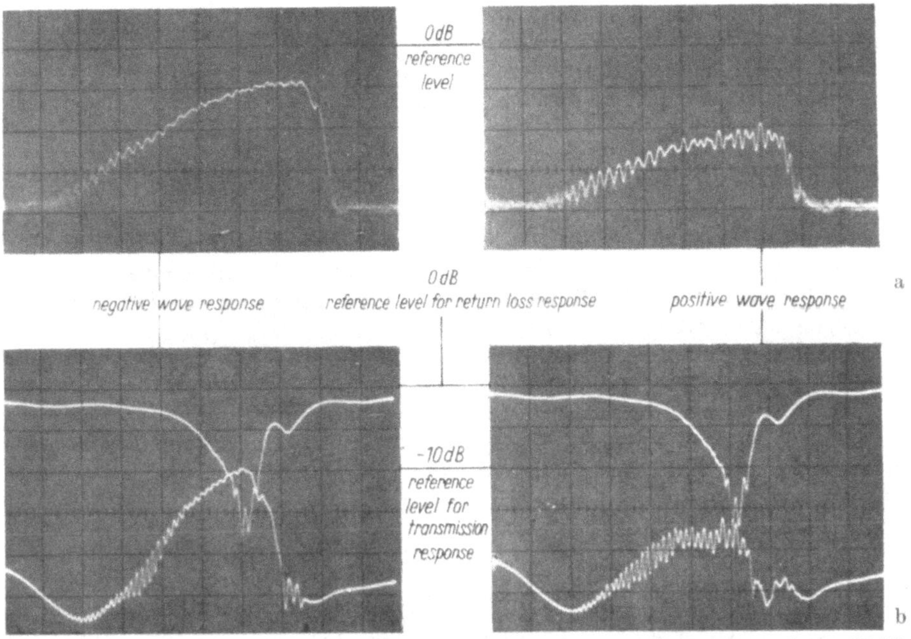

Fig. 9. Transmission and return loss response of two delay lines. Ordinate scale 10 dB/div., abscissae 100 MHz/div. $\varepsilon = 2.35$, $t = 105$ μm, 4 mm propagation path. a) Centre frequency = = 2.9 GHz, $d = 38$ μm, $\theta = 10°$, $H = 540$ Oe. b) Centre frequency = 8.5 GHz, $d = 18.4$ μm, $\theta = 20°$, $H = 2350$ Oe

It is therefore seen that the experimental results are in agreement with the theoretical predictions.

An estimate of the propagation loss in an 18.4 μm film may be obtained by reference to Fig. 9b which provides not only transmission but return loss as a function of frequency. At mid band (8.6 GHz) the insertion loss is about 9 dB while the return loss from the transducer is greater than 30 dB. These results indicate that the contribution to the insertion loss from transducer return loss may be neglected. Thus the propagation loss is estimated at a maximum of 6 dB which leads to a value of 120 dB/μs. This figure is misleading because the return loss measurement at mid band implies a transducer that is almost perfectly matched. In practice the untuned open circuit transducers radiate most of the input energy with only a small amount being transduced into the magnetostatic wave. This result is based on the observation that the delay line insertion loss is very insensitive to the level of return loss. It is therefore concluded that the propagation loss is considerably less than 120 dB/μs.

5. Conclusion

The general case of backward travelling magnetostatic volume waves propagating within a tangentially magnetised dielectric layered structure has been analysed in the magnetostatic limit. Justification for this approach is its recent success with magnetostatic surface [2] and forward volume [7] waves coupled with the fact that under typical experimental conditions only wave numbers in the range 40 to 10^3 cm^{-1} propagate. It has been shown theoretically and experimentally that under certain conditions, backward travelling volume waves are non-reciprocal. The experimental results have been shown to be in good agreement with theory through it should be noted that all the theoretical group delay results have been multiplied by a constant scaling factor (1.146). The origin of this discrepancy has not been located through it may well reside in the calibration of the time delay detecting system.

It should be noted that the derived general dispersion equation has a multiplicity of roots and that only the fundamental mode of propagation has been considered here. Higher order modes propagating at a given frequency and magnetic field, do so with different group velocities and wave numbers. For instance the first higher order mode has a group delay that is about three times that of the fundamental mode for the range of parameters reported herein. Further because the small valued wave numbers ($< 200\ \text{cm}^{-1}$) of higher order propagating modes are confined to a small region in frequency (< 10 MHz) the effect of the dielectric layered structure on propagation is negligible.

The backward travelling volume waves discussed, are similar to the surface wave in that non-reciprocal and non-dispersive propagation occur within a dielectric layered structure. However, unlike the surface wave, propagation of the volume wave occurs for all angles. From a device viewpoint, backward travelling volume waves should find applications at microwave frequencies (> 6 GHz) because of their increased bandwidth.

References

[1] R. W. DAMON and J. R. ESHBACH, J. Phys. Chem. Solids **19**, 308 (1961).
[2] W. L. BONGIANNI, J. appl. Phys. **43**, 2541 (1972).
[3] J. D. ADAM and J. H. COLLINS, Proc. IEEE **64**, 794 (1976).
[4] J. P. PAREKH, Proc. IEEE **61**, 1371 (1973).
[5] P. M. GRANT, J. D. ADAM, and J. H. COLLINS, IEEE Trans. Commun. **22**, 1410 (1974).
[6] C. VITTORIA and N. D. WILSEY, J. appl. Phys. **45**, 414 (1974).
[7] N. D. J. MILLER, phys. stat. sol. (a) **37**, 83 (1976).

(Received August 8, 1977)

phys. stat. sol. (a) 43, 601 (1977)

Subject classification: 1.2; 4; 21.1

II. Physikalisches Institut der Universität Wien[1])

Morphological Stability Study in Binary Alloys

II. Stability of Coherent Concentration Modulations[2])

By

H. P. Aubauer

The stability of a precipitated two phase structure is analysed with respect to changes in number and size of the precipitated domains, using the equilibrium interfacial concentration profile and width of part I of this paper, and keeping the precipitate volume fraction at a constant value. Thereby the total system energy may show an absolute minimum at a finite precipitate size, if the ratio η out of the interfacial self energy component and of the pair potential strength stays below a certain value in the case of ordering alloys. This is caused by an increase of the absolute value of the negative pair interaction energy component during coarsening, leading to a drop of the interface energy density. η is fitted to parts of the Fe–Al and the Cu–Al phase diagrams, leading to a transition region between classical one and two phase zones of the phase diagram, that is characterized by an unorthodox kinetic behaviour of the microstructure. This microstructure might interpret several experiments.

Die Stabilität einer ausgeschiedenen Zweiphasenstruktur wird bezüglich der Änderung der Anzahl und der Größe der ausgeschiedenen Bereiche untersucht, wobei die Gleichgewichtswerte des Konzentrationsprofils und der Randbereichsdicke des Teils I dieser Arbeit verwendet werden und der Volumenanteil der ausgeschiedenen Phase konstant gehalten wird. Dabei kann ein absolutes Minimum der gesamten Systemenergie bei einer endlichen Ausscheidungsgröße auftreten, wenn im Fall sich ordnender Legierungen das Verhältnis η aus der Selbstenergiekomponente der Zwischenphasenenergie und der Paarpotentialstärke unterhalb eines bestimmten Wertes bleibt. Es kann durch einen Anstieg des Absolutbetrages der negativen Komponente der Paarwechselwirkung während der Vergröberung entstehen, wobei dies die Zwischenphasenenergiedichte verringert. η wird an Teile des Fe–Al- und des Cu–Al-Phasendiagrams angepaßt, wobei dies zu einem Übergangsbereich zwischen klassischen Ein- und Zweiphasengebieten führt. Die Mikrostruktur dieses Bereiches zeigt ein unkonventionelles kinetisches Verhalten und könnte einer Vielfalt von experimentellen Ergebnissen zugrundeliegen.

1. Introduction

In order to interpret numerous experiments, which hint at the existence of stable coherent concentration modulations or a heterogeneous and stable state in imperfectly ordered alloys [1 to 7] (reviews in [8 to 12]), the stability of already precipitated two-phase structures has to be studied. As a difference to the usual concept of metastability this stable state forms from any nearby initial one [3] and corresponds to an absolute minimum of the total energy.

Such an analysis was performed in a group of works[3]) [13 to 16]. Khatchaturyan et al. [13, 14] demonstrated that the anisotropic elastic energy of an array of spherical

[1]) A-1090 Wien, Österreich.

[2]) Part I see phys. stat. sol. (a) 43, 261 (1977).

[3]) Hopper and Uhlmann [35] looked for an energy minimum within the Cahn-Hilliard formalism: Applying the two-dimensional Euler equation instead of that one for one dimension, as was done by the authors in their equation (2), an explicit dependence of ∇c on r follows in their equation (4), as was suggested (Aubauer [36]). In spite of the incorrect application of a one dimensional calculation (their equations (1) to (4)) to a two-dimensional problem (their Fig. 1), basically their conclusions remain valid.

domains shows a minimum with respect to the precipitate volume fraction as well as with respect to their relative positions. Nevertheless this is no absolute minimum, since the sum of the total interfacial and anisotropic elastic energy can be reduced further by an increase of the precipitate size (see also Appendix A). Considering uphill diffusion, as in nucleation theories [17, 18], this energy gain leads to unlimited coarsening [19].

Arguments for the existence of an absolute minimum of the total energy in isotropic alloys were presented in an earlier model[4]). As a difference to it in the present work the equilibrium values of the concentration profile and the width of the interfacial zone, determined in Part I of this paper [20] (to be referred to as I), were used, as they change with the precipitate size. Instead of a phenomenological value [16], a microscopic discrete or lattice model is used for the interfacial energy density [20], leading to its precipitate size dependence. Finally the total free energy was minimized with a variational procedure not only with respect to the precipitate volume fraction [13, 14], but also relatively to their size and their number. So the conditions of an absolute total energy minimum at a finite precipitate size were studied here, not asking for the diffusion process leading to this thermodynamically stable equilibrium (the latter was analysed in a further paper [19]). Reproducing qualitatively the results of the previous model [16] in case of ordering, the precipitate size and number are found to depend essentially on the ratio η of an effective interface energy (corrected for the elastic energy), and the strength of the interface potential. η is derived within the frame of the regular solution theory from phase diagrams generally in Section 4.2 and an application to so far experimentally studied alloy systems (Fe–Al and Cu–Al) is made in Sections 4.3 and 4.4. Since the first-order effect of a possible size stability was the main concern here, no equilibrium study of the precipitate shape was made. Deviations of their relative positions from a regular arrangement are discussed in Appendix B.

2. Total Free Energy

The total energy F of a monodisperse system of M spherical precipitates of constant composition c_P, being embedded in a matrix with a composition c_M was derived in I. For a special model of the excess free energy density

$$\Delta g = \gamma\,\Delta c^4(1/4 - Y_k^2)^2; \qquad \Delta c = c_P - c_M\,, \tag{1}$$

and the pair potential Ω (I), it reads

$$F = F_\infty + M\Sigma; \qquad F_\infty = N[Zg(c_P) + (1 - Z)\,g(c_M)]; \qquad Z = (\bar{c} - c_M)/\Delta c;$$

$$\Sigma = F_0 \sum_{k=(R-D)/a}^{R/a} [(V_k/a^3)\,\eta\,\Delta g\,\gamma^{-1}\,\Delta c^{-4} + 4\pi^2 \sum_{l=1}^{L} (Y_k - Y_l)^2\,h(k, l)];$$

$$F_0 = \Omega_0(Na^3)^2; \qquad \eta = \gamma\,\Delta c^2/(\Omega_0 N a^3); \qquad V_k = (4\pi a^3/3)\,[k^3 - (k-1)^3]\,. \tag{2}$$

Thereby a denotes the lattice constant, Y_k a normalized local and $\bar{c}$ the average concentration value. F is expressed here in terms of a dimensionless ratio η, that essentially represents the strength of an effective interface energy (corrected by the elastic energy) relative to the strength Ω_0 of the interacting pair potential Ω. The dimensionless geometrical energy factor $h(k, l)$ is defined in I. If R signifies the precipitate radius and D the interface width, then due to material conservation, the precipitate volume fraction Z is related with R, D, and Y_k (see Appendix C).

---

[4]) De Fontaine [37] suspected an error in the elastic energy expression of [16] and presented an alternative formulation which is equivalent to that one used there for a small precipitate volume fraction (Kirchner [38]). An expression that is not restricted to a linear variation of the atomic volume misfit, is presented in [39].

3. Minimization of Total Energy *F* with Respect to the Precipitate Radius *R*

F depends on the interfacial concentration profile (defined by the parameters Y_k), on the interface width D, the precipitate radius R, and the ratio η. For the further work the equilibrium values of Y_k and D, determined in I were used. Thus F only depends on the precipitate radius R and the material constant η.

If there exists a value R_0 among all other precipitate sizes R_i, that minimizes the total energy F of a monodisperse system (equation (2)), so that all competing R_i lead to a larger energy: $F(R_0) < F(R_i)$; $i \neq 0$, then it can be shown that there exists no polydisperse system whose energy is lower than $F(R_0)$. Then this monodisperse state also represents the stable configuration of a polydisperse system, and diffusion leads to a particle size distribution function that is sharply peaked at R_0 [19].

In order to study the existence of a stable equilibrium precipitate size, the total energy F, or $M\Sigma = F - F_\infty$ of (2) was minimized with respect to R, keeping the precipitate volume fraction Z equal to its equilibrium lever rule (equations (2), (C1)). So an increase of the precipitate size corresponds to a decrease of their number M. With the exception of this variation of M, the procedure is analogous to that in the previous work [20].

The total interface energy $(F - F_\infty)$ can be seen as the sum of a positive self-energy component of interfacial volume elements (first term within cornered brackets of (2)) and an interaction term which is negative in case of ordering (last term within cornered brackets). Considering a coarsening process at constant Z, the first term diminishes, while the absolute value of the second one rises. Thus if the strength η of the first self-energy term relative to the second interaction term exceeds a critical η_∞, a finite coarsening driving force exists for every finite precipitate size, causing a coarsening process which is only limited by a lack of concentration fluctuations or the finite sample size (Fig. 1).

If, however, the ratio η drops below a value η_∞, the driving force for coarsening vanishes at a stable precipitate size $R = R_0$, changing its sign for $R > R_0$. Then coarsening stops at $R = R_0$, even in the precence of large concentration fluctuations [19] (Fig. 1).

The increase of R_0 with the precipitate fraction Z (Fig. 1) appears to be plausible, if one considers the positive sign of the interaction energy between pairs of atoms within different precipitates (see Appendix B). This interaction partially cancels the negative precipitate-matrix one and rises with the fraction Z, since this goes along with a decrease of their distance.

The total energy F of (2) also can be minimized analytically for a small width D as well as for a concentration profile, that is approximated by a closed form function. Thereby the results agree qualitatively with one another and those of Fig. 1.

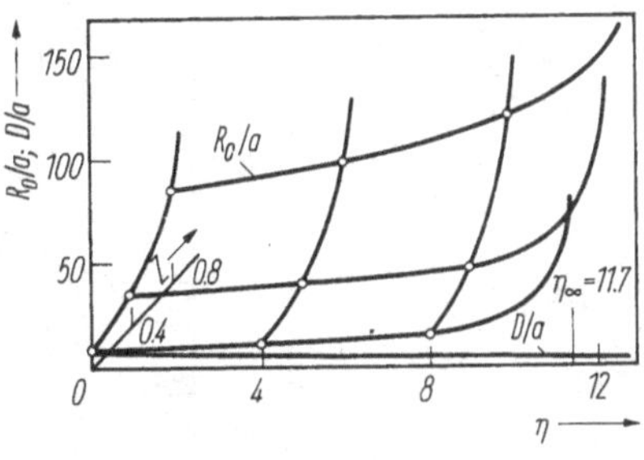

Fig. 1. Normalized equilibrium values of the precipitate size R_0 as a function of the ratio η and the volume fraction Z. For $\eta < \eta_\infty$ the stabilization breaks down

4. Discussion and Application

4.1 General discussion

During coarsening the first self-energy term of (2) decreases due to the drop of the volume fraction of interfacial zones (hatched in Fig. 2). This is illustrated in Fig. 2, sketching two stages of the coarsening process at different times $t_0 < t_1$. The simultaneous rise of the absolute value of the second negative interaction energy term is caused by an increase of the number of pairs between unequal atoms within the precipitate (concentration c_P) and in the matrix (c_M) respectively, having two origins: Firstly the increase of the volume fraction of homogeneous material with compositions c_P and c_M, respectively, and secondly the drop of the interface curvature. The latter can be shown by a simple geometric reflection, using the cosine law and the scetch of Fig. 2b in I.

Thus the first self-energy term in (2) gives a positive and the second term a negative contribution to the coarsening driving force. The ratio η of their weight can be derived from the excess free energy density Δg (see next section). Δg is related to the homogeneous energy density $g(c)$ by [20]

$$\Delta g(c_e, c) = g(c) - g(c_e) + (c_e - c)\, dg/dc|_{c=c_e}; \qquad c_e = c_P, c_M . \tag{3}$$

On the other hand, Δg can be found from the phase diagram, at least approximately, as demonstrated in Sections 4.2 to 4.4.

The change of energy during coarsening is demonstrated quantitatively by Fig. 3, showing in Fig. 3a the change of a *single* precipitate energy $F/(MF_0)$ and in Fig. 3b, c the variation of the *total* system energy $(F - F_\infty)/F_0$ with the precipitate size R. While an instable case is plotted with full lines ($\eta = 16$), two stable cases are shown by dashed lines ($\eta = 4, 11$). The variation of the total energy for constant Z in Fig. 3b, c is less familiar than that of single precipitates in Fig. 3a, and demonstrates a drop of the coarsening driving force even in the case of unlimited coarsening $\eta = 16 > \eta_\infty = 11.7$. For $\eta < \eta_\infty$ this drop is so strong that the driving force obtains a value equal to zero at $R = R_0$. For an infinite precipitate size ($R = \infty$), the total energy F equals F_∞.

All of this also can be discussed in the terminology of an energy density within the interfacial zone, that decreases with increasing precipitate size (dropping interface curvature). If this effect outweights that of the overall decrease of the interfacial area during coarsening, the latter stops.

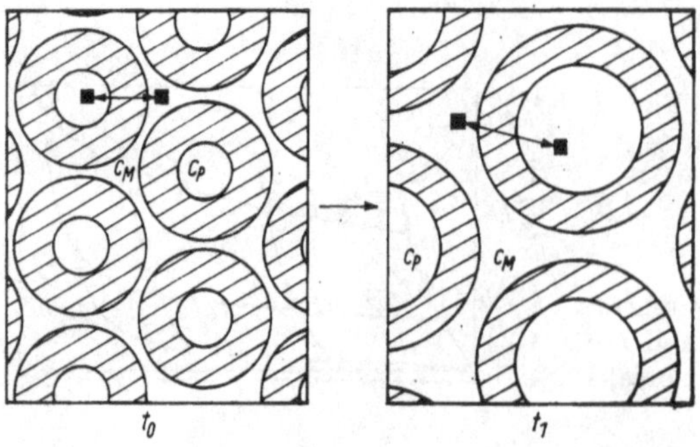

Fig. 2. Decrease of the hatched volume fraction of interfacial zones during coarsening, leading to an increase of pairs of unequal atoms, whereby one of them is located within the matrix and the other one inside the precipitates respectively: time $t_0 < t_1$; Z = constant

Fig. 3. The dependence of the energy on the precipitate domain size at a volume fraction $Z = 0.2$: a) Normalized energy $F/(M_0F_0)$ of a *single* precipitate domain as a function of normalized precipitate domain size R/a, whereby $F_0 = F_0Na^3$. b) Dependence of normalized *total* energy $(F - F_\infty)/F_0$ of all precipitates on R/a. c) Total energy variation with inverse normalized precipitate size a/R

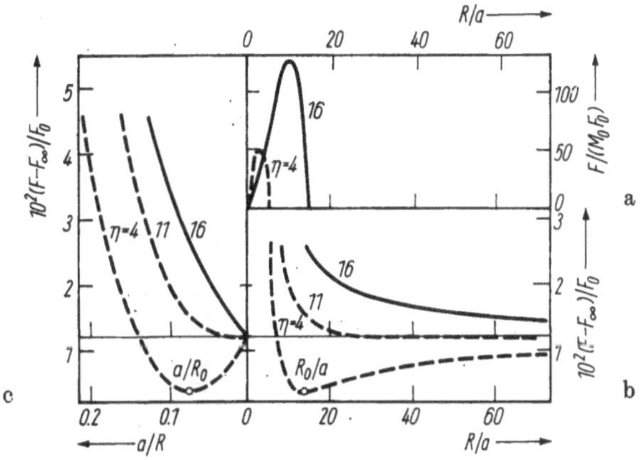

4.2 Relation to phase diagram

Using (3) and (6), the constant γ which is contained in the essential ratio $\eta = \Delta c^2/(\Omega_0 Na^3)$, is derivable from the excess free energy density Δg^r of a real alloy system. Thereby it can be found within the accuracy of this model, if the integral over Δg of (6) (hatched area in Fig. 1b of I):

$$\int_{c_M}^{c_P} \Delta g(c)\, dc$$

is set equal to that of a known excess free energy density Δg^r

$$\Delta g^r = \begin{cases} \Delta g_\alpha(c); & c_M < c \leqq c_0 \\ \Delta g_\beta(c); & c_0 < c \leqq c_P; \end{cases} \qquad g_\alpha(c_0) = g_\beta(c_0);$$

$$\int_{c_M}^{c_P} \Delta g(c)\, dc = \int_{c_M}^{c_0} \Delta g_\alpha\, dc + \int_{c_0}^{c_P} \Delta g_\beta\, dc\,. \tag{4}$$

The density Δg^r usually is derived by fitting theoretical models of it to a real phase diagram.

It is assumed in this work that the origin of the experimentally detected heterogeneous short-range order [1 to 21] might be a two-phase miscibility gap between an ordered phase whose free energy density $g_\beta(c)$ possibly can be written as

$$g_\beta(c) = \beta(c - c_\beta)^2 \tag{5}$$

and a solid solution phase whose energy density $g_\alpha(c)$ might be expressed within the frame of regular solution theory [21]

$$g_\alpha(c) = \gamma_0 + \gamma_1 c + \omega c(1 - c) + \varepsilon(c - \bar{c})^2 + kT[c \ln c + (1 - c) \ln (1 - c)];$$

$$\omega = 4\pi N \int_0^\infty d\varrho\, \varrho^2 \Omega(\varrho); \qquad \varepsilon = \delta^2(2\mu/3)\,[1 + 4\mu/(3K)]^{-1}\,. \tag{6}$$

Here γ_0, γ_1 are constants, containing the pure metal free enthalpies, the third term denotes the chemical energy contribution, caused by a pair potential Ω, the fourth indicates the elastic energy share (δ denotes the atomic volume misfit and $K(\mu)$ signifies the elastic bulk (shear) modulus), and the last one the entropy component.

With the aid of the lever rule $\Delta g(c_M) = 0$, the temperature dependent limits $c_P(T)$ and $c_M(T)$ of the miscibility gap can be derived from the energy model of (5) and (6). Thus this energy expression can be fitted to a real phase diagram by adjusting the appearing parameters, so that there is an optimum agreement between the theoretically derived solubility limits $c_P(T)$ and $c_M(T)$ and the real ones.

In Appendix D the excess densities Δg_α and Δg_β are derived by a substitution of (5) and (6) into (3). By substituting these values of $\Delta g_{\alpha,\beta}$ and Δg of (1) into (4), η is found. Thereby η depends on the concentration splitting $\Delta c = c_P(T) - c_M(T)$ of the miscibility gap. This gap has to close at the disordering temperature T_c: $c_P(T_c) = c_M(T_c)$. Moreover c_P and c_M are adjusted to the values of the real phase diagram at two temperatures T in the next two sections.

4.3 Iron-rich Fe–Al alloy

During the first period of analysis quite different versions of the iron-rich side of the Fe–Al equilibrium diagram were published [22, 23]. Even if the recently general agreement as to the first-order character of the transition between the α-solid solution and the Fe_3Al ordered phase was obtained [23], the theoretically determined two-phase field ($\alpha + Fe_3Al$) appears to extend towards lower Al concentrations [24] (< 10 at% Al) than the experimentally determined one [25] (< 20 at% Al) (see also Fig. 4a). Numerous other experiments on the other hand hint at the existence of a heterogeneously ordered phase [1, 23, 26] below 20 at% Al and within the hatched area of Fig. 4a. So the left side of the hatched region was taken as a border between the disordered α-one phase field and the $\alpha + Fe_3Al$ two-phase field, as suggested earlier [1], and the energy densities of (12), (13) were fitted to it at two temperatures ($T = 297$ K; $c_M = 0.037$, $c_P = 0.25$, $T = 813$ K: $c_M = 0.22$, $c_P = 0.26$). This way the fully drawn limits of the $\alpha + Fe_3Al$ two-phase field follow (see Fig. 4a). The gap closes above the eutectoid temperature (≈ 813 K) at $T_c = 823.5$ K[5]). Evaluating the elastic energy constant E_0 defined in Appendix D, using the appropriate atomic volume misfit [27] and lattice constants [28] ($E = 0.22$), the temperature change of $\eta(T)$ can be determined. η increases with temperature and attains its limiting value of η_∞ above the eutectoid temperature (at $T = 843$ K). Nevertheless the drop of the precipitate volume fraction Z with temperature dominates the temperature dependence of the equilibrium precipitate size R_0 and their number M_0, as shown in Fig. 4b for two alloy compositions $\bar{c} = 0.106$ (10.6 at%) and $\bar{c} = 0.19$ (19 at%). Thereby R_0 is found from Fig. 1, using the known values of η and Z. M_0 is evaluated from this by considering material conser-

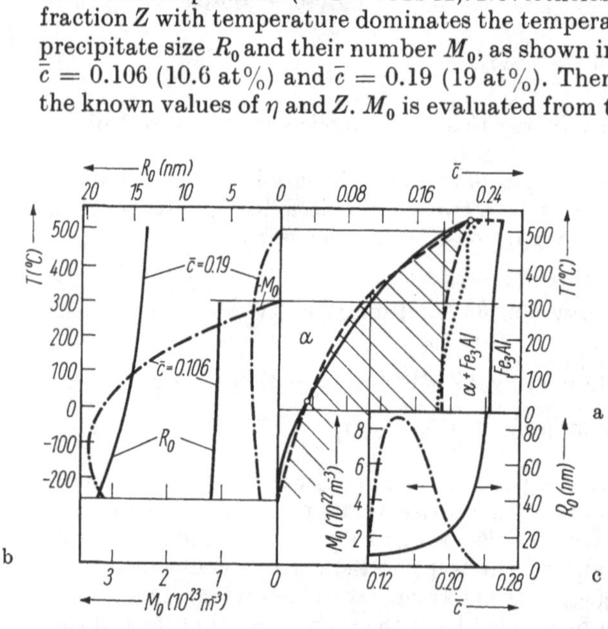

Fig. 4. a) The dashed lines denote the experimental and the full ones the calculated iron-rich Fe-Al phase diagram. The anomalous field is hatched and the dotted line indicates the equality of the equilibrium radius R_0 with that of coherency breakdown R_c. This dotted line is equivalent to a kinetic boundary between the classical two-phase field $\alpha + Fe_3Al$ and the hatched one, being characterized by a thermodynamic limitation of precipitate coarsening. b) The temperature dependence of the equilibrium precipitate size R_0 (full lines) and their number M_0 (dashed-dotted lines) is shown for two alloy compositions: $\bar{c} = 10.6$ at% Al (or $\bar{c} = 0.106$) and $\bar{c} = 19$ at% Al. c) R_0 (full) and M_0 (dashed-dotted) as a function of $\bar{c}$ for $T = 300$ °C

[5]) Thereby the Fe_3Al field is extrapolated to temperatures above 813 K.

vation of equation (C1) in Appendix C. R_0 slightly drops with T, while M_0 is zero for $Z = 0$ ($\bar{c} = c_M$) and goes through a maximum at lower temperatures. In Fig. 4c the values of R_0 and M_0 are shown as they change with $\bar{c}$ at a temperature of $T = 573$ K. M_0 is a factor 10 below the experimental values [1] for $0.16 < \bar{c} < 0.18$, while R_0 is a factor of two above it. Nevertheless, in agreement with experiment [1], a slight dependence of these values on the average composition $\bar{c}$ and no change with the ageing time arises.

A transition between the hatched two-phase field, characterized by precipitate domains of a stable size and the classical two-phase field $\alpha + Fe_3Al$ might be caused by a breakdown of coherency [29 to 31] due to the equilibrium radius R_0 rising with $\bar{c}$, and becoming as large as the critical [6]) radius R_c. The corresponding relief of coherency strains leads to a breakdown of stabilisation. The critical radius R_0 for the breakdown of coherency is indirectly proportional to the atomic volume misfit [29 to 31], or proportional to $(\Delta c)^{-1}$. According to the most simple criterion by Brooks [30] R_c has the order of $3/\Delta c$ nm in Fe–Al. The equality of R_c with the equilibrium radius R_0 is indicated by the dotted line in Fig. 4a, agreeing qualitatively with the border between the classical two phase field $\alpha + Fe_3Al$ and the hatched anomalous region [25, 26] in Fig. 4b.

4.4 *Copper-rich Cu–Al alloy*

Among the alloy systems where a stable, heterogeneous short-range order is suspected, also Cu-rich Cu–Al was studied in experimental detail [2 to 7, 32, 33]. The anomalous region is indicated by the hatched field in Fig. 5a [2, 9]. Since one of the experiments [5] indicates a disordering temperature of 543 K at an alloy composition $\bar{c} = 0.15$, (5) and (6) were fitted to this value and the eutectoid triple point at 636 K: c_M (543 K) $= 0.15$, c_P (543 K) $= 0.216$; c_M (636 K) $= 0.196$, c_P (636 K) $= 0.23$, leading to the full lines for the proposed two-phase field.

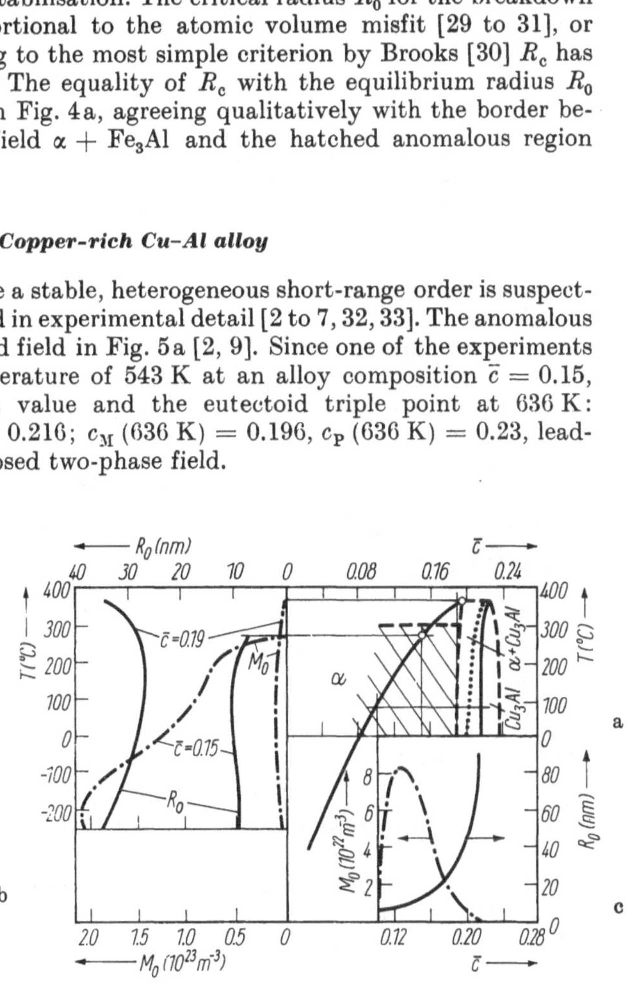

a

b

c

Fig. 5. The analogue to Fig. 4 for copper-rich Cu–Al alloys and a variation of M_0 and R_0 with temperature (Fig. 5b) at $\bar{c} = 15$ at% Al and $\bar{c} = 19$ at% Al as well as a change of R_0 and M_0 with composition $\bar{c}$ (Fig. 5c) at $T = 75$ °C

[6]) According to Brooks [30], the critical radius R_c is related to the length of the interface required to build a misfit equal to the Burgers vector $\boldsymbol{b}$ of an interface dislocation: $R_c > 3b/(2\sigma)$, whereby δ is the atomic volume misfit. δ for the $\alpha + Fe_3Al$ field has the order of $\delta = 0.138\Delta c$, giving $R_c = 3.1/\Delta c$ nm (on the Cu-rich side of Cu–Al, δ is of the order of $\delta \approx 0.051$: $R_c = 10.6/\Delta c$ nm).

This way a critical temperature at $T_c = 648.5$ K and a temperature, at which η attains the value η_∞ of 670 K are found. All of this agrees with an experimentally detected breakdown of the microstructure stability at alloy compositions above 19.2 at% Al [7] for $T = 473$ K ($\bar{c} > 0.192$). Nevertheless below $\bar{c} = 0.15$ the (fully drawn) disordering temperatures appear to be too small. Using the special Cu–Al value of the elastic energy constant ($E = 0.23$), the Cu–Al system at 348 K turns out to have a microstructure that is similar to that one of the Fe– Al system at 573 K (see Fig. 5c). In accordance with electron microscopic observations [2], the equilibrium density M_0 of the precipitate zones lies around 6×10^{22} m^{-3}, between $0.1 < \bar{c} < 0.15$, while R_0 changes from around 5 nm at $\bar{c} = 0.1$ to 10 nm at $\bar{c} = 0.15$. Also the dotted line, denoting the equality of R_0 with the critical radius of coherency breakdown R_c, $R_c = 10.6/\Delta c$ shows a rough agreement with the border between the classical $\alpha + Fe_3Al$ two-phase field and the anomalous one.

5. Summary

It is attempted to show that the total energy of a monodisperse system of spherical precipitates might have a minimum with respect to the precipitate size at a finite value of this size, if the ratio η of the self-energy and a negative pair interaction energy stays below a certain value. A negative pair interaction energy only appears in the case of ordering.

η is fitted to two real phase diagrams, deriving possibly stable equilibrium values R_0, M_0 of the precipitate size R and number M, as they change with temperature and the average alloy composition $\bar{c}$. Thus it appears to be possible that at low temperatures and precipitate volume fractions a transition region appears between the classical one- and two-phase regions of the binary diagram, that is characterized by a kinetic behaviour of the precipitates, which deviates strongly from that of classical theories [34].

Acknowledgements

The author is indebted to Prof. K. Lintner, Prof. G. Schöck, and Prof. G. Thomas for valuable communications.

Appendix

A. Total anisotropic elastic interaction energy

The elastic interaction energy E_{int} of a precipitate pair of equal volume v, being a large distance l apart, in edge-face configuration and infinitesimal approximation [14] can be written as

$$E_{\text{int}} = -kv^2/l^3\,;$$

whereby the constant k contains the elastic constants and the constrained elastic mismatch.

The total interaction energy $E_{\text{int}}^{\text{tot}}$ of M precipitates within a sample volume V, interacting with z nearest neighbours only, thus follows as

$$E_{\text{int}}^{\text{tot}} = (Mz/2)\,E_{\text{int}} = -(kz/2)\,(Mv)\,(v/l^3)\,.$$

For large M ($M \gg 1$), $E_{\text{int}}^{\text{tot}}$ stays unchanged during precipitate coarsening, as long as the precipitate volume fraction keeps constant, since both Mv and v/l^3 are proportional to this precipitate volume fraction (Ml^3 is a constant, proportional to the total volume V). A possible stabilizing effect due to the relative precipitate volumes is compensated by that of the change of their distance l.

B. Size of pair interaction energy between precipitates

The interaction energy E_{int} between precipitates originates from pairs of sites, where by one site $\boldsymbol{r}_n$ is located within one precipitate and the other one $\boldsymbol{r}_m$ within the other precipitate. E_{int} equals the difference between the total energy of (1) in I and the total energy, in which each of the two precipitates is surrounded only by matrix material with composition c_{M}. Assuming both of them equally sized

$$E_{\text{int}} = \sum_{n}^{\text{prec. 1}} \sum_{m}^{\text{prec. 2}} [(c_n - c_m)^2 - (c_n - c_{\text{M}})^2] \, \Omega(|\boldsymbol{r}_n - \boldsymbol{r}_m|)$$
$$= \sum_{m}^{\text{prec. 2}} (c_m - c_{\text{M}}) \sum_{n}^{\text{prec. 1}} (c_m + c_{\text{M}} - 2c_n) \, \Omega(|\boldsymbol{r}_n - \boldsymbol{r}_m|) \, .$$

Since $c_n \gtreqqless \bar{c}_{\text{M}}$ as well as $c_m \gtreqqless c_{\text{M}}$ and in case of ordering $\Omega(|\boldsymbol{r}_n - \boldsymbol{r}_m|) < 0$, $E_{\text{int}} > 0$. This can be confirmed, using concrete examples.

C. Material conservation

Taking into account the spherical symmetry of Fig. 2 of I, the equation of material conservation (equation (4) of I) changes to

$$M = 3VZ/[4\pi R^3 \mu(R)] \, ,$$
$$\mu(R) = (1 - D/R)^3 + (a/R)^3 \sum_{n=(R-D)/a}^{R/a} [n^3 - (n-1)^3] \, (Y_n - Y_m)/\Delta c \, . \quad \text{(C1)}$$

V denotes the total volume here.

D. Temperature dependence of the ratio η within the regular solution theory

The dependence of η, defined in (1) on the concentration splitting Δc follows from a substitution of g_α and g_β in (5) and (6) into $\Delta g(c_{\text{M}}) = 0$ and of the excess densities Δg_α and Δg_β, which are found that way, as well as Δg of (1) into (4), whereby it is assumed, that $c_0 = c_{\text{P}}$ and $v \gg 1$

$$\gamma \, \Delta c^5/30 = (\varepsilon + 4\pi N a^3 2\Omega_0) \Delta c^3/3 +$$
$$+ (kT/2) \{c_{\text{P}}^2 \ln (c_{\text{P}}/c_{\text{M}}) - (1 - c_{\text{P}})^2 \ln [(1 - c_{\text{P}})/(1 - c_{\text{M}})] - \Delta c\} \, ; \quad \text{(D1)}$$
$$\Delta c(T) = c_{\text{P}}(T) - c_{\text{M}}(T) \, .$$

With the substitution of the excess densities Δg_α and Δg_β into the lever rule $\Delta g(c_{\text{M}}) = 0$, the relation between the concentration splitting $\Delta c = c_{\text{P}}(T) - c_{\text{M}}(T)$ and T can be found

$$- (\varepsilon + 4\pi N a^3 2\Omega_0) \, \Delta c^2/k + T[c_{\text{P}} \ln (c_{\text{M}}) + (1 - c_{\text{P}}) \ln (1 - c_{\text{M}})] =$$
$$= T_{\text{c}}[c_{\text{P}} \ln c_{\text{P}} + (1 - c_{\text{P}}) \ln (1 - c_{\text{P}})] \, . \quad \text{(D2)}$$

Moreover this equation (D2) fullfills the condition that at the temperature T_{c} the miscibility gap between the phases α and β closes $[c_{\text{M}}(T_{\text{c}}) - c_{\text{P}} = 0]$. Substituting Ω_0 of (D2) into (D1), γ can be found and replacing γ of (2) by this expression the essential effective interface energy η arises as

$$\eta[\Delta c(T)] = (20\pi/3) \, (1 + E)^{-1} \, \{1 + (1.5/\Delta c) \, [\Delta c^2(0)/\Delta c^2(T) - 1] \times$$
$$\times [\Delta c - c_{\text{P}}^2 \ln (c_{\text{P}}/c_{\text{M}}) + (1 - c_{\text{P}})^2 \ln [(1 - c_{\text{P}})/(1 - c_{\text{M}})]] \, [c_{\text{P}} \ln c_{\text{M}} + (1 - c_{\text{P}}) \ln (1 - c_{\text{M}})]^{-1}\} \, ;$$

$$E = \frac{\varepsilon \delta^2 \, \Delta c^2(0)}{k T_{\text{c}}[c_{\text{P}} \ln (c_{\text{P}}) + (1 - c_{\text{P}}) \ln (1 - c_{\text{P}})]} \, .$$

Thereby (D2) can be used to derive the pair potential strength Ω_0 from the phase diagram. δ denotes the atomic volume misfit between the phases.

References

[1] H. Warlimont and G. Thomas, Metal Sci. J. **4**, 47 (1970).
[2] W. Gaudig and H. Warlimont, Z. Metallk. **60**, 488 (1969).
[3] L. Trieb, K. Siebinger, and H. P. Aubauer, Scripta Metall. **7**, 245 (1973).
L. Trieb and G. Veith, Acta metall, to be published.
[4] G. Veith, L. Trieb, W. Püschl, and H. P. Aubauer, phys. stat. sol. (a) **27**, 59 (1975).
[5] M. Zehetbauer, L. Trieb, and H. P. Aubauer, Z. Metallk. **67**, 431 (1976).
[6] G. Veith, L. Trieb, and H. P. Aubauer, Scripta Metall. **9**, 737 (1975).
[7] B. G. Lifshitz and M. P. Ravdel, Dokl. Akad. Nauk. SSSR **43**, 1033 (1953).
[8] H. Warlimont and H. P. Aubauer, Z. Metallk. **64**, 484 (1973).
[9] S. K. Das and G. Thomas, in: Order–Disorder Transformations in Alloys, Springer Verlag, Berlin/New York 1974 (p. 332).
[10] W. Gaudig, P. Okamoto, G. Schanz, G. Thomas, and H. Warlimont, Proc. III. Bolton Landing Conf., Claitors 1970 Baton Rouge La., USA (p. 347).
[11] H. Thomas, Z. Phys. **129**, 219 (1951).
[12] H. P. Aubauer and H. Warlimont, Z. Metallk. **65**, 297 (1974).
[13] A. G. Khachaturyan and V. M. Airapetyan, phys. stat. sol. (a) **26**, 61 (1974).
A. G. Khachaturyan, Soviet Phys. – J. exper. theor. Phys. **31**, 98 (1970).
[14] L. M. Brown, R. H. Cook, R. Ham, and G. R. Purdy, Scripta metall. **7**, 815 (1973).
[15] A. G. Khachaturyan, Soviet Phys. – Solid State **13**, 8, 2024 (1972).
[16] H. P. Aubauer, Acta metall. **20**, 165 (1972).
[17] B. V. Derjaguin, J. chem. Phys. **61**, 3665 (1974).
[18] J. E. McDonald, Amer. J. Phys. **31**, 31 (1963).
[19] W. Püschl and H. P. Aubauer, to be published.
[20] H. P. Aubauer, phys. stat. sol. (a) **43**, 261 (1977).
[21] M. Hillert, in: Phase Transformations, Ed. H. I. Aaronson, Amer. Soc. Metals, Cleveland (Ohio) 1968 (p. 181).
[22] A. Taylor and R. M. Jones, J. Phys. Chem. Solids **6**, 16 (1958).
H. J. McQueen and Kuczynzki, Trans. MS AIME **215**, 619 (1959).
P. S. Rudman, Acta metall. **8**, 321 (1960).
G. Erez and P. S. Rudman, J. Phys. Chem. Solids **18**, 307 (1961).
F. Lihl and H. Ebel, Arch. Eisenhüttenwesen **7**, 483 (1961).
A. Lawley and R. W. Cahn, J. Phys. Chem. Solids **20**, 204 (1961).
[23] R. G. Davies, J. Phys. Chem. Solids **24**, 9852 (1963).
[24] N. S. Golosov, A. M. Tolstik, and L. Ya. Pudan, J. Phys. Chem. Solids **37**, 273 (1976).
[25] G. Lütjering and H. Warlimont, Acta metall. **12**, 1460 (1964).
[26] H. Saito and H. Morita, J. Japan. Inst. Metals **30**, 930 (1966).
[27] A. J. Bradley and A. H. Jay, Proc. Roy. Soc. **A136**, 210 (1962).
[28] H. J. Leamy, Acta metall. **15**, 1827 (1967).
[29] J. Friedel, in: Les Dislocations, Gauthier Villars, Paris 1956.
[30] H. Brooks, in: Metal Interfaces, Amer. Soc. Metals, Cleveland (Ohio) 1952 (p. 20).
[31] J. M. Brown, G. R. Woolhouse, and U. Valdre, Phil. Mag. **17**, 781 (1968).
[32] V. E. Panin and E. K. Zerkova, Dokl. Akad. Nauk. SSSR **29**, 1024 (1959).
[33] W. Köster and H. P. Rave, Z. Metallk. **52**, 158 (1961).
[34] J. M. Lifshitz and J. Slyozov, J. Phys. Chem. Solids **19**, 35 (1961).
C. Wagner, Z. Elektrochem., Ber. Bunsenges. phys. Chem. **65**, 581 (1961).
[35] R. W. Hopper and D. R. Uhlmann, Scripta metall. **7**, 467 (1973).
[36] H. P. Aubauer, Scripta metall. **6**, 1061 (1972).
[37] D. De Fontaine, Scripta metall. **7**, 463 (1973).
[38] H. O. K. Kirchner, Acta metall. **22**, 553 (1974).
[39] H. P. Aubauer, phys. stat. sol. (a) **35**, 487 (1976).

(Received August 8, 1977)

phys. stat. sol. (a) 43, 611 (1977)

Subject classification: 1.2; 23

Materials Research Laboratory and Department of Physics, University of Illinois, Urbana

Density of Solid Argon at Melting [1])

By

A. T. MACRANDER and R. K. CRAWFORD[2])

Lattice parameters for crystalline argon along its melting line are reported with an estimated experimental uncertainty of 40 ppm. The X-ray data span the range of melting temperatures from 86.7 to 124.3 K. An upper limit of roughly 3×10^{-3} for vacancy concentrations along the melting line is obtained by comparing these data to bulk densities.

Es werden die Gitterparameter für kristallines Argon entlang seiner Schmelzkurve mit einer geschätzten experimentellen Unsicherheit von 40 ppm mitgeteilt. Die Röntgendaten umfassen einen Bereich von Schmelztemperaturen von 86,7 bis 124,3 K. Eine obere Grenze von ungefähr 3×10^{-3} für die Leerstellenkonzentration entlang der Schmelzkurve wird durch Vergleich dieser Werte mit den Volumendichten erhalten.

1. Introduction

Density determinations of solid argon along its melting line have been made primarily be means of weighing or piezometric techniques [1 to 5]. These techniques appeal to investigators interested in melting phenomena because the liquid state as well as the solid state can be studied. Unfortunately, densities determined by these techniques are limited to uncertainties on the order of 0.1%. X-ray density determinations can be made only of the solid, but they can be made with far greater precision. Indeed, we report here lattice parameters (l.p.'s) of argon along its melting curve with 40 ppm experimental uncertainty.

The volume change upon melting, $\Delta V = V_l - V_s$, is an important quantity in any study of melting transitions. Since for argon V_s has until now been determined by means of bulk density measurements only (except very near the triple point), our X-ray diffraction measurements improve the accuracy of ΔV for argon. Macroscopic voids and other inhomogeneities are a hazard in any experiment measuring bulk densities. No experimental check besides the reproducibility of the data has existed for argon, but by comparing our X-ray densities to bulk densities some inferences about the samples of the experimenters measuring bulk densities can be made.

Of course the differences between bulk densities and X-ray densities may be due to vacancies, and we obtain an upper limit on the concentration of vacancies if we take them to be responsible for the entire difference. These limits are of obvious interest to those constructing theories of melting around vacancies or vacancy clusters. Furthermore, these limits are of interest because vacancy concentration measurements in the rare gas solids have proved puzzling since they do not scale according to the law of corresponding states. An order of magnitude more vacancies have been found in krypton single crystals at its triple point [6] than were found in either argon [7] or neon [8] single crystals ar their triple points. A triple-exchange interaction has

[1]) Supported in part by U.S. Energy Research and Development Administration contract EY-76-C-02-1198 and by the Research Corporation.

[2]) Present address: Argonne National Laboratory, Argonne, Illinois, USA.

been proposed by Jansen [9] which decreases the enthalpy of formation of vacancies calculated for krypton to the observed value, but there is still much doubt about the strength of this force.

2. Experimental Method

2.1 X-ray camera

Our X-ray measurements were made on a much used oscillating back reflection camera first described in 1965 [10]. Data were taken on five separate argon crystals each with a different orientation. Both cobalt and nickel Kα radiation were Bragg-reflected and recorded on Ilford G film to yield our data. From three to five hours of exposure time were needed typically to produce defineable Kα_1 lines. For a few temperatures Kα_2 reflections were also recorded and these required at least ten hours of exposing. The X-rays had to pass through an effective thickness of 1.75 mm of beryllium (half this going in and half going out) in penetrating through a window and through the X-ray pressure cell. The loss in intensity due to absorption in the beryllium was about the same as that due to the Debye-Waller factor of argon.

2.2 Pressure system

Pressures up to 1.6 kbar were generated thermally. Low pressure argon was first passed through a commercial noble gas purifier and then through a molecular sieve suspended above liquid nitrogen before it was condensed in a maraging steel cell of 6.75 cm^3 internal volume. This pressure generating cell was held at the temperature of liquid nitrogen to condense the argon, and the argon was then pressurized by warming the generating cell to room temperature. Since the remaining high pressure plumbing added only 0.46 cm^3 of volume, several kilobars of pressure were within our reach. After the argon flowed through the sieve, the major impurities present in it were neon and helium which we removed by pumping on the pressure generating cell for several minutes before pressurizing. Residues that remained inside the high pressure plumbing and X-ray cell were then the main impurities. Since we used 0.15 mm i.d. stainless steel tubing, pumping out these residues was impractical, and we relied instead on several successive pressurizations to dilute them. Independent analyses performed with a mass spectrograph and a gas chromatograph showed that our argon was indeed very clean. No rare gas impurities were found by these means. The smallest concentrations that could be detected by these analyses were 4 ppm for the rare gasses including He, 4 ppm for O_2, N_2, and CO_2, and 10 ppm for H_2. The only impurity definitely detected was water at a concentration of 3 ppm.

Strain gauges glued to a separate cell provided pressure sensing via a dc bridge. These gauges were sensitive to a few tenths of a bar for time periods on the order of an hour, although the readings drifted up by to 40 bar over several days. The gauges proved extremely useful not only during the thermal pressure generation but also during specimen growth.

2.3 Sample chamber

Central to our apparatus is a beryllium cell in which crystals were grown (see Fig. 1). The cell is clamped to the fill line end by a maraging steel clamp containing a stainless steel seal, and the clamp in turn is held firmly in place by a large piece of copper. In a well within this same piece of copper we placed a platinum thermometer which was accurate to 15 mK. Our long exposure times increased the uncertainty of the temperature for each film to more than 15 mK though, and we estimate the total uncertainty to be 30 mK. A copper post thermally shorted the top of the sample chamber to the bottom of the sample cell making the temperature at the bottom of the cell the

Fig. 1. Sample chamber. a beryllium sample cell, b copper support, c platinum thermometer well, d control thermometer well, e copper post, f incolonel post (1 of 4), g copper braid, h cooling coils, i fill line; j maraging steel clamp, k stainless steel seal

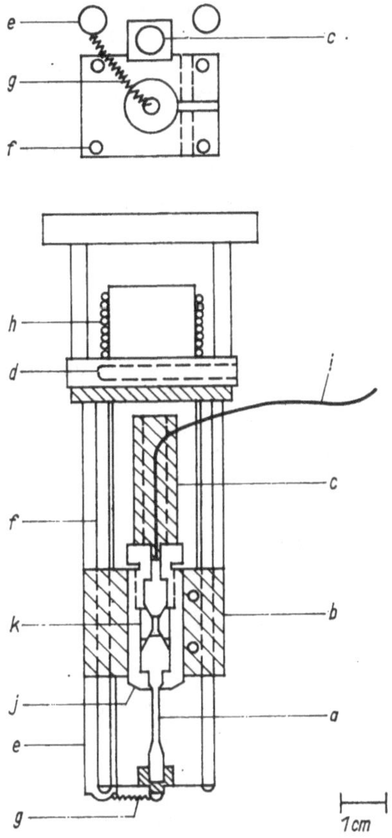

controlled temperature. Gradients between the bottom of the cell and the accurately measured temperature at the top of the cell were monitored by means of a Au + 200 ppm Fe thermocouple. Various temperature gradients across the sample cell were induced during those times when we attempted to grow single crystals since this gradient is an obvious parameter to be varied in obtaining the best crystal growing conditions. But once a useable sample was obtained a uniform temperature was of course greatly desired. Unfortunately, a persistent 80 mK gradient across the sample cell was present during the data taking, and we compensated the platinum thermo- meter readings accordingly.

2.4 Sample preparation

Growing argon single crystals proved to be possible in our beryllium cell, but the process was complicated because the liquid refused to crystallize until it had been supercooled (half a degree typically) below its melting point for all the temperatures at which we attempted to grow crystals. Simply cooling the liquid never produced useable single crystals, but, fortunately, did provide us with the means to do so. The samples that resulted by cooling until crystallization occurred were always polycrystalline showing many transmission Laue spots for all the sample cell gradients and cooling rates that we tried, but these samples could be melted down to serve as a seed with the help of an X-ray detector lined up on a suitably intense spot. The detector was needed to monitor the melting down process since it was quite easy to melt away the sample entirely. Generally, the parameter most closely correlated with crystal quality was the time taken to grow a crystal from a seed; longer times resulting in better crystals. An hour proved to be long enough to yield samples with a single crystalline portion of sufficient size for us to use.

2.5 Method for the taking of melting data

After a sample had been deemed useable and had been oriented (all were found to be f.c.c.) the fill line was allowed to plug, and the back reflection measurements were begun. By warming past the melting temperature that was observed when the fill line was open, l.p.'s along the melting line were obtained. The thermodynamic path for this procedure is shown in Fig. 2. Of course we had to be careful not to cause the plug in the fill line to slip, and, fortunately, from the shape of the diffraction spot we were able to discern when we could not safely warm further. Warming in this way generated melting data for four of the five crystals. For the fifth (crystal 4) we found that the l.p.'s fell on the melting line after cooling, and we concluded that the fill line had plugged with some liquid present in the cell. We cooled this crystal until it became obvious from the data that all the liquid had frozen.

3. Results, Comparisons, and Remarks (Including Remarks on Direct Vacancy Measurements)

3.1 Results

Our values for l.p.'s of argon at melting calculated using the wavelengths given by Bearden [11] are listed in Table 1. The X-ray camera that we used is most precise if after making an exposure of a Bragg reflection the incident and diffracted beams are interchanged for a second exposure. All but three of the values listed in the table entailed two such exposures and we estimate that they are precise to 40 ppm; the three values of lower precision, 80 ppm, were obtained from only one exposure. The shapes of the diffraction spots varied for the different crystals, and the greatest single source of error was the judgement of the center of such a diffraction spot. Other sources of error along with required corrections are discussed in [10]. In addition, our experimental arrangement required a special correction for X-ray absorption. Argon has a large

Table 1

Measured lattice parameters of argon crystals at melting along with their differences from the linear fit quoted in the text

crystal	radiation and reflecting plane used	temperature (K)	lattice parameter (Å)	$a_{\text{fit}} - a_{\text{meas}}$ (Å)
1	$\text{CoK}\alpha_1$*)	86.71	5.46271	+0.00039**)
	(442)	88.34	5.45944	−0.00002
2	$\text{CoK}\alpha_1$	93.56	5.44767	+0.00003
	(531)	94.84	5.44510	−0.00027
3	$\text{NiK}\alpha_1$*)	95.76	5.44290	−0.00015
	(620)	96.84	5.44038	−0.00004
		98.39	5.43709	−0.00024
		100.40	5.43251	−0.00017
4	$\text{CoK}\alpha_1$	111.98	5.40575	+0.00055**)
	(531)	113.90	5.40187	+0.00011**)
		116.67	5.39559	+0.00017
		119.08	5.39038	−0.00005
5	$\text{CoK}\alpha_1$	120.32	5.38764	−0.00007
	(531)	122.36	5.38305	−0.00009
		124.34	5.37869	−0.00016

*) Wavelengths used are those given in [11]: $\text{CoK}\alpha_1$, 1.788965 Å; $\text{NiK}\alpha_1$, 1.657910 Å.
**) Estimated uncertainty is 80 ppm. For the other values the estimated uncertainty is 40 ppm.

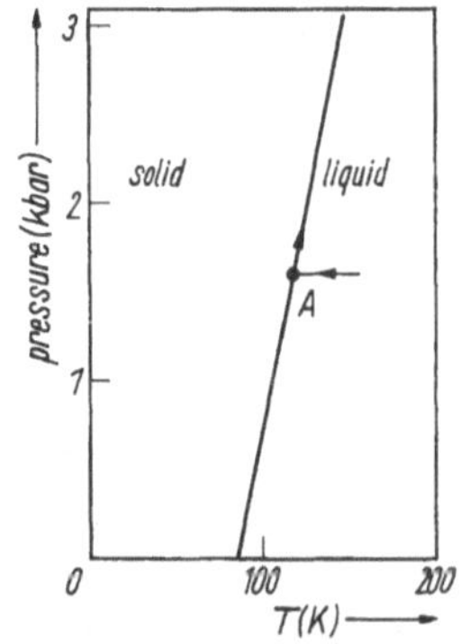

Fig. 2. High pressure phase diagram of argon showing the thermodynamic path followed in obtaining melting line data. Initially, a crystal is grown by cooling at constant pressure, then at A the fill line is allowed to plug and the crystal is warmed along the melting line to the desired melting temperature where X-ray diffraction measurements are made

X-ray absorption coefficient for which in our case, a cylindrical crystal bathed in the beam (laterally), an adjustment must be made in the way described by Buerger [12]. This adjustment amounted to a 15 ppm change in l.p.'s and was the largest that we made.

The data are shown in Fig. 3 along with a linear fit to the data:

$$a = 5.65798 - 0.00225T \ ,$$

where T is in degrees Kelvin and the resultant l.p. is in angstroms. We see that the data can be very well fit to a straight line. Also listed in the table are differences between the data and the above linear fit. In comparing our estimated uncertainties and these differences one notices that the l.p.'s differing the most from the linear fit are those of lower precision and that most of the high precision l.p.'s differ from the linear fit by considerably less than 40 ppm.

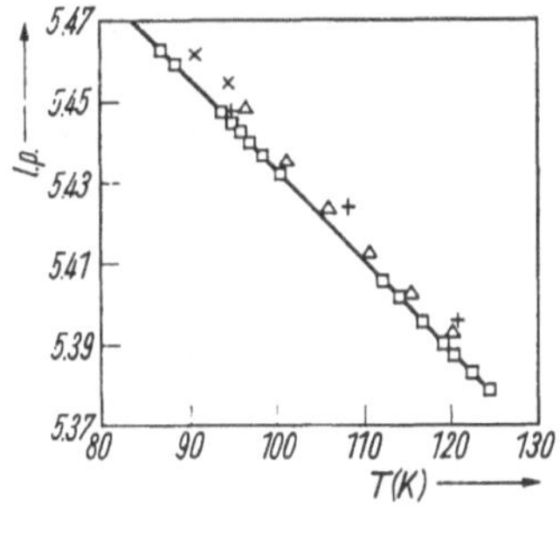

Fig. 3. Melting line of argon. The line is the linear fit to the data of the present work quoted in the text (□). The uncertainty of the present work is 0.004% which is less than the thickness of this line. The bulk density data of van Witzenburg and Stryland (△) and of Crawford and Daniels (+) yield lattice parameters precise to 0.04% or slightly less than the size of the symbols. Bronsveld and Stryland (×) quote an uncertainty of 0.01% which is about one-fifth the size of the symbols

3.2 Comparisons

Lattice parameters of free standing argon crystals were measured by Schwalbe with the same camera that we used [7]. These l.p.'s are shown on Fig. 4 along with his fit to them. Our data near the triple point is also shown along with the linear fit to display the comforting result that the intersection of the two fits occurs at a temperature very near the triple point temperature; the two fits intersect at 83.827 K and the triple point temperature of argon is 83.806 ± 0.001 K [13]. The temperature uncertainties of the data probably account for the difference; the temperature uncertainty of our melting line data is 30 mK and that of the data along the sublimation curve is better than 30 mK.

The results of experiments measuring bulk densities are shown on Fig. 3 where for ease in comparing the bulk results and X-ray results we have converted all densities

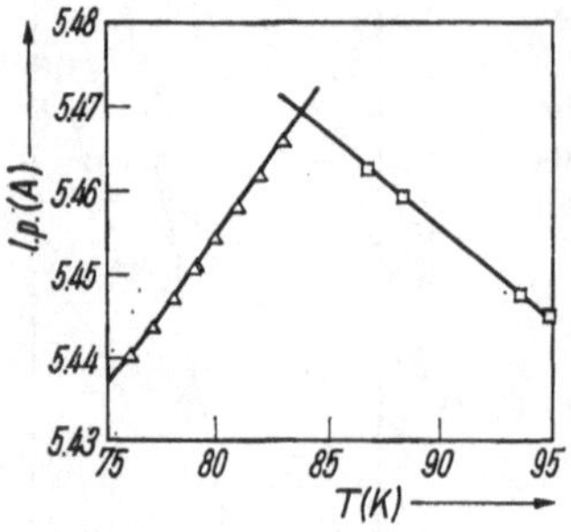

Fig. 4. Lattice parameters near the triple point of argon. A check on the data is obtained by comparing the intersection temperature of the two fits to the triple point temperature of argon
(□ present work, △ Schwalbe)

to lengths l using the formula $l = (4M/N\varrho)^{1/3}$ with $M = 39.948$ g/mol for the atomic weight of argon and $N = 6.0221 \times 10^{23}$. According to quoted uncertainties such lattice parameters obtained from the results of van Witzenburg and Stryland should be accurate to 0.04% as should be the l.p.'s obtained from the results of Crawford and Daniels. Also shown are two melting line l.p.'s reported by Bronsveld et al. [14] who quote an uncertainty of 0.01% for their X-ray diffraction data. Our experimental arrangement was somewhat similar to theirs but the two sets of X-ray data disagree. Perhaps a systematic error is inherent in their cylindrical camera since corrections to their raw data were small. The results of the present work all lie below the results of the experiments measuring bulk densities so that we may indeed interpret the difference as due to vacancies. Using the van Witzenburg data and the formula

$$\frac{n}{N} = 3\,\frac{(l - a)}{a}\,,$$

we obtain $(3 \pm 1) \times 10^{-3}$ for the concentration of vacancies at 120 K where the range of values represents the uncertainty of the bulk density quoted by van Witzenburg and Stryland. The scatter in the bulk data do not make it worthwhile to make this calculation for other temperatures, and we take $(3 \pm 1) \times 10^{-3}$ to be representative of the whole temperature range which we have investigated.

3.3 Remarks

After cooling three of the five crystals below the temperature at which they grew we attempted to make vacancy measurements using the method introduced for solid He by Balzer and Simmons [15]. Vacancy concentrations are then given by

$$\frac{n}{N} = 3\,\frac{\Delta l}{l} - 3\,\frac{\Delta a}{a}\,,$$

where l is a macroscopic length and a is the l.p..

If volume changes are small or can be calculated from the thermal expansion and Young's modulus of beryllium, it should be possible to measure vacancy concentrations with X-ray measurements alone. Preliminary data suggest an upper limit of 1×10^{-3} on vacancy concentrations at melting over the range of melting temperatures listed in the table. By way of comparison to this limit Schwalbe found the concentration of vacancies at the triple point to be $(1.7 \pm 0.6) \times 10^{-4}$ [16].

In light of the scatter in the bulk densities the limit of 1×10^{-3} is most easily reconciled with the higher limit, $(3 \pm 1) \times 10^{-3}$, found above if the bulk density data were somewhat less accurate than claimed.

In summary, we have measured lattice parameters of argon single crystals along the melting line by means of X-ray diffraction over a temperature range in which up to now only bulk densities had been measured. The results have shown that no great macroscopic inhomogeneities in the samples studied by van Witzenburg and Stryland or in the samples studied by Crawford and Daniels need to be invoked to explain the difference between bulk densities and X-ray densities; reasonable vacancy concentrations result if one takes most of the difference to be due to vacancies.

Acknowledgement

With pleasure we acknowledge many helpful discussions with Prof. R. O. Simmons.

References

[1] R. K. Crawford and W. B. Daniels, J. chem. Phys. **50**, 3171 (1969).
[2] W. van Witzenburg and J. C. Stryland, Canad. J. Phys. **46**, 811 (1968).
[3] V. M. Cheng, W. B. Daniels, and R. K. Crawford, Phys. Letters **43A**, 109 (1973).
[4] S. M. Stishov, I. N. Makarenko, V. A. Ivanov, and V. I. Fedosimov, Soviet Phys. – J. exper. theor. Phys., Letters, **11**, 13 (1970).
[5] S. M. Stishov and V. I. Fedosimov, Soviet Phys. – J. exper. theor. Phys., Letters **14**, 217 (1971).
[6] D. L. Losee and R. O. Simmons, Phys. Rev. **172**, 934 (1968).
[7] L. A. Schwalbe, Ph. D. Dissertation, University of Illinois, 1973.
[8] W. E. Schoknecht, Ph. D. Dissertation, University of Illinois, 1971.
[9] L. Jansen, Phys. Rev. **135**, A1292 (1964).
[10] D. N. Batchelder and R. O. Simmons, J. appl. Phys. **36**, 2864 (1965).
[11] J. A. Bearden, Rev. mod. Phys. **39**, 78 (1967).
[12] M. J. Buerger, X-Ray Crystallography, Wiley, New York 1942 (p. 404).
[13] H. H. Chen, R. A. Aziz, and C. C. Lin, Canad. J. Phys. **49**, 1569 (1971).
[14] P. M. Bronsveld, S. K. Kumra, and J. C. Stryland, Canad. J. Phys. **51**, 25 (1973).
[15] R. Balzer and R. O. Simmons, in: Low Temperature Physics, LT13, Vol. 2, Ed. D. K. Timmerhaus, W. J. O'Sullivan, and E. F. Hammel, Plenum Press, New York 1974 (p. 115).
[16] L. A. Schwalbe, Phys. Rev. B **14**, 1722 (1976).

(Received June 28, 1977)

phys. stat. sol. (a) 43, 619 (1977)

Subject classification: 5; 22.8.2

Laboratory of Magnetism, Ghent State University

A Mössbauer Study of Ca^{2+}-Containing Magnetites

By

J. DE SITTER, A. GOVAERT, E. DE GRAVE, D. CHAMBAERE, and G. ROBBRECHT

The substitution of Ca^{2+} ions in magnetite is studied, mainly by Mössbauer spectroscopy. The Ca^{2+} ions are located at the tetrahedral sites and the conduction electrons can be described by a band model. The magnetite structure can at most contain 6.67 wt% of Ca. The presence of a Ca^{2+} ion at an A-site decreases the hyperfine field at the neighbouring B-site by about 30 kOe.

Es wird die Substitution von Ca^{2+}-Ionen in Magnetit hauptsächlich mit Mößbauerspektroskopie untersucht. Die Ca^{2+}-Ionen sind auf den Tetraederplätzen lokalisiert und die Leitungselektronen lassen sich durch ein Bändermodell beschreiben. Die Magnetitstruktur kann höchstens 6,67 Gew% Ca enthalten. Die Anwesenheit von Ca^{2+}-Ionen auf den A-Plätzen verringert das Hyperfeinfeld an den benachbarten B-Plätzen um etwa 30 kOe.

1. Introduction

The magnetite phase in basic iron ore sinters can contain some Ca^{2+} ions, substituting for part of the ferrous ions. Previously these so-called calciomagnetites have been studied by means of chemical, microscopic, and X-ray methods [1]. Some Mössbauer effect measurements have also been reported, but no detailed analysis of the spectra has been made [2]. In this work we determined the solubility range of Ca^{2+} in the magnetite structure. We also payed attention to the distribution of the Ca^{2+} ions among the octahedral and tetrahedral lattice sites. By careful analysis of the room temperature Mössbauer spectra some fundamental aspects of the substitution of diamagnetic impurities in magnetite are studied.

2. Experimental Results

Samples having nominal composition $Ca_xFe_{3-x}O_4$ with x ranging from 0.0 to 0.4 were prepared by the following technique: CaO was dried at 900 °C and then weighted together with α-Fe_2O_3 in stoichiometric amounts. The powders were throughly mixed in an agate mortar, pressed into pellets, fired for 4 h at 1320 °C in a CO_2 atmosphere and quenched to room temperature. In order to obtain a better homogeneity of the samples, the firing procedure was repeated.

X-ray analysis was made using FeKα radiation. For $x < 0.2$ only a single spinel pattern was found without any broadening of the reflections and showing a good splitting of the Kα_1 and Kα_2 components at higher angles, indicating these materials to be pure and homogeneous spinels. For $x > 0.2$ some additional reflections indicated the presence of a secondary phase, which could be identified as being $Ca_4Fe_9O_{17}$ [3]. No trace of any other Ca or Fe compound was found. Fig. 1 shows the lattice parameter of the spinel phase as a function of composition x using a least-squares fit [4]. The lattice constant increases linearly from 8.40 Å at $x = 0.0$ to 8.47 Å at $x = 0.20$ indicating that the substitution takes place on one site, or in a constant ratio on both lattice sites. For $x > 0.20$ the lattice parameter remains constant, showing that no more Ca^{2+} ions enter the spinel phase. From a linear extrapolation of the two parts of Fig. 1 we suggest that the maximum amount of Ca^{2+} ions which can substitute for ferrous ions in the magnetite structure, corresponds to $x = 0.19$.

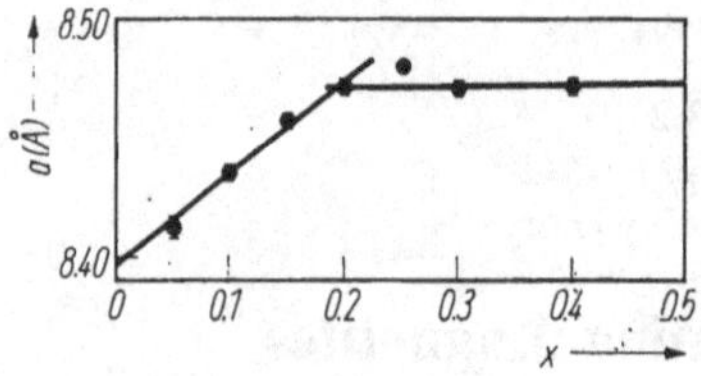

Fig. 1. The lattice parameter of the spinel phase in the series $Ca_xFe_{3-x}O_4$

From measurements of the spontaneous magnetization at room temperature, we can conclude that, at least for small x-values, the Ca^{2+} ions preferably occupy the tetrahedral sites, as indicated by an increase in spontaneous magnetization. For larger values of x, the Néel temperature becomes too low to determine the site occupancy from magnetization measurements at room temperature. Assuming that the calcio-magnetite phase does not change anymore for $x > 0.20$ and that the secondary phase is paramagnetic at room temperature, as will be affirmed by our Mössbauer effect measurements, it is possible to calculate the Fe/Ca ratio in this secondary phase from the reduction of the spontaneous magnetization and from the stoichiometry of the starting mixtures. The determined value of 2.21 is in good agreement with the value of 2.25, calculated from the theoretical composition $Ca_4Fe_9O_{17}$.

Mössbauer spectra at room temperature were obtained with a transducer, operating in a constant acceleration mode. The source was 10 mCi ^{57}Co in Pd metal. The spectrometer was calibrated using a natural iron absorber. The absorber thickness was held at about 10 mg Fe/cm^2 for all samples.

Fig. 2 shows the Mössbauer spectra of some of the samples. For $x < 0.20$ the spectra are composed of two subspectra, each showing six peaks. The sharper pattern can be assigned to a ferric oxidation state of iron. The second pattern is very strongly broadened.

A superposition of two Lorentzian-shaped six-lines patterns does not provide an acceptable description of the data, except for $x = 0.0$. There is however one clear point: the intensity of the broadened spectrum, assigned to the B-sites, increases significantly, relative to the sharp one, assigned to the A-sites. This is a strong indication that the Ca^{2+} ions occupy the A-sites. A further, detailed analysis of the spectra will be discussed below.

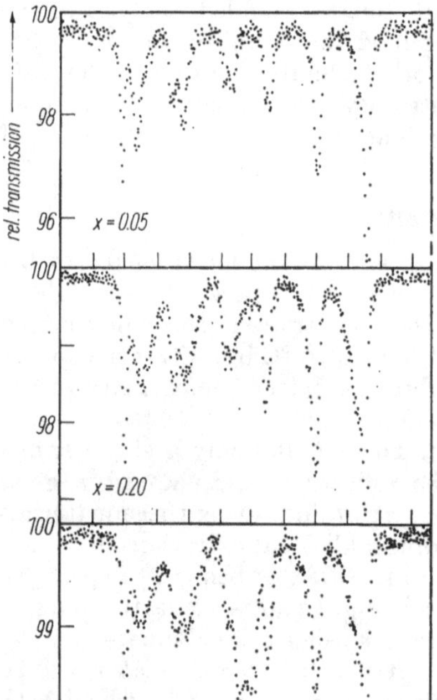

Fig. 2. Room temperature Mössbauer spectra of some samples in the series $Ca_xFe_{3-x}O_4$

For $x > 0.20$ a supplementary component appears in the spectra due to a paramagnetic phase. The magnetically split part of the spectrum does not change anymore by a further increase of x. After stripping off the contribution of the magnetite like phase, we attempted to analyse the remaining spectrum. This analysis showed mainly the presence of two quadrupole doublets, having a quadrupole splitting of 0.95 and 0.41 mm/s respectively, and isomer shifts of 0.08 and 0.01 mm/s relative to the $^{57}Co/Pd$ source. These values are typical for ferric iron and very close to data for $CaFe_2O_4$ [5].

We did not find a clear evidence for the presence of ferrous iron, although some asymmetry in the intensities of the components of the outer doublet might indicate a third site occupied by iron ions. These are characterized by an isomer shift of 0.55 mm/s and a zero quadrupole splitting, and could possibly be Fe^{2+} in a covalently bounded surrounding.

3. Analysis of the Mössbauer Spectra

The spectra of the calciomagnetites were analysed, making the following assumptions: 1. The conduction electron mechanism has a band character, so that all B-site Fe ions have equal effective charge [6]. 2. The Ca^{2+} ions are located at the tetrahedral sites.

Under these conditions the asymmetric broadening of the B-lines can be assigned to a distribution of the hyperfine field at the B-site, caused by the random distribution of Ca^{2+} ions at the A-sites [7]. In a first neighbour model, the hyperfine field at each B-site Fe ion depends on the number of Ca^{2+} ions at the neighbouring A-sites, due to the contribution of supertransferred hyperfine fields to the total hyperfine field.

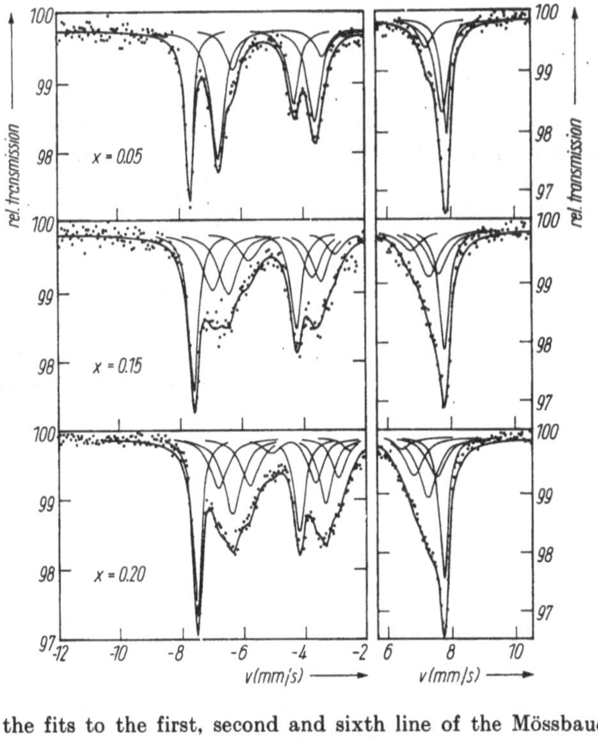

Fig. 3. Results of the fits to the first, second and sixth line of the Mössbauer spectra of some $Ca_xFe_{3-x}O_4$ samples

The relative intensity, as a function of the number n of such neighbours, can be expressed as:

$$P(n, x) = \binom{6}{n} x^n (1 - x)^{6-n} , \tag{1}$$

where x is the concentration of Ca^{2+} ions. Although this model is not completely correct, we adopt it because it is the simplest one that accounts in a reasonable way for the observed spectra. The validity of the two a priori conditions will be ascertained by the obtained results.

The experimental data were fitted to this model in the following way. The fine structure of the B-site absorption peaks was only taken into account for the first, second and sixth absorption line, because these are the best resolved lines, and give sufficient information to determine the hyperfine parameters. The relative intensities of the B-site lines were fixed by (1) and we also accepted the 3:2:1 rule for the intensities of the different lines of each subspectrum. The line widths of the several components of each B-line were assumed to be equal. No restrictions were made on the positions of the lines. In this way we obtained reasonable fits to the experimental data, as is illustrated in Fig. 3.

4. Discussion

4.1 Intensities

The total intensities of the A- and B-site subspectra were free parameters in the fits. If all iron ions on the B-sites contribute to the broadened pattern, and all A-site iron ions to the sharp one, and if we neglect thickness effects, the site occupancy ratio can be written as:

$$\frac{n_B}{n_A} = \frac{I_B}{I_A} \frac{f_A}{f_B} , \tag{2}$$

where f_A and f_B stand for the respective Mössbauer fractions and n for the number of iron ions at each site. The ratio f_B/f_A was calculated from the area ratio, for a pure magnetite sample, prepared in exactly the same way as the calciomagnetites. The determined site occupancy ratios are given in Fig. 4. The straight line indicates the theoretical value of this ratio if both a priori conditions were completely met. The agreement between the experimental results and the proposed model is reasonable.

Moreover, if one accepts a pairwise hopping model for the conduction mechanism in magnetite and in the calciomagnetites, the introduction of some A-site Ca^{2+} ions would lead to the creation of isolated ferric ions at the B-sites. In the Mössbauer spectrum, these ions should contribute to the sharp component, resulting in a decrease

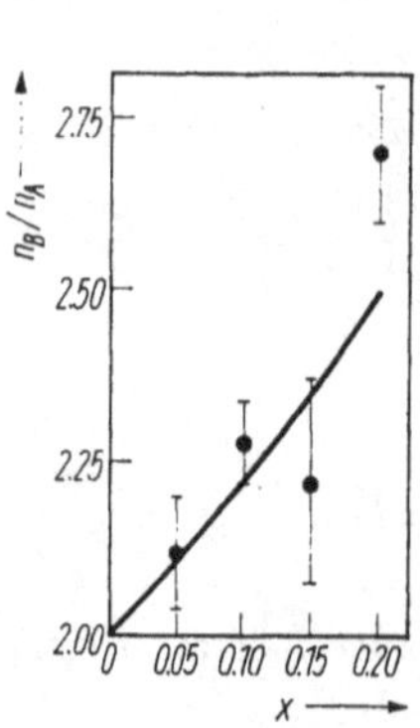

Fig. 4. The site occupancy ratio in the $Ca_xFe_{3-x}O_4$ samples, as calculated from the intensity ratio of A- and B-site subspectra

in the area ratio I_B/I_A with increasing x. A decrease in I_B/I_A, or at least a slower increase, is also expected if Ca^{2+} ions should occupy the B-sites. So the behaviour of the intensity ratio of both spectra gives evidence of the correctness of our initial assumptions.

4.2 Linewidths

Although our fitting procedure gives a good agreement between the experimental data and the theoretical fit, some residual broadening of the B-components needs to be explained. For the A-site line a full width at half maximum of 0.30 mm/s was found, independent of the value of x. The widths of the first and the sixth B-line are 0.65 mm/s and for the second line 0.50 mm/s. All these values slightly increase with increasing x. Several mechanisms can cause this broadening. Part of it can be explained by an anisotropy effect caused by the trigonal symmetry at the B-sites [8]. This effect explains the difference in widths between the inner and outer lines. We should also keep in mind that the samples were obtained by quenching from high temperatures. Quenching results also in a broadening of the lines [9], due to local variations in the quadrupole splitting caused by the introduction of thermal vacancies. This effect vanishes at the A-sites because of their cubic symmetry. Some broadening can also be caused by the influence of local deformations and electric polarisation on the mobility of the conduction electrons [10]. Finally, part of this broadening can also be ascribed to fluctuations in the molecular field acting on the B-site ions, caused by the disordered distribution of next-nearest and further neighbours.

4.3 Isomer shifts

Fig. 5 shows that the isomer shift difference between the A-site ferric ions and the iron ions on the B-sites, which have no Ca^{2+} ions as first neighbours, decreases with increasing x. This proves that the B-site iron ions become more ferric. The isomer shift can be written as:

$$\delta = A\varrho_{3d} + C ,$$

where ϱ_{3d} represents the effective number of 3d-electrons. Neglecting the difference in covalency between the A- and the B-sites, the parameter A can be calculated from the difference in isomer shift between the A- and the B-sites in Fe_3O_4. In this way a theoretical value for the isomer shift difference as a function of the parameter x can be calculated. This is plotted as the straight line in Fig. 5, and shows the same tendency as the experimental data. The larger decrease, than theoretically expected, can be attributed to a decrease in the conduction electron concentration due to the increase in lattice parameter, and to changes of covalency by the introduction of the large Ca^{2+} ions. An analogous result is found in Cd an Zn substituted magnetites [6].

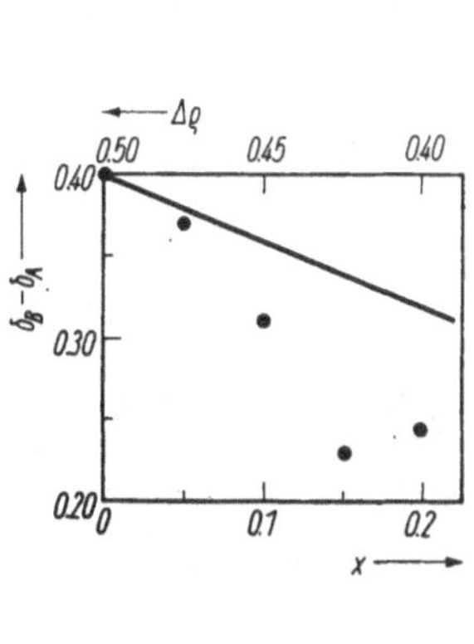

Fig. 5. Isomer shift difference between the A- and the B-sites in the $Ca_xFe_{3-x}O_4$ series

4.4 Hyperfine field

The evolution to a more ferric character of the B-site iron ions should, at first sight, give rise to an increase of the magnetic hyperfine field at the B-sites. However, the introduction of diamagnetic Ca^{2+} ions lowers the Néel temperature. Due to these two competing effects, no further affirmation of the band model for the conduction electrons can be obtained from hyperfine field data at room temperature.

In our analysis of the spectra, the broadening of the B-site lines is ascribed to changes in the supertransferred hyperfine field caused by the Ca^{2+} ions at the A-sites. The observed decrease in hyperfine field is about (30 ± 4) kOe for each A-site Fe^{3+} nearest neighbour replaced by Ca^{2+}, in good agreement with the values obtained on $MgFe_2O_4$ [11]. The somewhat higher value for the calciomagnetites is probably due to a lower value of the Néel temperature. Ok and Evans [6] give also a qualitative explanation for this larger value based on the difference between insulating and conducting ferrites.

5. Conclusions

From X-ray diffraction, spontaneous magnetization, and Mössbauer measurements, valuable information is obtained on the substitution of Ca^{2+} ions for the ferrous ions in the magnetite structure. The maximum amount of Ca^{2+}, that can enter the spinel structure, corresponds to the composition $Ca_{0.2}Fe_{2.8}O_4$. For higher amounts of Ca^{2+}, a secondary phase is formed which can be identified as $Ca_4Fe_9O_{17}$. The ferric ions in that structure behave similar to those in $CaFe_2O_4$. The Mössbauer spectrum gives also an indication of the presence of a third iron site, having an isomer shift of 0.55 mm/s relative to $^{57}Co/Pd$ and a zero quadrupole splitting. A careful but straightforward analysis of the room temperature Mössbauer spectra indicates that all Ca^{2+} ions are on tetrahedral sites. The asymmetric broadening of the B-site lines is explained by variations of the supertransferred hyperfine field, depending on the number of Ca^{2+} ions that are first neighbours of a given B-site iron ion. The decrease is about 30 kOe for each neighbour.

From the intensity ratio between the A- and B-site subspectra and from isomer shift data, we obtain evidence of the band character of the conduction electrons in the calciomagnetites.

Acknowledgements

The authors are grateful to the Belgian foundations IWONL and FKFO for financial support of this work.

References

[1] E. Mazanek and M. Wyderko, Polska Akademik Nauk – Metallurgia **22**, 11 (1974) and references therein.
[2] A. Z. Hrynckiewicz, D. S. Kulgawczuk, E. S. Mazanek, A. J. Pustówka, J. A. Sawicki, and M. E. Wyderko, phys. stat. sol. (b) **43**, 401 (1971).
[3] S. B. Homquist, Nature (London) **218**, 604 (1960).
[4] E. De Grave, J. De Sitter, and R. Vandenberghe, Appl. Phys. **7**, 77 (1975).
[5] E. F. Bertaut, J. Chappert, A. Apostolov, and U. Seminov, Bull. Soc. Franc. Minér. Crist. **89**, 206 (1966).
[6] H. N. Ok and B. J. Evans, Phys. Rev. **B14**, 2956 (1976).
[7] B. J. Evans, Mössbauer Effect Methodology **4**, 139 (1968).
[8] A. M. Van Diepen, Physica **86-88B**, 955 (1977).
[9] E. De Grave, C. Dauwe, A. Govaert, and J. De Sitter, to be published.
[10] A. M. Van Diepen and F. K. Lotgering, Physica **86-88B**, 961 (1977).
[11] E. De Grave, Internat. Conf. Mössbauer Spectroscopy Bucharest 1977.

(Received June 30, 1977)

phys. stat. sol. (a) 43, 625 (1977)

Subject classification: 12.1; 1.3; 21; 21.3

Laboratoire de Physique des Vibrations, U.E.R. des Sciences, Limoges (a) and Laboratoire de Métallurgie Physique, U.E.R. des Sciences, Poitiers (b)

Variations of the Elastic Constants of Aluminium–Magnesium Single Crystals with Guinier-Preston Zones

By

C. GAULT (a), A. DAUGER (b), and P. BOCH (a)

Ultrasonic measurements of elastic constants are performed in a single crystal of aluminium–magnesium alloy of 12 at% Mg during isothermal ageing at room temperature after quenching. They show that the shear constants C and C' increase in the first stage of ageing, whereas the bulk modulus B is at first nearly constant before increasing again. This is attributed to the development of Guinier-Preston zones, and these variations are compared with those observed in polycrystals. Then, an attempt is made to analyse the causes of these variations. Measurements of the elastic constants versus concentration are made in solid solution Al–Mg (from 4.5 to 12 at%), ordering to evaluate the matrix depletion effect during the formation of G. P. zones. The contribution of effect to the enhancement of the elastic constants of the zones should explain the variations observed.

Des mesures de constantes élastiques par une méthode ultrasonore ont été effectuées dans un monocristal d'alliage aluminium–magnésium de teneur 12% at. durant des vieillissements isothermes à température ambiante après trempe. Elles ont montré que les constantes de cisaillement C et C' augmentent dès le premier stade du vieillissement, alors que le module de rigidité B commence par rester sensiblement constant avant d'augmenter ensuite. Ceci est attribué au développement des zones de Guinier-Preston, et ces variations sont comparées avec celles observées dans un polycristal du même alliage. Une tentative d'analyse des causes de ces variations est ensuite entreprise. Des mesures des constantes élastiques en fonction de la teneur ont été effectuées pour la solution solide Al–Mg (de 4,5 à 12% at.), ce qui a permis d'évaluer l'effet d'appauvrissement de la matrice lors de la formation des zones G.P. La contribution de l'effet d'ordre dans les zones au renforcement des constantes élastiques des zones peut être responsable des variations observées.

1. Introduction

The elastic moduli of alloys vary with concentrations and structures. In the Al–Mg case, we have shown [1] that the Young's and shear moduli E and G of 12 to 16 at% supersaturated solid solutions change during isothermal ageing. This has been attributed to precipitation phenomena: the formation of Guinier-Preston (G.P.) zones for ageing at low temperatures and the precipitation of β' and β phases [2]. The elastic moduli have been measured in polycrystals using an ultrasonic method [3] and they have been found to increase with the formation of G.P. zones, but to decrease with β' and β development. Thus, measurements of ultrasonic moduli are a useful tool to study the kinetics of formation of G.P. zones, which is easily differentiated from the formation of the other phases.

Using a high angle X-ray scattering method, Dauger [4] has first shown the existence of G.P. zones in Al–Mg alloys, for concentrations above 9 at% Mg. Then, from neutron scattering [5] and electron microscopy [6] experiments, a model has been elaborated: G.P. zones are ordered domains of 25 at% Mg in a depleted matrix (8 at% Mg). The existence of a resistivity maximum for ageing at low temperatures (up to 40 °C) of Al–Mg 13.5 at%, connected with ultrasonic measurement results, has suggested that the demixtion of the solid solution should be a spinodal decomposition [1] in this temperature range. Further, modulated structures similar to those observed by electron microscopy in alloys such as Cu–Ti [7] have been investigated in Al–Mg [6] and have enhanced this hypothesis. But this point is still controverted [8] and it is of great interest to have, besides resistivity measurements, another technique to investigate the kinetics of the early stage of the demixtion of the solid solution, in order to determine the temperature limits of the spinodal decomposition, and nucleation and growth domains.

In this paper, measurements are extended to single crystals and an attempt is made to analyse the causes of the variations observed.

2. Experimental

In polycrystals, the measurements of the ultrasonic velocities have been made by a "pulse echo overlap" method described elsewhere [3]. The frequency is 6 to 20 MHz, and the sensitivity about 10^{-4}; velocities are measured with an accuracy of 10^{-3}. This method is limited by the size of specimens which must have a length superior to 10 mm.

As far as we can conjecture the growth of single crystals of Al–Mg alloys, for increasing concentrations, is a difficult problem which has not been solved yet. Using a compressional strain-anneal grain growth method, we have obtained crystals large enough to be cut after orientation into samples whose lengths and diameters are only a few millimeters. The thinness of these specimens has led to perform a "phase comparison" method for ultrasonic measurements [9]. A fused silica rod is used as a buffer between the transducer and the specimen. The superposition of the echoes at the buffer–specimen interface and in the specimen is observed when a long rf pulse is applied to the transducer. The block diagram of the apparatus is given on Fig. 1. At given frequencies of the waves, the echoes in the specimen are in phase, but out of phase with the echo at the interface buffer–specimen. These frequencies are measured, and the velocity may be calculated. Here the frequency varies about 20 MHz, and the precision is quite the same as for the "pulse echo overlap" apparatus, for specimen lengths superior to 1.5 mm.

3. Results on Isothermal Ageing at Room Temperature

3.1 *Variations of the elastic constants in an Al–Mg single crystal*

The single crystal used here is a 12 at% Mg specimen with two sets of plane parallel faces in ⟨100⟩ and ⟨110⟩ directions of 4 mm length.

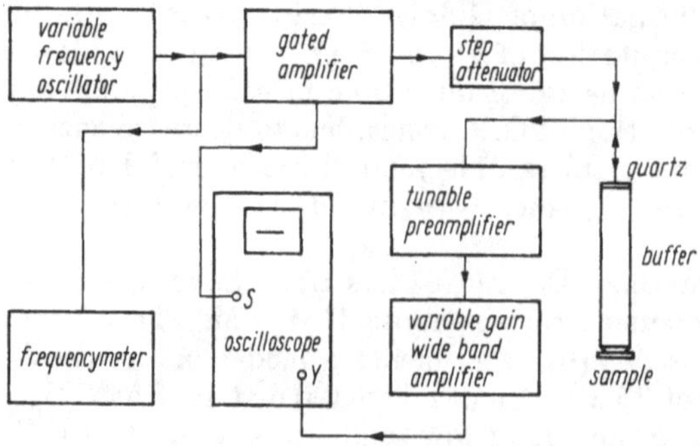

Fig. 1. Block diagram of the phase comparison apparatus

The longitudinal and shear wave velocities $V_{L\langle 100\rangle}$ and $V_{T\langle 100\rangle}$ are measured for propagation in the ⟨100⟩ direction, and $V_{L\langle 110\rangle}$ in the ⟨110⟩ direction. They are related to the elastic constants and the density ϱ by:

$$C_{11} = \varrho V^2_{L\langle 100\rangle}\,, \quad (1)$$

$$C_{44} = \varrho V^2_{T\langle 100\rangle}\,, \quad (2)$$

$$C_{11} + C_{12} + 2C_{44} = \varrho V^2_{L\langle 110\rangle}\,. \quad (3)$$

Then, the Zener elastic constants may be calculated:

$$C = C_{44}\,, \quad (4)$$

$$C' = \frac{C_{11} - C_{12}}{2}\,, \quad (5)$$

$$B = \frac{C_{11} + 2C_{12}}{3}\,. \quad (6)$$

The isothermal ageing is made at room temperature (20 °C) after quenching in air from 450 °C [1].

The three velocities $V_{L\langle 100\rangle}$, $V_{T\langle 100\rangle}$, and $V_{L\langle 110\rangle}$ increase with ageing time and the corresponding variations of the shear constants C and C' and of the bulk modulus B have been calculated and are plotted on Fig. 2. The shear constants increase, while the bulk modulus has a slight minimum in the first period of ageing, and then increases. As for polycrystalline alloys these variations are attributed to the formation of G.P. zones.

3.2 Comparison of the shear modulus variations in a single crystal and a polycrystal

The Hashin-Shtrikman relations evaluate the shear modulus G of an isotropic polycrystalline aggregate, from the elastic constants of the crystal [10]. These calculations lead to lower and upper bounds G_1 and G_2 for G, the average value of which is

$$\bar{G} = \frac{C + C'}{2}(1 + \varepsilon)\,,$$

with

$$\varepsilon = \frac{1}{2(C + C')}\left[\frac{3}{\dfrac{5}{C - C'} + \dfrac{12(B + 2C')}{5C'(3B + 4C')}} + \frac{2}{\dfrac{5}{C' - C} + \dfrac{18(B + 2C)}{5C(3B + 4C)}}\right]^{-1}.$$

In Al–Mg 12 at% in solid solution state, the experimental results are

$$C = 2.915 \times 10^{10}\ \text{Pa}\,,$$

$$C' = 2.284 \times 10^{10}\ \text{Pa}\,,$$

$$B = 6.949 \times 10^{10}\ \text{Pa}\,,$$

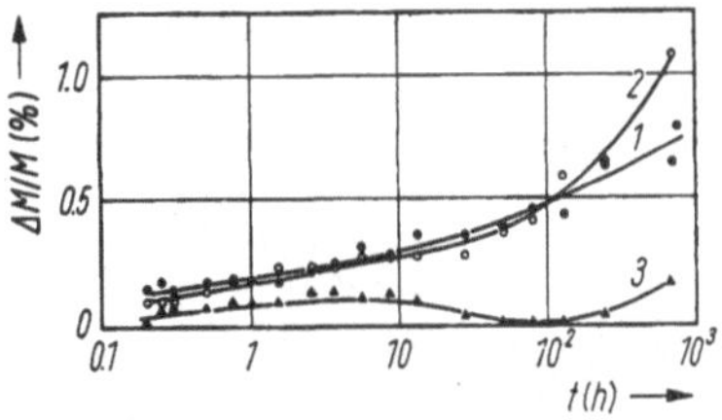

Fig. 2. Relative variations of the elastic constants of an Al–Mg 12 at% single crystal with ageing time at 20 °C after quenching: (1) $\Delta C/C$, (2) $\Delta C'/C'$, (3) $\Delta B/B$

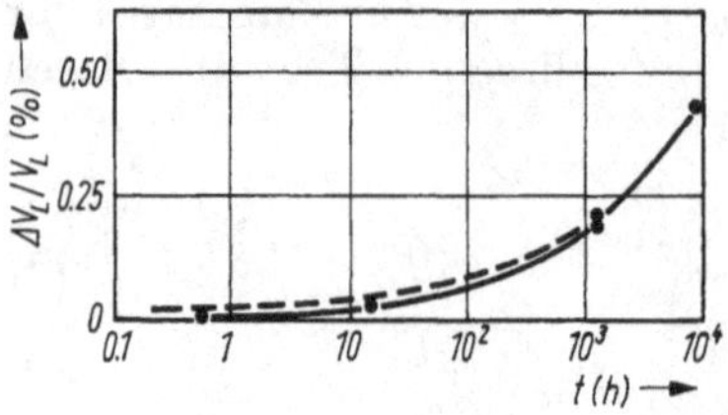

Fig. 3. Relative variation of the longitudinal velocity in an Al–Mg 12 at% polycrystal with ageing time: the full line is experimental result, and the dashed line is calculated from relation (10)

and so

$$\varepsilon = 1.02 \times 10^{-2} .$$

In first approximation ε may be neglected and:

$$\overline{G} \approx \frac{C + C'}{2} . \tag{7}$$

In polycrystals, G and B are related to V_{L} and V_{T} by

$$G = \varrho V_{\mathrm{T}}^2 , \tag{8}$$

$$B = \varrho \frac{3V_{\mathrm{L}}^2 - 4V_{\mathrm{T}}^2}{3} . \tag{9}$$

Then, using relations (7) to (9), it is possible to calculate the approximate average longitudinal wave velocity $\overline{V}_{\mathrm{L}}$ in a polycrystal from the elastic constants of the single crystals:

$$\overline{V}_{\mathrm{L}} \approx \left[\frac{2(C + C') + 3B}{3\varrho}\right]^{1/2} . \tag{10}$$

This leads to a very satisfactory value, $\overline{V}_{\mathrm{L}} = 6370$ m/s, compared with the experimental value measured in a polycrystal of same origin, $V_{\mathrm{L}} = 6367$ m/s. Fig. 3 shows the evolution of the longitudinal velocity in an Al–Mg 12 at% polycrystal at 20 °C. The full line is experimental measurement, and the dashed line is calculated from the experimental variations in the single crystal of Fig. 2 reported in the relation (10). The two curves are in very good agreement.

3.3 Results in polycrystals

The sample is a 13.5 at% polycrystalline alloy. It is cubic shaped and 20 mm long. The measurements of longitudinal and shear wave velocities in three orthogonal directions of propagation confirmed that this sample had a good isotropy, and thus it was possible to evaluate G and B.

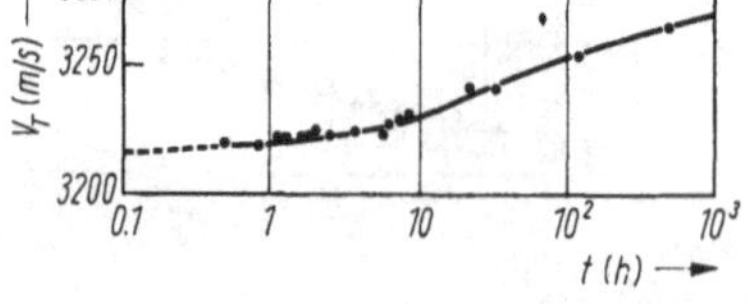

Fig. 4. Variations of the ultrasonic velocities in an Al–Mg 13.5 at% polycrystal with ageing time at 20 °C after quenching

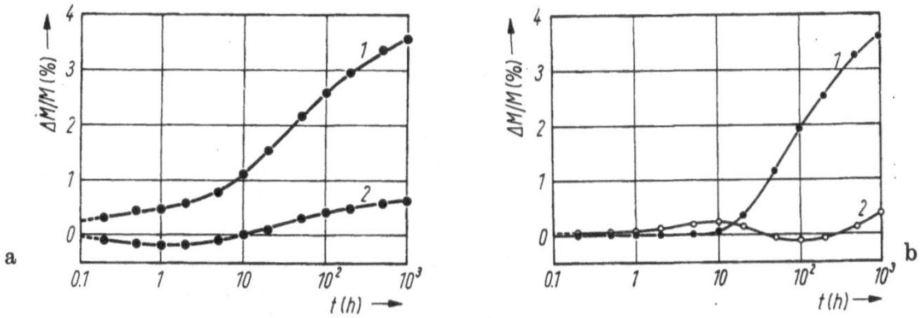

Fig. 5. a) Relative variations of the elastic moduli $\Delta G/G$ (curve 1) and $\Delta B/B$ (curve 2) in an Al–Mg 13.5 at% polycrystal with ageing time at 20 °C after quenching. b) The same after reversion

Fig. 4 shows the evolution at 20 °C of V_L and V_T in this sample as-quenched. They both increase with ageing time and this has been attributed to the development of G.P. zones [1]. From these curves the variations of G and B have been plotted (Fig. 5a), and it is shown, as for single crystals, that the increase of the bulk modulus is much smaller than the increase of the shear modulus (the variations are superior to those observed in the 12 at% single crystal because of the difference of concentration of the two alloys). The slight minimum of B appears after 2 h ageing though it appeared after 80 h ageing in the 12 at% single crystal, and so it seems to be characteristic of the decomposition rate of the solid solution. Indeed on Fig. 5b it appears later for an isothermal ageing at 20 °C after reversion (10 min at 100 °C), when the demixtion of the solid solution is slower than after quenching [1]. Then, the behaviour of B, found in single crystal, is confirmed by polycrystal experiments.

4. Discussion

The quantitative study of the variations of elastic moduli due to G.P. zones requires the knowledge of two cumulative effects: 1. The variation of the elastic constants of the solid solution with concentration, to evaluate the consequence of matrix depletion when G.P. zones are developed, and 2. the proper effect of G.P. zones which are considered as a second phase with elastic properties different from those of the matrix.

4.1 Matrix depletion

We have studied the influence of the addition of magnesium on the elastic constants of Al–Mg 4.5; 7.7; 12 at% Mg as-quenched. The velocities $V_{L\langle 100\rangle}$, $V_{T\langle 100\rangle}$, and $V_{L\langle 110\rangle}$ have been measured for the three alloys, and the elastic constants C_{11}, $C_{44} = C$, and B

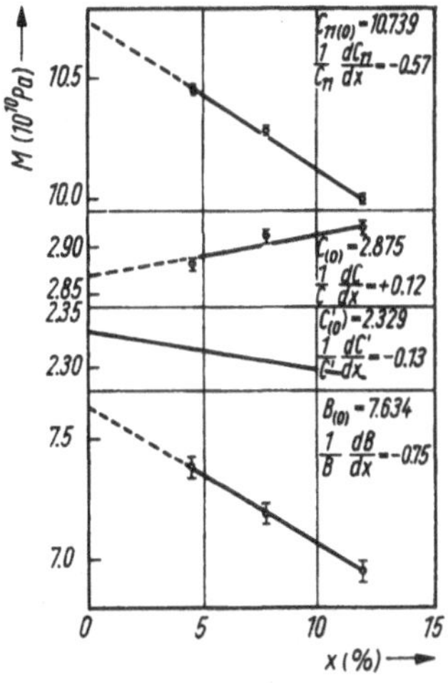

Fig. 6. Variations of the elastic constants of the aluminium–magnesium solid solution with atomic concentration

have been calculated. The velocities being measured with an accuracy of 0.1%, the precision is 0.2% for C_{11} and C_{44} and 0.6% for B (neglecting the error in ϱ, which is calculated from the variations of the lattice constant with concentration).

The results and the corresponding errors are plotted on Fig. 6, which shows that C_{11}, C_{44}, and B vary linearly with concentration (up to 12 at%).

As a consequence C' also varies linearly with concentration and its variation calculated from the three others is plotted on the same graph. The extrapolated values for pure aluminium are in good agreement with those of the literature [11].

These results prove that the increase of elastic constants during the formation of G.P. zones is not only caused by the variation of concentration of the matrix. Indeed G.P. zones are considered as ordered domains of 25 at% Mg, the matrix being a depleted solid solution of 8 at% [5]. Then the effect of matrix depletion at equilibrium is, for Al–Mg 12 at%:

$$\left(\frac{\Delta C}{C}\right)^* = -0.48\%\,, \qquad \left(\frac{\Delta C'}{C'}\right)^* = +0.52\%\,, \qquad \left(\frac{\Delta B}{B}\right)^* = +3\%\,.$$

These values, compared with the evolutions of Fig. 2, clearly show that the variations of elastic constants with G.P. zones do not depend only on matrix depletion, but that the proper effect of G.P. zones must be taken into account.

4.2 Analysis of the variations of the elastic constants of Al–Mg solid solution with concentration

When an elastic constant M of an alloy varies linearly with concentration x, the effect of solute may be written [12]:

$$\frac{1}{M_0}\frac{\partial M}{\partial x} = \frac{1}{M_0}\left(\frac{\partial M}{\partial x}\right)_a + \frac{1}{M_0}\left(\frac{\partial M}{\partial \log a}\right)_{x=0}\frac{1}{a}\frac{\partial a}{\partial x}\,. \tag{11}$$

The left-hand side of this relation is measured experimentally (Fig. 6). The first term of the right-hand side is the intrinsic effect of the solute assuming that the lattice constant does not change with concentration, the second term is the extrinsic effect due to the difference between the atomic volume of solvent and solute. This term may be calculated from measurements of the variation of M with pressure, and of a with concentration:

$$\left(\frac{1}{M_0}\frac{\partial M}{\partial \log a}\right)_{x=0} = -3B_T\left(\frac{1}{M_0}\frac{\partial M}{\partial P}\right)_{x=0}, \tag{12}$$

where B_T is the isothermal bulk modulus of aluminium given by [12]:

$$B_T = \frac{B}{1 + \dfrac{9\alpha^2 B T a^3 N}{4C_p}}$$

Table 1

M	E_m	$\frac{1}{M_0}\frac{\partial M}{\partial P}$ (10^{-10} Pa)	E_e	$E_i = E_m - E_e$
C_{11}	−0.57	0.66	−1.59	+1.02
$C = C_{44}$	+0.12	0.81	−1.95	+2.07
C'	−0.13	0.71	−1.71	+1.58
B	−0.75	0.64	−1.54	+0.79

α being the linear expansion coefficient, B the adiabatic bulk modulus (measured experimentally), T the temperature, a the lattice constant, N the Avogadro number, and C_p the molar specific heat at constant pressure.

For aluminium at 20 °C: $B = 7.634 \times 10^{10}$ Pa, $\alpha = 23.2 \times 10^{-6}$, $a = 4.05$ Å, $C_p = 25$ J/K and so $B_T = 7.317 \times 10^{10}$ Pa. For Al–Mg: $1/a\ \partial a/\partial x = 0.110$.

The variations of the elastic constants of aluminium with pressure have been measured by a "pulse echo overlap" method in the range of hydrostatic pressure 1 to 3000 bar. The results are given in Table 1 with the extrinsic effect E_e calculated from relation (12), and the intrinsic effect E_i deduced from relation (11) and experimental results (E_m) of Fig. 6:

$$E_m = \frac{1}{M_0} \frac{\partial M}{\partial x},$$

$$E_e = -3B_T \left(\frac{1}{M_0} \frac{\partial M}{\partial P}\right)_{x=0} \frac{1}{a} \frac{\partial a}{\partial x},$$

$$E_i = E_m - E_e .$$

This shows that the intrinsic effect causes the elastic constants of Al–Mg to increase with concentration, while the extrinsic effect causes them to decrease.

4.3 *Contribution of the proper effect of G.P. zones to the elastic constant variations*

G.P. zones being domains of 25 at% Mg they should have the elastic constants of a 25 at% solid solution. It is not possible to measure them experimentally because the maximum concentration for the existence of the solid solution Al–Mg is 17 at%. We can only assume that the signs of the intrinsic and extrinsic effects are always the same.

The interpretation of high angle X-ray scattering measurements involves that the lattice constant in G.P. zones is inferior to that expected for a hypothetic 25 at% solid solution because of ordering effect in the zones. This contributes to decrease the negative extrinsic effect in G.P. zones, thus to increase the elastic constants of G.P. zones above those extrapolated for a 25 at% solid solution.

Assuming that the depleted matrix with G.P. zones is a two-phase medium, this enhancement of the elastic constants of the zones may be the cause of the augmentation of elastic constants of Al–Mg alloys in the G.P. zone state.

Acknowledgements

The authors are grateful to Dr. E. Dedieu and Mr. H. Desointre of the Laboratory of "Métallurgie Physique" of Poitiers for the orientation and cutting of single crystals.

References

[1] C. Gault, A. Dauger, and P. Boch, phys. stat. sol. (a) **31**, 179 (1975).
[2] M. Bernole, R. Graf, and P. Guyot, Phil. Mag. **28**, 771 (1973).
[3] C. Gault, J. Vergnol, A. Dauger, J. P. Villain, and P. Boch, Mém. sci. Rev. Métall. **12**, 921 (1975).
[4] A. Dauger, J. P. Guillot, and J. Caisso, C.R. Acad. Sci. (France) **273**, 915 (1971).
[5] A. Dauger, E. K. Boudili, and M. Roth, Scripta metall. **10**, 1119 (1976).
[6] E. K. Boudili, M. F. Denanot, and A. Dauger, in Scripta metall. **11**, 543 (1977).

[7] D. E. LAUGHLIN and J. W. CAHN, Acta metall. **23**, 329 (1975).
[8] G. LASLAZ and P. GUYOT, Acta metall. **25**, 277 (1977).
[9] H. J. McSKIMIN, J. Acoust. Soc. Amer. **34**, 404 (1962).
[10] Z. HASHIN and S. SHTRIKMAN, J. Mech. Phys. Solids **10**, 335 (1962).
[11] G. N. KAMM and G. A. ALERS, J. appl. Phys. **35**, 327 (1964).
[12] D. FARGEOT, J. C. GLANDUS, and P. BOCH, phys. stat. sol. (a) **35**, 687 (1976).

(Received July 4, 1977)

phys. stat. sol. (a) 43, 633 (1977)

Subject classification: 10.2; 19; 22.8

Solid State Physics Laboratory, University of Groningen[1])

Paramagnetic Color Centers in Calciumchlorapatite Doped with F^- Ions

By

H. W. DEN HARTOG and H. BENNING

The results are given of EPR experiments on electron excess color centers in X-rayed synthetic calciumchlorapatite which is doped with fluoride ions. A comparison of the results obtained with those given in earlier papers on color centers in calciumchlorapatite and fluorapatite enables one to propose models for some of the electron excess centers observed. The centers observed in the material consist of a) an electron trapped at an anion vacancy neighbored by two equivalent (or approximately equivalent) Cl^- ions, b) an electron trapped at an anion vacancy neighbored by one Cl^- and one F^- ion, c) an electron trapped at an anion divacancy neighbored by two (nearly) equivalent F^- ions.

Es werden Ergebnisse von Elektronenspinresonanzmessungen an Elektronenüberschußzentren in röntgenbestrahlten synthetischen, mit F^--Ionen dotierten Kalziumchlorapatitkristallen mitgeteilt. Ein Vergleich mit früheren Ergebnissen für Kalziumchlorapatit und Fluorapatit ergibt Modelle für einige Elektronenüberschußzentren: Die beobachteten Zentren sind a) ein in einer Cl^--Anionlücke eingefangenes Elektron; zwei (fast) äquivalente Chlorionen sind die Nachbarn, b) ein in einer Cl^--Anionlücke eingefangenes Elektron; die Nachbarn sind ein Cl^-- und ein F^--Ion, und c) ein in einer Cl^--Doppellücke eingefangenes Elektron; die Nachbarn sind zwei (fast) äquivalente F^--Ionen.

1. Introduction

In previous papers we have studied the defect structure of calciumchlorapatite (ClAP) after X-irradiation [1 to 5]. It was found that impurities affect the color centers formed appreciably. Most of the experimental work published so far deals with ClAP doped with impurities of a valency differing from that of the ions of the host crystal. In the present paper we study the effect of doping with F^- ions.

Calciumchlorapatite can be considered to be a mixture of $Ca_3(PO_4)_2$ and $CaCl_2$. The Cl^- ions in ClAP are located in monoionic chains; the interionic distance within the chain is about 3.5 Å (see Fig. 1); the separation between two chains of Cl^- ions is 9.5Å . During X-irradiation color centers are produced in these chains of halide ions. It is well known that flux-grown ClAP is non-stoichiometric and this results in the presence of Ca^{++} and Cl^- vacancies [6 to 8]. The Ca^{++} vacancies are distributed equally over the Ca I and Ca II sites [6]; the Cl^- vacancies are found at the Cl^- chains. When electrons created by X-irradiation are trapped at the anion vacancies F-centers are formed. It has been shown however that more complicated electron excess centers are formed because of the coagulation of defects which takes place at temperatures higher than 77 K [4, 5].

It is obvious that apart from the electron excess centers also hole excess centers should be created because of electroneutrality. These hole centers have been studied by Knottnerus et al. [1, 2]; some of these centers are perturbed by O^{2-} impurities and others should be considered as intrinsic defects [2]. The crystals used for the

[1]) 1 Melkweg, Groningen, The Netherlands.

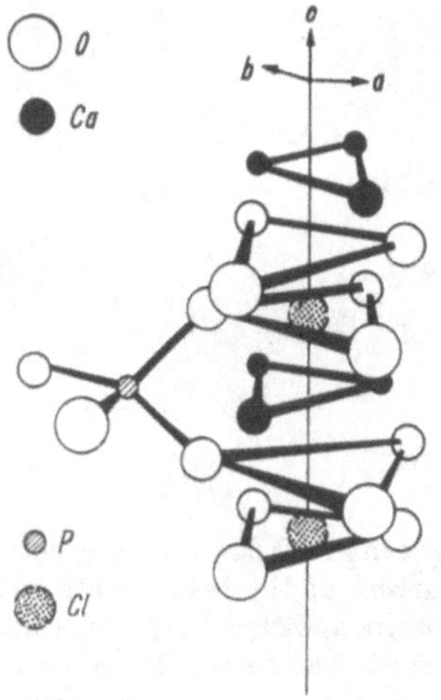

Fig. 1. Three-dimensional view of the environment of the Cl^- ion column in ClAP single crystals

present investigations contain O^{2-} impurities; therefore, $(OCl)^{2-}$ centers as studied by Knottnerus et al. [1] have been observed.

The purpose of the present investigation is to study the magnetic properties of F-like color centers perturbed by fluoride ions. In general the magnetic hyperfine interactions are much larger for the fluorine nucleus than for the chlorine. Therefore, it was expected that hfs interaction between the F-electron and the ^{19}F nuclei would produce well resolved EPR spectra. The experimental results show that this is indeed the case. In addition, we have been able to draw conclusions with regard to the models to be proposed for the different color centers observed in calciumchlorapatite crystals doped with fluorine ions. Also we will give a new model for a color center in pure ClAP observed by Knottnerus et al. [3].

We have found that for small X-ray doses at room temperature (RT) there are two types of paramagnetic electron excess centers: first, we have the F-center, which is an electron trapped at a negative ion vacancy; the neighboring ions occupy positions which are approximately equivalent. Secondly, we observed a perturbed F-center, which consists of a Cl^- vacancy and a trapped electron its nearest halide neighbors are one Cl^- and one F^- ion.

For higher X-ray doses we have observed a third electron excess center which appears to be composed of an electron trapped by two anion vacancies. The complex is neighbored by two equivalent F^- ions. It turns out that this more complicated defect is the predominant one after annealing at temperatures between 50 and 100 °C.

2. Experimental Procedures

The samples used in this investigation were prepared in a $CaCl_2$ flux. A detailed description of this method has been given by Knottnerus [5]. The size of the samples varied between 2 and 6 mm. The EPR experiments were carried out with a Varian V4500 spectrometer. The magnetic field strength was determined with a Bruker NMR oscillator B-NM12 field marker.

The X-irradiation was obtained from a Philips PW1008 X-ray generator running at 40 kV and 18 mA using a W anode. The bleaching experiments with white light were carried out with a 1000 W halogen lamp. The annealing experiments were performed in a stove which was dark in the inside. From time to time the optical absorption properties of the samples were checked with a Beckman DK2A automatically recording spectrophotometer.

3. Experimental Results

Immediately after X-irradiation the EPR spectrum of ClAP: F consists of at least nine lines. For small doses, apart from the EPR lines due to the hole excess centers studied by Knottnerus et al. [1], we observed three extra lines associated with electron excess centers (see Fig. 2). When the X-ray dose is increased the EPR spectrum becomes more complicated; additional lines due to electron excess centers are ob-

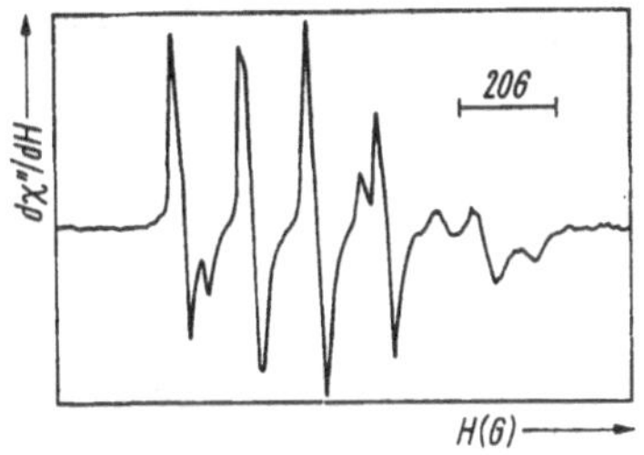

Fig. 2. EPR spectrum of an X-irradiated ClAP: F single crystal measured immediately after the irradiation. The magnetic field is perpendicular to the c-direction

served. The electron excess centers observed after subjecting the sample to small doses of X-rays show a symmetrical EPR spectrum; the detailed shape of the spectrum (for $\boldsymbol{H}_0 \perp \boldsymbol{c}$) is shown in Fig. 3. This spectrum can be considered to be composed of two spectra: the central portion consisting of a partly resolved line and one peak at both sides of the central line. The latter EPR lines are separated by 18.2 G. It was found that the spectrum observed at small X-ray doses is only slightly orientation-dependent. A rotational diagram of these lines as a function of the angle between $\boldsymbol{H}_0$ and the c-axis is given in Fig. 4. This rotational diagram also shows the behavior of the EPR lines due to hole excess centers of the type $(OCl)^{2-}$ studied by Knottnerus et al. [1]. The lines of the spectrum shown in Fig. 3 behave axially symmetric for rotation about the crystallographic c-axis.

Table 1
Magnetic properties of electron excess centers in ClAP: F

center	$g_{\parallel}$	$g_{\perp}$	$A_{\parallel}$ (G)	$A_{\perp}$ (G)
Ia	—	2.000	—	2
Ib	—	2.000	—	18.2
II	1.997	2.003	6.1	12.2

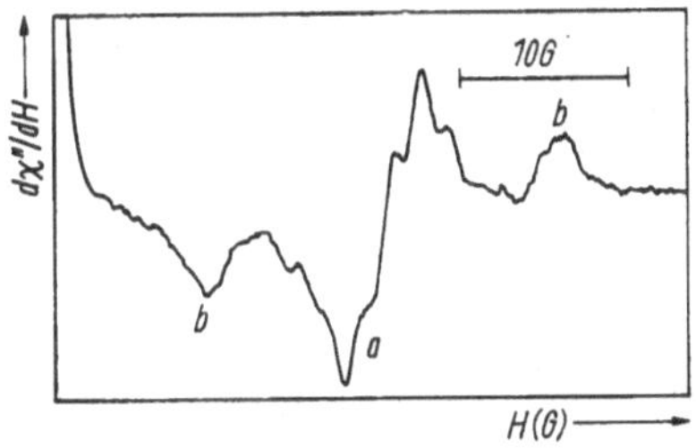

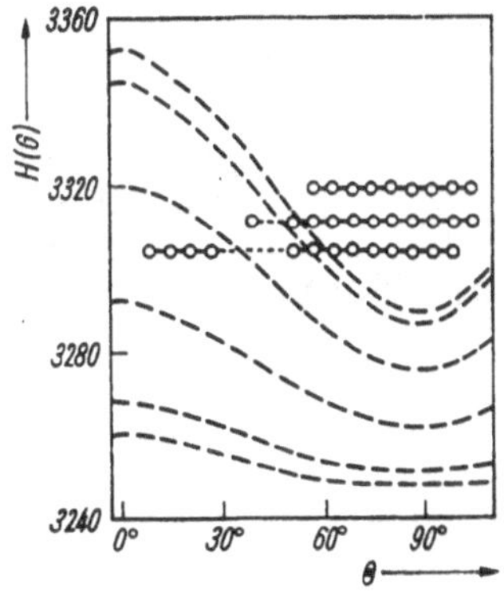

Fig. 3 Fig. 4

Fig. 3. EPR spectrum of the electron excess centers formed during the first stage of X-irradiation ($\boldsymbol{H} \perp \boldsymbol{c}$)

Fig. 4. Rotational diagram of the EPR lines observed in ClAP: F after a short period of X-irradiation. The broken lines correspond with the EPR lines du to the $(OCl)^{2-}$ centers studied by Knottnerus et al. [1]. The crystal was rotated about the a-axis

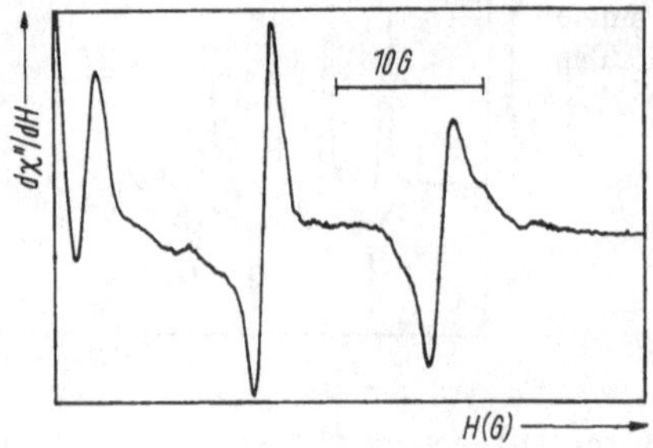

Fig. 5

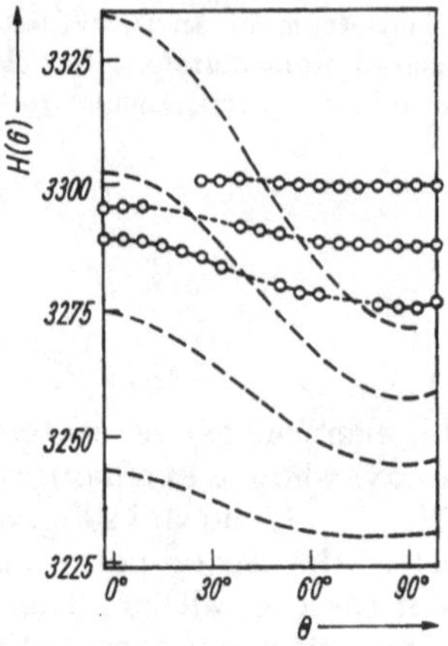

Fig. 6

Fig. 5. EPR spectrum of type II for $\boldsymbol{H} \perp \boldsymbol{c}$. The spectrum was obtained from a sample which has been irradiated for 120 h with X-rays and which was subsequently bleached with white light

Fig. 6. Rotational diagram of the EPR lines observed in a sample showing spectrum II. The behavior of the lines due to the $(OCl)^{2-}$ centers has been indicated by broken curves. The crystal was rotated about the a-axis

In this paper we refer to the central line, which probably consists of seven equidistant lines of intensities 1:2:3:4:3:2:1, as spectrum Ia; the two additional lines are denoted by Ib. The magnetic properties of the centers investigated in the present paper have been given in Table 1.

When a ClAP: F sample is irradiated for many hours three extra EPR lines (spectrum II) appear in the same region as spectra Ia and b. This makes the new absorption pattern rather complicated. It is however possible to obtain rather simple spectra after annealing at temperatures ranging between 50 and 100 °C. After about 1 h in the dark at 100 °C spectra Ia and b have disappeared whereas spectrum II is still present. Another more effective method to reduce the lines of spectra Ia and b is to illuminate the sample with white light. Spectrum II consists of three equidistant lines; the intensities of the lines have a ratio of 1:2:1 (see Fig. 5). The separation between the lines for $\boldsymbol{H}_0 \perp \boldsymbol{c}$ is 12.2 G. Spectrum II turns out to vary more strongly during rotation of the sample with respect to the external magnetic field than spectrum I (see Fig. 6). In Fig. 6 the behavior of the spectrum due to $(OCl)^{2-}$ centers has also been indicated. The hyperfine splitting of spectrum II for $\boldsymbol{H}_0 \parallel \boldsymbol{c}$ is 6.1 G.

4. Discussion

4.1 Spectrum Ia

As shown in Section 3 immediately after a short period of X-irradiation in ClAP: F two different types of paramagnetic color centers of the electron excess type are present. First, we have spectrum Ia which is associated with a color center showing interaction between the unpaired electron and two equivalent Cl nuclei ($I = 3/2$). For a center of this type one expects an EPR signal consisting of seven hyperfine lines; using statistics one can easily find that the intensity ratio of the lines is 1:2:3:4:3:2:1. The most simple center that can be constructed in the material is an F-like center (see Fig. 7). When the ions surrounding the vacancy are in positions which are the same as in the perfect lattice, the Cl^- ions neighboring the F-center are

Fig. 7. Three-dimensional schematic representation of the F-center in ClAP

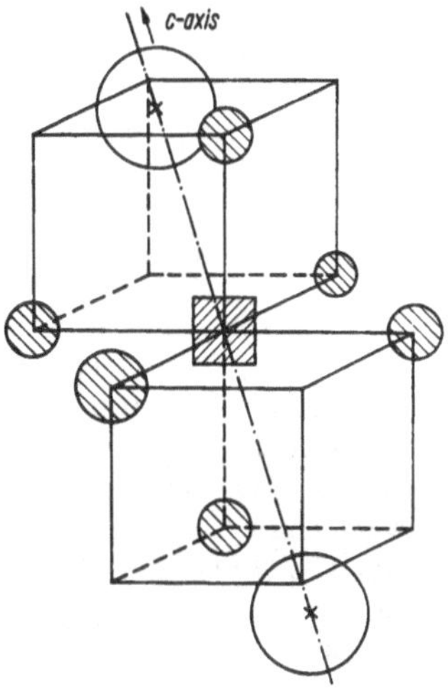

inequivalent. We therefore propose that these ions have relaxed towards new positions which are equivalent or approximately equivalent.

A spectrum similar to Ia has been observed by Knottnerus et al. [3] in pure ClAP, however, the splitting between the hyperfine lines was 1 G indicating that the center associated with the EPR spectrum described above is different from this one. It should be noted that the resolved EPR signal found by Knottnerus et al. [3] was obtained after one night storage in the dark at RT. The seven-lines spectrum given in the present paper is obtained immediately after X-irradiation, suggesting that it is connected with a color center of a more simple nature than the one found by Knottnerus et al. This implies that the resolved spectrum found by these authors is probably due to a center consisting of one unpaired electron spin trapped at a defect more complicated than a simple vacancy. Because of the defect structure of ClAP [2, 4] we tentatively propose that the seven-lines spectrum observed by Knottnerus et al. [3] is due to an electron trapped at a divacancy. The present interpretation is supported by the fact that the hfs constant observed for the F-center studied in the present paper is twice as large as the one given by Knottnerus et al.; in the latter case the electron is distributed over two vacancies.

4.2 Spectrum Ib

The other center which occurs in the first stage of X-irradiation is connected with spectrum Ib. This spectrum consists of two lines separated by 18.2 G; the lines have equal intensities indicating that the unpaired electron interacts with one nucleus with nuclear spin 1/2. We now recall that in the samples used in this investigation we have introduced about 1 mol% fluorapatite and as a result many of the Cl^- ions in the monoionic chains have been replaced by F^- ions. This implies that some of the anion vacancies present in the samples will be surrounded by one or two F^- ions. The statistical probability of finding an F-center neighbored by one Cl^- and one F ion is about 2%. Spectrum Ib could be well explained by a color center of this type, because ^{19}F has a nuclear spin 1/2. Statistically one would expect a signal of type Ib which has an intensity 50 times smaller than that of signal Ia. From Fig. 3 we see that the ratio of the intensities of signals Ia and b is appreciably larger. Probably the formation of F-centers surrounded by one Cl^- and one F^- ion is favored as compared to that of centers with two Cl^- neighbors.

The interpretation given above is in agreement with the results on F-centers in pure fluorapatite published by Piper et al. [9] who found $A_{\perp} = 17.2$ G.

4.3 Spectrum II

As shown in Section 3 spectrum II can be obtained after very long periods of X-irradiation, after annealing at temperatures higher than RT and by optical bleaching.

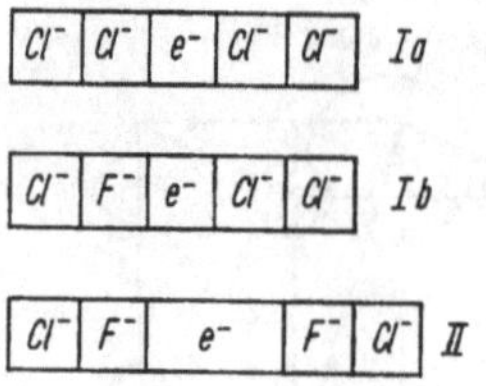

Fig. 8. Schematic representation of the centers corresponding with the EPR signals Ia, Ib, and II

This suggests that the center under consideration is of a complicated nature. From the fact that the spectrum consists of three equidistant lines with intensity ratio 1:2:1 we conclude that the unpaired electron interacts with two equivalent ^{19}F nuclei. The first possible model is an F-center neighbored by two equivalent F^- ions. This model is however less attractive because the hyperfine splitting A is appreciably smaller than the one found for the F-center neighbored by one Cl^- and one F^-. The model we propose here is similar to the one we have given for the F-like center observed by Knottnerus et al. [3] in pure ClAP. Spectrum II is associated with a color center consisting of an electron trapped at a divacancy surrounded by two equivalent F^- ions. A survey of the color centers investigated in this paper has been given schematically in Fig. 8.

Acknowledgements

The authors wish to thank P. Wesseling for growing the crystals. We are grateful to Dr. D. I. M. Knottnerus for stimulating discussions. This work is part of the research program of the Stichting voor Fundamenteel Onderzoek der Materie (FOM) and has been made possible by financial support from the Nederlandse Organisatie voor Zuiver Wetenschappelijk Onderzoek (ZWO).

References

[1] D. I. M. Knottnerus, H. W. den Hartog, and W. van der Lugt, phys. stat. sol. (a) **13**, 505 (1972).
[2] D. I. M. Knottnerus and H. W. den Hartog, phys. stat. sol. (a) **29**, 183 (1975).
[3] D. I. M. Knottnerus, H. W. den Hartog, and A. Lanjouw, phys. stat. sol. (a) **32**, 287 (1975).
[4] D. I. M. Knottnerus, A. Lanjouw, and H. W. den Hartog, phys. stat. sol. (a) **41**, 95 (1977).
[5] D. I. M. Knottnerus, Thesis, Groningen 1976.
[6] P. E. Mackie, J. C. Elliott, and R. A. Young, Acta cryst. **B28**, 1840 (1972).
[7] J. S. Prener, J. Electrochem. Soc. **114**, 77 (1967).
[8] J. S. Prener, J. Solid State Chem. **3**, 49 (1971).
[9] W. W. Piper, L. C. Kravitz, and R. K. Swank, Phys. Rev. **138**, 1802 (1965).

(Received July 19, 1977)

phys. stat. sol. (a) 43, 639 (1977)

Subject classification: 17; 22.5.2

Moscow State University

Thermostimulated Exoemission from LiF

By

N. I. Konyushkina, I. V. Krylova, V. N. Opekunov, and A. A. Predvoditelev

The surface of LiF cleaned by prolonged bombardment with Ar^+ ions or electrons in high or ultrahigh vacuum does not exhibit thermostimulated exoemission. The adsorption of water vapours restores the emissivity. The point defects (colour centres or magnesium impurities) do not form exoemission centres in the absence of adsorbed water.

Die Oberfläche von LiF, die durch ausgedehnten Beschuß mit Ar^+-Ionen oder Elektronen im Hoch- oder Ultrahochvakuum gereinigt wurde, zeigt keine thermisch stimulierte Exoelektronenemission. Die Adsorption von Wasserdampf stellt die Emissionsfähigkeit wieder her. Die Punktdefekte (Farbzentren oder Magnesiumstörstellen) bilden keine Exoemissionszentren bei Abwesenheit von adsorbiertem Wasser.

1. Introduction

At present for thermostimulated exoemission (TSE) from solids two basic models are considered. Exoemission from oxides is assumed to be due to electrons emitted directly from capture levels formed by adsorbed centres [1, 2]. This hypothesis has been corroborated by experiments which proved that oxides without adsorbed layers under ultravacuum do not show emissivity [3, 4]. For alkali halides TSE is considered as quasistationary thermoemission from the conduction band. It is assumed that the local capture levels are formed by colour centres (F-, F′-, V-centres or their aggregates) [5].

For LiF it has been established that TSE peaks result from thermal dissociation of bulk M-, F-, F′-colour centres, and of surface σ- and Σ-centres [6, 7]. Gordann and Scharmann [8] consider that these surface centres in LiF are not due to adsorption. Besides the stable TSE peaks (at 25, 117 to 130, and 290 °C), attributed to bulk M-, F-, F′-centres, Golzapfel [7] observed TSE peaks (at 190 to 200 °C) which vanished on reheating (σ-centres). He proposed that these peaks are due to the presence of a sorption layer, whose nature is not known. Assuming that exoemission is a sensitive method for detecting the colour centres and LiF is an effective material in dosimetry [9], we made an attempt to use TSE for investigating the radiation damage caused by Ar^+ ion bombardment at different stages of surface erosion. At high doses of bombarding ions which give rise to visible colour (F- and M-)centres [10] we found that the crystals do not exhibit emissivity. A comprehensive study was undertaken to investigate the effect of adsorbed layers on TSE from single crystals with surfaces cleaned under ultrahigh vacuum.

2. Experimental

The experiments were carried out in vacuum ($\approx 10^{-5}$ Torr), produced by an oil-diffusion pump, and in a heated oil-free ultravacuum (10^{-8} to 5×10^{-9} Torr) system equipped with sorption magneto-discharge pump operating on titanium getter. The composition of residual gases was determined by mass-spectrometer of type Omegatron. Hydrogen prevailed over other gases in the ultravacuum system ($p \leqq 10^{-8}$ Torr), the pressure of other gases (CO, CO_2, H_2O) being 10^{-10} to 10^{-11} Torr. The emission was recorded by a secondary electron multiplier. Excitation by electron bombardment was carried out directly in the exoemission measurement chamber (600 to 1500 eV, 10 to 15 μA).

LiF single crystals were grown from the melt by the Kyropoulos method. Bombardment with Ar^+ ions was carried out at a current density 0.02 to 45 μA/cm² and an energy 20 keV.

3. Results

To estimate the influence of the surface microrelief on TSE the results of [10] have been used, where for studying the initial stages of surface erosion under ion bombardment the vacuum surface decoration method was employed.

It has been established that Au crystallization centres are formed on aggregates of vacancies (≈ ten vacancies). The defects caused by single Ar^+ ions (dose $\Phi \leq \leq 10^{12}$ ions/cm², distance between the incident ions ≈ 100 Å) do not give rise to Au crystallization centres.

Thermostimulated exoemission takes place at the stage of the formation of single defects. TSE from LiF after bombardment with different doses of ions, is shown in Fig. 1. Curve 1 was obtained after the bombardment of a freshly cleaved surface with a minimum possible dose, i.e. $\Phi = 1.6 \times 10^{10}$ ions/cm² (current density 0.02 μA/cm², $t = 0.12$ s). This corresponds to a mean distance between the incident ions of ≈ 700 Å. The distinct TSE peak at 230 °C indicates the presence of single defects (according to [7] these are Σ-centres). The activation energy of electron ejection calculated as in [11] was ≈ 1 eV. On reheating TSE does not take place, but the curve is easily reproduced at subsequent electron bombardment. Curve 2 corresponds to TSE after ion bombardment with the dose $\Phi = 1.3 \times 10^{12}$ ions/cm² (current 0.16 μA, $t = 1.2$ s); this corresponds to a distance between incident Ar^+ ions of ≈ 45 Å and to the lower sensitivity of the surface decoration method in the radiation defect investigations (the stage of formation of vacancy aggregates). TSE curve 2 has a large number of overlapping peaks, which are indicative of a large number of point defects having different electron binding energies. In this case the number of defects having large thermal dissociation energy increases significantly: the high-temperature emission intensity (at 300 °C) exceeds the intensity of the main peak at 230 °C. For an ion bombardment dose $\Phi \approx 1.5 \times \times 10^{15}$ ions/cm², when the distance between the incident Ar^+ ions is approximately equal to the interionic distance in a crystal ($\lambda = 2.5$ Å), the emissivity decreases abruptly. In this case the monoatomic steps of the original relief totally get destroyed, but new steps connected with the formation of cavities are formed [10]. Optical methods permit to detect the colour centres (F- and M-centres) only at very

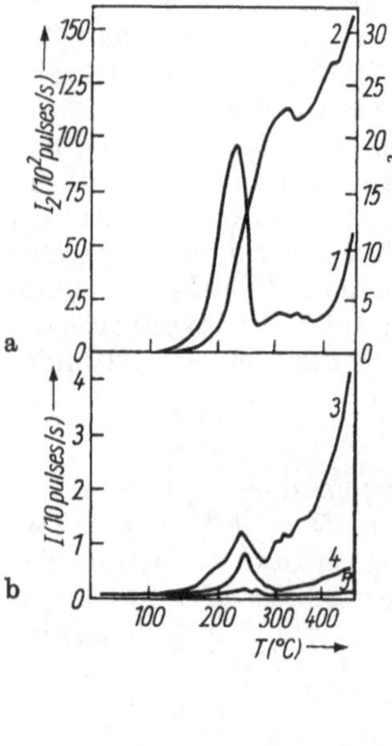

Fig. 1. TSE from LiF crystals excited by Ar^+ ions. (1) $\Phi = = 1.6 \times 10^{10}$, (2) 1.3×10^{12}, (3) 3.3×10^{15}, (4) 3×10^{16}, (5) 1.2×10^{18} ions/cm²

Fig. 2. Net counts (20 to 430 °C) vs. dose of exciting radiation. (1) bombardment with Ar^+ ions, (2) subsequent electron bombardment

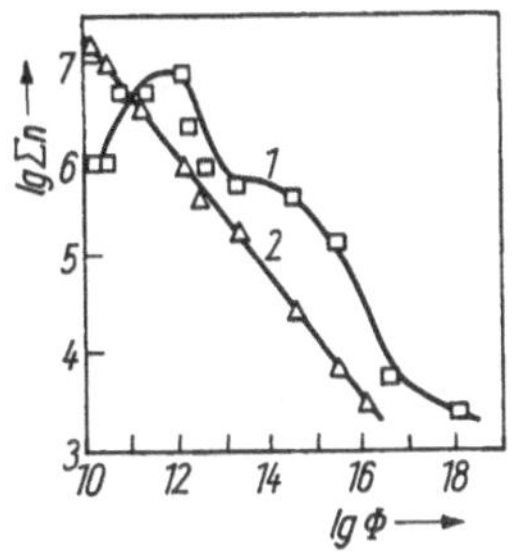

high doses of Ar^+ ions $\Phi \approx 10^{18}$ ions/cm² (4.5 μA/cm², 10 h). These centres are stable and get destroyed only on heating the crystal to $T \geqq 600$ °C. Exoemission from such la crystal is not detected and not excited by a subsequent electron bombardment.

Fig. 2 shows the dependence of the number of emitted electrons on the ion bombardment dosage. The emissivity increases only in the small dose ($\approx 10^{12}$ ions/cm²) range prior to the formation of vacancy aggregates, when the surface decoration method does not permit to investigate the process of defect formation.

Since the colour centres produced by ion bombardment are stable at $T \leqq 600$ °C, it is difficult to explain the observed relationship within the framework of the general band model of exoemission [5] according to which the output of the electrons takes place from the conduction band by overcoming the surface potential.

3.1 TSE after electron bombardment

In Fig. 3 TSE curves are shown for the cleaved surface of a single crystal excited by equal doses ($\approx 10^{15}$ cm^{-2}) of bombarding Ar^+ ions and electrons. The shape of the curve — number, temperature position, and ratio of peak intensities — does not depend upon the nature (charge, mass) of bombarding particles.

If a crystal, bombarded by Ar^+ ions and heated to 400 °C, is excited by electrons, the total number of emitted electrons is determined by the number of defects caused by ion bombardment (Fig. 2, curve 2). Thus, the ion bombardment gives rise to stable defects which can be filled up with electrons and on subsequent heating can be emptied. If the dose of bombarding electrons ($\approx 10^{15}$ electrons/cm²) exceeds the preliminarily bombarding ion dose ($\Phi < 10^{12}$ ions/cm²), then the electron bombardment gives rise to additional defects of the same nature ($T_{max} = 230$ °C).

3.2 TSE and thermodesorption

The analysis of gas evolution attended by thermostimulated emission, shows that the bombardment by Ar^+ ions results in a considerable dehydration of crystal surfaces (Fig. 4). Maxima of the thermodesorption rate in this process were found to correspond to TSE maxima. Thus, for the crystals of alkali halides (see also [12]) as for oxides [1]

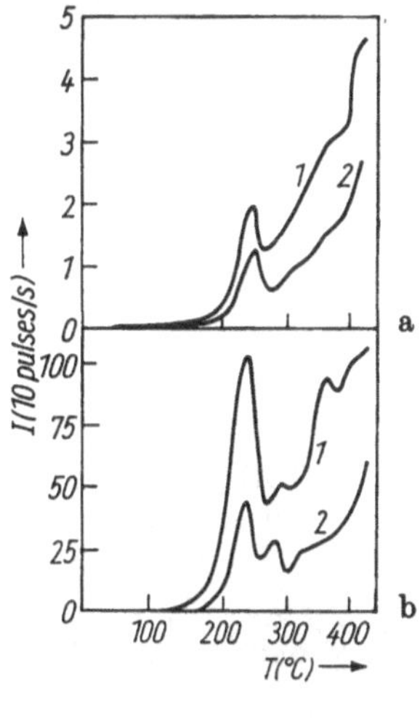

Fig. 3. a) TSE from LiF single crystals bombarded by (1) Ar^+ ions or (2) electrons with doses $\Phi \approx 10^{15}$ particles/cm²; b) TSE from a LiF crystal bombarded by (1) Ar^+ ions ($\Phi = 3.3 \times 10^{12}$ ions/cm²) and (2) subsequent electron bombardment ($\Phi \approx 10^{15}$ electron/cm²)

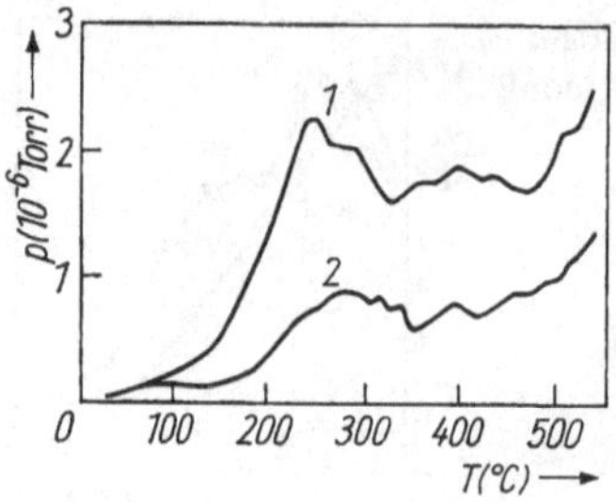

Fig. 4. Gas evolution in linear heating of a LiF crystal. (1) original specimen, (2) bombarded by Ar^+ ions ($\Phi = 3.1 \times 10^{15}$ ions/cm²)

the TSE centres serve as centres of activated desorption of water. Such centres exist even in original (unexcited) samples, and in the presence of adsorbed water they determine the "self-exciting" thermostimulated emission process.

3.3 TSE in ultrahigh vacuum

Exoemission from LiF in ultrahigh vacuum was investigated after vacuum heat treatment, prolonged electron bombardment, and adsorption of water vapour. The results of these studies are shown in Fig. 5. On heating the sample for 6 h at 300 °C the intensity of thermostimulated exoemission decreases to about one quarter. The emissivity is totally suppressed after a subsequent prolonged electron bombardment (10 μA, 600 eV, $\Phi = 2 \times 10^{17}$ electrons/cm²). Despite the high dose of exciting radiation, TSE at subsequent heating in the temperature range from 20 to 350 °C does not exceed the "background level". Contact with atmospheric air does not lead to the restoration of emissivity of the sample. After exposing the samples, which have lost their emissivity, to water vapour, "self-exciting" TSE was observed (Fig. 6, curve 2). After excitation by electron bombardment (dose $\approx 10^{15}$ electrons/cm²) the form of the "self-exciting" TSE curve is restored, and an additional low-temperature peak (at $\approx$ 100 °C) appears. This emission vanishes at subsequent prolonged electron bombardment.

3.4 Effect of Mg impurities on TSE from LiF

In [6] it has been shown that in LiF crystals doped with Mg, In, and O_2 after they have been excited by X-rays, low-temperature peaks (at 160 and 190 °C) appear in

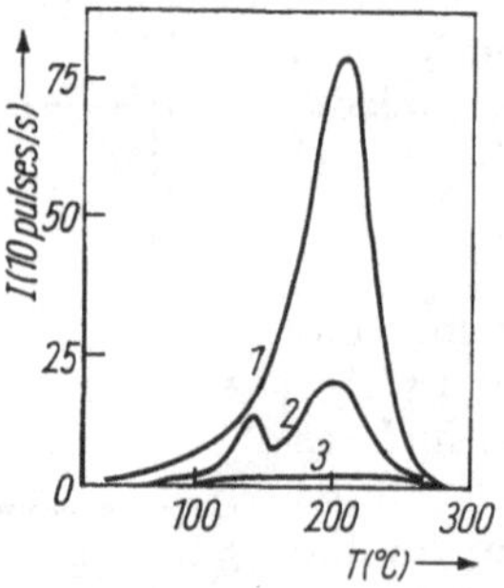

Fig. 5

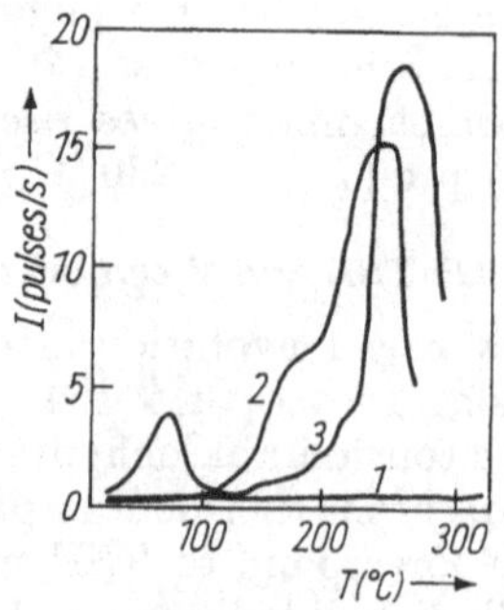

Fig. 6

Fig. 5. TSE from a LiF crystal bombarded by low electron dose ($\Phi \approx 10^{14}$ electrons/cm²) (1) original specimen, (2) after 6h heating at 300 °C, and (3) by high electron dose ($\Phi \approx 10^{17}$ electrons/cm²)

Fig. 6. Curve 1 is curve 3 from Fig. 5. "Self-exciting" TSE from LiF crystals (2) after adsorption of water (saturated vapour) and (3) after electron bombardment ($\Phi \approx 10^{14}$ electrons/cm²) addition to the 230 and 330 °C peaks which, according to the authors, belong to different types of F-centres. These peaks, according to [6], characterize F-centres localized near complexes of impurities and vacancies.

In our experiments for the freshly cleaved surface of the LiF:Mg single crystal an intensive TSE of negative ions was observed (Fig. 7) with a maximum of 180 °C. After excitation by electron bombardment the TSE intensity was considerably less ($\approx$ 5 times), but it increased on subsequent heating and bombardment. Unlike in LiF, heating of a LiF:Mg crystal in vacuum for 6 h at 300 °C led to an increase in TSE excited by electron bombardment. Prolonged electron bombardment (for 30 min, $\Phi = 2 \times 10^{17}$ electrons/cm^2) totally suppressed the emissivity, while adsorption of water vapour restored it again (Fig. 7).

4. Discussion

Exoemission from LiF crystals excited by electron bombardment, after their cleavage in ultrahigh vacuum ($p \leqq 10^{-8}$ Torr) was studied by Gordann and Scharmann [8]. For a dose of 5×10^{14} bombarding electrons they found a TSE peak at 120 °C which was attributed to the formation of surface states due to the emergence of dislocations. The TSE peak at 310 °C which appears in specimens excited by X-rays and which is associated with F- and M-centres, was not observed in [8]. It is concluded in [8] that exoemission from LiF is not governed by the presence of adsorbed layers. No special purification of the sample was done in this case.

The basic results of our studies — total suppression of emissivity after cleaning the cleaved surface of LiF by vacuum heat treatment with subsequent prolonged electron bombardment and its restoration after the adsorption of water vapour — demonstrate that the adsorption layers play a decisive role in TSE from alkali halides. It is also known that for cleaning certain surfaces from adsorbed layers, vacuum heat treatment is insufficient and emissivity is suppressed only after prolonged bombardment by Ar^+ ions or after electron bombardment at elevated temperatures [2, 3]. Linke has shown [13] that cleavage of LiF single crystals in vacuum is accompanied by a considerable evolution of gas. Gas evolution is especially intense in the crystals grown in air. Thus, the cleavage of a LiF single crystal in vacuum does not ensure to obtain an atomically pure surface. The gases (vapours) liberated from the bulk upon cleavage could be very well retained by the surface.

The very intensive self-exciting ionic emission from a LiF:Mg crystal is caused by the dehydration of a highly hydrated cleaved surface due to the great affinity of Mg ions to water. As distinct from undoped LiF, repeated heating, bombardment at small doses, and vacuum heat treatment at 300 °C of LiF:Mg increases the intensity of TSE. A considerable amount of water in the surface layers of LiF:Mg, apparently, hinders the formation of exoemission centres, as shown in [14] for SiO_2 and in [15] for oxidized aluminium, as a result of absorption of energy of exciting radiation in adsorbed layers. The high affinity to water is responsible for the rather intensive

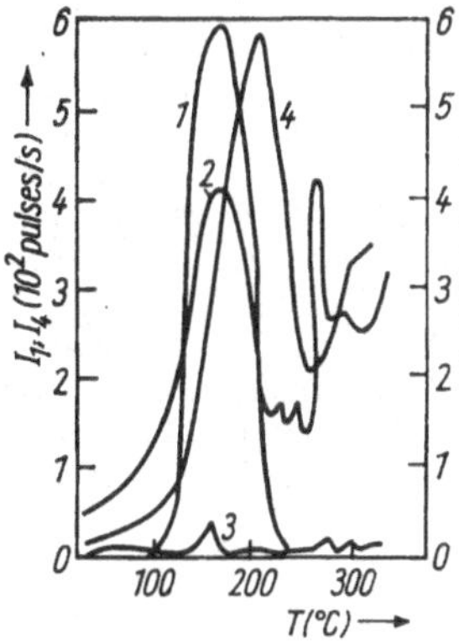

Fig. 7. TSE from LiF:Mg crystal. (1) "self-exciting" TSE negative ions, (2) after electron bombardment ($\Phi \approx 10^{15}$ electrons/cm^2), (3) after prolonged bombardment ($\Phi \approx 10^{17}$ electrons/cm^2), (4) after adsorption of water vapour and additional electron bombardment ($\Phi \approx 10^{15}$ electrons/cm^2)

emission from LiF:Mg after adsorption of water on the surface which had lost its emissivity.

These investigations have shown that the presence of adsorbed layers (of water) is necessary to excite TSE from LiF crystals. The desorption centres act as emission centres. The TSE peak (at $T_{max} = 230$ to 260 °C), which coincides with the maximum of gas desorption rate, is observed for a large number of oxides [1]. The coincidence of TSE peaks for oxides and LiF at the same conditions of excitation and registration is shown in [16]. The TSE peak position determines the activation energy of the electron output E_{act} which is equal according to our estimates obtained as in [11] to 1.1 eV for $T_{max} = 230$ °C. The weak dependence of E_{act} on the material of the emitter indicates that the output of electrons is controlled by the same stage in a complicated chain of surface transformations. According to the physico-chemical mechanism of the exoemission [17] the output of electrons is the result of ternary impact — recombination of two active particles in the mobile adsorption layer plus active centre. The active emission centre is the adsorption complex, for example the point lattice defect (vacancy) with OH^-, H^-, O^- etc. ions (OH^-, U-colour centre etc.) localized near this defect.

In accordance with our results (Section 3.1) active centres may be only in the form of point defects; the formation of vacancy aggregates (under Ar^+ ion doses $\Phi \geqq 10^{12}$ ions/cm^2 according to investigations by means of the surface decoration method [10]) and the subsequent destruction of the surface — the formation of cavities — lead to a decrease in the emissivity of the crystals. The same result, the decrease of emissivity under high doses of γ-radiation ^{60}Co ($\Phi > 2 \times 10^5$ rd), is obtained by investigations of TSE from Li_2SO_4 + graphite [18]. The decrease of emissivity up to total disappearance under high doses of radiation is explained by the strong destruction of the outermost atomic layers, which causes the break of the emission centres.

The investigations have shown that the measurement of the exoemission is a sensitive method for the detection of point defects when these defects are isolated and do not form vacancy clusters (colloid complex). These surface point defects at the initial stage of surface erosion cannot be detected by other known methods.

References

[1] I. V. KRYLOVA, phys. stat. sol. (a) 7, 359 (1971).
[2] M. EULER and A. SCHARMANN, phys. stat. sol. (a) 34, 297 (1976).
[3] I. V. KRYLOVA, V. I. SVITOV, and N. I. KONYUSHKINA, Zh. fiz. Khim. 50, 933 (1976).
[4] M. EULER and A. SCHARMANN, Z. Phys. B22, 373 (1975).
[5] A. BOHUN, PTB-Mitteilungen 5, 320 (1970).
[6] A. A. URUSOVSKAYA and G. G. KNAB, phys. stat. sol. (a) 30, 59 (1975).
[7] G. HOLZAPFEL, Z. angew. Phys. 29, 107 (1970).
[8] P. GORDAN and A. SCHARMANN, Z. Phys. 217 309 (1968).
[9] M. FRANK, Isotopenpraxis 2, 369 (1966).
[10] M. M. MELAMED and A. A. PREDVODITELEV, Kristallografiya 20, 807 (1975).
[11] M. BALARIN and A. ZETZSCHE, phys. stat. sol. 2, 1670 (1962).
[12] N. I. KONYUSHKINA and I. V. KRYLOVA, Fiz. tverd. Tela 15, 1925 (1973).
[13] E. LINKE, Z. angew. Phys. 29, 241 (1970).
[14] I. V. KRYLOVA and Z. L. KRYLOVA, Khim. vys. Energii 11, 268 (1968).
[15] W. SCHLENK, phys. stat. sol. (a) 33, 217 (1976).
[16] H. KIRIHATA and J. AKUTSU, phys. stat. sol. (a) 35, K73 (1976).
[17] I. V. KRYLOVA, Z. Chem. 13, 415 (1973).
[18] T. SASAKI, J. appl. Phys. 47, 2241 (1976).

(Received April 1, 1977)

phys. stat. sol. (a) 43, 645 (1977)

Subject classification: 1.2; 8; 14.3; 18.1; 22.6.1

Department of Inorganic Chemistry, Arrhenius Laboratory, University of Stockholm[1])

Magnetic, Electrical, and Thermal Studies of the $V_{1-x}Re_xO_2$ System with $0 \leqq x \leqq 0.15$

By

Ö. Sävborg and M. Nygren

The insulator–metal transition of VO_2 and VO_2 doped with Re is studied by means of DTA, X-ray diffraction, magnetic susceptibility, and electrical conductivity measurements. A linear decrease of the transition temperature of 18 K/at% Re is observed. The low-temperature phase is found to be temperature-independent paramagnetic for all compositions studied and semiconducting properties are observed for $x \leqq 0.09$ in $V_{1-x}Re_xO_2$. The properties of the (V, Re)O_2 alloys are compared with previous and some new findings of the $V_{1-x}Me_xO_2$ systems with Me = W, Mo, and Nb.

La transition métal–isolant de la solution solide $V_{1-x}Re_xO_2$ ($0 \leqq x \leqq 0,15$) a été étudiée par ATD, diffraction X et par des mesures de susceptibilité magnétique et de conductivité electrique. La température de transition décroit de 18 K/at% en Re. La phase $V_{1-x}Re_xO_2$ de basse température présente un paramagnetisme indépendant de la température dans tous le domaine de composition étudié. C'est un semiconducteur pour $x \leqq 0,09$. Les propriétés de la solution solide (V, Re)O_2 sont comparées avec les travaux antérieurs et nos travaux encours sur les systèmes $V_{1-x}Me_xO_2$ (Me = W, Mo, Nb).

1. Introduction

Vanadium dioxide undergoes a first-order metal–insulator transition at 340 K. The high-temperature metallic form possesses the rutile structure [1], named R below, while the low-temperature semiconducting modification (hereafter called M_1) is of the MoO_2-type [2], which can be regarded as a monoclinic distortion of the rutile structure ($a_{M_1} = 2c_R$, $b_{M_1} = a_R$, $c_{M_1} = a_R - c_R$). In the monoclinic phase the V atoms are displaced from their rutile positions in such a way that zig-zag chains of alternating longer and shorter metal–metal distances are formed. The V^{4+} ions (d^1-configuration) thus occur in pairs, which explains the nonmagnetic character of the insulating phase.

Several VO_2–MeO_2 systems and also the VO_2–VOF system have been studied in order to elucidate the influence of substitution on the phase relations and in particular on the metal–insulator transition temperature. It has thus been found that cations, with a formal charge of +5 or +6, e.g. Nb [3], Mo [4], and W [5] or anions such as F [6], lead to a rapid decrease of the transition temperature. The phase diagrams of the $V_{1-x}Me_xO_2$ systems with Me = Cr [7], Al [8], and Ti [9] are more complex and beside the M_1-phase they display two or in the case of the $V_{1-x}Ti_xO_2$ system, one more insulating phase. It has been pointed out that the magnetic and electrical properties of the latter phases cannot be properly described without taking into account the interatomic correlation energy while pure and doped VO_2, possessing the M_1-type of structure, is a borderline case for classical band transport description [7, 10].

This paper presents data for the $V_{1-x}Re_xO_2$ system obtained by means of DTA, X-ray, magnetic susceptibility, and electrical conductivity studies. These data will be compared with previous and some new findings for the $V_{1-x}Me_xO_2$ systems with Me = Nb, Mo, and W.

[1]) S-106 91 Stockholm, Sweden.

2. Experimental

The starting materials consisted of V_2O_5 (Fisher Sci. Co. p.a.) and NH_4ReO_4 (Koch-Light 99%). V_2O_3 and ReO_2 were prepared and characterized as described in [4, 11]. Powder samples of the $V_{1-x}Re_xO_2$ system were synthesized by heating appropriate amounts of thoroughly mixed V_2O_3, V_2O_5, and ReO_2 at 1275 K for 7 d in sealed evacuated silica tubes placed in a metal cylinder.

In some cases the upper part of the silica tube protruded just above the metal cylinder. After the heat treatment a few crystals up to 0.5 mm in length were found at the top of the protruding tubes. These crystals were used for the conductivity measurements described below.

The powder samples of the $V_{1-x}Me_xO_2$ systems with Me = Nb, Mo, and W, used for some complementary measurements were prepared and characterized as in [3 to 5].

The X-ray powder patterns were recorded at ambient temperature in a Guinier-Hägg focusing camera with $CuK\alpha_1$ radiation ($\lambda = 1.54051$ Å) and with KCl as internal standard ($a = 6.2919$ Å).

The DTA studies were performed in the temperature region 100 to 400 K with an equipment permitting simultaneous analysis of five samples [12].

The magnetic susceptibility measurements were performed with an automatized apparatus applying the Faraday method. $HgCo(SCN)_4$ was used as standard. The measurements covered the temperature region 78 to 700 K [13].

The electrical conductivity measurements were performed with the four-probe technique on rectangular rod-shaped crystals ranging from 0.1 to 0.5 mm in length. The logarithm of the conductivity was continuously recorded as function of the inverse temperature from 100 to 500 K.

3. Results

3.1 Thermal and X-ray analysis

The transition temperature, T_t, determined by DTA is plotted versus x in Fig. 1. The T_t-values given are the average of the data obtained from raising and lowering the temperature. The hysteresis was found to be less than 3 K for lower x-values and increased to 14 K for $x = 0.09$.

The linear relation between x and T_t for $x \leqq 0.07$ gives a decrease of the T_t of 18 K/at% Re. For $x = 0.09$ one observes a transition temperature considerably lower than one might expect when applying the linear relation between x and T_t. Samples with the compositions $V_{0.90}Re_{0.10}O_2$ did not exhibit any DTA peak above 100 K although the magnetic susceptibility measurements indicate the presence of a transition around 115 K (see below).

The latent heat of the $M_1 \rightarrow R$ transition of pure VO_2 is 1020 cal/mol [14]. If the DTA peak area is assumed to be proportional to the latent heat of the transition one observes a decrease of the latent heat for increasing Re content as shown in Fig. 1.

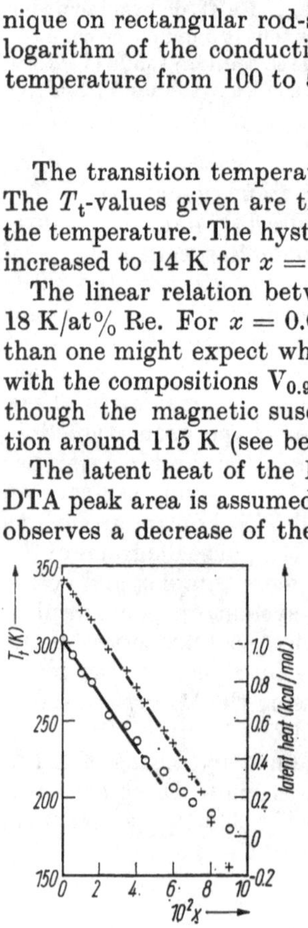

Fig. 1. The transition temperature (+) and latent heat (○) of the $M_1 \rightarrow R$ transition plotted vs. x in $V_{1-x}Re_xO_2$

Table 1

Lattice parameters of the $V_{1-x}Re_xO_2$ system

x	lattice parameters at 295 K				
	a (Å)	b (Å)	c (Å)	β	V (Å^3)
0.00	5.752 (1)	4.526 (1)	5.379 (1)	122.61 (1)	117.96 (4)
0.01	5.751 (1)	4.530 (1)	5.386 (1)	122.58 (1)	118.27 (4)
0.02	5.746 (1)	4.535 (1)	5.389 (1)	122.51 (1)	118.42 (4)
0.03	4.561 (1)		2.849 (1)		59.26 (1)
0.05	4.564 (1)		2.849 (1)		59.35 (2)
0.10	4.578 (1)		2.847 (1)		59.67 (1)
0.15	4.591 (1)		2.845 (1)		59.96 (2)

At lower concentrations of Re the latent heat decreases with approximately 130 cal/per mol and at% rhenium.

The X-ray powder patterns show an extension of the homogeneity range of the M_1-phase up to $x = 0.02$, while the rutile structure prevails for $0.03 < x \leq 0.15$. The lattice parameters for some representative samples are given in Table 1.

3.2 *Magnetic susceptibility measurements*

The molar magnetic susceptibility of some representative samples are given as a function of T^{-1} in Fig. 2. The χ_M-values are not corrected for the underlying diamagnetism. The data presented are those obtained when the samples were heated from 78 to 700 K.

The low-temperature modification of pure VO_2 is temperature-independent paramagnetic. The χ_M-value (65×10^{-6} cgs e.m.u./mol) is in good agreement with our previous findings [4, 15]. Re-doped VO_2 is also temperature-independent paramagnetic

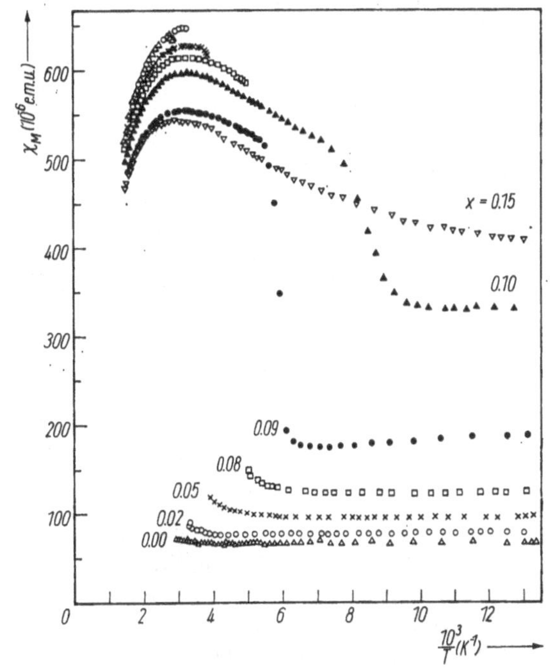

Fig. 2. The molar magnetic susceptibility of $V_{1-x}Re_xO_2$ samples plotted vs. the inverse temperature

in the temperature region $T_t > T \geqq 80$ K. The observed χ_M-values increase for increasing Re contents in the M_1-phase area. For $x \leqq 0.09$ one observes a rather sharp discontinuity in the susceptibility curve at the T_t while a somewhat more gradual increase of susceptibility values is observed for $x = 0.10$. In the R-phase area the susceptibility values decrease with increasing x-value. (The susceptibility values of $V_{0.91}Re_{0.09}O_2$ in the R-phase deviate from those expected, the reason of which is unknown.)

3.3 Electrical conductivity measurements

The logarithm of the conductivity of VO_2 is plotted versus T^{-1} in Fig. 3a. The data for pure VO_2 are in good agreement with our previous findings [4, 15].

The composition of the Re-doped crystals with $x \leqq 0.07$ is determined from their transition temperatures by applying the linear relation between x and T_t described above. The Re contents of the crystals with $x \leqq 0.07$ are estimated from a comparison of the observed T_t with those obtained for powder samples of similar compositions by means of DTA and magnetic susceptibility measurements. The accuracy in x might be as low as ± 0.01 in this part of the phase diagram as dT_t/dx is less well defined due to the considerable hysteresis in the T_t.

From Fig. 3a and b one can observe that the activation energy decreases from 0.45 eV for pure VO_2 to 0.004 eV for $x \approx 0.09$ in $V_{1-x}Re_xO_2$.

The activation energies given in Fig. 3b are calculated from the slope of the straight lines in Fig. 3a. An extrapolation of these lines to $T = \infty$ gives an intersection on the conductivity axis of the order 40000 $(\Omega m)^{-1}$.

According to Fig. 3b one would expect an insulator–metal transition, induced by the variation of composition, to take place around $x = 0.10$. A temperature-induced metal–metal transition would then be found for $x > 0.10$. In order to confirm the existence of such metal–metal transitions in the (V, Re)O_2 system similar to those observed in the (V, Mo)O_2 system [4] the conductivity measurements must, however, be extended below the liquid nitrogen temperature, the lowest temperature applied in this study.

4. Discussion

The above described properties of Re-doped VO_2 will below be compared with previous and some new findings for the $V_{1-x}Me_xO_2$ systems with Me = Nb, Mo, and W.

It has been shown that beside Re also Nb, Mo, and W, when substituting in VO_2, lower the M_1–R transition temperature linearly with increasing content of substituent [3 to 5]. Formally one might assume that a redistribution of the d-electrons of the substituent element and the vanadium atoms takes place in these materials leading to the formation of V^{3+} ions. A decrease of the T_t of approximately 12 K/at% thus formed V^{3+} ions, named x' below, is then found. Thus for Nb and Mo substituted material $x = x'$ while for W-doped VO_2 $2x = x'$. If this relation between transition temperature depression and contents of V^{3+} formed is applied to our case one obtains a formal charge distribution of $V^{4+}_{1-5x}V^{3+}_{3x}Re^{5+}_xRe^{6+}_xO_2$ $(2x = 3x')$.

The c/a ratios of VO_2 doped with Nb, Mo, W, and Re are plotted versus x' in Fig. 4. The lattice parameters were recorded at room temperature and the c/a ratio of the M_1-phase is calculated as $\frac{1}{2}\,(a^3_{M_1}/V_{M_1})^{1/2}$. Within the M_1-phase area $(x' < 0.03)$ the c/a ratio exhibits a tendency to decrease with increasing x'-values independent of the type of doping element. The c/a ratio of pure VO_2 is, however, smaller than that of the doped alloys with $x' < 0.01$. In the metallic phase $(x' > 0.03)$ an increase of the c/a ratio with increasing x'-values is observed for (V, Me)O_2 with Me = W and Nb, while the c/a for Mo- and Re-doped VO_2 decreases with increasing x'-values. If one assumes that for each doping element there exists a linear relation between c/a and x' within the R-phase

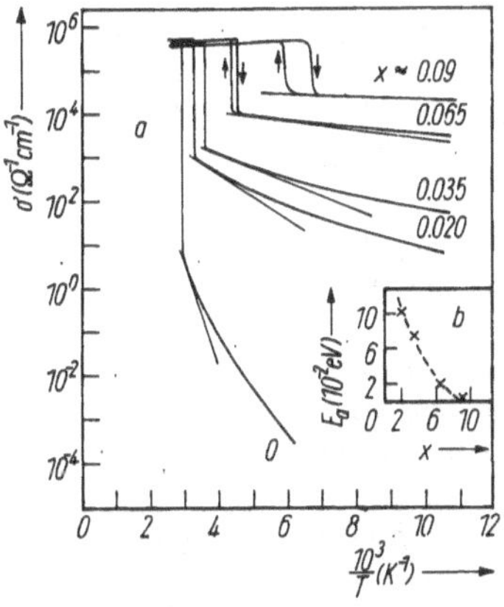

Fig. 3

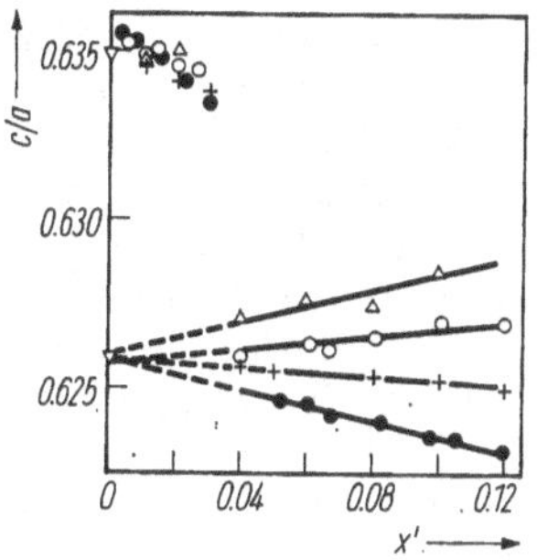

Fig. 4

Fig. 3. a) Electrical conductivity σ and b) activation energy of $V_{1-x}Re_xO_2$

Fig. 4. The c/a ratios of the $V_{1-x}Me_xO_2$ systems plotted vs. x' (see text). ▽ VO_2 (this work and [23]), ● $V_{1-x}Re_xO_2$ (this work), ○ $V_{1-x}W_xO_2$ (this work and [5]), + $V_{1-x}Mo_xO_2$ (this work and [4]), △ $V_{1-x}Nb_xO_2$ (this work)

area and compositional range studied, an extrapolation of these lines towards $x' = 0$ gives c/a ratios which are close to the one of pure VO_2 just above the $M_1 \rightarrow R$ transition temperature.

The latent heat of the $M_1 \rightarrow R$ transition decreases with increasing content of dopant. Thus we have observed a decrease of the latent heat of (75 ± 15) cal/mol and x' for VO_2 moderately doped with W, Nb, Mo, and Re, respectively. However, preliminary studies of the $V_{1-x}Mo_xO_{2+y}$ system with $x = 0.01$ and 0.05 and $y = -0.005$, 0.000, and 0.005 for each x-value have shown that the latent heat varies not only with the x-value but also with the stoichiometry of the samples. This implies that the latent heat value per x' given above might also be affected by the actual oxygen stoichiometry of the various samples.

The magnetic and electrical properties of the M_1-phase of Re-doped VO_2 are very similar to those observed for Mo-doped VO_2 [4]. Thus the magnetic susceptibility is found to be temperature-independent paramagnetic for both Mo- and Re-doped VO_2. The electrical conductivity measurements show that the activation energy decreases with increasing content of dopant in both cases. For Mo-doped VO_2 an insulator to metal transition due to a change in the composition takes place in the region $0.05 < x < 0.10$ while the corresponding value for the $V_{1-x}Re_xO_2$ system seems to be around $x = 0.10$. Assuming this metal–insulator transition to be an Anderson transition a minimum metallic conductivity value of the order 30000 $(\Omega m)^{-1}$ is expected (see Mott [16]) which compares favourably with the experimental values around 40000 $(\Omega m)^{-1}$ found for Re- and Mo-doped VO_2. The observed properties might, however, also be interpreted in terms of a narrow impurity band which is pushed towards the conduction band for increasing x-values.

The small activation energy observed in the conductivity might thus be due either to Anderson localization and thus to thermally activated hopping or to a small energy gap. The two alternatives might be distinguished by means of Seebeck coefficient measurements. Thus, if there is an energy gap the thermopower, $|\alpha|$, will decrease with increasing temperature while in the hopping case $|\alpha|$ will increase with the temperature (see e.g. Mott [16]). Such measurements are in progress.

The magnetic and electrical properties of the M_1-phase of Nb-doped VO_2 are very similar to those of $(V, W)O_2$, but differ from those of Mo- and Re-substituted VO_2. Thus the $(V, Me) O_2$ alloys with Me = Nb and W exhibit temperature-dependent paramagnetism and the activation energy is found to be almost independent of the composition in both cases [3, 15, 17].

As mentioned above the physical properties of the M_1-phase of Nb-, W-, and Mo-doped VO_2 might qualitatively be interpreted within the framework of a one-electron band model [18 to 20] as might also the properties of the $V_{1-x}Re_xO_2$ system.

If experimental errors in e.g. l and A in $\sigma = l/\varrho A$ are accounted for there is no significant difference between the conductivity values of the various $(V, Me)O_2$ alloys in the R-phase area. The magnetic susceptibility data, however, exhibit certain differences.

The molar magnetic susceptibility values can be converted to squared Bohr magneton numbers per metal atom according to

$$\mu_{\text{eff}}^2 = \frac{3k_B T \chi_M}{N_A \mu_B^2 \mu_0},$$

where μ_B is the Bohr magneton and $\mu_0 = 1$ (cgs e.m.u.) $= 4 \times 10^{-7}$ (SI units).

In Fig. 5 μ_{eff}^2 is plotted versus T for VO_2 and VO_2 doped with W, Mo, and Re. All samples but $V_{0.85}Re_{0.15}O_2$ exhibit an $M_1 \rightarrow R$ transition. At the transition temperature one observes a discontinuous increase of the μ_{eff}^2-value while below the T_t an almost linear increase of the μ_{eff}^2-values with the temperature is found. Within the R-phase area a nonlinear increase of the μ_{eff}^2-value is observed. For $V_{0.85}Re_{0.15}O_2$ which is metallic over the entire range of temperature one finds a continuous increase of the μ_{eff}^2-value with the temperature. Similar μ_{eff}^2 versus T curves have been reported in a study of the magnetic properties of the TiO_2–VO_2 system [21]. The shape of the latter curves and those in Fig. 5 suggests that spin pairing also occurs in the R-phase area at lower temperatures. The analysis of the magnetic susceptibility data of the TiO_2–VO_2 system revealed a spin exchange coupling constant of $-300\ k_B T$ between adjacent vanadium atoms. Such an analysis can be accomplished provided that for instance the number of valence electrons per vanadium atom does not exceed one. However, this condition is most probably not fulfilled in either of the MeO_2–VO_2 systems discussed here.

Pouget and Launois [22] have suggested that the increase in the magnetic susceptibility of Nb-doped VO_2 in the R-phase might be related to the variation of the lattice

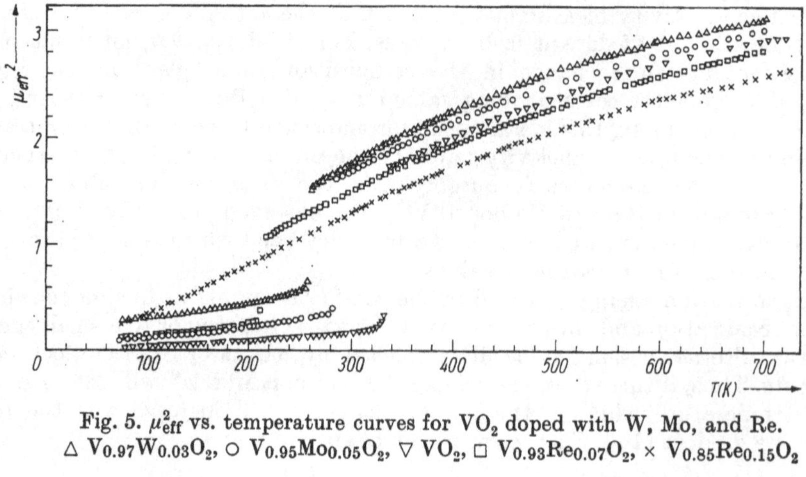

Fig. 5. μ_{eff}^2 vs. temperature curves for VO_2 doped with W, Mo, and Re. △ $V_{0.97}W_{0.03}O_2$, ○ $V_{0.95}Mo_{0.05}O_2$, ▽ VO_2, □ $V_{0.93}Re_{0.07}O_2$, × $V_{0.85}Re_{0.15}O_2$

parameters with x and T. Starting from Goodenough's one-electron band scheme [18] they argue that an increase of c_R and a_R will decrease the $t_{||}$ and π^*-bandwidths and an increase of the c_R/a_R ratio will destabilize the $t_{||}$-band with respect to the π^*. Both effects contribute to an increase of the electron–electron correlation energy and ultimately lead to formation of clusters of moments. This model seems to be appropriate for VO_2 and VO_2 doped with W and Nb as an increase of both the μ^2_{eff}-values and lattice parameters with x and T has been observed (see Fig. 4, 5, and [23]). In the R-phase area Re-doped VO_2 exhibits lower μ^2_{eff}-values than those of pure VO_2 and the μ^2_{eff}-values decrease with increasing x and so does the c_R/a_R ratio. In the case of Mo-doped VO_2 one observes an increase of the μ^2_{eff}-values with increasing x while c_R/a_R decreases. The model proposed by Pouget and Launois seems thus to be appropriate for Re-doped VO_2 but cannot in a straightforward way be applied to the $(V, Mo)O_2$ system. In this connection it can also be noticed that for VO_2 the tendency of obtaining constant μ^2_{eff}-values at elevated temperatures does not seem to have any counterpart in the c_R/a_R ratios [23].

Zylbersztejn and Mott [10] prefer an alternative model for describing both the magnitude and temperature dependence of the magnetic susceptibility of the R-phase. Thus in metallic VO_2 the electrons in the π^*-band screen those in the narrower $t_{||}$-band and the magnetic susceptibility of the latter ones ($\chi_{||}$) are exchange-enhanced over the Pauli value ($\chi^p_{||}$) by a Stoner enhancement factor

$$\frac{\chi_{||}}{\chi^p_{||}} = \frac{1}{1 - U_{eff} N_{||}} = S\,,$$

where U_{eff} is the effective intra-atomic coulomb repulsion and $N_{||}$ the number of $t_{||}$ electronic states per vanadium atom at the Fermi level. Thus their analysis of pure VO_2 revealed a Stoner enhancement factor of 3.2 with $U_{eff} = 0.13$ eV and $N_{||} = 5.25$ states/eV assuming an occupation of 0.3 electrons per vanadium in the π^*-band and 0.7 electrons in the $t_{||}$. If this model is applied to the magnetic properties of the various $(V, Me)O_2$ systems described above it seems that each $V_{1-x}Me_xO_2$ composition has its own Stoner enhancement factor. Thus U_{eff} and/or more probably $N_{||}$ varies not only with Me but also with x.

Acknowledgements

The authors wish to thank Prof. A. Magnéli and Dr. L. Kihlborg for their interest in this work. Dr. T. Hörlin and Dr. T. Niklewski are also acknowledged for many challenging discussions on various parts of this paper.

This investigation has been supported by the Swedish Natural Science Research Council.

References

[1] S. Westman, Acta chem. Scand. **15**, 217 (1961).
[2] G. Andersson, Acta chem. Scand. **10**, 623 (1956).
[3] P. Lederer, H. Launois, J. P. Pouget, A. Casalot, and G. Villeneuve, J. Phys. Chem. Solids **33**, 1969 (1972).
[4] T. Hörlin, T. Niklewski, and M. Nygren, Mater. Res. Bull. **8**, 179 (1973).
[5] M. Nygren and M. Israelsson, Mater. Res. Bull. **4**, 881 (1969).
[6] M. L. F. Bayard, T. G. Reynolds, M. Vlasse, H. L. McKinzie, R. J. Arnott, and A. Wold, J. Solid State Chem. **3**, 484 (1971).
[7] J. P. Pouget, H. Launois, T. M. Rice, P. Dernier, A. Goosard, G. Villeneuve, and P. Hagenmuller, Phys. Rev. B **10**, 1801 (1974).
[8] M. Drillon and G. Villeneuve, Mater. Res. Bull. **9**, 1199 (1974).

[9] T. Hörlin, T. Niklewski, ahd M. Nygren, Acta chem. Scand. **A30**, 6196 (1976).
[10] A. Zylbersteijn and N. F. Mott, Phys. Rev. B **11**, 4383 (1975).
[11] Ö. Sävborg, Mater. Res. Bull. **11**, 275 (1976).
[12] T. Hörlin, T. Niklewski, and M. Nygren, Chem. Commun. Univ. Stockholm No. 9 (1975).
[13] B. Blom and T. Hörlin, Chem. Commun. Univ. Stockholm No. 5 (1977).
[14] E. J. Ryder, F. S. L. Hsu, H. J. Guggenheim, and J. E. Kimzler, quoted by C. N. Berglund and H. J. Guggenheim, Phys. Rev. **185**, 1022 (1969).
[15] T. Hörlin, T. Niklewski, and M. Nygren, Mater. Res. Bull. **7**, 1515 (1972).
[16] N. F. Mott, Metal–Insulator Transitions, Taylor and Francis, London 1974.
[17] G. Villeneuve, J. C. Launey, and P. Hagenmuller, Solid State Commun. **15**, 1683 (1975).
[18] J. B. Goodenough, J. Solid State Chem. **3**, 49 (1971).
[19] E. Caruthers, L. Kleinman, and H. I. Chang, Phys. Rev. B **7**, 3753 (1973).
[20] E. Caruthers and L. Kleinman, Phys. Rev. B **7**, 3760 (1973).
[21] T. Hörlin, T. Niklewski, and M. Nygren, J. Physique **10**, C4-69 (1976).
[22] J. P. Pouget and H. Launois, J. Physique **10**, C4-49 (1976).
[23] K. V. K. Rao, S. V. N. Naida, and L. Iyengar, J. Phys. Soc. Japan **23**, 1380 (1967).

(Received June 2, 1977)

phys. stat. sol. (a) 43, 653 (1977)

Subject classification: 9; 21.6

National Chemical Laboratory, Poona

On the Diffusion of Iron in Silver [1])

By

S. BHARATI, S. BADRINARAYANAN, and A. P. B. SINHA

In the temperature region 780 to 940 °C the diffusion of ^{59}Fe in silver single crystals is investigated. The diffusion coefficient, $D = 1.90 \exp(-49.36/RT)$ cm^2/s, agrees with the earlier work of Mullen. The calculations of the correlation factor f_i for the Ag–Fe system using Mullen's results and from the model of LeClaire and the T_m-model, further support the vacancy mechanism to be operative in this system.

Die Diffusion von ^{59}Fe in Silber-Einkristallen wird im Temperaturbereich von 780 bis 940 °C untersucht. Der Diffusionskoeffizient, $D = 1{,}90 \exp(-49{,}36/RT)$ cm^2/s, stimmt mit dem in der früheren Arbeit von Mullen überein. Berechnungen des Korrelationsfaktors f_i für das Ag–Fe-System mit den Mullenschen Ergebnissen sowie den Modellen von LeClaire und dem T_m-Modell bestätigen den Leerstellenmechanismus als in diesem System wirkenden Mechanismus.

1. Introduction

In their studies on impurity diffusion in silver metal, Hirano et al. [1] showed that the frequency factor and activation energy for impurity diffusion of iron in silver are small compared to the corresponding values for self-diffusion in silver. This was interpreted on the basis of the low solid solubility of iron in silver. The solute flux through the lattice is considered to be reduced to a point where diffusion along dislocation short-circuiting paths becomes the controlling factor in comparison to diffusion through the lattice. It is generally accepted that diffusion in the f.c.c. lattice takes place through the vacancy mechanism. Accordingly one expects the activation energy for impurity diffusion of iron in silver to be close to the self-diffusion activation energy of silver as reported by Mullen [2]. Hence the purpose of the present paper is to reinvestigate the result on the diffusion of Fe in Ag and to find out the reason for the inconsistency in the existing data. This paper reports the result of such an investigation in which the diffusion of ^{59}Fe in Ag has been carried out in the temperature interval 780 to 940 °C.

2. Material and Method

Silver single crystals about 10 mm in diameter used in this experiment were grown in vacuum by the Bridgman technique from 99.999% pure silver obtained from Bhabha Atomic Research Center, Trombay, Bombay. The samples which did not have the correct density of 10.52 g/cm^3 were rejected. One face of each diffusion sample was mechanically polished with 400 and 600 mesh silicon carbide powder followed by a polish by fine alumina paste and etched in a solution composed of 100 ml of the saturated solution of chromic acid and 5 ml of 5% hydrochloric acid. The diffusion samples were pre-annealed for about 100 h at a temperature 50 °C below the melting point to remove the effects of cold work during the polishing procedure. The radioactive isotope ^{59}Fe was obtained as $FeCl_3$ in HCl solution from the Bhabha

[1]) Communication No. 2145 from the National Chemical Laboratory, Poona 411008, India.

Atomic Research Center, Bombay. A thin layer of the radioactive isotope was electrodeposited on the polished surface of the silver sample from a solution which consisted of a saturated ammonium oxalate solution whose pH was reduced to four by the addition of oxalic acid. Two samples thus prepared were put in a graphite crucible with their radioactive faces separated by a thin silica flat of 1 mm thickness and sealed in a quartz tube containing spectroscopically pure argon at a pressure of about 100 Torr. In order to minimize the loss of iron isotope by evaporation, the entire space above the graphite crucible was filled with silica beads. The samples were annealed in the temperature interval 780 to 940 °C. The temperature for each diffusion anneal was maintained constant to better than ± 1 K by a platinum resistance proportional temperature controller. The annealing temperature was measured by means of a chromel–alumel thermocouple calibrated at the melting points of spectroscopically pure tin, lead, cadmium, and aluminium.

3. Procedure

The concentration gradient of ^{59}Fe in a diffused sample was determined by a serial sectioning technique as described earlier [3]. In order to ensure a reproducible counting geometry the chips from each section were dissolved in equal volume of 6N nitric acid before counting for the activity. The diffusivity of ^{59}Fe was determined by counting the 1.10 MeV γ-radiation using a scintillation spectrometer and a single channel pulse height analyzer. The concentration C of a radioactive isotope under our laboratory conditions of an instantaneous plane source in a semiinfinite cylinder can be expressed as follows:

$$C = \frac{C_0}{\sqrt{\pi D t}} \exp\left(-\frac{x^2}{4Dt}\right), \quad (1)$$

where C_0 is the initial concentration and t is the time of the diffusion anneal. The values of D were determined by the method of least squares from the slopes of the linear plots of the logarithm of C versus the square of the penetration distance.

4. Result

A good linear plot of the logarithm of the specific activity versus the square of the penetration depth was obtained in all the runs. These plots are shown in Fig. 1 for various diffusion anneal temperatures. The values of D are summarized in Table 1 along with the temperature and time of the diffusion anneal. The logarithms of the diffusion coefficients were found to vary linearly with the reciprocal of the absolute temperature (Fig. 2) and hence can be expressed as

$$D = 1.90 \exp\left(-\frac{49.36}{RT}\right) \text{cm}^2/\text{s} . \quad (2)$$

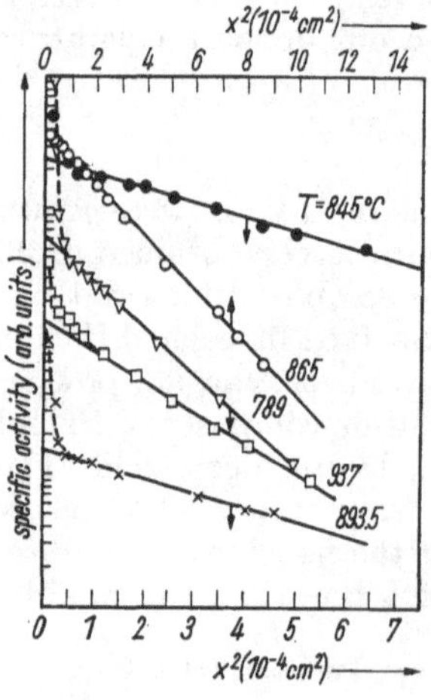

Fig. 1. Penetration plots for the diffusion of iron in silver single crystals

Fig. 2. Arrhenius plot for the diffusion of ^{59}Fe in silver; ○ Mullen [2], ● present work

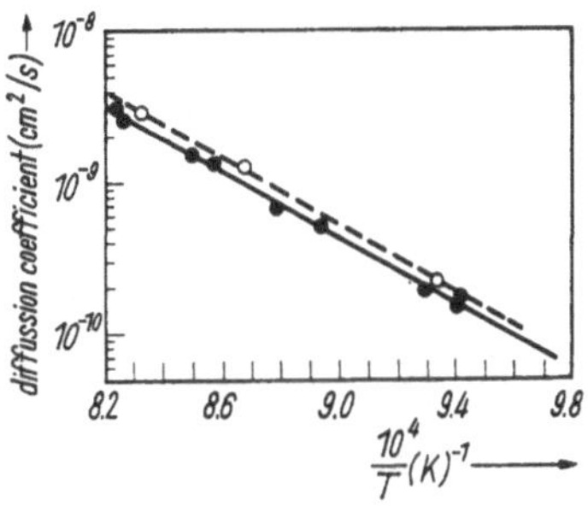

5. Discussion

The fact that plots of the logarithm of specific activity versus the square of the penetration depth up to a penetration depth of 400 μm were linear indicated a negligible contribution from grain boundary diffusion. However, from the above plots, a sharp drop in the activity within an initial penetration depth of about 40 μm near the surface can be seen. Similar behaviour has been observed by many workers in their studies [1, 4 to 8] on the intermetallic diffusion. This has variously been attributed to (i) the formation of an oxide layer [4] on the surface of the sample, (ii) a very low solubility of the solute in the solvent metal, and (iii) the near surface effect, the mechanism of which is not clearly understood. The possibility of the formation of a silver oxide layer can be ruled out in this case because silver oxide is not stable beyond 300 °C. The solid solubility of the Fe metal in Ag is very low and it is likely that a portion of the iron tracer may have remained undissolved and undiffused within a layer of a few microns which cause the sudden drop in activity. It may be pertinent to point out that the phenomenon of the sharp drop in the activity near the surface of the solvent metal is confined not only to the systems with very low solute solid solubilities but has also been observed where the solid solubilities of the diffusing impurities in the solvent are known to be fairly high [6].

The value of the activation energy and the frequency factor for the diffusion of ^{59}Fe in Ag obtained by us are in good agreement with earlier results of Mullen [2] and Bernardini and Cabane [9], but are significantly different from that of Hirano et al.

Table 1

Diffusion coefficient for the diffusion of iron in silver

sample No.	temperature (°C)	time of annealing (s)	diffusion coefficient (cm^2/s)
1	940.0	3.60×10^4	3.00×10^{-9}
2	937.0	1.80×10^4	2.54×10^{-9}
3	903.5	9.00×10^4	1.47×10^{-9}
4	893.5	7.56×10^4	1.33×10^{-9}
5	865.0	1.62×10^5	6.84×10^{-10}
6	865.0	1.62×10^5	6.42×10^{-10}
7	845.0	2.16×10^5	5.02×10^{-10}
8	845.0	2.16×10^5	5.36×10^{-10}
9	802.5	1.67×10^5	1.88×10^{-10}
10	790.0	5.47×10^5	1.53×10^{-10}
11	789.0	3.65×10^5	1.76×10^{-10}

Table 2

Jump frequency ratios $\frac{w_4}{w_0}$, $\frac{w_3}{w_1}$, $\frac{w_2}{w_1}$ for the diffusion of iron in silver

sample No.	temp. (°C)	$\frac{w_4}{w_0}$	$\frac{w_3}{w_1}$	$\frac{w_2}{w_1}$	f_i	model used
1	725	0.1	0.015	0.05	0.95	Howard and Manning [15]
2	725	0.5	0.30	0.21	0.89	
3	725	1.0	10.0	3.45	0.88	
4	750	—	0.90	0.33	0.91	LeClaire [13]
5	725	—	—	—	0.80 ± 0.08	experimental value [2]

[1] who carried out their analysis of the ^{59}Fe concentration gradient in Ag by the residual activity method of Gruzin [10]. A recent communication on the diffusion of cobalt in Ag [11] has clearly brought out the disadvantages in carrying out the diffusion analysis by the indirect Gruzin's technique particularly in cases where the solid solubility of the solute in the solvent metal is very low. The same analysis holds good for the diffusion of Fe in Ag also.

Our values are in good agreement with the theory of Lazarus [12]. The difference between the activation energies for self- and impurity diffusion can be calculated using a) the electrostatic model of LeClaire [13] or b) the T_m-model [14]. The calculated values of ΔQ by these models are 4.8 and 5.5 kcal/mol, respectively, and are in good agreement with our experimental value of 4.2 kcal/mol. Activation energy for self-diffusion has been taken to be 45.2 kcal/mol [17].

It is possible to calculate a range of possible values of f_i and the jump frequency ratios when $D_i/D_s(0)$ and b are known using the method of Howard and Manning's [15] calculation. Using Lidiard's equation [16] the value of b has been estimated to be -4.6 [13]. Howard and Manning's calculations are made at a low temperature viz. at 725 °C since the divacancy contribution at this temperature is small. Present values of D_i are used and $D_s(0)$ values are taken from Rothman et al. [17].

The allowed range of values for the jump frequency ratios are tabulated in Table 2. The ratios computed from LeClaire's model are also tabulated in Table 2 for comparison.

6. Conclusion

Our values for the frequency factor and the activation energy for the diffusion of iron in silver agree well with those of Mullen. Good agreement with the models of Lazarus, LeClaire, and the T_m-model is an additional proof that iron diffuses in Ag mainly through vacancy migration. Further, the repulsive nature of the interaction between the solute and vacancy is suggested by the ratio of w_4/w_0, w_3/w_1, and w_2/w_1 from Howard and Manning's calculation. Such behaviour is expected from LeClaire's model also for any electro-negative impurity diffusing in the f.c.c. solvent lattice.

Acknowledgement

Our thanks are due to Dr. H. B. Mathur for his continued interest during the course of this investigation.

References

[1] K. Hirano, M. Cohen, and B. L. Averbach, Acta metall. **11**, 463 (1963).
[2] J. G. Mullen, Phys. Rev. **121**, 1649 (1961).

[3] S. Badrinarayanan and H. B. Mathur, Indian J. pure appl. Phys. **8**, 324 (1970).
[4] T. Hirone and H. Yamamoto, J. Phys. Soc. Japan **16**, 455 (1961).
[5] C. B. Pierce and D. Lazarus, Phys. Rev. **114**, 686 (1959).
[6] D. L. Styris and C. T. Tomizuka, J. appl. Phys. **34**, 1001 (1963).
[7] K. Hirano, R. P. Agarwala, and M. Cohen, Acta metall. **10**, 857 (1962).
[8] T. S. Lundy and J. F. Murdock, J. appl. Phys. **33**, 1671 (1962).
[9] J. Bernardini and J. Cabane, Acta metall. **21**, 1561 (1973).
[10] P. L. Gruzin, Dokl. Akad. Nauk SSSR **86**, 289 (1952).
[11] S. Badrinarayanan and H. B. Mathur, Indian J. Chem. **11**, 465 (1973).
[12] D. Lazarus, Phys. Rev. **93**, 973 (1954).
[13] A. D. LeClaire, Phil. Mag. **7**, 141 (1962).
[14] G. Neumann and W. Hirschwald, phys. stat. sol. (b) **55**, 99 (1973).
[15] R. E. Howard and J. R. Manning, Phys. Rev. **154**, 561 (1967).
[16] A. B. Lidiard, Phil. Mag. **5**, 1171 (1960).
[17] S. J. Rothman, N. L. Peterson, and J. T. Robinson, phys. stat. sol. **39**, 635 (1970).

(Received July 27, 1977)

phys. stat. sol. (a) **43**, 659 (1977)

Subject classification: 1.5; 2; 22.1.1; 22.6

Institute of Semiconductor Physics,
Academy of Sciences of the USSR, Siberian Branch, Novosibirsk

Structure of SiO_2 Pyrolitic Films on Ge Substrates

By

O. P. Pchelyakov, E. B. Gorokhov, S. V. Pokrovskaya, L. V. Sokolov, and S. I. Stenin

SiO_2 pyrolitic films are studied by ellipsometry, transmission electron microscopy (TEM), and electron diffraction techniques. During the condensation the films are shown densified starting from interface by a heterogeneous mechanism which results in the formation of more porous layer near the film surface. Between the SiO_2 film and Ge substrate a region with higher concentration of GeO_2 appears. Synthesized in oxygen flow this region is hundreds Angströms thick and a phase separation occurs. All the films have dome-like defects as potential centres of spherulite crystallization. On the interface polycrystalline diamond-like germanium and in some cases the germanium high pressure phase, Ge III, is found.

Пиролитические плёнки SiO_2 изучены методами эллипсометрии, просвечивающей электронной микроскопии и дифракции электронов. В процессе конденсации плёнки уплотняются по гетерогенному механизму, начиная от границы раздела. В результате вблизи поверхности плёнок образуется более рыхлый слой. Между плёнкой SiO_2 и подложкой Ge возникает область с повышенным содержанием GeO_2. Если синтез проводился в потоке кислорода, толщина этой области составляла сотни ангстрем и в ней наблюдалось фазовое расслоение. Во всех плёнках присутствовали куполовидные дефекты, являющиеся потенциальными центрами сферолитной кристаллизации. На границе раздела обнаружен поликристаллический алмазоподобный германий, а в ряде случаев фаза высокого давления германия, Ge III.

1. Introduction

When analyzing the electrophysical properties of Ge MIS structures, one assumes a definite structure of dielectric layers and metal–dielectric and dielectric–semiconductor interfaces as well [1]. However, because of the limited experimental structure data the models considered do not always correspond to a real structure of the systems investigated. The present paper is an experimental study of SiO_2 pyrolitic films on Ge by ellipsometry, TEM, and electron diffraction techniques.

2. Experimental

SiO_2 pyrolitic films were deposited on the (111) surface of Ge substrates by tetraethylorthosilicate pyrolysis [2] in argon flow at a synthesis temperature $T_s = 670$ °C (A-films) and in oxygen flow at $T_s = 500$, 620, and 670 °C (O-films). The growth rate was 80 to 100 Å/min for A-films and 10 to 15, 60 to 80, and 150 to 170 Å/min for O-films. For comparison films on Si substrate were grown under the same conditions. The refractive index (n) and the film thickness (d) of the dielectric after synthesis and etching in P-etchant [3] were measured by ellipsometry using nomograms for a two-layer film in some cases [1].

The presence of GeO_2 in the films was determined by transmission IR-spectroscopy with step-by-step removing the films in P-etchant.

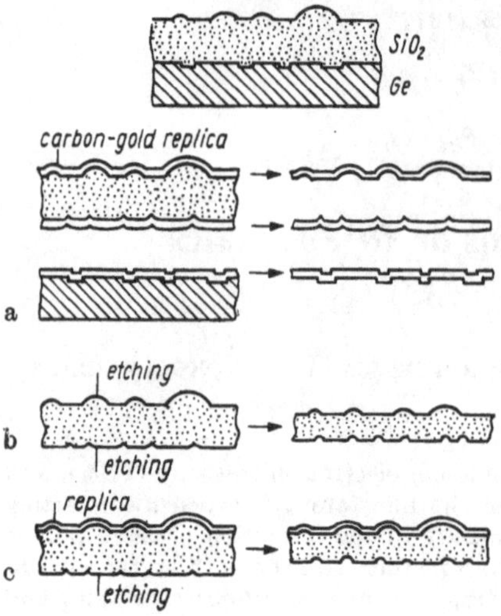

Fig. 1. Electron microscopic techniques: a) gold shaded carbon replica technique; b) transmission technique; c) combined technique

For the investigation of inhomogeneities in the film and at the film-substrate interface a set of electron microscopic techniques was used (Fig. 1). Inhomogeneities in the film were revealed by structure-sensitive etchants on the basis of hydrofluoric acid. Our experience has shown that HF vapour reveals inhomogeneities in an amorphous phase and the buffer etchant [4] crystallites. The SiO_2 films were step-by-step thinned in one of the etchants. The relief of the dielectric surface was examined by a gold-shaded carbon replica (Fig. 1a). In order to study the inner surface of the dielectric with the substrate a sample was sticked with the free surface on a glass plate with epoxy. The substrate was then dissolved in hydrogen peroxide with KOH added. The further procedure was as usual. The relief of the Ge surface under the dielectric was investigated after removing the films in buffer etchant (Fig. 1a). Free films treated by HF were studied by TEM (Fig. 1b). For a comparison of the film relief with its bulk TEM was combined with replica techniques [5] (Fig. 1c).

Crystalline phase analysis of films and film-substrate interface was performed by microdiffraction and RHEED.

3. Results

Ellipsometric data obtained by step-by-step thinning of films have shown that the dissolving rate (V) is not equal throughout the thickness. All the films have three layers with different etching rate (Fig. 2a). The upper part is etched at a higher rate equal to 15 to 17 Å/s (curves 3, 4) for A-films and 10 to 11 Å/s for O-films (curves 1, 2, 5).

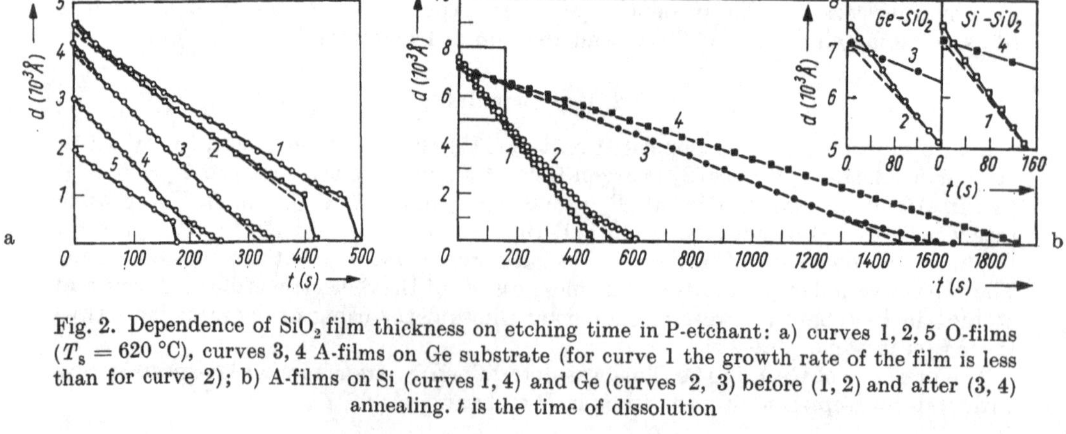

Fig. 2. Dependence of SiO_2 film thickness on etching time in P-etchant: a) curves 1, 2, 5 O-films ($T_s = 620$ °C), curves 3, 4 A-films on Ge substrate (for curve 1 the growth rate of the film is less than for curve 2); b) A-films on Si (curves 1, 4) and Ge (curves 2, 3) before (1, 2) and after (3, 4) annealing. t is the time of dissolution

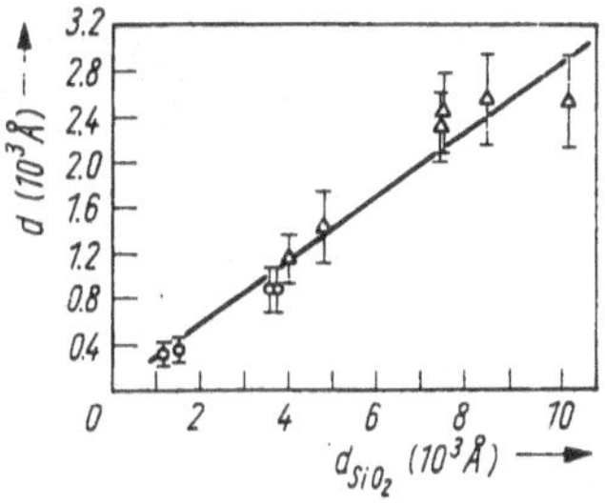

Fig. 3. Dependence of the thickness of the upper layer on SiO_2 thickness △ A-films; ○ O-films

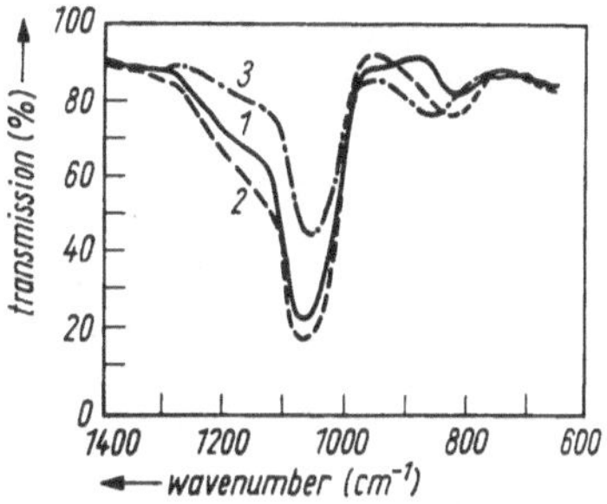

Fig. 4. Transmission IR-spectra of SiO_2 films on Ge: (1) A-film; (2) O-film; (3) O-film after removing $0.5 d_{SiO_2}$

The thickness of this part is about $0.25 d_{SiO_2}$ (Fig. 3). The bulk of the film ($\approx 0.6 d_{SiO_2}$) is etched at the rate of 14 to 16 and 8 to 10 Å/s, and the lower layer ($\approx 0.15 d_{SiO_2}$) at 7 to 8 and 6 to 7 Å/s for A-films and O-films, respectively. O-films have in addition one extremely rapidly etched layer whose thickness increases with increasing synthesis time (Fig. 2a). The exact value of V of this layer is unknown but is assumed to be about 200 Å/s.

It should be noted that ellipsometric results for O-films coincided with the theoretical nomogram only when a two-layer film model was taken into account: a dielectric with $n = 1.55$ to 1.57 and SiO_2. The SiO_2 films on Si substrate are characterized by the absence of the lower slowly etched layer (Fig. 2b). The thermal treatment of A-films on Ge and Si in vacuum or argon flow at 800 °C for 2 h results in (Fig. 2b): a) an overall decrease of d_{SiO_2} by 4 to 5%, b) the disappearance of the upper rapidly etched layer, c) a decrease of V in the middle part of films on Ge and Si by 4 to 5 and the lower slowly etched layer on Ge by 2 to 3 Å/s.

The transmission IR spectra (Fig. 4) of all the films show an absorption peak near 1075 cm^{-1}. A-films on Ge and all films on Si have also a peak at 805 cm^{-1} (curve 1). O-films on Ge have a somewhat broad asymmetric 825 cm^{-1} peak (curve 2) instead of the 805 cm^{-1} one. If $0.5 d_{SiO_2}$ is removed then this peak, keeping the asymmetry, is shifted to 860 cm^{-1} (curve 3).

No marked relief has been found on the Ge surface after removing O-film in HF. If the substrate is dissolved in hydrogen peroxide then a characteristic picture of phase separation appears (Fig. 5) on the inner surface of the film. On the Ge surface under an A-film triangular pits are found. Their density increased with increasing d_{SiO_2} from 5×10^8 at $d_{SiO_2} = 0.1$ μm to 2×10^9 cm^{-2} at $d_{SiO_2} = 1$ μm. Heating the

Fig. 5. Relief of the inner surface of O-film on Ge

sample with thin A-films in argon at 670 °C resulted in a higher density of pits, while vacuum heating did not change the relief of the substrate. An inner A-film surface had no marked relief.

The electron diffraction study indicated polycrystalline diamond-like Ge at the interface of all the specimens, but its amount was markedly less for A-films. In some O-film samples ($T_s = 500$ °C) another polycrystalline phase was found. Its interplanar spacings coincided satisfactorily with those of the high-pressure phase, of the Ge III (see Table 1).

Table 1
Interplanar spacings for the polycrystalline phase near the O-film–Ge interface

hkl	d_{obs} (Å)	d_{cal} (Å)	$(I/I_0)_{cal}$	d_{obs} (Å) [6]
—	5.24*)	—	—	—
—	4.91*)	—	—	—
011	4.52	4.519	8	4.518
111	3.60	3.594	17	3.593
012	3.00	3.008	45	3.005
021	2.74	2.729	100	2.728
112	—	2.682	22	2.681
121	2.51*)	2.479	16	2.478
022	2.26	2.259	0	—
113	2.02	2.034	2	2.032
031		1.902	12	1.900
130	1.87*)	1.875	5	1.873
222	1.79	1.797	35	1.796
004	} 1.75*)	1.745	11	
123		1.749	4	1.744
230	1.65	1.645	11	—
223	1.55	1.557	0.1	1.543
033	1.50*)	1.506	5	1.504
040	} 1.48*)	1.482	5	1.483
232		1.488	0.7	

d_{cal} and $(I/I_0)_{cal}$ are calculated for electron diffraction at a tetragonal body-centered lattice with $a_0 = 5.93$ Å, $c_0 = 6.98$ Å, $z = 12$.
*) Observed not in all electronograms.

Removing the SiO_2 film from these samples and long-time storage under usual conditions leads to a transition of Ge III to usual polycrystalline Ge I. Vacuum heating of specimens with uncovered interface at 350 °C for 1 h transforms Ge III to a twinned single crystal of Ge I.

Dome-like defects on the surface of both A- and O-films on different substrates (Ge, Si, fused quartz) are observed. Fig. 6a shows a replica of the film surface shifted relative to the same part of the film bulk by about 0.4 μm. The inner surface of this film was etched by HF vapour. A darker part in the film bulk is well seen to correspond to each dome of the replica. The micrograph indicates areas with different chemical stability inside the dome-like inclusions. Surface replicas of films etched in HF vapour also revealed that the central parts of most defects are dissolved more rapidly (Fig. 6b). Etching the film surface in buffer etchant is accompanied by a shape change a considerable part of the dome-like defects which gradually transform to cylindrical ones (Fig. 6c). The density of "cylinders" was lower than that of domes

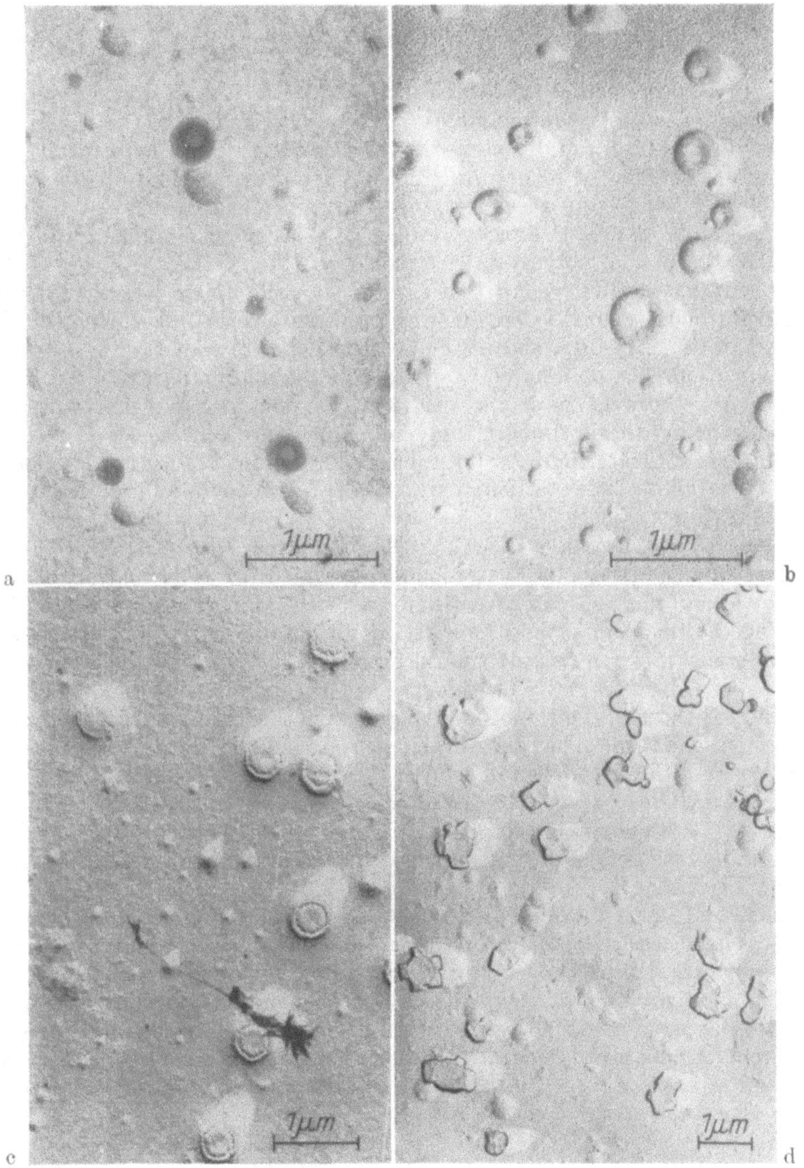

Fig. 6. Dome-like defect structure in SiO_2 films. a) Superimposed image of the replica of the film surface and film bulk. b) Replica of the inner film surface treated in HF vapour. c) Cylindrical defects. d) Crystallites around cylindrical defects

of film surface. If thinning is continued, then the "cylinders" are surrounded by particles having distinct crystallographic facets (Fig. 6d). Vacuum heating of these films at 800 °C for 1 to 2 h leads to the formation of spherolites in the dielectric bulk near the interface. The density of crystals is 10^6 cm^{-2} independent of annealing time. An analysis of several sections of reciprocal lattice obtained by microdiffraction from spherolites showed α-quartz.

4. Discussion of the Results

The structural peculiarities of the Ge–SiO_2 pyrolitic film system can be divided into two groups. The first group includes features independent of the substrate type: the presence of the upper rapidly etched layer and a layer with constant etching rate as well as the existence of structural microinhomogeneities in the film bulk. The second group includes effects due to the influence of the substrate: slowing down the A- and O-film etching near Ge, the presence of a layer with anomalously high dissolvability in O-films and polycrystalline phases at the interface.

First we consider the first group. The existence of the upper two layers in [1, 7, 8] was explained to be due to the densification processes proceeding during film growth. The analysis of the data obtained in the present paper reveals the macromechanism of the densification. The etching constant rate of the middle part of the film implies that in this area the density of the substance condensed has a constant value independent of the synthesis time at a given temperature (cf. curves 1, 5 and 3, 4 in Fig. 2a). The growth rate increase (cf. curves 1, 2 in Fig. 2a) leads to a decrease of the density of a given layer. Films synthesized in oxygen flow have higher density compared to A-films.

The process of densification to a constant value can proceed according to two mechanisms: homogeneous and heterogeneous. The first densification takes place throughout the bulk of the non-densified part of the film simultaneously. The time for achieving the final density is the same at all points of the film. Because of this fact the thickness of a porous part should be constant during the SiO_2 condensation. However, as seen from Fig. 3, the thickness of the non-densified portion increases with the SiO_2 thickness. This fact is explained in the model of the heterogeneous densification mechanism. The densification is considered to start at the interface and the velocity of the densification front towards the surface of the film is constant and less than the growth rate, in the films studied their ratio is ≈ 0.75 (Fig. 3). Film annealing at temperatures equal to or higher than T_s leads to further densification: the etching rate decreases, the upper porous layer disappears and film shrinkage takes place (Fig. 2b).

The presence of dome-like defects indicates a higher growth rate in those places. Fig. 6 shows that these defects are pre-crystallization areas in which the structure is obviously more similar to the crystalline modification of α-quartz, since just this phase is the final product of crystallization. The complicated constitution of pre-crystallization areas (Fig. 6a, b) can be due either to different chemical composition or to different structures of separate regions. The concentration of spherolites does not rise with annealing time. It follows that all crystallization centres are formed during the synthesis.

As for the second group of effects the formation of the layer in O-films with anomalously high dissolvability is of considerable interest (Fig. 2a). As was shown above [1, 7] this layer contains 60 to 75% GeO_2. The dependence of the given layer thickness on synthesis time at 620 °C is practically the same as for thermal oxidation of Ge at 600 °C [9]. The latter proves again the conclusion that free oxidation of Ge through SiO_2 takes place, which is also visualized by transmission IR spectra (Fig. 4). The 1075 and 805 cm^{-1} peaks observed for A-films on Ge and for all films on Si are characteristic of IR absorption by Si–O bond vibrations. The asymmetric peak in the 825 to 860 cm^{-1} range for O-films is due to the superposition of the bond vibrations Si–O (805) and Ge–O (860 cm^{-1}) [9]. A comparison of curves 2 and 3 (Fig. 4) shows that Ge–O bonds are localized near the interface. In A-film spectra the GeO_2 contribution to absorption is not large because of its little amount. Recently the thermal oxidation

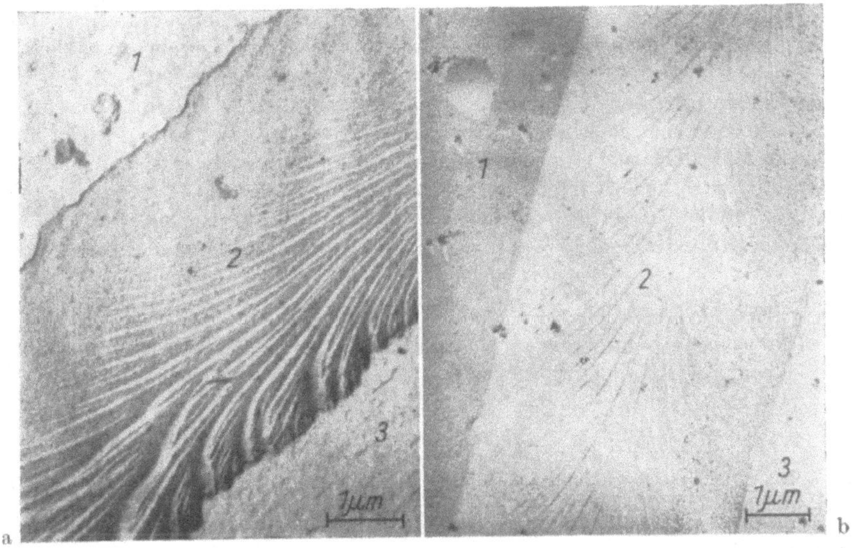

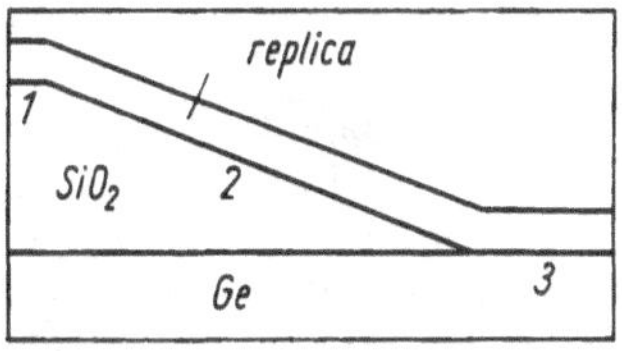

Fig. 7. Relief of the fracture surface of a) A- and b) O-films on Ge; c) fracture scheme

of Ge through the SiO_2 layer in oxygen at 600 °C was once more experimentally confirmed using secondary ion mass-spectrometry [10].

Electron microscopic data show the phase decomposition in a layer enriched with GeO_2 (Fig. 5). Because of the absence of binodal curves for GeO_2–SiO_2 glasses there is no possibility to discuss the phase separation mechanism in detail and the temperature of the GeO_2–SiO_2 glass separation as well as the composition of the phase separation products remain unknown.

At the interface of A-films with Ge besides the initial 20 to 30 Å thick oxide layer also thermal oxide appears due to oxidation of Ge by traces of oxygen-containing agents in the gas mixture. The oxidation obviously proceeds in local centres on the surface producing etch pits. The density increase of pits on the Ge surface while annealing in argon and the constancy of their density while vacuum annealing proves the presence of traces of oxygen-containing components in the carrier gas.

The formation of the polycrystalline phases Ge I and Ge III at the interface may be explained as follows. A fraction of GeO_2 formed on the Ge surface transforms to GeO by disproportion reaction so that near the interface at the synthesis temperature the two oxides GeO_2 and GeO coexist. A further cooling results in a reverse reaction $2\,GeO \rightarrow GeO_2 + Ge$ with formation of a crystalline Ge phase. If the Ge crystallization is governed by the rule of Ostwald steps, then under some conditions one can fix an intermediate state. In our case, for instance, it was achieved by decreasing T_s and rapid cooling the samples. It should be noted that in [11] while studying the phase separation of GeO_x glasses a metastable amorphous metallic Ge phase was found.

Layers of low etching rate in A- and O-films on Ge cannot be explained using our data.

In conclusion we give micrographs of replicas of cross-sectional fractures of A- and O-films on Ge which clearly show the main peculiarities of the structure of the systems studied (Fig. 7). In the left-top part of the micrographs (Fig. 7a, b) a SiO_2 surface with dome-like defects is seen. A porous layer of the film has no marked microstructure. One can make sure that its thickness is 0.25 of the total thickness of SiO_2. The next layer is the bulk of SiO_2 with step-like fracture. In a specimen with A-film (Fig. 7a) the layer in contact with the Ge surface has etch pits. On fracture of O-film sample (Fig. 7b) between the smooth Ge surface and the layer with step-like fracture there is an area of phase separation GeO_2–SiO_2 which is ≈ 0.25 of d_{SiO_2} in agreement with the data of Fig. 2a.

Acknowledgements

The authors wish to thank Dr. I. G. Neizvestny for helpful discussions, and V. A. Ivanchenko and N. Yu. Tchernysheva for assistance during this work.

References

[1] A. V. RZHANOV, Svoistva structur metal–dielektrik poluprovodnik, Izd. Nauka, Moscow 1976.
[2] E. B. GOROKHOV, E. A. DEMYANOV, I. G. NEIZVESTNY, and S. V. POKROVSKAYA, Izv. Akad. Nauk SSSR, Ser. neorg. Mater. **11**, 270 (1975).
[3] W. A. PLISKIN and H. S. LEHMANN, J. Electrochem. Soc. **112**, 1013 (1965).
[4] F. P. PRESS, Photolitographiya v proizvodstve poluprovodnikovykh priborov, Izd. Energiya, Moscow 1968.
[5] O. P. PCHELYAKOV, Zav. Labor. **40**, 1125 (1974).
[6] C. H. BATES, F. D. DACHILLE, and R. ROY, Science **147**, 360 (1965).
[7] E. B. GOROKHOV, V. A. IVANCHENKO, I. G. NEIZVESTNY, S. V. POKROVSKAYA, O. P. PCHELYAKOV, S. I. STENIN, and E. M. TRUKHANOV, Elektronnye protsessy na poverkhnosti poluprovodnikov i na granitse razdela poluprovodnik–dielektrik, Inst. Fiz. Poluprov., Novosibirsk 1974 (p. 33).
[8] E. M. TRUKHANOV, E. B. GOROKHOV, and S. I. STENIN, phys. stat. sol. (a) **33**, 435 (1976).
[9] V. N. DROZDOV, T. I. KOVALEVSKAYA, A. V. RZHANOV, and K. K. SVITASHEV, Mikroelektronika **2**, 46 (1973).
[10] K. L. WANG and H. A. STORMS, J. appl. Phys. **47**, 2539 (1976).
[11] P. DE NEUFVILL and D. TURNBULL, Disc. Faraday Soc. **50**, 182 (1970).

(Received June 6, 1977)

phys. stat. sol. (a) 43, 667 (1977)

Subject classification: 10.1; 12.1; 21.1

Department of Materials Science and Engineering, Faculty of Engineering, Osaka University[1])

Slip Geometry and Anomalous Strengthening in β-Cu(ZnMn) Single Crystals

By

M. YAMAGUCHI and Y. UMAKOSHI

Single crystals of ternary β-Cu(ZnMn) alloys containing 2 and 7 at% Mn are deformed in compression in the temperature range from 293 to 573 K. The slip geometry, morphology, and yield stress are determined as function of orientation and temperature. It is shown that addition of manganese results in a considerable change in slip behaviour for orientations near the [011]–[$\bar{1}$11] boundary in β-brass. These results are discussed in detail in terms of the separation of 1/2⟨111⟩ superpartials.

Einkristalle der ternären β-Cu(ZnMn)-Legierungen mit 2 und 7 at% Mn werden durch Kompression bei Temperaturen zwischen 293 und 573 K verformt. Die Gleitgeometrie, Morphologie und Fließspannung werden als Funktionen der Orientierung und Temperatur bestimmt. Es wird gezeigt, daß die Zugabe von Mangan eine beträchtliche Änderung im Gleitverhalten des β-Messings bei Orientierungen in der Nähe der [011]–[$\bar{1}$11] Grenze verursacht. Diese Ergebnisse werden ausführlich an Hand der Aufspaltung zwischen den 1/2⟨111⟩ Superpartialversetzungen diskutiert.

1. Introduction

In CsCl-type ordered alloys and intermetallic compounds with ⟨111⟩ slip vector, superdislocations are usually composed of two $1/2a_0$⟨111⟩ superpartial dislocations separated by a ribbon of antiphase boundary (APB) [1, 2]. In these alloys and compounds, it has been generally recognized that the operative slip planes and the nature of slip surfaces are sensitive to the separation of superpartials [3 to 5]. This is best seen in the case of β-brass where a continuous change in the separation is easily accomplished by the addition of third elements; in this case it has been demonstrated that as the separation decreases, there is an increasing tendency for slip to occur preferentially on {112} instead of {110} [6] and, in general, for slip line to become wavy [7, 8]. The transition in the slip plane is also shown to exert a considerable influence upon the characteristics of the well-known anomalous strengthening observed in β-brass at around 470 K [9].

As the separation increases on the other hand, each superpartial dislocation presumably tends to slip independently like unit-perfect dislocations in disordered b.c.c. lattices [5, 10]. For this reason, the operative slip planes and the nature of slip surfaces for larger separations should be different from those for smaller separations. Information on the alloys with large separations is, however, rather sparse, although it is also of importance for a complete understanding of how the variation of superpartial separation affects the slip plane and the nature of slip surfaces in CsCl-type ordered alloys with ⟨111⟩ slip vector.

One of the purposes of this investigation is to provide a detailed information on the orientation dependence of slip plane and slip line morphology in ternary β-Cu(ZnMn) alloy single crystals at room temperature and at 473 K where the anomalous strength peak is expected to occur. Manganese was chosen as the third element because the addition of manganese increases the superpartial separation in β-brass consider-

[1]) Suita, Osaka, Japan.

ably and also because in β-Cu(ZnMn) alloys containing more than 5 at% manganese the disordered state can be retained by rapid quenching [11] and therefore the β-Cu(ZnMn) alloys have the advantage that work on both ordered and disordered states can be done. The other purpose of this investigation is to examine the effect of manganese addition on the anomalous strengthening in β-brass.

2. Experimental Procedure

Two β-Cu(ZnMn) alloys were prepared under purified argon atmosphere using electrolytic grades of copper (99.99%), Zn (99.99%), and manganese (99.9%) as charged materials. Table 1 shows the compositions of the two alloys used in this study, together with the values of the critical temperature for ordering, T_c, measured by means of differential thermal analysis carried out at 3 K/min in a purified argon atmosphere.

Table 1

The compositions and T_c of alloys

alloy	Zn (at%)	Mn (at%)	T_c (K)
TM1	46.9	1.8	724
TM2	41.7	6.8	691

Single crystals of the alloys were grown from the ingots by the Bridgman technique and homogenized for 5 h at 773 K. The compression specimens were machined from the homogenized crystals and subsequently they were chemically polished to remove the surface damage. All of the machined specimens were annealed at 773 K for 30 min and then cooled very slowly in the furnace to produce complete order. Some of specimens of the TM2 alloy were again heat-treated for 10 min at 823 K and quenched into iced brine to allow mechanical testing in disordered conditions. Crystal orientations were determined using the back-reflection Laue method. The compression test was carried out at a strain rate of 1.2×10^{-4}/s on an Instron type machine. Test temperatures higher than 323 K were obtained by immersing the specimens in constant-temperature silicon oil baths. In general, about 1% compression strain was required before slip lines can be distingished clearly enough for trace analysis. The slip lines were observed using the Nomarski interference contrast technique. Operative slip planes were determined by analysis of the traces visible on the surface of crystals by rotation of the cylindrical cross-section. The compression axis is referred to the standard [001]–[011]–[$\bar{1}$11] unit triangle.

3. Experimental Results

3.1 Observed slip planes and slip line morphology at room temperature

3.1.1 Slowly cooled TM1 and TM2 alloys

Fig. 1 a and b show the results for slip plane variation with orientation in slowly cooled TM1 and TM2 alloys, respectively. Observed slip planes are plotted against orientation in the conventional form of ψ–χ relation, where χ is the angle between the maximum resolved shear stress (m.r.s.s.) plane and ($\bar{1}$01), and ψ is the angle between the observed slip plane and ($\bar{1}$01). As seen from Fig. 1a, the observed slip plane in the TM1 alloy is the ($\bar{1}$01) plane (i.e. $\psi = 0°$) for $\chi \lessapprox 20°$. For $\chi > 20°$, however, as χ increases, it quickly approaches from ($\bar{1}$01) to the m.r.s.s. plane. In the TM2 alloy, slip

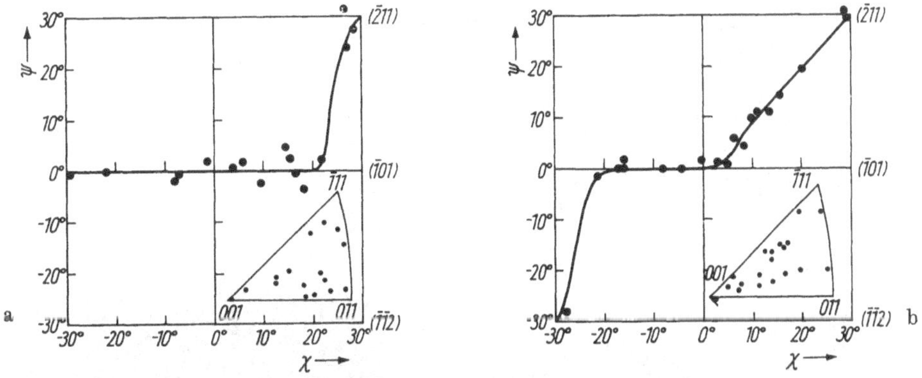

Fig. 1. $\psi(\chi)$ curve of the slowly cooled a) TM1 and b) TM2 alloy crystals deformed at room temperature

approximately follows the m.r.s.s. line for $\chi > 5°$. This shows the occurrence of pure noncrystallographic slip in this range of χ. For $\chi < 5°$ except for orientation near [001] (i.e. $\chi = -30°$), slip still occurs on (1̄01). For orientations with $\chi \approx -30°$, there is a remarkable difference between the two alloys. The observed slip plane in TM1 was (1̄01) but in TM2 it was (1̄1̄2). The authors [6] have previously shown that slip in binary β-CuZn occurs on {110} planes regardless of crystal orientation (i.e. $\psi = 0°$ over the entire range of χ). Therefore, it is seen that the effect of manganese addition on the ψ–χ relationship in β-brass is considerable.

Fig. 2 and 3 show the slip traces observed on top faces of single crystals of slowly cooled TM1 and TM2 alloys, respectively. The top face means the part of the cylindrical surface to which primary edge dislocations are parallel. It is found that in the two alloys both the nature and the distribution of slip vary with crystal orientation.

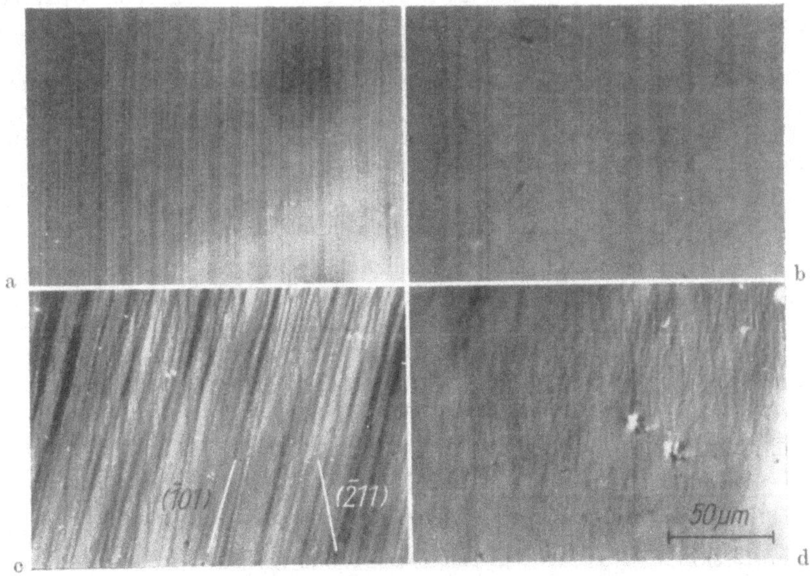

Fig. 2. Slip traces on the top face of slowly cooled TM1 alloy single crystals with orientations a) $\chi = -20°$, b) $-3°$, c) 22°, and d) 27° deformed at room temperature

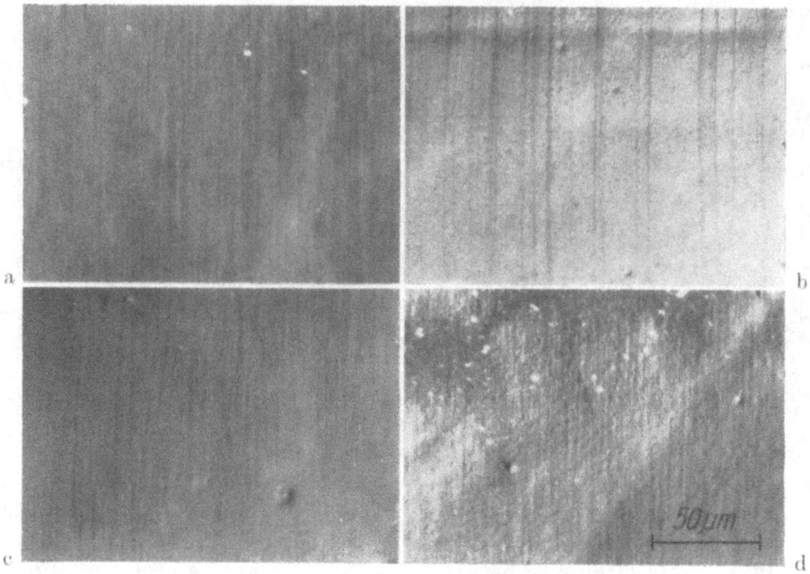

Fig. 3. Slip traces on the top face of slowly cooled TM2 alloy single crystals with orientations a) $\chi = -27°$, b) $-18°$, c) $23°$, and d) $30°$ deformed at room temperature

Slip lines corresponding to the slip on the ($\bar{1}01$) plane are always long and straight in both alloys (Fig. 2a, 2b and 3b), although slip traces in TM2 are much coarser than those in TM1. However, cross slip between these straight slip lines was frequently observed at the beginning of the transition from the planar slip on the ($\bar{1}01$) plane to wavy non-crystallographic slip. As seen in Fig. 2c, the cross slip traces are found to be nearly parallel to the trace of the ($\bar{2}11$) plane, the shear in [111] on which is in the twinning sense. This seems to suggest that the m.r.s.s. slip observed for $\chi > 20°$ in TM1 and for $\chi > 5°$ in TM2 may be interpreted on the basis of the composite slip on the ($\bar{1}01$) and ($\bar{2}11$) planes. Fig. 2d and 3c show the typical wavy slip traces corresponding to the m.r.s.s. slip in TM1 and TM2 alloys, respectively. Slip lines of crystals of TM2 with $\chi \approx -30°$ and $\chi = 30°$ are shown in Fig. 3a and 3d. They are fine, but fairly straight when compared with those in Fig. 2d and 3c.

3.1.2 Quenched TM2 alloy

The $\psi(\chi)$ curve for the TM2 alloy quenched from 823 K is shown in Fig. 4. The difference between the shape of $\psi(\chi)$ curves for slowly cooled and quenched alloys is obvious. The $\psi(\chi)$ curve in the figure shows large deviations in the observed slip plane from the m.r.s.s. plane toward ($\bar{2}11$) and this implies that slip on the twinning {112} planes is preferred for the deformation of the quenched TM2 alloy. Slip traces observed in the quenched TM2 alloy were generally fine and wavy. Fig. 5 shows those on the top face of orientation $\chi \approx 0°$. Slip traces were found to become less wavy as χ approaches 30°.

3.2 Observed slip planes and slip line morphology at 473 K

The compression test at 473 K was carried out only for slowly cooled alloys. Fig. 6a and 6b show the variation of the observed slip plane with orientation in TM1 and TM2 alloys. The basic feature of the $\psi(\chi)$ curve for the TM1 alloy is the same for

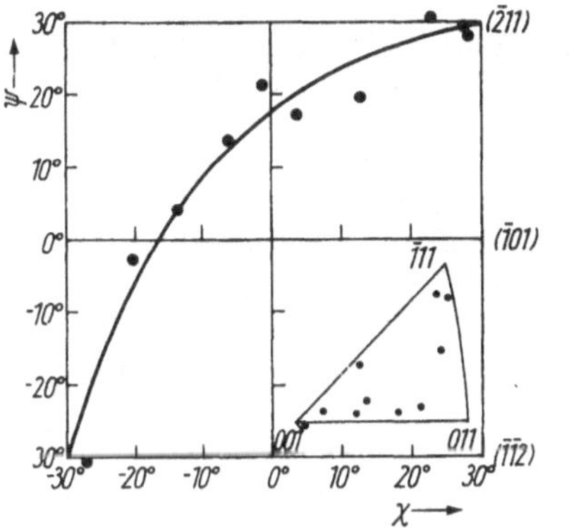

Fig. 4. $\psi(\chi)$ curve of the quenched TM2 alloy crystals deformed at room temperature

Fig. 5. Slip traces on the top face of the quenched TM2 alloy single crystal with orientation $\chi \approx 0°$ deformed at room temperature

both room temperature and 473 K. The difference between the two temperatures is that at 473 K the slip planes start deviating from the $(\bar{1}01)$ plane at smaller values of χ than for room temperature. The ψ–χ relationship in the TM2 alloy is changed considerably with temperature mainly for orientations $\chi > 0°$; at 473 K, the slip plane deviates from the m.r.s.s. plane and the twinning $(\bar{2}11)$ plane becomes more predominant. For $\chi < 0°$, however, we cannot see any significant differences in the feature of the $\psi(\chi)$ curve between the two temperatures.

The basic feature of the variation of the slip line nature with orientation was the same for the two test temperatures in both alloys; for orientations where slip occurs on {110} planes, slip traces were still long and straight and for other orientations they were generally wavy. As seen in Fig. 7a, however, slip lines for the planar slip on {110} planes at 473 K were much coarser than those at room temperature. Slip traces for the m.r.s.s. slip on the other hand became gradually fine as temperature increased. This is clearly seen from Fig. 7b and 7c which show slip traces observed in TM1 and TM2 alloys at 473 K, respectively, for the orientations corresponding to Fig. 2c and Fig. 3d.

3.3 Temperature and orientation dependence of yield stress

The yield stress was determined for the two slowly cooled alloys over the temperature range 293 to 573 K using single crystals with some different orientations.

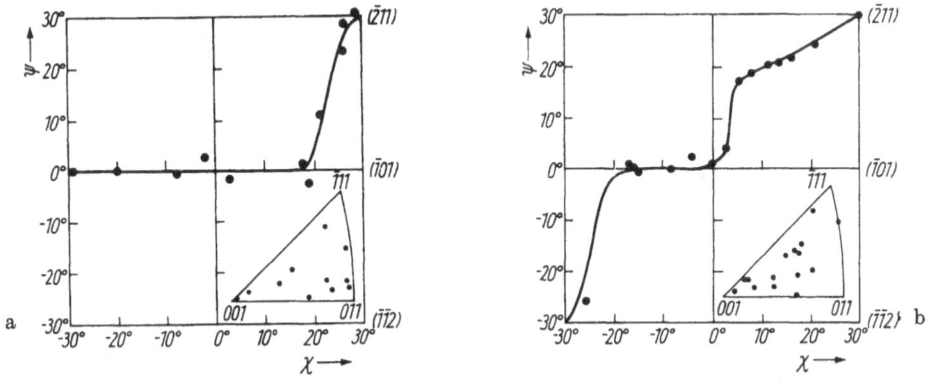

Fig. 6. $\psi(\chi)$ curve of the slowly cooled a) TM1 and b) TM2 alloy crystals deformed at 473 K

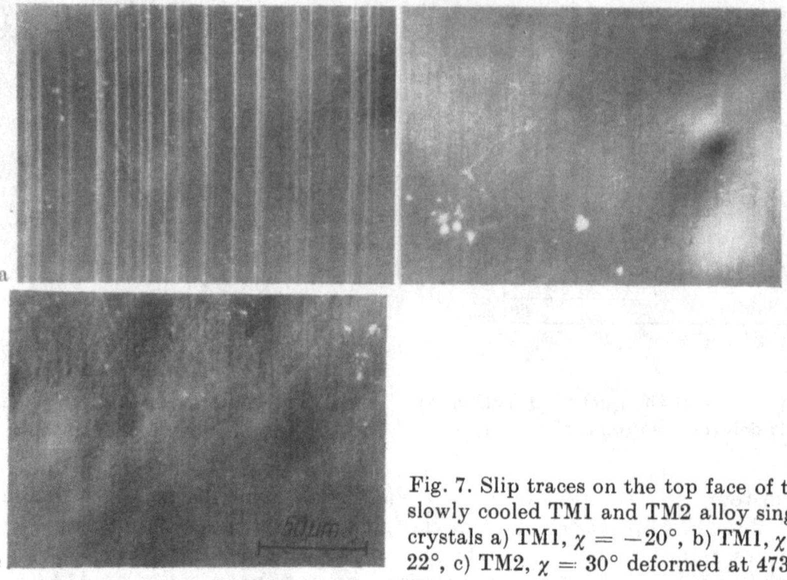

Fig. 7. Slip traces on the top face of the slowly cooled TM1 and TM2 alloy single crystals a) TM1, $\chi = -20°$, b) TM1, $\chi = 22°$, c) TM2, $\chi = 30°$ deformed at 473 K

The results are shown in Fig. 8. All the specimens exhibited an anomalous peak in yield stress with increasing temperature as previously noted in binary β-brass by Ardley and Cottrell [1] and Brown [12]. This behaviour in ternary β-brass alloys containing manganese has been previously observed by Shea and Stoloff [8], however, using polycrystalline specimens. As seen from the figure, the yield stress reaches a peak at around 520 K for TM2 and at around 510 K for TM1 alloy. The peak temperature is virtually independent of crystal orientation in the two alloys. This is in sharp contrast to the fact that in the binary β-CuZn [13] and ternary β-(CuNi)Zn [9] containing 15 at% Ni the peak temperature is highly sensitive to the crystal orientation. However, there exists a marked difference in yield stress increase between the orientations where the planar {110} slip occurs and other orientations; the relative yield stress increase $\left(= (\sigma_{peak} - \sigma_{373K})/\sigma_{373K}\right)$ for the former orientations is roughly two times higher than that for the latter orientations for both TM1 and TM2 alloys.

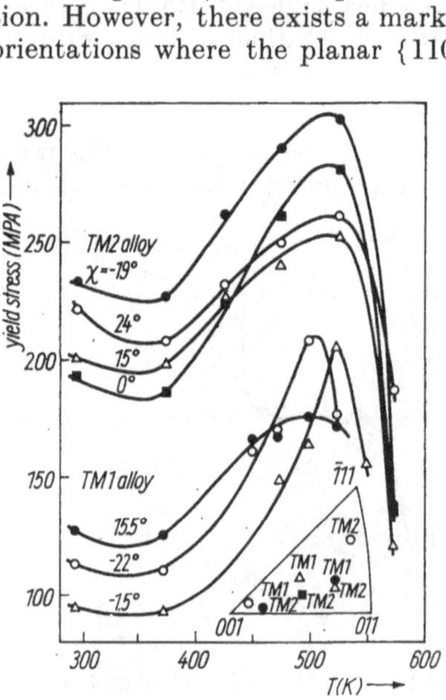

Fig. 8. Temperature dependence of yield stress in the slowly cooled TM1 and TM2 alloy single crystals. Orientations of crystals are shown on the (001) stereographic projection (yield stress in MPa)

4. Discussion

4.1 *Orientation and temperature dependence of the slip plane*

Recently, the operative slip planes and the nature of slip surfaces in ternary β-brass alloys such as CuAuZn [7], CuNiZn [6, 8], and CuZnMn [7, 8] have been studied, although systematic data on single crystals are available only for β-(CuNi)Zn alloys [6]. In general, the addition of third elements which decrease the superpartial separation enhances wavy slip, while the elements which increase the separation tend to confine slip to a single set of planes. However, at large separations each superpartial dislocation should be able to slip independently [5, 10] and therefore the occurrence of wavy slip would be again increased.

The separation has been shown to be proportional to (Ga_0^4/T_cS^2) [5], where G is the shear modulus, a_0 is the lattice parameter, and S is the degree of long-range order. In the Cu–Zn–Mn system [14] the β-phase exists along the line connecting the equiatomic points on Cu–Zn and Cu–Mn binary systems in the composition range studied. This implies that manganese atoms are thought to replace zinc atoms in the lattice. Therefore, the maximum degree of order attainable in the β-Cu(ZnMn) alloys studied is presumably about the same as that in binary β-CuZn alloys. This assumption seems to be supported also by the following facts: 1. the prominent anomalous strength peaks which are quite characteristic of ordered alloys are observed in both TM1 and TM2 alloys and 2. the $\psi(\chi)$ curve for the slowly cooled TM2 alloy is found to be completely different from that for the quenched alloy in which quenched-in disorder is thought to be achieved [11]. On the other hand, T_c for TM1 and TM2 are 10 and 43 K, respectively, lower than that for the binary β-brass containing about 48 at% Zn. a_0 and G increase with the addition of manganese as shown previously by Shea and Stoloff [8], so that the separation of superpartials in the β-Cu(ZnMn) alloys is expected to increase considerably with increasing manganese content not only due to the decrease in T_c but also to the decrease in a_0 and G.

The m.r.s.s. slip observed in TM1 and TM2 alloys may be the result of continual cross slip of screw dislocations on {110} and the twinning {112} since cross slip from {110} to the twinning {112} plane is frequently observed in the specimen with orientations where the transition from the planar {110} slip to the wavy m.r.s.s. slip takes place. However, it is unlikely that such a continuous cross slip occurs in such a way that the trailing superpartial follows a path atomically identical with that of the leading superpartial. Each superpartial dislocation presumably moves independently during the continual cross slip. In the TM1 alloy the deviation in slip planes from {110} occurs for orientations only near the [011]–[$\bar{1}$11] boundary where the resolved shear stress for the cross slip ($\bar{2}$11) plane reaches a maximum. This seems to suggest that the separation in the TM1 alloy is close to the value of separation, above which each superpartial dislocation can move independently. Cross slip onto the ($\bar{1}\bar{1}$2) plane on the other hand may not occur easily because slip occurs in the anti-twinning direction on ($\bar{1}\bar{1}$2), so that the planar {110} slip occurs predominantly for $\chi < 0°$. An APB trail is expected to be left by each superpartial dislocation during continuous cross slip. This may be a cause of strengthening. In fact the critical resolved shear stress (CRSS) for the planar {110} slip in TM1 was 49MPa at room temperature, while that on the twinning {112} planes for the specimen with $\chi = 27°$ and $\psi \approx 30°$ was 62MPa. The CRSS on the twinning {112} planes for the single crystal of slowly cooled TM2 with $\chi \approx 30°$ and $\psi \approx 30°$ was found to be 102MPa. This is about 20% larger than that for the quenched specimen. The superpartial separation in TM2 would be much larger than that in TM1. This is in good agreement with the fact that the m.r.s.s. slip in TM2 occurs over a much wider range of χ than in TM1.

At elevated temperatures, the separation may increase mainly due to the increase in a_0 and the decrease in S. This is the reason why the deviation in slip planes from {110} occurs at smaller χ at 473 K than at room temperature in the TM1 alloy. In the TM2 alloy, deviation in the observed slip plane towards the twinning {112} planes such as that found at 473 K would be enhanced with increasing temperature since as seen from Fig. 5 the twinning {112} plane is preferred for the deformation of the TM2 alloy in the disordered condition.

4.2 Anomalous strengthening

Since Ardley and Cottrell [1] showed that a peak in yield stress occurred at around 470 K in β-CuZn, some theories [1, 8, 12, 13] have been proposed to interpret this anomalous peak. Brown [12], for example, has interpreted it in terms of a mechanism in which an additional stress is shown to be required to move a superdislocation in ordered lattices since the APB created by slip is significantly different from that under thermal equilibrium. Recently Umakoshi and Yamaguchi [13] have shown that the anomalous strengthening in β-CuZn and β-(CuNi)Zn alloys can be explained using an analogous mechanism proposed by Takeuchi and Kuramoto [15] in order to interpret that in Ni_3Ga. In the mechanism, jogs created by thermally activated cross slip are considered to act as a dragging point for the srew superdislocation which undergoes planar slip. These mechanisms obviously should hold for the alloys and compounds which deform by slip of superdislocation, and the strengthening arising from these mechanisms would be highest when the slip is confined to a single set of planes. The prominent yield stress peaks observed in both TM1 and TM2 alloys for the orientations where the planar {110} slip takes place may be interpreted in terms of these mechanisms. However, the peaks are found to occur even for the orientations where each superpartial dislocation presumably moves independently although the peaks are much broader and less prominent than those related to the planar {110} slip. The afore-mentioned mechanisms obviously may not be applicable to these less prominent peaks. Therefore, the anomalous strengthening in the β-Cu(ZnMn) alloys studied should be explained by a combination of mechanisms such as those mentioned above and a mechanism which is not related to the dislocation structure, e.g. the mechanism proposed by Ardley and Cottrell [1], in which the anomalous strength peak is considered to be a result of the maximum in the shear modulus in ⟨111⟩ direction.

Acknowledgements

The authors are grateful to Prof. T. Yamane for his encouragement, to Y. Namba, K. Wada, and Y. Takahashi for their help in experiments.

References

[1] G. W. ARDLEY and A. H. COTTRELL, Proc. Roy. Soc. **219**, 328 (1953).
[2] M. J. MARCINKOWSKI and H. CHESSIN, Phil. Mag. **10**, 837 (1964).
[3] P. A. FLINN, Trans. MS AIME **218**, 145 (1960).
[4] M. J. MARCINKOWSKI and R. M. FISHER, J. appl. Phys. **34**, 2135 (1963).
[5] N. S. STOLOFF and R. G. DAVIES, Acta metall. **12**, 473 (1964).
[6] M. YAMAGUCHI and Y. UMAKOSHI, Acta metall. **24**, 1061 (1976).
[7] H. M. CLARK, Phil. Mag. **16**, 853 (1967).
[8] M. M. SHEA and N. S. STOLOFF, Metallurg. Trans. **5**, 755 (1974).
[9] Y. UMAKOSHI and M. YAMAGUCHI, Scripta metall. **11**, 211 (1977).
[10] M. J. MARCINKOWSKI and N. BROWN, Acta metall. **9**, 764 (1961).
[11] M. GONDA, T. TAKAHASHI, and H. WATANABE, J. Japan Inst. Metals **39**, 939 (1975).
[12] N. BROWN, Phil. Mag. **4**, 693 (1959).
[13] Y. UMAKOSHI, M. YAMAGUCHI, Y. NAMBA, and K. MURAKAMI, Acta metall. **24**, 89 (1976).
[14] J. R. GRAHAM, J. R. LONG, C. E. ARMANTROUT, and A. H. ROBERSON, Inst. Mineral. Met. Engr. **185**, 675 (1949).
[15] S. TAKEUCHI and E. KURAMOTO, Acta metall. **21**, 415 (1973).

(Received June 17, 1977)

phys. stat. sol. (a) 43, 675 (1977)

Subject classification: 14.3.4; 13.3; 22.5.1

Central Laboratory of Photographic Processes, Bulgarian Academy of Sciences, Sofia[1])

Ionic Field Effect in Epitaxial AgBr Layers

By

V. ZHELEV and J. MALINOWSKI

Reversible variations in the conductance of thin AgBr layers due to a capacitively applied electric field normal to the surface are measured. A suitable etching of the layers is made leading to changes in surface properties and to the appearance of the field-effect signal. The effect of etching is discussed in terms of a healing of the surface disorder. The classical semiconductor physics theory of Garrett and Brattain is adapted to describe the surface conductance vs. surface potential dependence in silver halides. Observation of a minimum in the conductance curve makes possible the calculation of the surface potential and the charge in surface states. The results obtained support the model of Poeppel and Blakely indicating a finite number of surface sites.

Es werden reversible Leitfähigkeitsänderungen beim kapazitiv angelegten elektrischen Feld senkrecht zur Oberfläche dünner Silberbromidschichten gemessen. Es wird festgestellt, daß eine geeignete Ätzung, die eine Veränderung der Oberflächeneigenschaften verursacht, für die Erscheinung des Feldeffektsignals (FE) notwendig ist. Es wird angenommen, daß die Ätzung zu einer Verminderung der Oberflächenfehlordnung führt. Die in der Halbleiterphysik klassische Theorie von Garrett und Brattain wird angepaßt, um die Abhängigkeit der Oberflächenleitfähigkeit vom Oberflächenpotential bei den Silberhalogeniden zu beschreiben. Das Auftreten eines Minimums in der Leitfähigkeitskurve ermöglicht die Berechnung des Oberflächenpotentials und der Ladung an den Oberflächen. Die erhaltenen Ergebnisse unterstützen das Modell von Poeppel und Blakely, bei dem eine begrenzte Zahl der Oberflächenstellen angenommen wird.

1. Introduction

According to current views, latent image formation involves the participation of electronic and ionic charged carriers. Latent image specks built up on illumination of contemporary photographic materials are located predominantly at the surface of the silver halide grains. Therefore, the eventual role of a space charge (SC) region in the primary photographic process has been suggested [1, 2]. A number of techniques have been employed for SC studies of silver halide crystals. By photolysis experiments Saunders et al. [3] established the existence of a SC layer in AgBr single crystals. Using their results, Trautweiler [4] evaluated the surface potential and related surface characteristics. Wakabayasi and Sonoike [5, 6] and later Blakely et al. [7 to 10] made use of the Kelvin probe technique to determine directly the surface potential in samples with different crystallographic orientation. Measurements of the surface conductance of epitaxial silver halide layers by Baetzold and Hamilton [11, 12] and by Starbov et al. [13, 14] also contributed to the evaluation of the basic surface parameters. Little is known, however, about the nature and behaviour of the ionic surface states. As well established [15], the surface states on solid surfaces may be introduced or removed by adsorption of ambient molecules or by appropriate treatment of the surface. Since the charge in the SC region is electrically compensated by the localized charge on the surface itself, the potential and other surface characteristics depend strongly on the surface preparation. Hoping to obtain additional information about ionic surface states in silver halides, we have attempted to apply in this work the field

[1]) 1000 Sofia, Bulgaria.

effect (FE) method widely used in semiconductor studies. With a capacitively applied transverse electric field it should be possible to induce *ionic* charge in the subsurface region. This leads to reversible variations in the surface conductance accompanied by a corresponding change in the surface potential. If the surface conductance variation as a function of the voltage applied passes through a minimum, a comparison with the corresponding theoretical curve gives immediately the values of the SC potential and the immobile charge on the surface for any voltage applied [16]. Thin epitaxial AgBr layers are convenient objects for FE experiments in view of the substantial contribution of the subsurface to the total conductance [11, 13, 14].

This paper is a first attempt to determine the applicability of the FE method to ionic crystals, in particular silver bromide, and to report data obtained at the present stage of investigations.

2. Fundamental

The theoretical expressions for the surface conductance and surface SC as a function of the potential are obtained following the classical works by Garrett and Brattain [17] and Kingston and Neustadter [18].

Evaluation of the theoretical curves requires information on the bulk properties of the material. It has been shown [13] that the bulk conductivity of large single crystals and epitaxial AgBr layers prepared in this laboratory is identical. However, the FE experiment requires Ohmic electric contacts made in this study by vacuum deposition of silver. Silver dissolves partially in AgBr increasing thus the concentration of interstitial Ag^+ ions [19]. Therefore, in our experiments the epitaxial AgBr layers should be regarded as single crystals with excess concentration of majority carriers.

The characteristics of the SC region depend on the value of the surface potential Φ. If the reference zero $\Phi = 0$ is taken in the interior of the crystal at an infinite distance from the surface, the concentration of interstitials and vacancies in the SC region is correspondingly [4]

$$n_\mathrm{i}(x) = n_\mathrm{i}' \exp\left(-\frac{e\Phi}{kT}\right), \tag{1a}$$

$$n_\mathrm{v}(x) = n_\mathrm{v}' \exp\left(\frac{e\Phi}{kT}\right), \tag{1b}$$

where n_i' and n_v' are the bulk concentrations of interstitials and vacancies, the other symbols having their conventional meaning. The Poisson equation for the x-direction perpendicular to the surface in dimensionless variables can be written as

$$\frac{\mathrm{d}^2\psi}{\mathrm{d}x^2} = \frac{-\varrho(x)\, e}{\varepsilon\varepsilon_0\, kT}. \tag{2}$$

Here $\psi = e\Phi/kT$ is the dimensionless potential, $\varrho(x)$ is the net charge density, ε is the dielectric constant (13.1 for AgBr), ε_0 is the permittivity in vacuum, and e the electronic charge. The dissolution of silver from the contacts gives excess interstitial Ag^+ ions and electrons [20]. At room temperature the electrons are readily trapped and do not contribute substantially to the conductivity [21]. Therefore, the net charge density is given by

$$\varrho(x) = en_\mathrm{i}(x) - en_\mathrm{v}(x) - en_\mathrm{e}\,, \tag{3}$$

where n_e is the density of the trapped electrons. Since the total bulk charge equals zero

$$n_\mathrm{i}' = n_\mathrm{v}' + n_\mathrm{e} \tag{4}$$

and for $\varrho(x)$ one obtains

$$\varrho(x) = en_i'(e^{-\psi} - 1) + en_v'(1 - e^{\psi}) . \tag{5}$$

Equation (2) can be rewritten as

$$\frac{d^2\psi}{dx^2} = \frac{K^2}{2} [\gamma(1 - e^{-\psi}) + \gamma^{-1} (e^{\psi} - 1)] . \tag{6}$$

Here

$$K = \left[\frac{2e^2 n_0}{\varepsilon\varepsilon_0 kT}\right]^{1/2}$$

is the inverse Debye length of the SC layer, n_0 is the intrinsic concentration of Frenkel pairs in pure AgBr, and the parameter $\gamma = n_i'/n_0 = n_0/n_v'$ is used for convenience.

The integration of (6) at the boundary conditions

$$\left(\frac{d\psi}{dx}\right)_{x\to\infty} = 0 \quad \text{and} \quad (\psi)_{x=0} = \psi_s$$

gives

$$\frac{d\psi}{dx} = \pm\, KF(\psi, \gamma) . \tag{7}$$

Since for AgBr $\Phi < 0$, the positive sign in (7) must be taken.

For the very surface at $x = 0$ one has

$$\frac{d\psi}{dx} = KF(\psi_s, \gamma) . \tag{7a}$$

Now for the excess concentration of interstitials and vacancies in the SC region one can write

$$\Delta n_i = \int\limits_\infty^0 [n_i(x) - n_i'] \, dx = -\frac{n_i'}{K} \varphi(\psi_s, \gamma) \tag{8}$$

and

$$\Delta n_v = \int\limits_\infty^0 [n_v(x) - n_v'] \, dx = \frac{n_v'}{K} \Phi(\psi_s, \gamma) . \tag{9}$$

The numerical solutions of the functions $F(\psi_s, \gamma)$, $\varphi(\psi_s, \gamma)$ and $\Phi(\psi_s, \gamma)$ have been tabulated [22].

Using (8) and (9), the excess conductance $\Delta\sigma$ can be calculated as an explicit function of the surface potential. One obtains

$$\Delta\sigma = \frac{e\mu_i}{K} [- n_i'\varphi(\psi_s, \gamma) + bn_v'\Phi(\psi_s, \gamma)] , \tag{10}$$

where $b = \mu_v/\mu_i$. μ_i and μ_v are the mobilities of interstitials and vacancies assumed to be equal in the SC region and the bulk.

The total SC per unit area Q_{sc} is readily obtained by integration of the expression

$$Q_{sc} = \int\limits_\infty^0 \varrho(x) \, dx .$$

With the help of (2) and (7a) the integration yields

$$Q_{sc} = -\frac{2en_0}{K} F(\psi_s, \gamma) . \tag{11}$$

3. Experimental

The epitaxial ⟨111⟩ AgBr layers were prepared and annealed as described in [14]. Sufficiently good Ohmic current electrodes were obtained by vacuum deposition of silver on the substrate prior to the evaporation of AgBr. The samples did not show any detectable open circuit voltage. Ohmic behaviour was checked by applying for 2 min 2.5 V dc. It was found that the current through the sample decreased not more than 1%. Besides, using a probe of electrolytically sharpened silver point, it was found that the voltage varied linearly between the electrodes. Using Ohmic contacts, the values of the conductance obtained here from dc measurements are similar to those from ac short pulses as described in [14]. Thus it can be concluded that at the specified conditions the flow of charge is connected with nearly reversible electrode processes at a negligible electrode polarization.

The geometry of the samples used for the FE experiment is shown in Fig. 1. The field electrode represents a narrow metal strip deposited on a 16 μm thick mylar sheet. The strip was carefully positioned symmetrically to the gap separating the current electrodes. The mylar spacer was pressed on the sample by means of a spring loaded rubber. No variation in the conductivity on pressing was detected.

Immediately before measurements the samples were immersed at 20 °C for 20 s in a solution containing 1% NH_4Br and NH_4CNS, rinsed with running deionized water and dried. The procedure etches smoothly away approximately 10 to 20 Å from the AgBr layer [24, 25]. This procedure changes significantly the surface properties of the layers and leads to the appearance of a FE signal which could not be detected in unetched samples. This points out that etching heals substantially the surface reducing thus the shielding due to surface disorder. Additional etching does not introduce any further changes in the surface properties of the layer.

However, it cannot be definitely excluded that due to the etching insoluble complexes or sorbed ions remain on the surface. If present, they might be responsible for the changes in the surface properties, which of course would require further elucidation.

The FE measurements were carried out in atmosphere of dry nitrogen using the circuit arrangement shown in Fig. 2, each sample being previously checked for leakage. The polarities of the current source and the electrometer input were reversed simultaneously every 5 s. The transverse high voltage was varied in increments. Usually at each voltage step an asymmetry in current values arised disappearing after 2 to

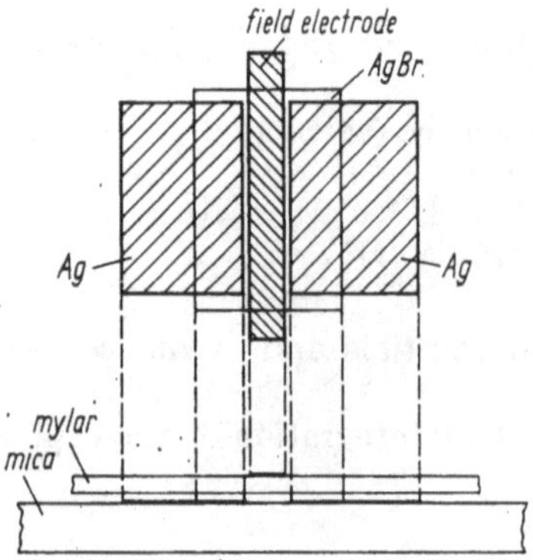

Fig. 1. Field-effect sample, spacing between the current electrodes 130 μm, field-electrode width 80 μm

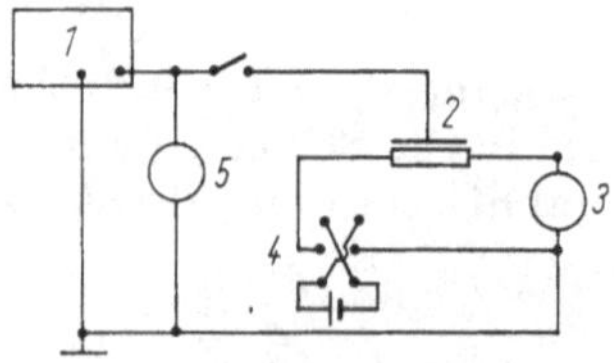

Fig. 2. Field-effect circuit: 1 dc power supply, 2 sample, 3 vibrating-capacitor electrometer, 4 commutator of the current source 1.5 V, 5 voltmeter

3 min. The reading was taken after the current values became equal in both directions. Since the field parallel to the film is small compared to the transverse field, the latter was almost constant over the whole surface. On switching off the transverse field the current reaches again its initial value.

4. Results and Interpretation

4.1 Effect of etching

Fig. 3a is a plot of the total conductance per unit area C as a function of reciprocal temperature for three values of the layer thickness d for etched and unetched samples. In general, etching decreases the conductance, the measured variation becoming smaller with thicker layers. At $d \approx 4.5$ µm etching causes practically unmeasurable changes. From the slope and the intercept of curve $C(d)$ one obtains the bulk and surface conductance, respectively [11]. Fig. 3b illustrates the drastic change in the dependence $C(d)$ on etching. The upper curves, related to the intrinsic region, show clearly that etching alters the surface component only, the bulk conductivity (the slope) remaining essentially the same. One can therefore conclude that this is due to a decrease in the concentration of excess interstitials in the SC region, the bulk concentration being unaffected.

As reported earlier [13], epitaxial [111] AgBr layers prepared in this laboratory exhibit deviations from the linear dependence $C(d)$ at small thicknesses d. This is also illustrated in Fig. 3b. As seen, etching eliminates the deviations and restores the linear behaviour observed by Baetzold and Hamilton [11].²)

The temperature dependence of the bulk and surface components of the conductivity as well as the corresponding activation energies are plotted in Fig. 4. $\Delta E_s = 0.37$ eV corresponds well to the values found in previous works: 0.36 [11] and 0.32 eV [13].

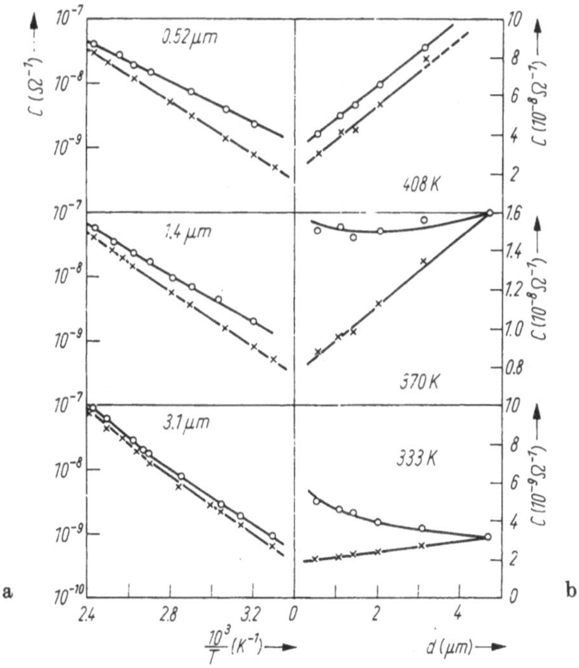

Fig. 3. The effect of etching on a) temperature dependence, b) thickness dependence of the total conductance: ○ before etching, × after etching

²) These deviations may be due to the different experimental conditions: in [11] the measurements have been made in situ, in ultrahigh vacuum, while our measurements have been carried out in nitrogen atmosphere after the samples have been exposed to the ambient. This may probably create significant variations in the surface states subsequently eliminated by etching.

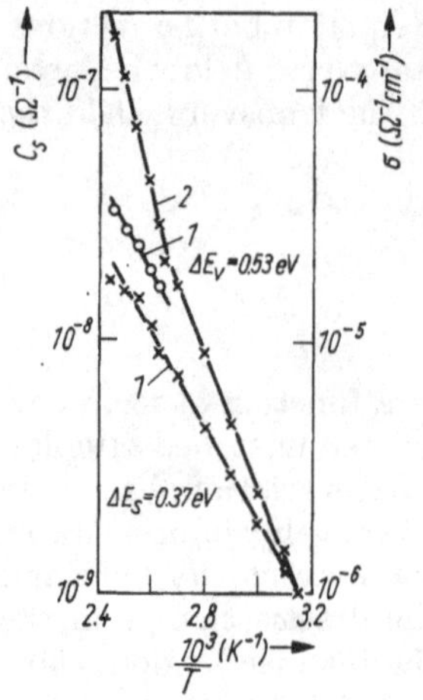

Fig. 4. Temperature dependence of (1) surface conductance and (2) bulk conductivity. Notation of the experimental points same as in Fig. 3

This is an indication that the mechanism of generation of interstitials from surface sites is affected neither by the presence of stoichiometric Ag excess nor by etching. Alternatively, the value of $\Delta E_v = 0.53$ eV is in good agreement with 0.49 eV found in [13] for similar layers deliberately fogged with silver.

4.2 Field effect measurements

Typical data for a sample 2 μm thick are shown in Fig. 5. The surface conductance versus field dependence exhibits a flat but clearly expressed minimum at 2400 V. Dielectric breakdown limits the field voltage up to 3100 V. The magnitude of the observed change in the conductance reaches 13% of the equilibrium value. When the field electrode is negatively biased, the curve surprisingly exhibits saturation at relatively low fields. The observation of a minimum allows a direct comparison of the experimental with the theoretical curves of relations (10) and (11). The numerical calculation of these relations requires the values of the bulk concentration and mobility of all carriers. The bulk conductivity of the AgBr layers is

$$\sigma = e(n_i'\mu_i + n_v'\mu_v) .$$

On the other hand, the interstitials are in equilibrium with the vacancies according to the law of mass action

$$n_i'n_v' = n_0^2 .$$

This equation system may be solved taking $\sigma = 8 \times 10^{-7}\ \Omega^{-1}\ cm^{-1}$ equal to the bulk conductivity of AgBr layers with stoichiometric excess of silver [19] and the values of n_0, μ_i, and μ_v as calculated by Müller [23]. Thus for $T = 300$ K one obtains $n_i' = 10^{16}\ cm^{-3}$ and $n_v' = 7.3 \times 10^{13}\ cm^{-3}$, the other possible solutions having no adequate physical meaning.

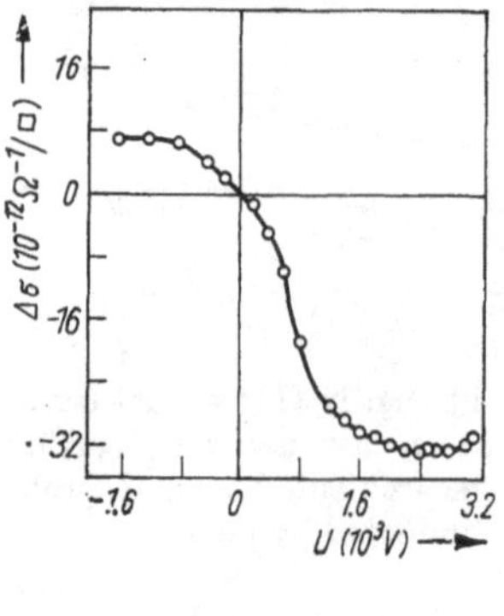

Fig. 5. Experimental FE curve for AgBr layers at room temperature. The change in the sample conductance per square area is evaluated relative to the value of σ at $V = 0$

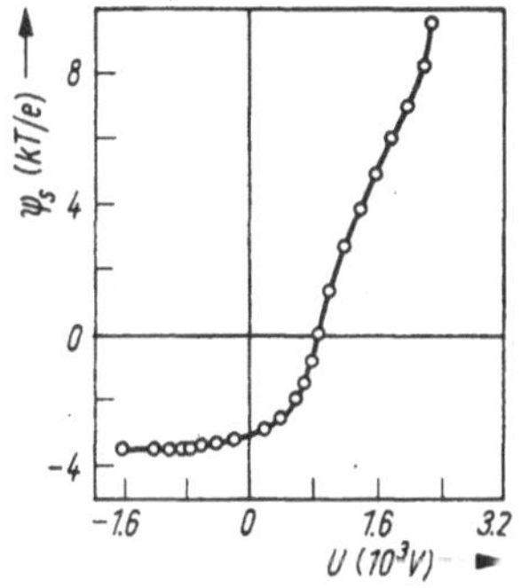

Fig. 6. Dependence of surface potential on field voltage

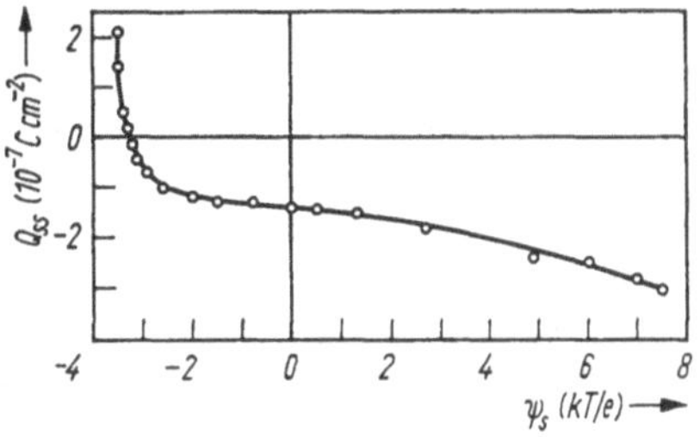

Fig. 7. Dependence of immobile surface charge on surface potential

In this way one obtains the dependence of the surface potential on the field electrode voltage as plotted in Fig. 6. For the equilibrium value of the surface potential one derives $\Phi_s = -0.08$ V. On the positive side the surface potential can be varied to a large extent. On the negative side, however, it remains practically constant, reaching the maximal value of $\Phi_s^m = -0.09$ V still at $-$ 800 V.

The charge induced in the sample was calculated from the relation $Q_i = C_g V$, where C_g is the geometric capacitance per unit area of the FE condenser and V is the field electrode voltage. This is subdivided into $Q_i = Q_{sc} + Q_{ss}$, where Q_{sc} is the charge of the SC region and Q_{ss} is the immobile charge located at the surface sites. A plot of Q_{ss} versus the surface potential is shown in Fig. 7, obtained by using (11) and the data in Fig. 6. In accordance with Fig. 7 on the positive side Q_{ss} changes slightly with the surface potential while on the negative side it changes the sign and rises very steeply.

In general, Ag^+ ions can be bound with different energies at various surface sites on the AgBr crystal. The ultimate purpose of a study like this is to determine the concentration and energy distribution of the surface states relative to some reference level. At discrete distribution of the surface states the theoretical dependence of Q_{ss} on surface potential may be evaluated using Fermi distribution functions

$$Q_{ss} = \sum_i \frac{|e|\, N_{sa}^i}{1 + \exp\left[(\Delta G_v - \mu_{Ag})/kT - \psi_s\right]} - \sum_i \frac{|e|\, N_{sd}^i}{1 + \exp\left[\psi_s - (\mu_{Ag} - \Delta G_i)/kT\right]}$$

for acceptor-like and donor-like states, respectively. By an appropriate choice of the state concentrations N_{sa}^i, N_{sd}^i and the energies ΔG_v, ΔG_i it is possible to fit the theoretical curve to the experimental points in Fig. 7. Unfortunately, the problem does not have a single-valued solution. An unambiguous model for the surface states cannot be obtained only on the basis of FE experiments. Different sets of energy and concentration values assumed lead to an acceptable coincidence with the experimental results [16].

5. Discussion

Usually the surface sites in silver halides are identified with surface kinks [26]. A kink may act as a source or a sink for the mobile interstitials. In AgBr the free energy of formation of an interstitial ΔG_i is less than the corresponding value for vacancy formation ΔG_v. Therefore, the surface kinks generate preferably interstitial Ag^+ ions building up the space charge.

As already mentioned, etching of AgBr layers proved to be of major importance for the observation of a FE signal. This seems to imply that the unetched surface contains a considerably large number of surface sites, shielding completely the bulk from external fields. On this basis Trautweiler [4] attempts to explain the observation that a strong adsorption of ions does not affect the surface potential [3]. An investigation of adsorption on etched surfaces could help to check the expectation that adsorbed species might be able to correspond directly with the SC region.

Fig. 6 and 7 illustrate an unusual peculiarity of the AgBr surface, namely that the surface potential could be varied under the effect of an external field in one direction only. This occurs at a positively biased field electrode when a field-assisted emptying of the donor-like surface sites takes place. In equilibrium, $\psi_s = -3\ kT/e$, a negative charge $Q_{ss} = -4.4 \times 10^{-8}$ C cm^{-2} is associated with surface sites corresponding to a density of the immobile vacancies $N_s = 2.7 \times 10^{11}$ cm^{-2}. From (11) one obtains $Q_{sc} = 3.9 \times 10^{-8}$ C cm^{-2} for the density of the net charge. This fair agreement implies that the negative charge in the SC region may be reasonably neglected. On applying a positive voltage at the field electrode a forced generation of mobile Ag^+ ions from occupied surface sites begins and for example if ψ_s equals zero (at about + 900 V), an addititional mobile charge $+9{,}6 \times 10^{-8}$ C cm^{-2} appears in the bulk. Nevertheless, the SC region has been depleted in majority carriers which are drawn to the contacts and hence are neutralized forming metallic silver. A complete emptying of the surface sites is probably not achieved, otherwise the curve in Fig. 7 should saturate at higher voltages.

From the data in Fig. 7 one can conclude that both donor-like and acceptor-like surface sites are present. In fact, at a negatively biased field electrode Q_{ss} equals zero at $\psi_s = -3.25\ kT/e$, when the donor-type surface sites, being empty in equilibrium, have now become occupied. Further increasing of the negative voltage results only in a change in ψ_s up to $-3.5\ kT/e$. Since $+ Q_{ss}$ continues to increase, one can conclude that acceptor-like surface sites begin to operate immobilizing the induced positive charge. This could explain why the surface potential retains the value of -0.09 V.

The data presented in Fig. 7 support the model suggested by Blakely et al. [27, 28] which assumes that in silver halides there is a finite number of surface sites partially occupied at equilibrium. Indeed, it is seen that only a small part of them is thermally emptied at normal conditions. Obviously, when induced charge appears in the subsurface region the charge equilibrium is disturbed leading to a change in the occupancy of surface sites.

The evaluated value of the surface potential $\Phi = -0.08$ V is several times smaller than the values found by other authors [4, 11, 14, 26]. However, such a value is to be expected in the present case for two reasons. First, etching substantially decreases the density of surface states which in accordance with theoretical expectations [27] leads to a decrease in the surface potential. A comparison of the density of negatively charged surface states 1.4×10^{13} cm^{-2} found in [14] corresponding to the value evaluated in the present work indicates a decrease by two orders of magnitude. Second, the enhanced concentration of majority carriers in the non-stoichiometric material studied involves in turn a reduced SC region, hence a lower surface potential.

It is taken for granted, however, that the arguments just exposed are not unambiguous, suggesting only a possible explanation of the data obtained. Further studies of the same material by different experimental techniques are needed to prove this point of view.

6. Conclusion

The present work to the best of our knowledge is the first observation of FE in ionic crystals. The theoretical treatment applied to electronic semiconductors was adapted to make possible an interpretation of the results obtained. Difficulties connected with electrode polarization, surface shielding as well as other factors inherent in studying high-Ohmic materials by this method are believed to have been successfully overcome. Information about the surface potential and surface sites is obtained for AgBr possessing by necessity an excess of stoichiometric silver. It is understood that the problems now discussed are not completely resolved. Nevertheless, the results shown allow an insight into the AgBr surface from a novel point of view. This might well be relevant for the elucidation of the role of ionic surface sites in the primary photographic process in silver halides.

Acknowledgement

Thanks are due to N. Starbov for the many helpful discussions.

References

[1] K. Lehovec, J. chem. Phys. **21**, 1123 (1953).
[2] L. Slifkin, W. McGowan, A. Fukai, and J. S. Kim, Photogr. Sci. Engng., **11**, 79 (1967).
[3] V. I. Saunders, R. W. Tyler, and W. West, Photogr. Sci. Engng. **12**, 90 (1968).
[4] F. Trautweiler, Photogr. Sci. Engng. **12**, 98 (1968).
[5] H. Wakabayasi, J. Phys. Soc. Japan **15**, 2000 (1960).
[6] S. Sonoike and H. Wakabayasi, J. Phys. Soc. Japan **14**, 1664 (1959).
[7] J. M. Blakely and S. Danyluk, Photogr. Sci. Engng. **18**, 489 (1974).
[8] S. Danyluk and J. M. Blakely, Surface Sci. **41**, 359 (1974).
[9] S. Danyluk, Ph.D. Thesis, Cornell Univ., Ithaca, N.Y. 1973.
[10] R. B. Poeppel, Ph.D. Thesis, Cornell Univ., Ithaca, N.Y. 1969.
[11] R. C. Baetzold and J. F. Hamilton, Surface Sci. **33**, 461 (1972).
[12] R. C. Baetzold, J. Phys. Chem. Solids **35**, 89 (1974).
[13] N. Starbov, A. Buroff, and J. Malinowski, J. Photogr. Sci. **23**, 44 (1975).
[14] N. Starbov, A. Buroff, and J. Malinowski, phys. stat. sol. (a) **38**, 161 (1976).
[15] See for example, H. Gatos, Lecture Notes, Internat. School of Crystal Growth, Mt Fuji District, Japan 1974.
[16] A. Many, Y. Goldstein, and N. B. Grover, Semiconductor Surfaces, North Holland Publ. 1965.
[17] C. G. B. Garrett and W. H. Brattain, Phys. Rev. **99**, 376 (1955).
[18] R. V. Kingston and S. F. Neustadter, J. appl. Phys. **26**, 718 (1955).
[19] V. Zhelev and J. Malinowski, to be published.
[20] F. A. Kröger, J. Phys. Chem. Solids **26**, 901 (1965).
[21] B. Ilschner, J. chem. Phys. **28**, 1109 (1958).
[22] G. E. Picus, Ed., Fiz. Poverhknosti Poluprovodnikov, Izd. Inostrannoj Lit. Moskva 1959.
[23] P. Müller, phys. stat. sol. **12**, 775 (1965).
[24] G. Karadjov, Dissertation, Sofia, 1969.
[25] J. Eneva and J. Malinowski, J. Photogr. Sci. **22**, 273 (1974).
[26] Y. T. Tan, Progr. Solid State Chem. **10**, 193 (1976).
[27] R. B. Poeppel and J. M. Blakely, Surface Sci. **15**, 507 (1969).
[28] J. M. Blakely and S. Danyluk, Surface Sci. **40**, 37 (1973).

(Received June 3, 1977)

phys. stat. sol. (a) 43, 685 (1977)

Subject classification: 20.3; 13.5.1; 22.3

Department of Electronics, Faculty of Engineering, Toyama University[1])

Photoluminescence in GaSe

By

T. MATSUMURA, M. SUDO, C. TATSUYAMA, and S. ICHIMURA

Photoluminescence of GaSe is measured between 4.2 and about 40 K. The spectra of many samples are classified into three types. The first type has only one broad peak at about 2.04 eV, and the second type has two sharp lines at 2.096 and 2.088 eV followed by some lines at the lower energy side. The last one contains a direct free exciton line at 2.108 eV followed by fine structure. The variety of the spectra is interpreted by the change of the relaxation rates of the excited electrons into the direct minima and into the indirect minima of the conduction band which may be perturbed by the fine cracks on the surface of the sample.

Die Photolumineszenz von GaSe wird zwischen 4,2 und etwa 40 K gemessen. Die Spektren vieler Proben werden in drei Typen klassifiziert. Der erste Typ hat nur eine einzige breite Linie bei etwa 2,04 eV, der zweite Typ zwei scharfe Linien bei 2,096 und 2,088 eV, gefolgt von einigen Linien auf der niederenergetischen Seite. Der letzte Typ enthält eine direkte Freie-Exzitonen-Linie bei 2,108 eV, gefolgt von einer Feinstruktur. Die Verschiedenheit der Spektren kann erklärt werden durch die Änderung der Relaxationsrate der angeregten Elektronen in das direkte Minimum und in das indirekte Minimum des Leitungsbandes, das gestört werden kann durch die feinen Spalten auf der Oberfläche der Probe.

1. Introduction

It has been reported by many authors [1 to 11] that luminescence spectra of GaSe at low temperatures contain many sharp lines in the near edge emission, and the spectra are considerably different from sample to sample. It may be probably due to stacking disorders in terms of the weak bonding of the Van der Waals force between layers, and due to the large stress on the surface of the sample induced by cleavage.

The energy difference between the direct and the indirect gaps of GaSe is very small [12, 13], so it has been interpreted in some different ways, whether the emission lines are to be associated with the direct minima or with the indirect minima of the conduction band.

Abdullaev et al. [3, 4] have reported about the Davydov splittings of the direct free exciton emission for both cases of radiation polarization $\boldsymbol{E} \parallel c$ and $\boldsymbol{E} \perp c$. They have treated GaSe as a molecular crystal consisting of planar quasimolecules in terms of the presence of weak bonding between the layers. But the free exciton level of GaSe may also be split by stacking disorders [10, 14, 15]. They have also reported about the phonon-assisted indirect emission.

Cingolani et al. [2] have observed a bound-to-bound emission followed by LO phonon replicas investigating photoluminescence of p- and n-type GaSe crystals.

Voitchovsky et al. [7, 8] have interpreted the dependences of emission spectra on temperature, excitation intensity, and impurity content as a result of a balance between the two recombination channels. According to them, the energy of the indirect gap of GaSe is smaller than the direct gap by about 20 meV, so the direct ex-

---

[1]) Takaoka, Toyama, Japan.

citon state with binding energy about 20 meV resonates with the bottom of the indirect conduction band. In the case of low temperatures, low excitation intensity, and high impurity content, the direct exciton emission is not observed, but for the condition of relative high temperature, high excitation intensity, and low impurity content, electrons at the bottom of the indirect conduction band are transferred resonantly into the direct free exciton state and yield the exciton line.

On the other hand, Kuroda and Nishina [9] have measured time-resolved photoluminescence spectra of $GaSe_{1-x}S_x$ and interpreted GaSe to be a direct-gap material. According to them, the bottom of the direct conduction band of GaSe lies 5 meV below the indirect one. The direct–indirect gap reversal occurs at $x \gtrapprox 0.01$. The luminescence lines have been attributed to the indirect bound excitons and their phonon replicas.

As described above the recombination kinetics of GaSe have been interpreted in some different ways.

In the present work, we report on our new results observed in our measurements on photoluminescence in GaSe. We find the competition between the intensity of the line at about 2.04 eV and of other lines on the higher energy side. The intensity of the line at 2.04 eV becomes stronger when there are many fine cracks on the surface of the sample. The competition is interpreted by the change of the energy difference between the direct and the indirect gaps induced by the presence of fine cracks. The phonon replicas in the spectra are also reported.

2. Sample Preparation and Experimental Methods

Most of crystals used in the present measurements were grown from the melt by the Bridgman method, and some samples were grown by the closed tube sublimation method. In the Bridgman method, the bottom of a quartz ample was bent according to Schmid et al. [6], and the ample was lowered slowly in a vertical furnace. When all the melt in the ample crystallized into one single crystal ingot, the c-axis of the crystal was almost parallel to the axis of the ample. In this case, the ingot was easily cleaved, and the cleaved surface was as smooth as a mirror although containing a few fine cracks on the surface. But when the crystal in the ample consisted of some grains, it was sometimes difficult to cleave. In these samples, the cleaved surface was not smooth and contained many fine cracks. In the closed tube sublimation method [16], GaSe single crystals sealed in a quartz ample were located at the high temperature side in a temperature-gradient horizontal furnace, and GaSe vapour was transported to the low-temperature side by the diffusion in the ample. In this case, single-crystal platelets were got on the wall of the ample. All crystals were found to consist of ε- and γ-modifications by the energy position corresponding to the direct free exciton absorption [12].

In measurements of photoluminescence, the samples were immersed in liquid He in a cryostat with quartz windows. The cleaved surface of the sample was excited by 365 nm line obtained by using an interference filter from a high pressure mercury lamp with power of 1 kW. The emission light from the surface was analyzed by a double grating monochromator (Jasco CT-80DS) and detected by a photomultiplier (HTV R-666). The excitation light was chopped and the output of the photomultiplier was amplified by a lock-in amplifier (PAR HR-8) and was recorded. The spectra at higher temperature beyond 4.2 K were measured on the way in which the temperature of the sample was rising after liquid He was evaporated. The rise of the temperature in a scanning of a spectrum was about 4 to 6 K. A thermocouple (Au + 2.1% Co-to-Cu) was used for the temperature measurement.

3. Experimental Results

3.1 Photoluminescence at 4.2 K

The photoluminescence spectra of many samples have been investigated at 4.2 K. The shapes of these spectra vary depending not only on the sample but also on the exciting position of the same sample. But these spectra are classified into the following three types paying attention to the main peaks in the spectra:

Type A. In the spectra of this type, the broad band with its peak at about 2.04 eV is observed as shown in Fig. 1. The samples exhibiting type A spectra are not cleaved easily, and contain many fine cracks on the cleaved surface. The energy position of the peak of this band agrees with the position of the line c_9 in the following types B and C.

Type B. The typical spectrum of this type is shown in Fig. 2a whose special feature is recognized by two sharp lines c_1 at 2.095 eV and c_2 at 2.088 eV followed by some lines c_i on the lower-energy side. Type B spectra are observed mainly in the samples obtained from the top of the ingot grown by the Bridgman method. These lines were designated as $c_1, c_2, \ldots, c_9$ from the higher-energy side by Voitchovsky and Mercier [7]. According to them, these spectra exhibiting only c_i lines are a characteristic of highly doped crystals. Type B spectra are also observed in the samples grown by the closed tube sublimation method. But, in this case, the line c_9 is not observed as shown in Fig. 3.

Type C. Fig. 4a shows a typical spectrum of this type which consists of a sharp line E at 2.108 eV, and φ_i and a_i lines named also by Voitchovsky and Mercier [7]. Type C spectra are observed mainly in the samples obtained from the bottom of the ingot. According to Voitchovsky and Mercier [7], these spectra showing the lines E,

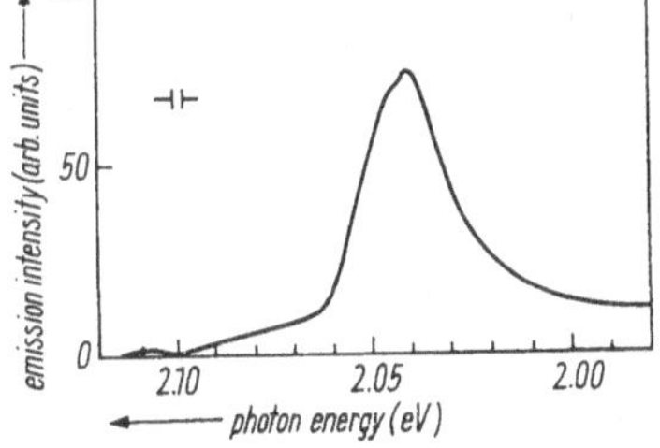

Fig. 1

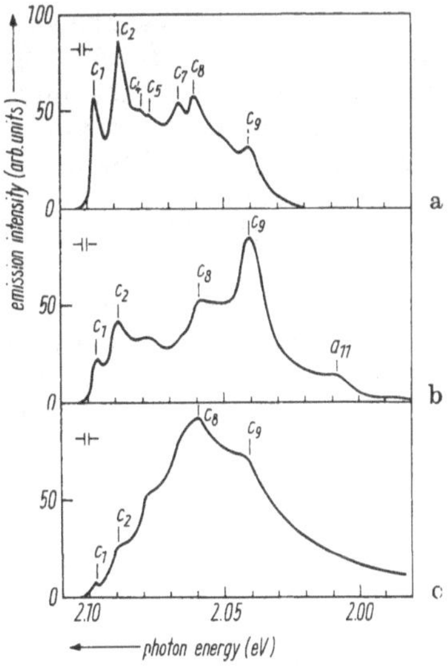

Fig. 2

Fig. 1. Photoluminescence spectrum of type A at 4.2 K. This type is observed in the samples containing many fine cracks on all surface

Fig. 2. Photoluminescence spectra of type B at 4.2 K. This type is observed mainly in the samples obtained from the top of the ingot grown by the Bridgman method. The spectra of a), b), and c) were measured in different samples. The lines c_1 and c_2 decrease with an increase in the line c_9

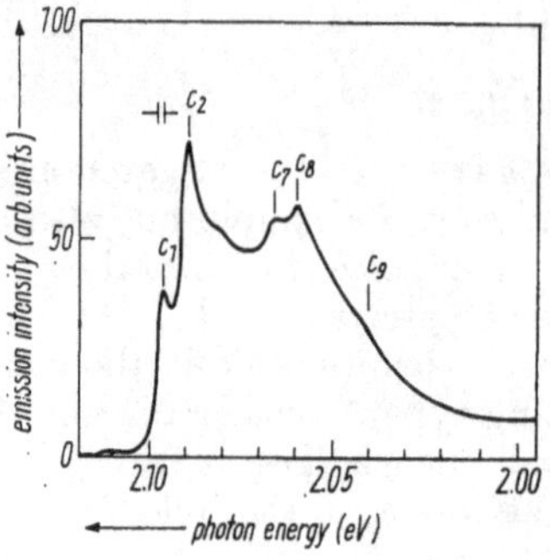

Fig. 3. The samples grown by the closed tube sublimation method show type B spectrum at 4.2 K, but the line c_9 is not observed

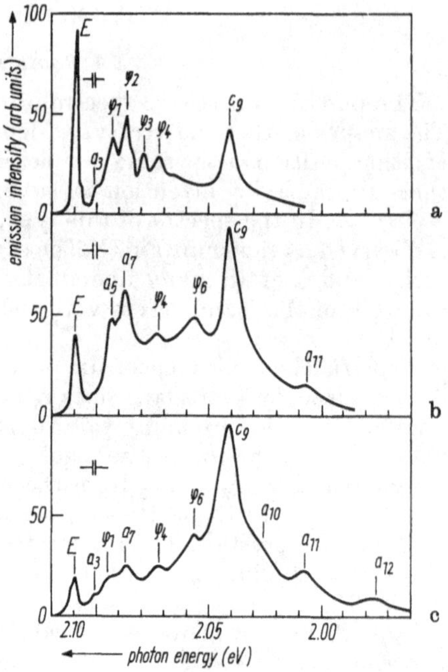

Fig. 4. Photoluminescence spectra of type C at 4.2 K. This type is observed mainly in the samples obtained from the bottom of the ingot grown by the Bridgman method. The spectra of a), b), and c) were obtained in the same sample, but by changing the exciting position on the surface of the sample. The high energy lines decrease with an increase in the line c_9

φ_i, and a_i are peculiar to non-doped pure crystals. The line E is due to the direct free exciton emission.

Type A spectra are almost the same in the samples containing many fine cracks on the surface. But, in type B and C spectra, the relative intensities between the lines vary drastically depending not only on the sample but also on the exciting position of the same sample. This behaviour for type B is shown in Fig. 2a, b, and c which have been obtained from different samples. In the spectrum of a, the intensities of the lines c_1 and c_2 are strong, and the intensity of the line c_9 is weak. In the spectra of b and c, on the other hand the intensities of the high-energy lines such as c_1 and c_2 become weak with an increase in the intensity of the line c_9. The similar behaviour for type C spectra is shown in Fig. 4 a, b, and c which have been obtained from the same sample changing the exciting position. In Fig. 5 a photograph of the surface of the sample of Fig. 4 is shown. We call the black parts shown by an arrow the fine cracks which may be induced by cleavage. The spectrum of a is obtained from the

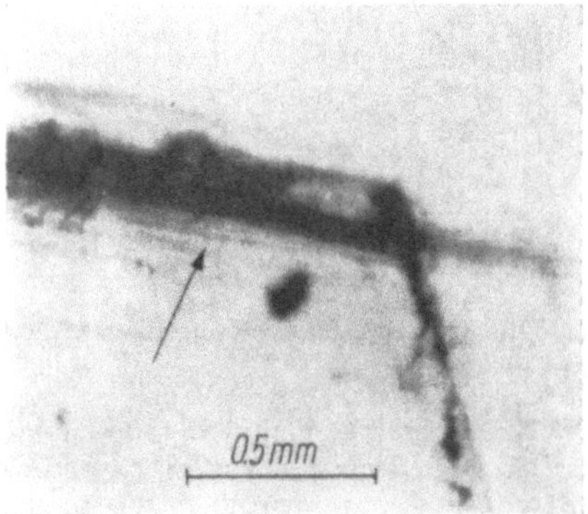

Fig. 5. The photograph of the surface of the sample of Fig. 4. The black part shown by an arrow is called fine cracks in this report. The Fig. 4c spectrum is obtained by exciting the position with more fine cracks than a) and b)

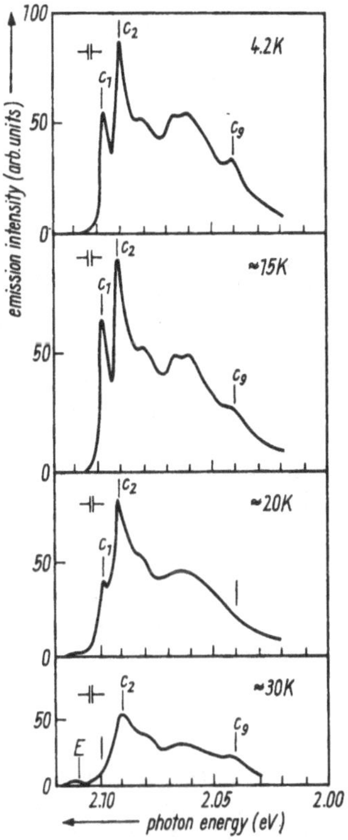

Fig. 6. The temperature dependence of type B spectrum. In this case, the line c_9 above about 15 K may be a two-phonon replica of the line c_1 (see text)

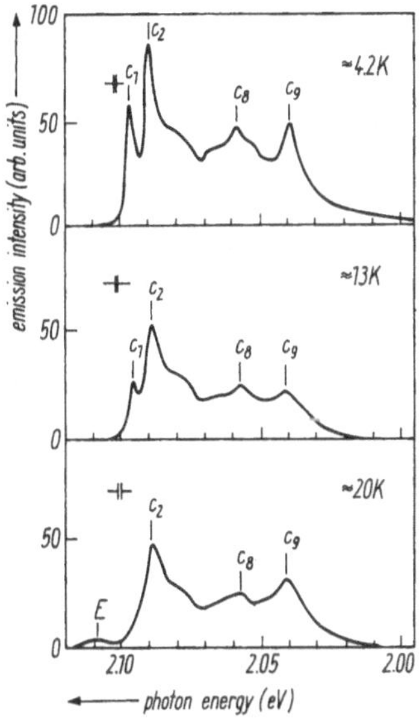

Fig. 7. The temperature dependence of type B spectrum, but in this case, the line c_9 does not disappear with an increase in temperature, although the line c_9 is quenched

position without fine cracks, while those of a and b are obtained by exciting the positions containing the fine cracks.

The change of the spectra of types A, B, and C suggests the competition between the line c_9 and other lines at the higher-energy side. The line c_9 become larger in the samples containing many fine cracks on the surface, and it is not observed in the samples grown by the closed tube sublimation method which need not to cleave. From these results, the origin of the line c_9 may be attributed to a radiative recombination through the level related to the fine cracks.

The reason that type B spectra are observed in samples obtained from the top of the ingot, while type C spectra in samples obtained from the bottom of the ingot may be as follows. In the case of crystal growth by the Bridgman method, the impurity content in the melt is higher at the upper side of the ample, since the crystallization begins firstly from the bottom and ends at the top of the ample, so highly doped crystal will be obtained at the top of the ample.

As shown in Fig. 4c, when the line c_i is strongly observed, the lines a_{11} and a_{12} are also detected. They are situated about 32 and 64 meV below the line c_9 at 2.041 eV, respectively. Furthermore, the line a_{10} is observed as a shoulder of the line c_9.

3.2 Temperature dependences of spectra

Temperature dependences of type B spectra are shown in Fig. 6 and Fig. 7. They have been obtained from the same sample, but from the different positions on the

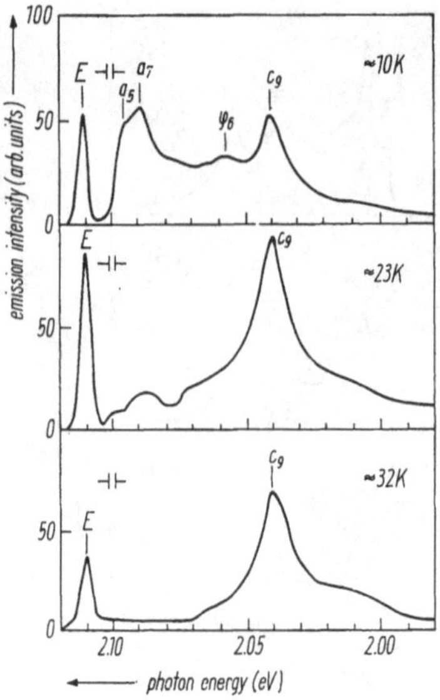

Fig. 8. The temperature dependence of type C spectrum, where a spectrum at 4.2 K is shown in Fig. 4b. The lines between the line E and the line c_9 decrease rapidly with an increase in temperature

surface. The behaviour of the line c_9 is different in both cases. In Fig. 6, the line c_9 disappears at about 20 K with a decrease of the line c_1. At about 30 K, the line c_1 cannot be seen, but the new line also named c_9 appears at the same energy position as the line c_9, and the direct free exciton line E becomes to be seen. In Fig. 7, the line c_9 is clearly seen, and it does not disappear in spite of the distinction of the line c_1 with an increase in temperature. The line E becomes to be detected at the temperature where the line c_1 disappears. The line c_9 in Fig. 7 and the new line c_9 at about 30 K in Fig. 6 may be ascribed to the fine cracks, but the line c_9 below 15 K in Fig. 6 may be probably due to a phonon replica of the line c_1, although it is situated at the same energy position as the line c_9 in other spectra.

In Fig. 8, temperature dependences of type C spectra are shown. The spectrum at 4.2 K of this sample has already been shown in Fig. 4b. The lines between the line E and the line c_9 decrease with an increase in temperature and disappear at about 30 K. At this temperature, only the lines E and c_9 are present.

In the above descriptions, the energy positions of the lines c_i, a_i, and φ_i have been determined by the reconstructions of the experimental spectra by the superposition of Gaussians. They almost agree with Voitchovsky and Mercier [7]. The line a_{12} has not been reported in [7].

4. Discussions

4.1 Model for the recombination processes

The common feature observed in many spectra is the competition between the line c_9 and the other lines. In the preceding chapter, the origin of the line c_9 has been attributed to the fine cracks on the surface induced by the cleavage of the sample. The increase of the fine cracks not only makes the density of the level exhibiting the line c_9 increase, but also may perturb the band itself. The energy difference between the direct and the indirect gaps of GaSe is very small [12, 13], so if the degrees of perturbation are different in both gaps, the relaxation rates of excited electrons into both direct and indirect minima should change remarkably depending on whether the fine cracks are present or not.

Besson et al. [17] have measured the pressure dependence of the direct free exciton absorption of GaSe. According to them, the indirect gap decreases faster than the direct gap by three times with an increase in the hydrostatic pressure. Therefore, if the perturbation by the fine cracks is similar to the hydrostatic pressure, the relaxation rate into the indirect minimum should increase with an increase in the density of the fine crack.

Fig. 9. A proposed model for the recombination kinetics in GaSe, where c_9 represents a level related to fine cracks, c_i and φ_i levels related to structural defects, DFE a direct exciton state, IFE an indirect exciton state, VB a valence band and A an acceptor level. In both direct (DCB) and indirect (ICB) conduction bands, solid lines represent a band for a sample without fine crack, and broken lines for a sample with many fine cracks and impurities. J is the total electrons relaxed into both minima of the conduction band. (A part of this figure is drawn after Voitchovsky and Mercier [7])

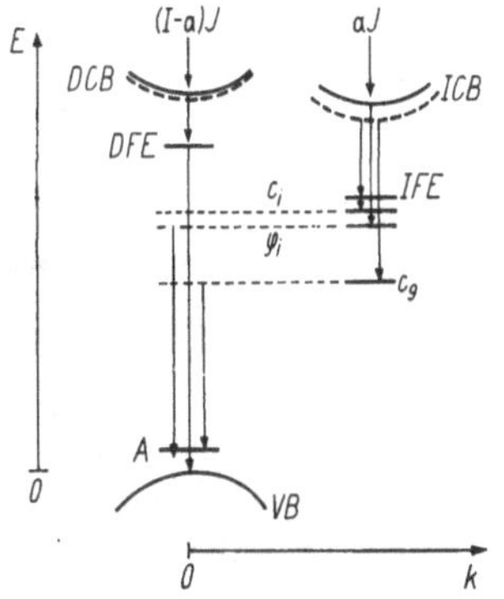

From these considerations, we may construct a model for the recombination kinetics in GaSe as shown in Fig. 9. This model is based on the assumptions that GaSe is an indirect gap semiconductor, but the energy difference between direct and indirect conduction minima is very small, probably a few meV. According to Schlüter [18], the indirect conduction minimum of GaSe lies at the M point in the Brillouin zone. In Fig. 9, c_9 represents the level related to the fine cracks, and c_i and φ_i represent unknown levels, probably related to structural defects and impurities. The conduction band written by a solid line represents for the sample without fine cracks, and the broken line represents for the sample with many fine cracks.

For the samples pure and without fine cracks, the level c_9 does not exist. In this case, most of excited electrons αJ may be relaxed into the indirect minimum, but some electrons $(1 - \alpha)\, J$ into the direct minimum, since the energy difference between both minima is very small. The electrons relaxed into the indirect minimum may recombine through the level φ_i. According to Voitchovsky and Mercier [19], the lifetime of the lines φ_i is very long, a few μs, and the associated phonon energy for the momentum conservation is about 3.4 meV. The electrons relaxed into the direct minimum form the direct exciton (DFE) and yield the exciton line E. For the samples containing many fine cracks, even if they are pure, the band may be perturbed [20] as written by broken line in Fig. 9, so the energy difference between direct and indirect minima may increase. As a result, the value of α may become larger in comparison with the samples without fine cracks, and the direct free exciton line E may decrease. The electrons relaxed into the indirect minimum recombine with holes, bound to the acceptor level (A) through the level c_9, emitting the line c_9. In this way the competition between the line c_9 and other lines in Fig. 4 a, b, and c may be interpreted.

For the cases of type B spectra, the interpretation is as follows. According to Schmid et al. [6], if the spectra contain only c_i lines, the impurity concentration in the crystals is more than 10^{19} cm^{-3}. Further the spectra are not affected by the chemical nature of the impurity, but are affected only by the degree of the impurity concentration. Therefore, they attributed the origin of the lines c_i to the structural defects induced by doping of impurity. But, if the crystal contains so much impurities, the band may also be perturbed [20, 21], and the value of α may become larger. In this case, most of the excited electrons may be relaxed into the indirect minimum, so that the direct exciton emission is not observed. The lines c_1 and c_2 may be ascribed to zero-phonon lines associated with the indirect bound excitons [22] as shown by Kuroda and Nishina [9]. If the crystal contains many fine cracks, the density of the level c_9 becomes larger, so the electrons at c_1 and c_2 may be trapped at the level c_9.

As a result, the intensities of the lines c_1 and c_2 may decrease with an increase in the line c_9.

As the temperature of the sample becomes higher, the population rate in the direct minimum increases, and the direct free exciton line E may become to appear, even if it is not observed at 4.2 K. On the other hand, the lines c_i, φ_i, and a_i may be thermally excited with an increase in temperature.

4.2 Phonon replicas

The emission lines situated at the same energy position as the line c_9 have been reported by some authors. Abdullaev et al. [4] have observed the line P whose energy position corresponds to the line c_9, and they assigned it to an indirect transition with one phonon emission with energy 38 meV. But it is not clear whether a phonon with energy 38 meV is present or not at the M point [23]. The line C_1 reported by Cingolani et al. [2] is situated at the same energy position, and they attributed this line to a bound-to-bound recombination followed by one and two LO phonon replicas C_2 and C_3. We have also reported the large emission line P_3 at about this energy position in the electroluminescence spectra using GaSe–SnO_2 heterojunction, and assigned it to a donor–acceptor pair recombination [11].

From the present experiments the line c_9 may be due to the recombination through the level c_9 induced by the fine cracks on the surface. But, it is not clear at the present time whether the level c_9 is merely a trapp level or a donor level. In either case, the momentum conservation may be relaxed by the presence of the fine cracks [20, 21], and a zero-phonon line c_9 may be observed.

As shown in Fig. 4c, when the line c_9 is observed strongly the lines a_{10}, a_{11}, and a_{12} are also observed at 17, 32, and 64 meV below the line c_9, respectively. The lines a_{11} and a_{12} may be due to one and two LO phonon replicas of the line c_9, respectively. In GaSe, there are two polar LO phonons. They are E′(LO) with energy 31.6 meV and A_2''(LO) with energy 30.6 meV [24 to 26]. The values of these phonon energy are very close, so it is not clear which phonon plays a dominant role for a_{11} and a_{12}. The line a_{10} may be attributed to one A_1'-LO phonon replica of the line c_9, since whose phonon energy is about 17 meV [24, 25, 27].

The line c_9 at above 15 K in Fig. 6 may be a phonon replica of the line c_1. The energy difference between the lines c_9 and c_1 is about 54 meV. The phonon with energy as large as 54 meV is not present in GaSe [23], so two phonons may assist. If we suppose a dipole transition, the assisted two phonons may be a M_1-phonon with energy 18 meV at the M point [23] and an A_1'-phonon with energy 38 meV at the Γ point [25] for ε-GaSe [28]. But if the crystal contains ε–γ mixtures, a M_3-phonon with energy 18 meV at the M point [23] may also be possible [28].

References

[1] A. CINGOLANI, A. MINAFRA, P. TANTALO, and C. PAORICI, phys. stat. sol. (a) **4**, K83 (1971).
[2] A. CINGOLANI, F. EVANGELISTI, A. MINAFRA, and A. RIZZO, phys. stat. sol. (a) **17**, 541 (1973).
[3] G. B. ABDULLAEV, G. L. BELENKII, KH. KHALILOV, and E. YU. SALAEV, Soviet Phys. – Semicond. 7, 559 (1973).
[4] G. B. ABDULLAEV, G. L. BELENKII, E. YU. SALAEV, R. A. SULEIMANOV, and V. KH. KHALILOV, Soviet Phys. – Solid State **16**, 11 (1974).
[5] A. MERCIER, E. MOOSER, and J. P. VOITCHOVSKY, J. Lum. 7, 241 (1973).
[6] PH. SCHMID, J. P. VOITCHOVSKY, and A. MERCIER, phys. stat. sol. (a) **21**, 443 (1974).
[7] J. P. VOITCHOVSKY and A. MERCIER, Nuovo Cimento **22B**, 273 (1974).
[8] A. MERCIER, E. MOOSER, and J. P. VOITCHOVSKY, Phys. Rev. B 12, 4307 (1975).
[9] N. KURODA and Y. NISHINA, phys. stat. sol. (b) **72**, 81 (1975).
[10] N. KURODA and Y. NISHINA, Nuovo Cimento **32B**, 109 (1976).

[11] C. Tatsuyama and S. Ichimura, Nuovo Cimento **38B**, 352 (1977).
[12] E. Aulich, J. L. Brebner, and E. Mooser, phys. stat. sol. **31**, 129 (1969).
[13] R. E. Nahory, K. L. Shaklee, R. F. Lehny, and J. C. DeWinter, Solid State Commun. **9**, 1107 (1971).
[14] J. L. Brebner and E. Mooser, Phys. Letters **A24**, 274 (1967).
[15] J. J. Forney, K. Maschke, and E. Mooser, Nuovo Cimento, **38B**, 418 (1977).
[16] V. L. Cardetta, A. M. Mancini, C. Manfredotti, and A. Rizzo, J. Crystal Growth **17**, 155 (1972).
[17] J. M. Besson, K. P. Jain, and A. Kuhn, Phys. Rev. Letters **32**, 936 (1974).
[18] M. Schluter, Nuovo Cimento **13B**, 313 (1973).
[19] J. P. Voitchovsky and A. Mercier, phys. stat. sol. (a) **18**, 545 (1973).
[20] J. I. Pankove, Optical Processes in Semiconductors, Chap. 1, Prentice-Hall, Inc., Englewood Cliffs, N.J. 1971.
[21] J. I. Pankove and P. Aigrain, Phys. Rev. **126**, 956 (1962).
[22] D. G. Thomas, M. Gerchenson, and J. J. Hopfield, Phys. Rev. **131**, 2397 (1963).
[23] S. Jandl, J. L. Brebner, and B. M. Powell, Phys. Rev. **B13**, 686 (1976).
[24] H. Yoshida, S. Nakashima, and A. Mitsuishi, phys. stat. sol. (b) **59**, 655 (1973).
[25] R. M. Hoff and J. C. Irwin, Phys. Rev. **B10**, 3464 (1974).
[26] T. C. Chiang, J. Camassel, Y. R. Shen, and J. P. Voitchovsky, Solid State Commun. **19**, 157 (1976).
[27] J. L. Brebner, S. Jandl, and B. M. Powell, Solid State Commun. **13**, 1555 (1973).
[28] C. Tatsuyama and S. Ichimura, to be published.

(Received July 8, 1977)

phys. stat. sol. (a) 43, 695 (1977)

Subject classification: 20.3; 22.9

Department of Structural Chemistry, Weizmann Institute of Sciences, Rehovot

A Metastable Excimer-Emitting Crystal Form of 9-Cyanoanthracene

By

Z. LUDMER

A new metastable crystal form of 9-cyanoanthracene showing a red excimer emission is described. The structure of this excimer is attributed a centric geometry.

Es wird eine neue metastabile Kristallform von 9-Zyanoanthrazen beschrieben, die eine rote Exzimer-Emission zeigt. Der Struktur dieses Exzimer wird eine zentrische Geometrie zugeschrieben.

1. Introduction

Several recent publications have demonstrated that a given compound can give rise to more than one excimer. Thus, Cohen et al. [1] showed that the two polymorphic forms of 9-chloroanthracene give rise to excimer emissions differing in their spectral distributions and sensitivity to temperature. Similar observations have been reported [2] for the two polymorphic forms of 2,4-dichloro-3′-methyl-trans-stilbene. Clark et al. [3] have found different excimer emissions from several crystal forms of acridine. Calculations on the excimer in perylene [4] suggest that the differences between such polymorphic forms are due to the different constraints on the excimer applied by the different structures.

Other constraints, too, can modify the properties of the excimer. Thus Froines and Hagerman [5] found that the syn- and anti-[2,2](1,4)naphthalenophanes both give excimer emission, but with different spectral distributions. Similar observations have been made [6 to 9] for various anthracenophanes.

Finally, Ferguson et al. [10, 11] report that the excimer emission from trapped pairs of anthracenes can vary with the conditions (wavelength, temperature, matrix) under which the pair is produced by photosplitting of the photodimer.

These examples show that the structure of the excimer, and therefore its properties, can be constrained by such factors as the structure of the crystal or other matrix, or by the covalent bonding between the components of the pair.

Considerable interest attaches to this phenomenon both because of the information it gives us about the excimer, and because of the increasing evidence that in many systems [2, 11] there is an intimate relationship between the structures of the excimer and the ease with which it can yield the photodimer. The excimer-emitting 9-cyanoanthracene crystal has been a central material in studies of solid-state photodimerization [12 to 16]. We therefore consider it desirable to report our observation of a second, metastable, form of 9-cyanoanthracene, and to describe some of its properties.

2. Experimental

The photodimer of 9-cyanoanthracene precipitated when a degassed ethanolic solution of the monomer was irradiated with long-wavelength UV. The precipitate was repeatedly washed with pure solvents until the wash fluid showed no monomer fluorescence.

Pure monomer was obtained from the cleaned photodimer as follows: the photodimer was dried and sealed off under vacuum, and then heated at 140 °C for several hours. This resulted in complete monomerization, the transformation being a single crystal to a single crystal one [17].

Sandwich pairs were prepared as described by Chandross and Ferguson [18] in a glass of ethanol/methanol (1/4) at 90 K. Solvents were spectrograde and were checked for absence of fluorescent impurities. Emission spectra were determined on an Aminco-Bowman spectrophotofluorometer, modified as previously described [19], with the samples being held in 3 mm cylindrical pyrex cells in a quartz dewar. Lifetimes were measured by means of a pulsed Molectron 500 kW nitrogen laser and a fast oscilloscope (Tektronix 7704).

3. Results

A near-saturated solution of pure 9-cyanoanthracene in chloroform was immersed rapidly in liquid nitrogen. The specimen became opaque. Under illumination with long-wavelength UV it gave two emissions, one green and one red, originating from different regions of the sample. In most cases these regions were so small that the spectra measured were a superposition of the two spectra (Fig. 1, curve 1); however, on occasion it was possible to obtain large enough areas ($\approx 0.2\ cm^2$) of red emission so that this could be measured on its own (Fig. 1, curve 2).

On warming the sample, first the emission in the red was lost and subsequently the green emission had vanished too, and only monomer emission was observed.

Slow cooling of the concentrated solutions gave a precipitate whose emission was found to be identical with that of the cyanoanthracene in its known crystal structure at the same temperature ($\lambda_{max} = 560$ nm at 77 K). On the other hand, quenching experiments with dilute solutions led to systems showing monomer emission only.

The emission of trapped sandwich pairs is shown, for comparison, in Fig. 2. There are thus established three long-wavelength emissions of 9-cyanoanthracene, having λ_{max} (77 K) at 505 nm (split pair), 560 nm (known crystal), and 580 nm (quenched concentrated chloroform solution). These have radiative decay times of, respectively, 170 [20], 160, and ca. 140 ns.

The red-emitting species does not undergo rapid photodimerization as does the split pair [20].

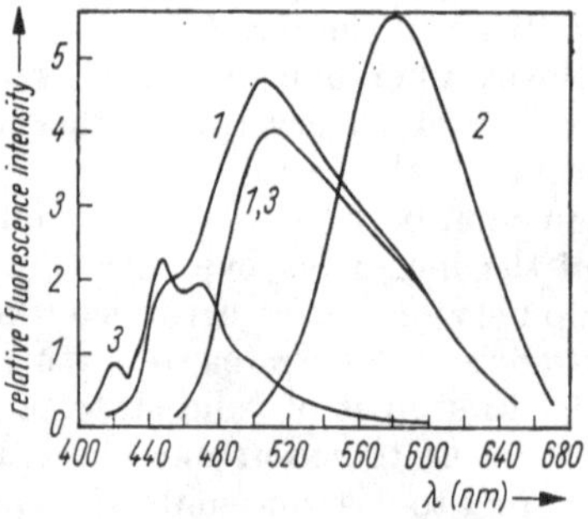

Fig. 1

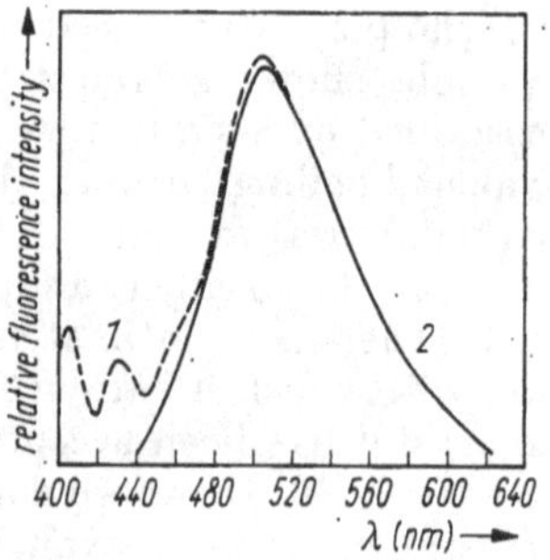

Fig. 2

Fig. 1. Emission spectra of chloroform solution of 9-cyanoanthracene: (1, 2) two zones of same frozen solution at 77 K; (3) thawed solution (read 1–3 instead of 1, 3)

Fig. 2. Emission spectrum of sandwich pair (produced by photolysis of dimer of 9-CNA in ethanol/methanol (1/4) at 77 K). Dashed curve: observed spectrum; full curve: observed spectrum minus monomer contribution

4. Discussion

We assign the red ($\lambda_{max} = 580$ nm) emission to an excimer of 9-cyanoanthracene, different from the two previously recognized, This assignment is based on the diffuseness and long radiative lifetime of the emission, Further, we find other possible assignments unacceptable. Thus, it cannont be due to a direct interaction with the solvent, since this emission is absent at low concentrations of 9-cyanoanthracene. Nor can it be attributed to an impurity because the emission emanates only from certain domains in the system.

Further, again because this red emission is localized in certain domains, we conclude that it derives from a metastable crystal form of the 9-cyanoanthracene. We can also suggest what the structure of this crystal form must be.

We have previously [1] shown that all known excimer-emitting anthracenes pack with adjacent molecules strongly overlapped, their planes parallel, and in "pair" or "stack" structures (Fig. 3). Further, by analysis of the temperature dependence of the emissions from these types of structure we concluded that

$$\frac{\omega_g}{\omega_e}\,(\text{pair}) > \frac{\omega_g}{\omega_e}\,(\text{stack}),$$

where ω_g and ω_e are the intrapair frequencies in ground and excited states, respectively. Since for a varity of excimers in both types of structure $\omega_e \approx 100\ \text{cm}^{-1}$, it follows that $\omega_g(\text{pair}) > \omega_g(\text{stack})$. Thus at the equilibrium ground-state geometry the potential energy surface is steeper for the pair than for the stack structures. And, since the fluorescence excitation spectra are similar for the two, the excimer emission for a given molecule lies at higher energies in stack than in pair structures (Fig. 3).

In keeping with this prediction we have found $E(\text{stack}) - E(\text{pair})$ to be 800 cm^{-1} for the polymorphic forms of 9-chloroanthracene, and to be 1000 cm^{-1} for those of (the non-anthracene) 2,4-dichloro-3′-methyl-trans-stilbene.

The known crystal form of 9-cyanoanthracene is of stack structure. Since the emission from this form is of energy about 700 cm^{-1} higher than that of the red

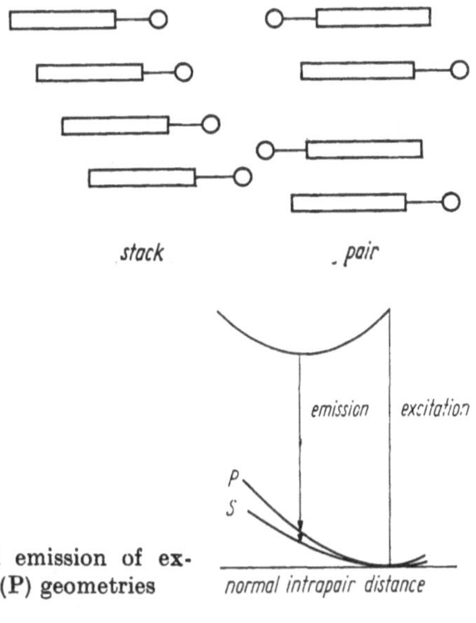

Fig. 3. Schematic representation of excitation and emission of excimer emitting crystals with stack (S) and pair (P) geometries

emission originating from the metastable crystal form, we conclude that the latter is of pair structure.

Ferguson and Miller [11] have reported that the photosplitting of the solid photodimer of 9-cyanoanthracene at 6 K gives monomer pairs showing no detectable emission, but undergoing re-photodimerization with unit quantum yield. The authors propose that the monomers in the pair lie antiparallel and with the anthracene skeletons fully eclipsed. Such a geometry is very unlikely for our pair structure; indeed, the photochemical stability of the red-emitting crystal form at 77 K indicates that the "paired" monomers are somewhat offset, as is found in other known pair structures [21]. This conclusion is in agreement with the interpretations of Ferguson et al.

References

[1] M. D. COHEN, Z. LUDMER, and V. YAKHOT, phys. stat. sol. (b) **67**, 51 (1975).
[2] R. COHEN, Z. LUDMER, and V. YAKHOT, Chem. Phys. Letters **34**, 271 (1975).
[3] B. P. CLARK, J. M. THOMAS, and J. O. WILLIAMS, Chem. Phys. Letters **35**, 251 (1975).
[4] M. D. COHEN, R. HABERKORN, E. HULER, Z. LUDMER, M. E. MICHEL-BEYERLE, D. RABINOVICH, R. SHARON, A. WARSHEL, and V. YAKHOT, J. chem. Phys. 1977, in the press.
[5] J. R. FROINES and P. HAGERMAN, Chem. Phys. Letters **4**, 135 (1969).
[6] T. HAYASHI, T. SUZUKI, N. MATAGA, Y. SAKATA, and S. MISUMI, Chem. Phys. Letters **38**, 599 (1976).
[7] T. HAYASHI, N. MATAGA, Y. SAKATA, S. MISUMI, M. MORITA, and J. TANAKA, J. Amer. Chem. Soc. **98**, 5910 (1976).
[8] J. FERGUSON, M. MORITA, and M. PUZA, Chem. Phys. Letters **42**, 288 (1976).
[9] F. C. DE SCHRYVER, J. HUYBRECHTS, N. BOENS, J. C. DEDEREN, M. IRIE, and K. ZACHARIASSE, X. Symp. Excited States in Organic Chemistry and Biochemistry, Jerusalem 1977.
[10] J. FERGUSON and A. W. H. MAU, Mol. Phys. **27**, 377 (1974).
[11] J. FERGUSON and S. F. H. MILLER, Chem. Phys. Letters **36**, 635 (1975).
[12] B. STEVENS, T. DICKINSON, and R. R. SHARPE, Nature (London) **204**, 876 (1964).
[13] H. BOUAS-LAURENT, R. LAPOUYADE, and J.-G. FAUGERE, C. R. Acad. Sci. (France) **256C**, 508 (1967).
[14] E. HELLER and G. M. J. SCHMIDT, Israel J. Chem. **9**, 449 (1971).
[15] D. P. CRAIG and P. SARTI-FANTONI, Chem. Commun., 742 (1966).
[16] M. D. COHEN, Z. LUDMER, J. M. THOMAS, and W. O. WILLIAMS, Proc. Roy. Soc. **A324**, 459 (1971).
[17] Z. LUDMER, to be published.
[18] E. A. CHANDROSS and J. FERGUSON, J. chem. Phys. **43**, 4175 (1965).
[19] M. D. COHEN, A. ELGAVI, B. S. GREEN, Z. LUDMER, and G. M. J. SCHMIDT, J. Amer. Chem. Soc. **94**, 6776 (1972).
[20] J. FERGUSON, A. W. H. MAU, and J. M. MORRIS, Austral. J. Chem. **26**, 91 (1973).
[21] J. C. J. BART and G. M. J. SCHMIDT, Israel J. Chem. **9**, 429 (1971).

(Received June 30, 1977)

phys. stat. sol. (a) 43, 699 (1977)

Subject classification: 7 and 14.3.2; 22.4.1

V. Kapsukas State University, Vilnius

Influence of the Surface Potential on the Acoustoelectric Interaction in Semiconductor–Piezoelectric Structures

By

V. KUNIGĖLIS, A. JUCYS, E. GARŠKA, and V. VITKAUSKAS

The influence of a semiconductor surface potential on the transverse accustoelectric field in a piezoelectric–semiconductor layer system is investigated both, theoretically and experimentally. It is shown that in the case of accumulation layer the sign of the transverse acoustoelectric field depends on the conductivity of the semiconductor. This is confirmed experimentally using CdS samples with surfaces treated in a special way to create the accumulation layer.

Теоретически и экспериментально исследовано влияние поверхностного потенциала полупроводника на поперечное акустоэлектрическое поле в слоистой структуре и показано, что оно может менять знак в зависимости от проводимости полупроводника, если потенциал такой, что поверхность обогащается электронами. Экспериментально это было подтверждено специальной обработкой поверхности образцов CdS созданием обогощающего потенциала.

1. Introduction

The influence of a surface potential on the acoustoelectric interaction was investigated earlier [1 to 4]. However, previous investigations both experimental and theoretical where mainly concerned with the attenuation of surface acoustic waves (SAW). In this paper we have calculated both the SAW attenuation constant and the transverse acaustoelectric field (TAE). The latter is shown to have a much stronger dependence upon the surface potential, and the change of the TAE sign may be well explained by the bending of the energy bands at the surface. This change of sign was treated previously in different ways [5, 6].

2. Theory

Here we shall deal with a structure consisting of a piezoelectric substrate and a semiconductor of certain thickness upon it. The SAW propagation on a surface of the piezoelectric substrate in x-direction creates an alternating potential which penetrates into the semiconductor. The potential is assumed to have the form $\varphi \sim e^{-\beta y + i(kx - \omega t)}$. The current caused by this potential in an n-type semiconductor is

$$\left.\begin{aligned} j_x &= -e\mu n_0(y)\,\frac{\partial \varphi}{\partial x} + e\mu n_s E_{0x} + eD_n\,\frac{\partial n_s}{\partial x}\,,\\ j_y &= -e\mu n_0(y)\,\frac{\partial \varphi}{\partial y} + e\mu n_s E_{0y} + eD_n\,\frac{\partial n_s}{\partial y}\,,\end{aligned}\right\} \tag{1}$$

where e is the electron charge, μ the carrier mobility, D_n the diffusion constant, n_0 the free electron density which depends on the surface potential, n_s the bunched charge, E_{0x} the external dc field, E_{0y} the surface field connected with the surface potential $\varphi_0(y)$. The dispersion relation for the coefficient $\varkappa = \beta/k$ is obtained from (1) and Poisson's equations

$$(1 - \varkappa^2)\left[1 + i\omega\tau\gamma + \omega\tau \frac{V_{dy}}{V_s}(\varkappa - a) + k^2 l_D^2 (1 - \varkappa^2)\right] + \varkappa b = 0\,, \tag{2}$$

where

$$b = \frac{1}{k\sigma}\frac{\partial\sigma}{\partial y}, \qquad \sigma = e\mu n_0(y)\,, \qquad a = \frac{1}{kV_s}\frac{\partial V_{dy}}{\partial y}, \qquad \gamma = 1 \pm \frac{\mu E_{0x}}{V_s},$$

$l_D = \sqrt{D_n\tau}$ is the Debye screening length, $\tau = \varepsilon_s/\sigma$ is the Maxwell relaxation time, and $V_{dy} = \mu\, \partial\varphi_0(y)/\partial y$. Positive values of V_{dy} correspond to electron accumulation on the crystal surface. When $V_{dy} = 0$ we obtain the solution for the case of a flat band

$$\varkappa_{01} = -\varkappa_{03} = 1\,, \qquad \varkappa_{02} = -\varkappa_{04} = \left(1 + \frac{1 + i\omega\tau\gamma}{k^2 l_D^2}\right)^{1/2},$$

which coincides with the results of other authors. Generally, the solution of the equation is rather complicated, so we have solved (2) in the case of a weak surface potential, such as $V_{dy}/V_s \ll 1$. Thus, neglecting terms of high order, we obtain

$$\varkappa_1 = -\varkappa_3 = 1\,, \qquad \varkappa_2 = \varkappa_{02} + \frac{\omega\tau}{2k^2 l_D^2}\frac{V_{dy}}{V_s}, \qquad \varkappa_4 = \varkappa_{04} + \frac{\omega\tau}{2k^2 l_D^2}\frac{V_{dy}}{V_s}.$$

Finally, the potential and the bunched charge are obtained in the form

$$\varphi = \sum_{i=1}^{4} A_i\, \mathrm{e}^{-\beta_i y + i(kx - \omega t)}$$

and

$$en_s = \sum_{i=1}^{4} \varrho_i\, \mathrm{e}^{-\beta_i y + i(kx - \omega t)}\,,$$

where $\varrho_i = -\varepsilon_s(k^2 - \beta_i^2)\,A_i$. The coefficient A_i can be found from the boundary conditions

$$\varepsilon_s \frac{\partial\varphi}{\partial y} = -\varepsilon_0 k\varphi \quad \text{at} \quad y = d\,,$$

$$j_y = 0 \qquad \text{at} \quad y = 0, d\,.$$

For sufficiently high conductivities $kl_D \ll 1$ and arbitrary values of $\omega\tau$ even in the case of a thick semiconductor, e.g. $kd \gg 1$, the ratio of amplitudes A_1 and A_2 is

$$\frac{A_1}{A_2} = i\omega\tau\varkappa_2\,,$$

and A_3, $A_4 \to 0$. The transverse acoustoelectric field is calculated using the formula

$$E_y^0 = \frac{\mu}{2\sigma d}\int_0^d \mathrm{Re}\,[en_s E_y^*]\,\mathrm{d}y\,,$$

and can be expressed as

$$E_y^0 = \frac{k^2\tau\mu}{2d}\,|\varkappa_2|^2\,|A_2|^2\left(\frac{1}{2} - \frac{\omega\tau}{k^3 l_D^3\,|\varkappa_2|^2}\frac{V_{dy}}{V_s}\right). \tag{3}$$

The basic conclusion which follows from (3) is that under influence of the surface potential the transverse AE field can change its polarity. According to [7] the absorption coefficient of acoustic waves is

$$\alpha = -\frac{K^2}{2} k \,\mathrm{Im} \frac{V_s Z}{\dfrac{i}{\varepsilon_p} + V_s Z},$$

where Z is the impedance of the semiconductor and ε_p the permitivity of the piezoelectric. Now we can find the absorption coefficient of acoustic waves which is equal to

$$\alpha = -\frac{K^2}{2} k \frac{\omega\tau\gamma \dfrac{\varepsilon_s}{\varepsilon_p}}{\left[\dfrac{\varepsilon_s}{\varepsilon_p} - \varkappa_2 \omega\tau \dfrac{V_{dy}}{V_s}\left(\dfrac{\varepsilon_s}{\varepsilon_p} + 1\right)\right]^2 + \left[\omega\tau\gamma\left(\dfrac{\varepsilon_s}{\varepsilon_p} + 1\right)\right]^2}. \tag{4}$$

The same value of the surface potential which is sufficient to change the polarity of the TAE field leads only to a slight increase of the longitudinal acoustoelectric current.

3. Experimental Results and Discussion

The measurements were made using the conventional experimental arrangement [5, 6] where a change of the gap between the faces of the semiconductor and the piezoelectric was possible. The working face of the semiconductor could be illuminated enabling a change of conductivity and strength of the TAE field in the photosensitivity range. The gap between the semiconductor and the piezoelectric was kept strictly constant and equal to 0.5 μm in all measurements. The Rayleigh wave pulse with the fundamental frequency of 20 MHz and duration of 5 μs was generated and detected by means of interdigital transducers deposited on the $LiNbO_3$ YZ-cut plate with dimensions $34 \times 10 \times 0.5$ mm³. The CdS samples were illuminated through a set of filters with maximum light intensity at $\lambda = 5400$ Å. The intensity of light with such a spectral distribution could be varied by a diaphragm. The specimens for measurements of photoconducting properties were cut from the same ingot as those for the measurements of the transversal acoustoelectric effect. The transverse acoustoelectric voltage $V_{ae}^{\perp}$ was observed connecting a pre-amplifier to the indium electrode deposited on the upper face of a CdS plate; the other transparent electrode was deposited on the face of the piezoelectric and grounded. The semidonductor surface near the $LiNbO_3$ plate with Rayleigh wave propagation was free of the electrode and was treated as described below. Six CdS samples for $V_{ae}^{\perp}$ measurements and two for a volume photoconductivity control wer cut from the same ingot, ground and polished to the dimensions $6 \times 5 \times$ $\times\ 2$ mm³. Before the measurements and deposition of the electrodes the samples were etched in a mixture of six parts (by volume) of fuming nitric acid, six parts of glacial acetic, and one part of water for 2 min at 20 °C, then washed in acetone and running distilled water. The surface potential $e\varphi_0/k_B T$ was measured using the method developed for the photosensitive CdS samples and described in [8]. As was shown in [8] and affirmed by our measurements of the surface potential, the strong illumination of CdS crystals leads to the creation of a positive surface potential. This happens because of the strong trapping of carriers by the surface states. In the other case the illumination of the semiconductor would decrease the surface-state potential because of the separation of the electron–hole pairs by the surface field [9, 10]. After the treatment described above the darkness surface potential was found to be $e\varphi_0/k_B T = -3$.

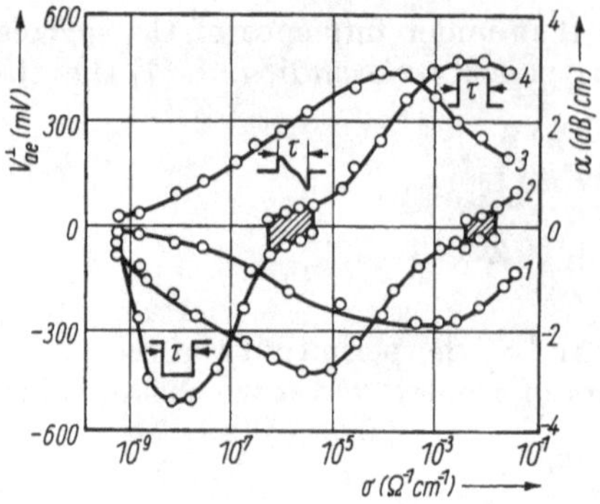

Fig. 1. Transverse acoustoelectric voltage $V_{ae}^{\perp}$ (curves 1, 2, 4) and SAW absorption coefficient α (curve 3) versus conductivity in the CdS–LiNbO$_3$ system. τ is the $V_{ae}^{\perp}$ pulse duration

The dependence of the transverse acoustoelectric voltage $V_{ae}^{\perp}$ upon the conductivity σ for sample No. 5 is shown in Fig. 1 (curve 2). Similar dependences $V_{ae}^{\perp}(\sigma)$ have been measured for the other samples. One can see that, when the conductivity is increased, the amplitude of $V_{ae}^{\perp}$ increases, then reaches its maximum negative value and begins to decrease. In the vicinity of zero ae fields the $V_{ae}^{\perp}$ pulse does not change its sign at once. The different parts of the pulse change their signs at different values of conductivity. For definite values of conductivity there exist both positive and negative parts of the $V_{ae}^{\perp}$ pulse. Hatched areas in Fig. 1 correspond to such a case. At high conductivities the $V_{ae}^{\perp}$ pulses are positive. The shapes of the $V_{ae}^{\perp}$ pulses at different conductivities are shown in Fig. 1. Then in samples No. 1 to No. 3 accumulated layers near the surfaces were produced. For this purpose the surface of the semiconductor was treated in gas discharge in helium atmosphere at a pressure of 4×10^{-2} Torr, the energy of ions being up to 400 to 500 eV, and the duration of the treatment 10 min. After such a treatment crystals showed all the properties of class 1 CdS samples in accordance with [11]. The darkness surface potential became $e\varphi_0/k_BT \approx 5$. The shape of the $V_{ae}^{\perp}$ dependence changed drastically, so that a distinct inversion of $V_{ae}^{\perp}$ versus σ appeared. This case is shown in Fig. 1 by curves 4 ($V_{ae}^{\perp}$ versus σ) and 3 (electronic attenuation α versus σ). Close to the surfaces of the rest samples (No. 4 to No. 6) the depletion layer was made by copper ion doping. This was performed by plunging the working parts of the samples into a water solution of Cu$_2$Cl$_2$ with concentration of about 8×10^{-4} M (pH = 2.5) for 5 min. After doping, the samples showed the properties of class 2 crystals [11]. The surface potential then was $e\varphi_0/k_BT \approx -11$. Curve 1 of Fig. 1 represents the typical $V_{ae}^{\perp}$ dependence for these samples (sample No. 6). As can be seen from Fig. 1, the value and the polarity of the transverse acoustoelectric voltage depend on surface conditions. When the surface potential is positive and, correspondingly, the surface layer is caused by electron accumulation, the transverse acoustoelectric field changes its polarity with changing conductivity, while the longitudinal acoustoelectric current keeps its polarity unchanged. This shows that there is no charge inversion in the surface layer as was assumed in [5] in order to explain similar results. The observed fact can be better explained using the expression (3), from which follows that for large $\omega\tau$ or low conductivities the second term is small and for all samples repulsion of electrons by the surface wave takes place. When the conductivity is increased the product $\omega\tau$ decreases, the second term grows up and the transverse acoustoelectric field changes its sign. This happens because the phase shift does not remain constant, when the conductivity is changed at a certain value of the surface potential. It may concluded that the investigation of the transverse acoustoelectric field seems to be a more sensitive tool to study a semiconductor surface than longitudinal acoustoelectric effect measurements.

References

[1] I. A. Victorov and A. A. Talashev, Acoust. J. (Soviet) **18**, 197 (1972).
[2] S. Takada, K. Hoh, H. Hayakawa, and N. Mikoshiba, Japan J. appl. Phys., Suppl. **42**, 21 (1973).
[3] S. Zemon and E. M. Conwell, Appl. Phys. Letters **17**, 218 (1970).
[4] I. A. Victorov and A. A. Talashev, Acoust. J. (Soviet) **21**, 288 (1976).
[5] P. Das, M. E. Motamedi, and R. T. Webster, Solid State Electronics **19**, 121 (1976).
[6] Yu. V. Gulyaev, A. M. Kmita, A. V. Medved, V. P. Plesky, N. N. Shibanova, and V. N. Fedorec, Zh. eksper. teor. Fiz., Pisma **21**, 353 (1975).
[7] K. A. Ingebrigtsen, J. appl. Phys. **40**, 2681 (1969).
[8] V. A. Tyagai, Yu. M. Shirshov, O. V. Snitko, and V. D. Yarinovskii, phys. stat. sol. **33**, 469 (1969).
[9] A. V. Rzhanov, Elektronye processy na poverkhnosty poluprovodnikov, Izd. Nauka, Moskva 1971.
[10] A. Many, Y. Goldstein, and B. Grover, Semiconductor Surfaces, John Wiley & Sons, New York 1965.
[11] E.-H. Weber, phys. stat. sol. **28**, 649 (1968).

(Received July 15, 1977)

phys. stat sol. (a) 43, 705 (1977)

Subject classification: 7 and 14.3; 16; 22.4.3

V. Kapsukas State University, Vilnius (a) and Charles University, Prague (b)

Investigation of the Bipolar Photoconductivity in CdTe

By

B.-P. KIETIS (a), R. REKSNYS (a), A. SAKALAS (a), and P. HÖSCHL (b)

A method for simultaneous determination of the densities and mobilities of both the polarities of the charge carriers in bipolar semiconductors is proposed. For this purpose the Hall effect for ohmic and acoustoelectric currents, as well as the critical electric field of the acoustoelectric current, are used. Experiments are performed on single crystals of CdTe under illumination near the conversion point of the Hall e.m.f. The drift mobility of holes is determined by measuring the critical field of the acoustoelectric current; the electron mobility and the densities of both types of carriers are calculated.

Предложена методика раздельного определения подвижностей и концентраций носителей тока в биполярном полупроводнике. Для этой цели применён эффект Холла для дрейфового и акустоэлектрического токов, а также измерения критического электрического поля акустоэлектрического тока. Эксперимент проведён на монокристаллическом CdTe в условиях инверсии типа проводимости при освещении. Определены дрейфовая подвижность дырок, а также подвижность электронов и концентрации обоих типов носителей при фотовозбуждении полупроводника.

1. Introduction

Photosensitive semiconductors and semi-insulators are of considerable interest in modern practice. It is not inconceivable that the conductivity (or photoconductivity) in such materials features a bipolar character. By way of illustration, the change in the conductivity type in GaAs from p-type to n-type and conversely is communicated by Bube et al. [1]. Conductivity type conversions in other materials also have been reported in [2, 3]. The typical monopolar semi-insulator n-CdS may have an appreciable bipolarity under some conditions [4]. Also it is common to believe that most if not all semiconductors are bipolar under strong photoexcitation.

In every case mentioned above it is wise to know the main parameters of the charge carriers — their mobilities and densities — in the given semiconductor. Except the method based on photomagnetic effect measurements [5], at present there is no method for solving this problem.

In this paper we suggest a simple method for the investigation of bipolar semiconductors. It is based on measurements of the ordinary Hall effect and the Hall effect for acoustoelectric (AE) current (HEAC), the latter one being combined with the investigation of the critical electric field of the AE current. An experimental illustration of the method is performed on photosensitive CdTe, which changes the type of conductivity from p- to n- under photoexcitation.

2. Set of Equations

The well-known method for the investigation of semiconductors is based on the Hall-effect and conductivity measurements. In the case of a monopolar semiconductor, these measurements are sufficient for the determination of the density and mobility of the majority charge carriers. If the semiconductor is bipolar, both types of carriers are important and thus additional measurements are nescessary for the determination of the main parameters of both types of carriers.

In our previous paper [4] it was noted that the HEAC is a very sensitive phenomenon with respect to bipolarity. In a wide range of physical conditions it can be characterized by the coefficient μ_{HEAC}, analogous to the Hall mobility for the drift current. The relation between μ_{HEAC} and the parameters of the carriers is as follows:

$$\mu_{HEAC} = A \frac{\mu_p^2 \sigma_p + \mu_n^2 \sigma_n}{\mu_p \sigma_p - \mu_n \sigma_n} . \tag{1}$$

Here μ_p, μ_n, σ_p, σ_n are mobilities and conductivities of holes and electrons, respectively, A the Hall factor of scattering, which is assumed to be equal for holes and electrons. By applying the formulae for the total bipolar conductivity σ of the semiconductor,

$$\sigma = \sigma_p + \sigma_n , \tag{2}$$

and for the ordinary Hall mobility μ_H,

$$\mu_H = A \frac{\mu_p \sigma_p - \mu_n \sigma_n}{\sigma} , \tag{3}$$

one can obtain two further equations. Finally, the drift mobility of holes μ_{pd} (or that of electrons) can be obtained from the critical electric field of the AE current [6]:

$$\mu_{pd} = V_s / E_{0p} , \tag{4}$$

where V_s is the sound velocity in the semiconductor, E_{0p} the critical electric field for the AE current, when the holes drift in the same direction as ultrasonic waves propagate. The relation (4) can serve as fourth equation alongside with (1), (2), (3), if only μ_{pd} coincides with μ_p [7, 8]. This condition is satisfied, when the period of the ultrasonic waves strongly differs from the time of hole trapping by levels in the forbidden energy gap of the semiconductor. Trapping may be readily eliminated by varying the frequency of applied ultrasonics. Then the set of equations (1), (2), (3), and (4) can by solved with respect of the unknowns μ_p, μ_n, σ_p, σ_n with the accuracy of the scattering factor A. The densities of holes (p) and electrons (n) are easily obtained from σ_p and σ_n. The factor A is commonly known for different scattering mechanisms. A source of error may occur in the vicinity of the Hall-e.m.f. conversion, when $\mu_p \sigma_p \approx \approx \mu_n \sigma_n$. Under such condition it is very useful to measure the coefficient of the acousto-magnetoelectric effect (AME) μ_{AME}:

$$\mu_{AME} = A \frac{\sigma_p \sigma_n (\mu_p + \mu_n)^2}{\mu_p \sigma_p - \mu_n \sigma_n} . \tag{5}$$

Equation (5) cannot be applied to the determination of the parameters of the carriers alongside (1), (2), (3), since

$$\mu_{HEAC} = \mu_H + \mu_{AME} , \tag{6}$$

i.e. (1), (3), and (5) are linearly dependent on each other. However, the experimental control of (6) partially allows to rule out the possible errors in the vicinity of the Hall-e.m.f. conversion point, where $\mu_H \to \pm 0$, μ_{HEAC} and $\mu_{AME} \to \pm \infty$.

3. Experimental Procedure

The measurements for obtaining the quantities μ_p, μ_n, n, p and the examination of (6) were performed using the device for Hall-effect and HEAC investigations [4] as well as the device for the investigation of the critical electric field of the AE current. The latter device comprises a high-voltage pulse generator and a recorder of the AE current. Samples of photosensitive compensated p-CdTe with dimensions $1 \times 2 \times \times 7$ mm³ and dark conductivity $\sigma_0 \approx 10^{-8}$ $(\Omega\ \text{cm})^{-1}$ were chosen for the experimental investigation. They had the sphalerite structure. The sound velocity along the [111] direction was calculated from the relation

$$\varrho V_s = \frac{1}{3}(C_{11} + 2C_{12} + 4C_{44})\,, \tag{7}$$

where $\varrho = 5.854$ gcm^{-3} is the density of CdTe, C_{11}, C_{12}, C_{44} the elastic moduli being equal to 5.351×10^{11}, 3.681×10^{11}, and 1.994×10^{11} Nm^{-2}, respectively [9]. The calculated sound velocity is 3.43×10^5 cms^{-1}. The directions of proper currents, voltages, and magnetic induction B are shown in Fig. 1. μ_{HEAC}, μ_{AME}, and μ_{pd} are expressed in the following way in terms of measuring quantities:

$$\mu_{\text{HEAC}} = \frac{lIU_{\text{HEAC}}}{wBI_aU}\,, \tag{8}$$

$$\mu_{\text{AME}} = \frac{lIU_{\text{AME}}}{wBI_aU}\,, \tag{9}$$

$$\mu_{\text{pd}} = \frac{V_s l}{U_{0\text{p}}}\,. \tag{10}$$

Here l, w are dimensions of the sample, I, I_a drift and AE currents, U, U_{HEAC}, U_{AME}, $U_{0\text{p}}$ drift voltage, e.m. forces of HEAC, AME, and critical voltage ($E_{0\text{p}} = U_{0\text{p}}/l$) of AE current, respectively.

Longitudinal ultrasonic waves of 10, 30, and 60 MHz frequency were applied. The drift mobility of holes $\mu_{\text{pd}} = 190$ cm²/Vs was obtained from critical field measurements. This value is independent of the frequency of ultrasonics and coincides with the theoretical hole mobility caused by optical phonons at room temperature. Therefore the trapping of holes was excluded and the condition $\mu_{\text{pd}} = \mu_{\text{p}}$ was satisfied. The scattering factor A was assumed to be equal to 1.3 (scattering by optical phonons). The electron mobility was obtained from the set of equations (1), (2), (3), (4):

$$\mu_{\text{n}} = \frac{\mu_{\text{H}}(\mu_{\text{HEAC}} - A\mu_{\text{p}})}{A(A\mu_{\text{p}} - \mu_{\text{H}})}\,. \tag{11}$$

μ_{H} and μ_{HEAC} are negative for n-type and positive for p-type conductivity. The hole density p was calculated from μ_{n} being known using the formula

$$p = \frac{(\mu_{\text{H}} + A\mu_{\text{n}})\,\sigma}{A(\mu_{\text{p}} + \mu_{\text{n}})\,e\mu_{\text{p}}}\,; \qquad e = 1.6 \times 10^{-19}\ \text{C}\,. \tag{12}$$

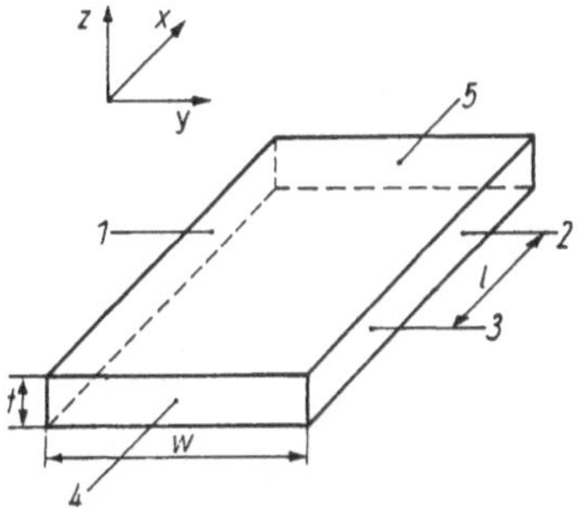

Fig. 1. Directions of the measured quantities with respect to the sample: $I = I_x$, $I_a = I_{ax}$, $B = B_z$, $V_s = V_{sx}$, $U = U_x$, $U_0 = U_{0x}$, $U_{\text{H}} = U_{\text{H}y}$, $U_{\text{HEAC}} = U_{\text{HEAC}y}$, $U_{\text{AME}} = U_{\text{AME}y}$

The electron density is

$$n = \frac{\sigma - e\mu_p p}{e\mu_n} . \tag{13}$$

4. Experimental Results

The dependence of the Hall mobility an photoconductivity in CdTe is shown in Fig. 2. The variation of μ_H is complicated by conductivity type conversion. Fig. 3 shows the dependences of μ_{HEAC} and μ_{AME} on σ. At conductivity type conversion μ_{HEAC} and μ_{AME} tend to infinity. One can see that in the vicinity of the conversion point μ_{AME} tends to be equal to μ_{HEAC}, $\mu_H \to 0$. μ_{AME} is never equal to zero, μ_{HEAC} is ever greater than μ_H. The relation (6) is always fulfilled. Fig. 4 illustrates the dependence of the drift mobility of holes, calculated from critical field measurements. The μ_n values calculated from (11) are presented, too. The dashed lines show the theoretical values of mobilities for scattering by optical phonons [10]. Fig. 5 illustrates the variation of the charge carrier densities at photoexcitation. The hole density is greater than the electronic one, although the sign of the Hall-e.m.f. shows n-type conductivity.

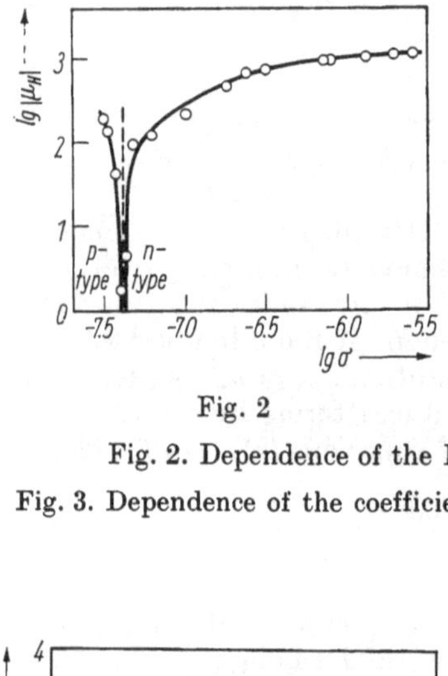

Fig. 2

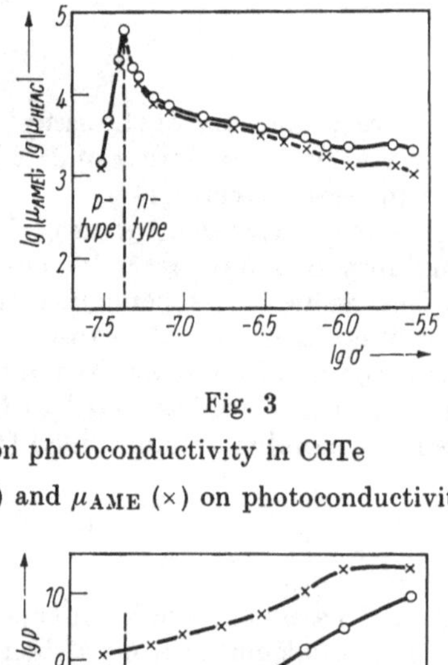

Fig. 3

Fig. 2. Dependence of the Hall mobility on photoconductivity in CdTe

Fig. 3. Dependence of the coefficients μ_{HEAC} (○) and μ_{AME} (×) on photoconductivity

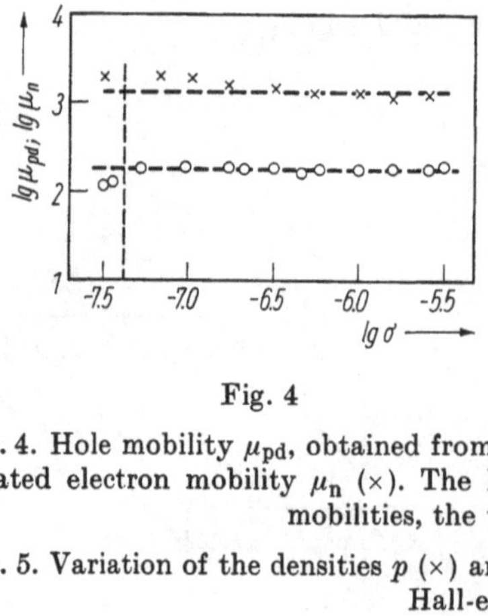

Fig. 4

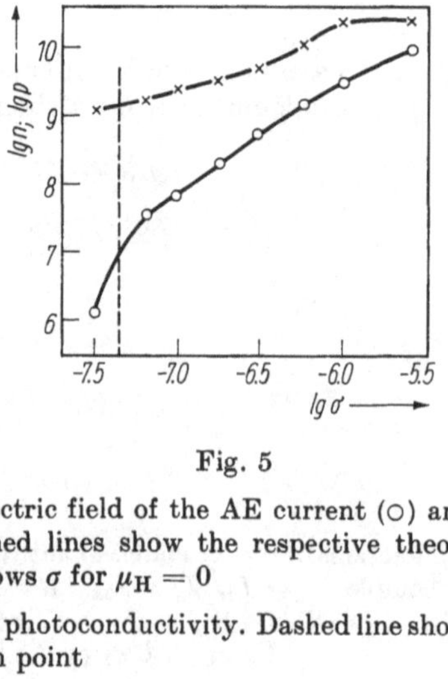

Fig. 5

Fig. 4. Hole mobility μ_{pd}, obtained from the critical electric field of the AE current (○) and calculated electron mobility μ_n (×). The horizontal dashed lines show the respective theoretical mobilities, the vertical line shows σ for $\mu_H = 0$

Fig. 5. Variation of the densities p (×) and n (○) versus photoconductivity. Dashed line shows the Hall-e.m.f. conversion point

5. Discussion

It is seen from Fig. 2 and 3 that bipolarity causes a great difference between μ_{HEAC} and μ_H. The relation (6) is in good agreement with experimental data also under conditions of Hall-e.m.f. conversion. The infinite μ_{HEAC} and μ_{AME} mean only that the e.m.f. transverse to the current direction is not equal to zero, when the AE current tends to zero due to bipolarity. The hole mobility obtained from the critical field measurements is in better agreement with the theoretical value than that obtained only from the Hall effect. It is very possible that the Hall mobility of both types of carriers is strongly reduced due to bipolarity. The calculated electron mobility is 1230 to 1700 cm^2/Vs, which may be thought to agree with the theoretical value 1320 cm^2/Vs. The error can be reduced by refinement of the chosen scattering factor A. The Hall mobilities for both types of carriers are in good agreement with the experimental data of other authors [11, 12]. The error of 20 to 30% for the microscopic mobility is not small, but for the densities of charge carriers in our experiments such accuracy is rather good. In the vicinity of the conductivity type conversion point the rated curves of μ_n, p, n are extrapolated.

Finally it should be pointed out that bipolar semiconductors can be successfully investigated by the proposed method in a wide range of physical conditions. The determination of the drift mobility from the saturation of U–I curves [13] requests relatively high conductivity and strong AE interaction. In our case the conductivity may be of the order of 10^{-10} $(\Omega\ cm)^{-1}$ and smaller; the AE interaction may be weak. Thus, the method can be applied to the investigation of non-piezoelectric materials.

References

[1] R. H. Bube, H. E. MacDonald, and J. Blanc, J. Phys. Chem. Solids **22**, 173 (1961).
[2] W. W. Tyler and R. Newman, Phys. Rev. **108**, 1409 (1957).
[3] S. Kubalkova, A. Sakalas, and J. Viščakas, Liet. Fiz. Rink. **15**, 617 (1975).
[4] B.-P. Kietis, R. Reksnys, and A. Sakalas, phys. stat. sol. (a) **39**, 601 (1977).
[5] V. V. Ostroborodova, L. I. Riabova, M. V. Simakin, M. M. Gubaeva, Izv. vuzov, Ser. fiz. **2**, 21 (1976).
[6] M. Fink and G. Quentin, phys. stat. sol. (a) **4**, 397 (1971).
[7] C. Fischler, J. appl. Phys. **41**, 1439 (1970).
[8] B. E. Cekvava, Fiz. Tekh. Poluprov. **7**, 1184 (1973).
[9] H. J. McSkimin and D. G. Thomas, J. appl. Phys. **33**, 56 (1962).
[10] S. S. Devlin, in: Physics and Chemistry of $A^{II}B^{VI}$ Compounds, Section II, North-Holland Publ. Co., Amsterdam 1967.
[11] B. Segal, M. R. Lorenz, and R. E. Halsted, Phys. Rev. **129**, 2471 (1963).
[12] S. J. Yamada, J. Phys. Soc. Japan **15**, 1940 (1960).
[13] R. W. Smith, Phys. Rev. Letters **9**, 97 (1962).

(Received July 11, 1977)

See Short Note by V. SIEGEL and W. RASP

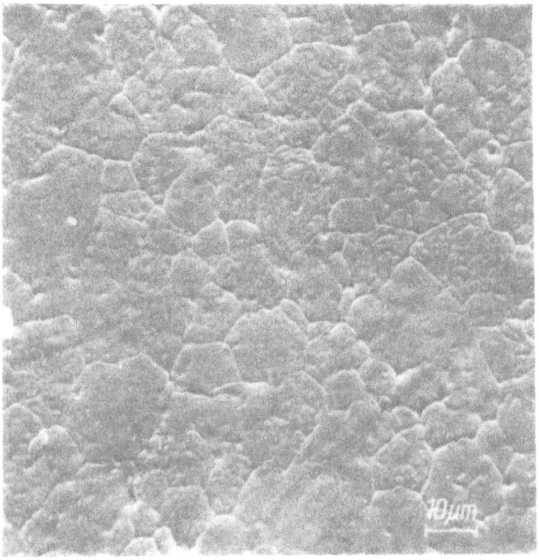

Fig. 2. Scanning electron microscope image of the surface of the original sintered BeO sample

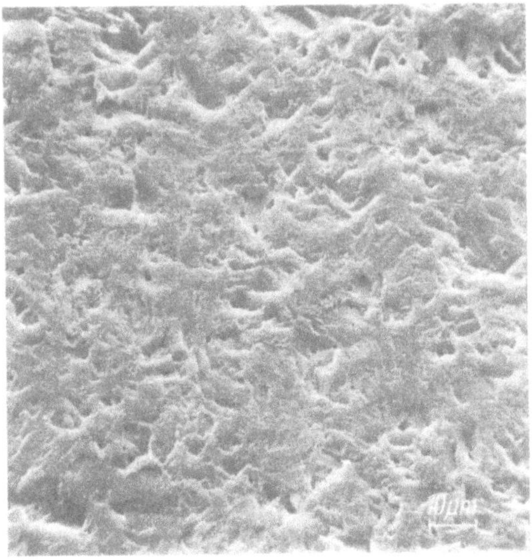

Fig. 3. Scanning electron microscope image of the surface of the lapped sintered BeO sample. Maximum grain size of the diamond powder used 3.0 μm

See Short Note by A. J. Morton

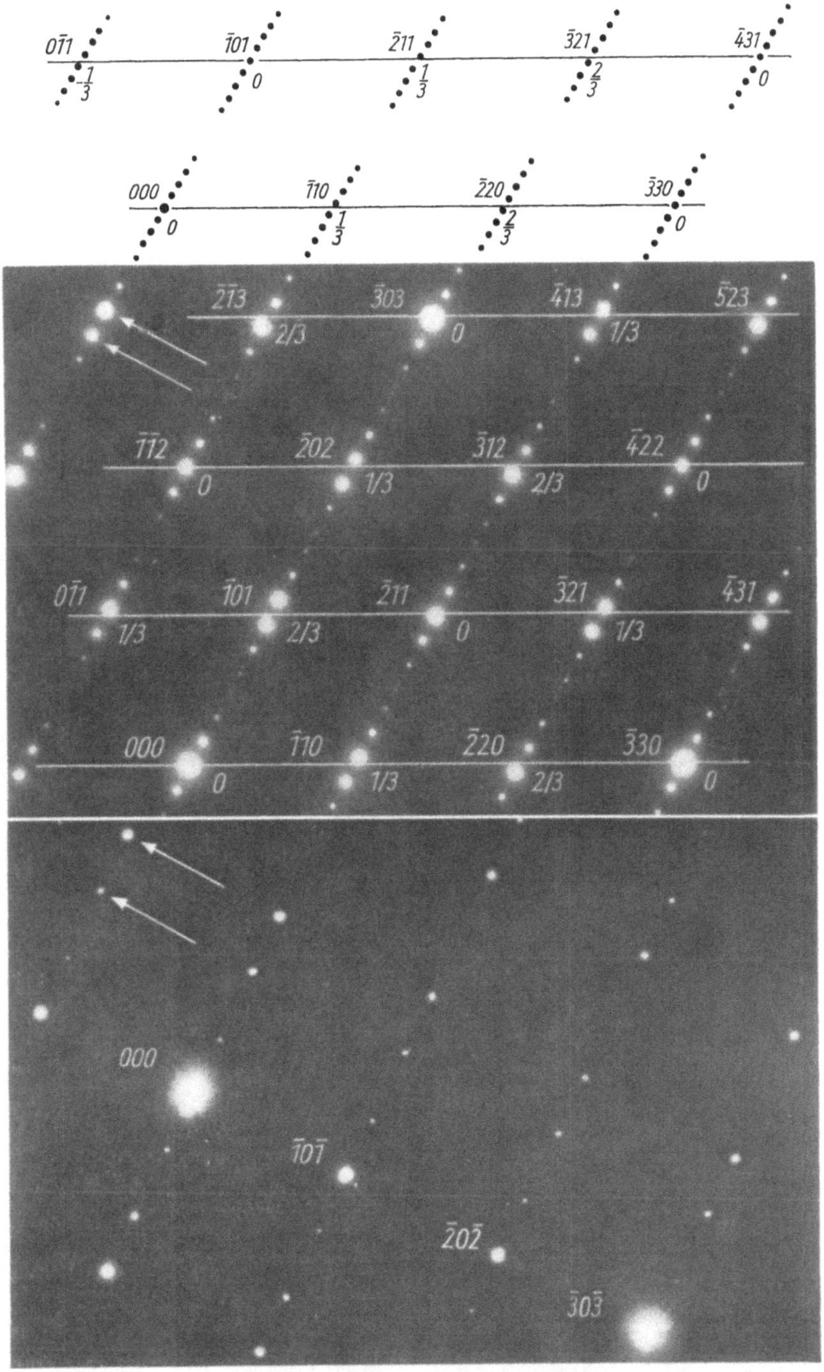

Fig. 1. a) Schematic [111] diffraction pattern of the planar IAPD structure in γ-brass drawn from observations of IAPD structure diffraction patterns; b), c) [111] and (010) diffraction patterns taken from a single area of the planar IAPD structure in γ-$NiZn_3$. In a) and b) displacements of the superlattice spots from positions of the basic structure spots are indicated as 0, 1/3, 2/3, 0, 1/3, etc. and in b) and c) attention is drawn to the different spacing of the superlattice spots

See Short Note by N. $\bar{O}$TSUKA et al.

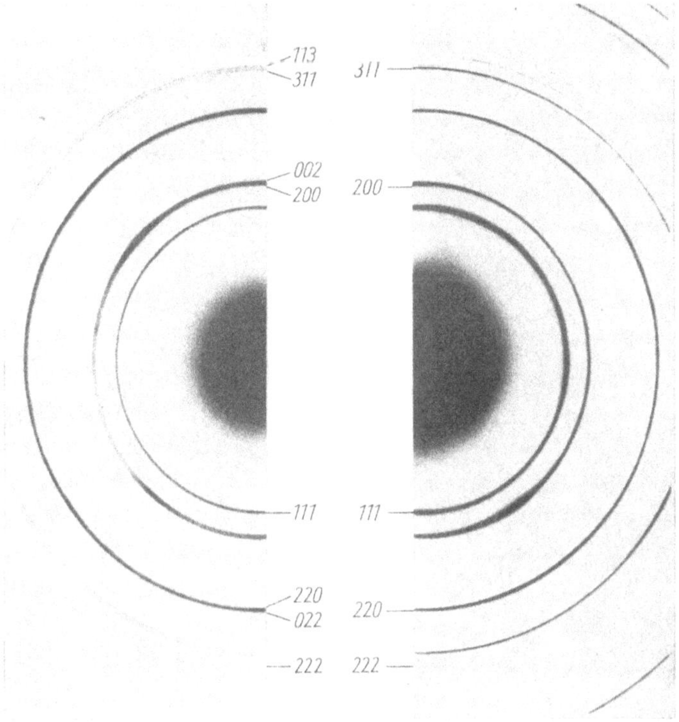

a b

Fig. 1. Electron diffraction patterns of a polycrystalline $Mn_6N_{5.32}$ film at a) room temperature and b) 430 °C

See Short Note by J. HÄRTWIG et al.

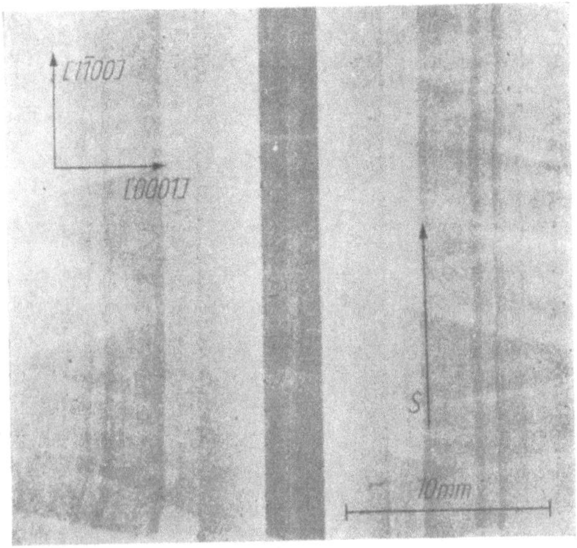

Fig. 2. Contrast photocopy of the crystal with visible light

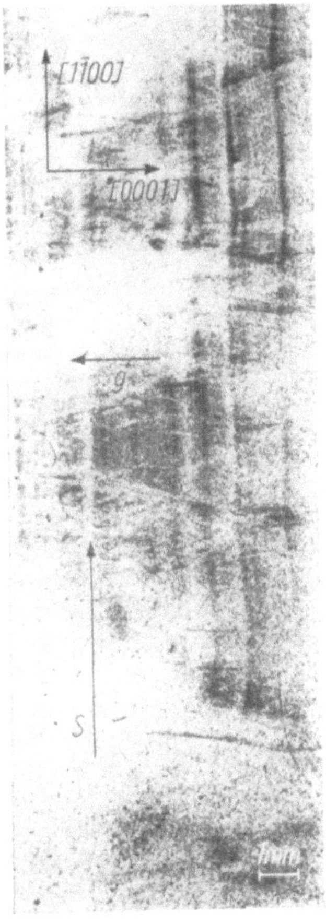

Fig. 3. Double crystal diffractometer topograph of a part of the specimen, (0003) reflection, $CuK_{\alpha 1}$ radiation, $\mu t \approx 3$

See Short Note by Németh-Sallay et al.

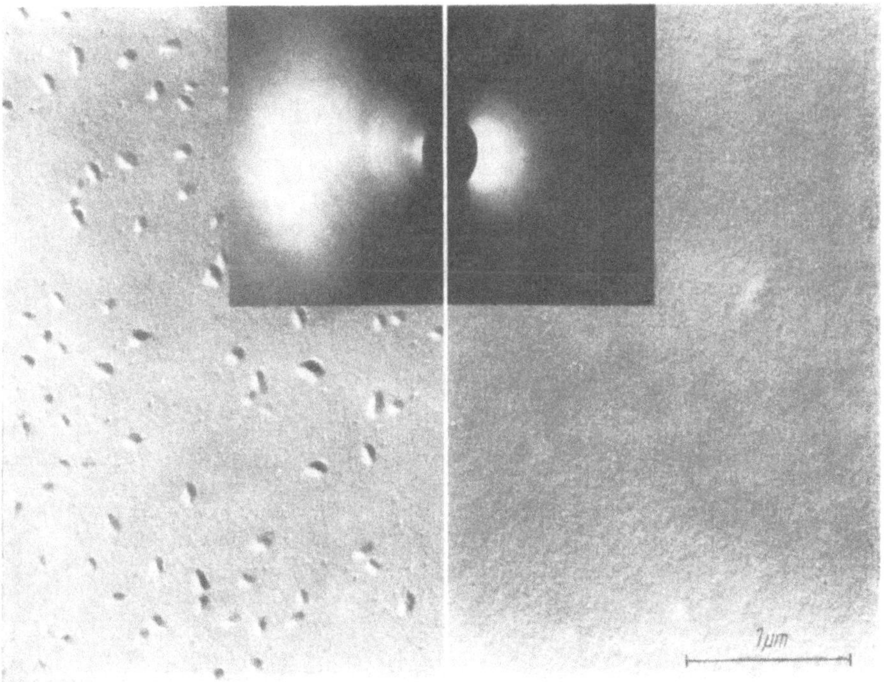

a b

Fig. 1

See Short Note by G. van Tendeloo et al.

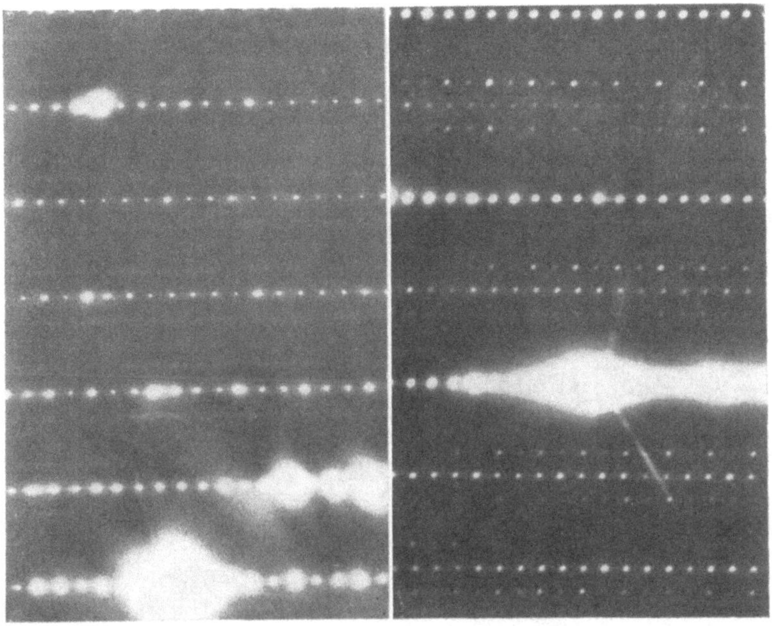

Fig. 2. Evolution of the electron diffraction pattern of TaS_3 with decreasing temperature. a) Continuous satellite lines parallel to the rows of Bragg spots reveal largely uncorrelated distortions of the chains (just below room temperature); b) the diffuse lines break up into discrete spots, revealing the formation of a superlattice as a result of the correlation between neighbouring chains (specimen holder at liquid nitrogen temperature)

Short Notes

phys. stat. sol (a) 43, K117 (1977)
Subject classification: 17; 16; 13.3; 22.6

Physikalisch-Technische Bundesanstalt, Braunschweig

The TSEE of Lapped and Unlapped BeO Ceramics

By
V. SIEGEL and W. RASP

For BeO and other oxides it is known that exoelectron emission is influenced by gas adsorption and chemical reactions at the surface /1 to 4/. According to Krylova et al. /5/ the thermally stimulated exoelectron emission (TSEE) is a result of stepwise recombination of active particles formed by excitation processes at the surface of a solid. Exoelectrons are regarded as released binding electrons during the desorption process. Studying the ability of BeO to TSEE, Euler et al. /2/ confirm that one of the TSEE maxima is due to the adsorption of oxygen. According to the investigations of Baazov et al. /6/, using γ-Al_2O_3 samples, the thermally stimulated exoelectron emission increases after adsorption of hydrogen with a simultaneous appearance of a new emission peak. The height of this maximum strongly depends on the mass of adsorbed hydrogen

On the other hand, it is known that a change in the constitution of the adsorbed surface layer leads to a distinct variation in the exoelectron yield. Thus, wetting BeO ceramic samples with methanol over a period of several hours, the exoelectron yield sharply decreases compared with the original exoelectron emission /7/. It should be noted that both the so-called surface and volume exoemission peaks are affected by surface methoxylation of the sample under investigation (Fig. 1). Therefore both types of trapping centres, which are responsible for the TSEE at about 270 and 325 ^{o}C, will be influenced by adsorbed layers as the methanol only forms a chemisorption coverage. Obviously, active surface sites for the exoelectron emission are exchanged or destroyed by the adsorption of the methanol molecules.

Usually the surface of a solid when stored in humid air is covered with different kinds of atoms and molecules The adsorption of water molecules on oxides is pro-

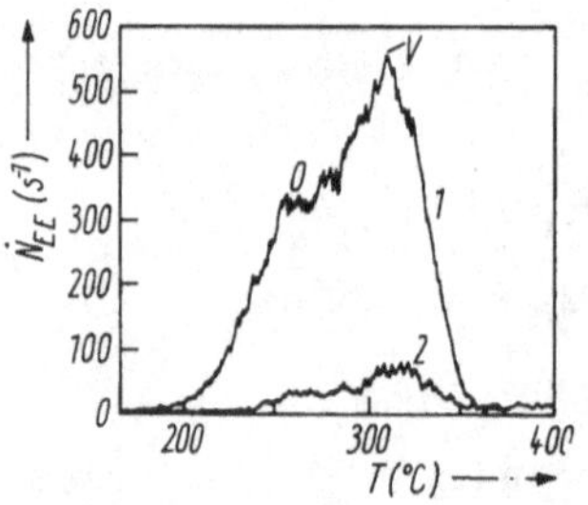

Fig. 1. Exoelectron emission $\dot{N}_{EE}$ of ceramic BeO before (1) and after (2) wetting with methanol. T heating temperature

moted by a consequent large decrease in the free energy. By measurements of the catalytic properties of different oxides it is known that imperfections on the surfaces, such as grain boundary layers, steps, cracks, and holes, intensify the adsorption of foreign molecules compared with an atomically flat surface.

Examining the surface of the commercially available BeO ceramics by a scanning electron microscope one observes on the surface a large number of the above-mentioned imperfections (Fig. 2, see on the photo pages before the Short Notes part) due to the production process of the samples. Since a direct relation exists between the number of active sites on the surface of a solid and the number of adsorbed species one would expect that a change in the number of these sites will also change the number of adsorbed molecules.

Calling to mind that for most oxides the centres of activated desorption serve as TSEE centres, a change of the constitution of these active adsorption centres must lead to an alteration of the exoelectron yield.

To change the constitution of the surface of the samples the lapping method with diamond powder is used. The two types of diamond powder used have a maximum grain size of 3.0 and 6.6 μm, respectively. After lapping for a constant time (Fig. 3, see on the photo pages before the Short Notes part) the disc-shaped BeO ceramics with a diameter of 5.5 mm and a thickness of 0.5 mm were cleaned by means of ultrasound in desealed water. The exoelectron emission before and after lapping the samples was measured using a windowless methane gas flow counter, operating in the proportionality region /8/. Throughout the measurements the rate of thermal stimulation was kept constant at about 1.4 K/s. Before each measuring cycle the samples were annealed in air at 500 °C.

The measurements of the TSEE of the original and lapped samples were carried out in the following way: Before the exposure to a constant amount of hydrogen ions at 20 °C the BeO discs were heated in stages up from room temperature to about

Fig. 4. The change of the exoelectron yield as a function of prior storage temperature for the original sintered BeO sample (above) and for the lapped sample (below)

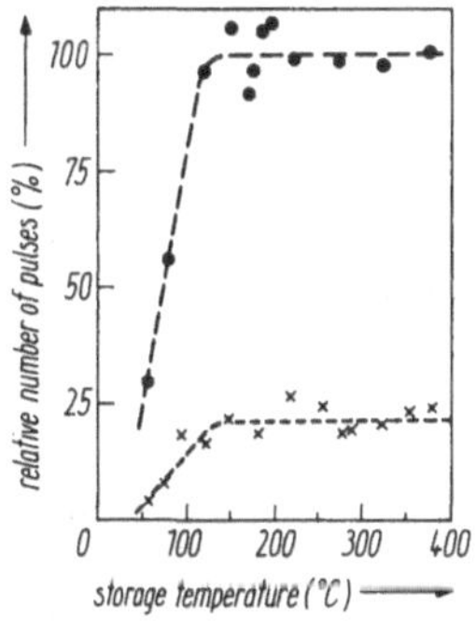

400 °C inside the counter in the methane atmosphere for the stepwise desorption of Van der Waals adsorbed water molecules and thermally weakest bound OH groups, bound by H bridges. By this means, more and more thermally stronger bound OH groups, bound by H bridges, are bared on the surface of the sample under investigation /9, 10/ and can be reached by the hydrogen ions.

The necessary hydrogen ions were generated by the dissociative ionization of the methane molecules by Compton and photo-electrons inside the counter. By switching a positive high voltage on the counting wire during the ion bombardment, the positive H^+ and H_2^+ ions are accelerated towards the sample surface and are partially adsorbed.

After this, the BeO sample was excited by γ-rays of a ^{60}Co source with constant photon fluence.

As can be seen from Fig. 4 (upper curve) the normalized exoelectron yield of the original sample increases with prior storage temperature reaching a quasi-saturation at about 130 °C. Following the conclusions in /9/, the Van der Waals adsorbed water molecules desorb in the temperature range from 30 to 130 °C, with a maximum loss of mass at about 70 °C. From desorption measurements /11/ it is known that in the temperature region above 130 °C, transformed OH groups bound by H bridges also become desorbed, the desorption rate of these groups as a function of increasing temperature, however, only showing a small slope covered by the large spread of the measured number of exoelectrons.

The lapped sample shows a similar course of desorption (Fig. 4, lower curve). In this case, however, the exoelectron yield only amounts to about 20% of the original value in the saturation region. Obviously, active adsorption sites are lost or destroyed on the surface of the sample during the lapping process as the number of H ions during bombardment was equal to that for the unlapped sample.

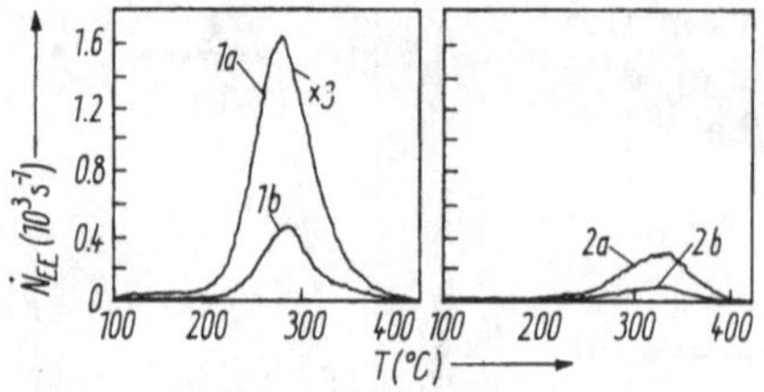

Fig. 5. The change of the TSEE maxima at about 270 and 325 °C for the unlapped BeO sample (curves 1a, 2a) and for the lapped sample (curves 1b, 2b)

The above results constitute direct evidence that the surface structure of the samples is drastically changed during the lapping process.

First, examining the exoelectron yield of the two emission maxima at about 270 and 325 °C before (Fig. 5, curves 1a and 2a) and after (Fig. 5, curves 1b and 2b) lapping the sample, one recognizes a large decrease of the exoelectron emission in both, the maxima at 270 and 325 °C, although the ion bombardment and excitation conditions by ionizing radiation were equal in both cases.

Furthermore, comparing the scanning electron microscope images (Fig. 2 and 3), one is forced to assume that the effective surface area of the lapped sample has become larger. On the other hand, the measurements of the exoelectron yield after the above mentioned procedure show a distinct decrease of the measured value of exoelectrons after lapping the sample indicating a diminution of the occupiable active adsorption sites. For this reason, one must search in an another direction for the explanation of the observed change in exoelectron yield. Examining the scanning electron microscope images in more detail, it is to be observed that the grain boundaries of the original sample have almost totally vanished after lapping the surface. Following the conclusions in /12, 13/ among other lattice defects, we note that particularly those places where the growth of individual crystal faces has stopped, as well as lattice disturbances in the surface, are active centres of adsorption. Obviously, grain boundaries are preferred adsorption sites from the point of view of energy, their disappearance showing the above-mentioned diminution of the exoelectron yield from sintered BeO.

Dr. Lauer is very thanked for the preparation of the scanning electron microscope images.

References

/1/ I.V. KRYLOVA, phys. stat. sol. (a) 7, 359 (1971).

/2/ M. EULER, W. KRIEGSEIS, and A. SCHARMANN, phys. stat. sol. (a) 15, 431 (1973).

/3/ M. EULER and A. SCHARMANN, phys. stat. sol. (a) 34, 297 (1976).

/4/ I.V. KRYLOVA and V.I. SVITOV, phys. stat. sol. (a) 37, K9 (1976).

/5/ I.V. KRYLOVA, V.I. SVITOV, and N.I. KONYUSHKINA, Zh. fiz. Khim. 50, 933 (1976).

/6/ D.I. BAAZOV, YU.A. KOLBANOVSKI, and L.S. POLAK, Izv. Akad. Nauk SSSR,Ser. fiz. 35, 361 (1971).

/7/ V. SIEGEL, W. RASP, and H.-H. KIRCHNER, phys. stat. sol. (a) 34, K203 (1976).

/8/ H.-J. KRIKS and W. SCHAARE, PTB-Mitt. 81, 10 (1971).

/9/ S.E. ERMATOV and T.S. KOSHAROV, Izv. Akad. Nauk Kazakh SSR, Ser. fiz.-mat. 12, 29 (1974).

/10/ S.E. ERMATOV, Zh. prikl. Spektroskopii 23, 14 (1975).

/11/ V. SIEGEL, W. RASP, and H.-H. KIRCHNER, to be published.

/12/ J.H. DE BOER, Adv. Catal. 8, 18 (1956).

/13/ E.H. TAYLER, Adv. Catal. 18, 111 (1968).

(Received July 13, 1977)

phys. stat. sol. (a) 43, K123 (1977)

Subject classification: 1.4; 1.3; 21.1

CSIRO Division of Tribophysics, University of Melbourne, Parkville[1)]

The Displacement Vector Associated with Inversion Boundaries in γ-Brass Type Phases - A Reassessment

By

A.J. MORTON

In two recent papers the author obtained by different techniques conflicting results for the displacement vector, $\vec{R}$, associated with inversion anti-phase domain (IAPD) boundaries in copper rich γ-brass. The image matching technique of Head et al. /1/ applied to domain boundary contrast gave displacements of $\frac{1}{3}[\bar{1}\bar{1}0]$ and $\frac{1}{3}[011]$ at adjacent boundaries parallel to $(\bar{1}01)$ planes which resulted in a displacement of $\frac{1}{3}[\bar{1}\bar{0}1]$ between like domains /2/. Previously /3/, an analysis due to van Landuyt et al. /4/ of the displacement of superstructure diffraction spots from the positions of those due to the basic structure was applied to diffraction patterns of the planar IAPD structure giving a displacement vector of $\frac{1}{3}[010]$ for domains parallel to $(\bar{1}01)$.

Because of the very fine spacing of the superlattice diffraction spots in the Cu-Zn γ-phase alloy (long period 78.7 → 160 Å) the pattern analysed in /3/ was a schematic [111] diffraction pattern drawn from observations of the [111] patterns of the planar IAPD structure. Current work to determine the mechanism of stabilisation of IAPD structures /5/ has shown that they are formed in Ni-Zn γ-phase alloys and in these alloys the period of the IAPD structure can be as small as 35 Å giving a much larger spacing of superlattice diffraction spots and allowing a more detailed and precise analysis of the diffraction patterns.

In Fig. 1 (see on the photo pages before the Short Notes part) the a) schematic [111] diffraction pattern used in the previous analysis and b) an actual diffraction pattern from a Ni-75 at% Zn γ-phase alloy are shown. The displacements of the superlattice spots from the basic structure positions are indicated by 0, 1/3, 2/3, 0, etc. where the displacements are in units of the superlattice spot spacing and

1) Parkville, Victoria, 3052, Australia.

are in the direction normal to the domain boundaries. The displacements in Fig. 1b have been determined by constructing lines parallel to the $\bar{1}10$ reciprocal lattice row passing through 000, $\bar{3}03$, and through the positions of the absent $\bar{1}01$ and $\bar{2}02$ normal γ-brass diffraction spots, i.e. equally spaced lines at 1/3 of the 000 - $\bar{3}03$ spacing. It is clear that the displacements in the two patterns are similar but that, in constructing the schematic pattern, the displacements shown for the whole row of diffracting vectors $0\bar{1}1$, $\bar{1}01$, $\bar{2}11$, ... have been misplaced. Thus those drawn in a) for $0\bar{1}1$ correspond to the actual displacement in b) for $\bar{1}01$; and that drawn for $\bar{1}01$ corresponds to $\bar{2}11$ and so on. In the notation of van Landuyt et al. the correct displacements $\vec{H} \cdot \vec{R}$ for the various diffracting vectors $\vec{H}$ are $\vec{H}$: $\bar{1}01$, $\bar{1}10$, $0\bar{1}1$, $\bar{2}11$, $2\bar{1}1$; $\vec{H} \cdot \vec{R}$: 2/3, 1/3, 1/3, 0, -1/3. These displacements lead to the result $\vec{R} = \frac{1}{3}[\bar{1}01]$ for the displacement vector at the domain boundaries in agreement with that obtained by the image matching technique. It should be noted that this displacement vector specifies the displacement between like domains in the IAPD array and is the result of two separate displacements occurring at consecutive plus/minus and minus/plus domain boundaries.

Another feature of the γ-NiZn diffraction patterns concerns the nature of the domains themselves. Fig. 1c is an [010] pattern taken from the same area as Fig. 1b and it is obvious that the superlattice spot spacing in the [010] patterns is twice that of the [111] patterns. This would seem to imply that the superstructure period is halved merely by rotating from the [111] beam direction to [010] around the domain normal. Since the domain thickness is not altered by such a rotation it is the inversion character of the domains which causes this effect. This is readily understood because the γ-brass structure, space group $I\bar{4}3m$, appears centrosymmetric when viewed in the [001] directions, i.e. the structure factors for all hk0 reflections are real. Therefore in the [010] beam direction the plus/minus character of adjacent domains is not detected and hence the long period associated with the pair of plus/minus domains disappears, the diffraction pattern merely reflecting the array of parallel displacement boundaries spaced at one domain thickness. Thus it is possible to infer that the superstructure is composed of inversion anti-phase domains simply from a study of diffraction patterns.

References

/1/ A.K. HEAD, P. HUMBLE, L.M. CLAREBROUGH, A.J. MORTON, and C.T. FORWOOD, Computed Electron Micrographs and Defect Identification, North Holland Publ. Co., Amsterdam 1973.

/2/ A.J. MORTON, phys. stat. sol. (a) 31, 661 (1975).

/3/ A.J. MORTON, phys. stat. sol. (a) 23, 275 (1974).

/4/ J. VAN LANDUYT, R. DE RIDDER, R. GEVERS, and S. AMELINCKX, Mater. Res. Bull. 5, 353 (1970).

/5/ A.J. MORTON, to be published.

(Received July 22, 1977)

phys. stat. sol. (a) 43, K127 (1977)

Subject classification: 1.2; 21.1

Department of Metallurgy, Tokyo Institute of Technology[1)]

Crystal Structure and Phase Transition of Mn_6N_5 Studied by Electron Diffraction

By

N. ŌTSUKA, Y. HANAWA, and S. NAGAKURA

The Θ-phase in the Mn-N system is tetragonal at room temperature; it is stable up to about 580 °C /1/. Its composition range extends from 45.8 to 47.9 at%N at 400 °C and the representative chemical formula is Mn_6N_5. The Mn atoms are known to take a face centered tetragonal arrangement, but the N atom positions have not yet been established. In the course of electron diffraction studies of interstitial compounds, the present authors succeeded in preparing Mn_6N_5 films and made a structural study. It was found that Mn_6N_5 transforms from tetragonal to cubic at (410 ± ± 20) °C. This is described in this note.

By applying the same specimen preparation method as was used for Mn_4N /2/, polycrystalline nitride films with a thickness of about 500 Å were produced by nitriding vacuum evaporated Mn films with NH_3 gas at 300 °C for 30 min. Electron diffraction investigations were made with 150 kV electrons. A high temperature specimen holder was used to investigate the phase transition. The specimen temperature was measured by means of a thermocouple fixed near to the specimen film. The temperature difference between the thermocouple and the specimen film was smaller than 20 °C.

Fig. 1a (see on the photo pages before the Short Notes part) reproduces an electron diffraction pattern of a nitride film at room temperature. All the reflections can be indexed on the basis of a face centered tetragonal lattice with parameters a = 4.214 and c = 4.144 Å and an axial ratio c/a = 0.983 (reference: TlCl, the lattice parameter of which is 3.8427 Å). The lattice parameter-composition relationship for the Θ-phase /1/ yields the specimen composition 47.0 at%N or $Mn_6N_{5.32}$. Unlike other compounds in the Mn-N system, superstructure reflections were observed in none of the specimens even after a prolonged heat treatment. The observed diffraction intensities were well explained by assuming that N atoms are

1) Oh-okayama, Meguro-ku, Tokyo 152, Japan.

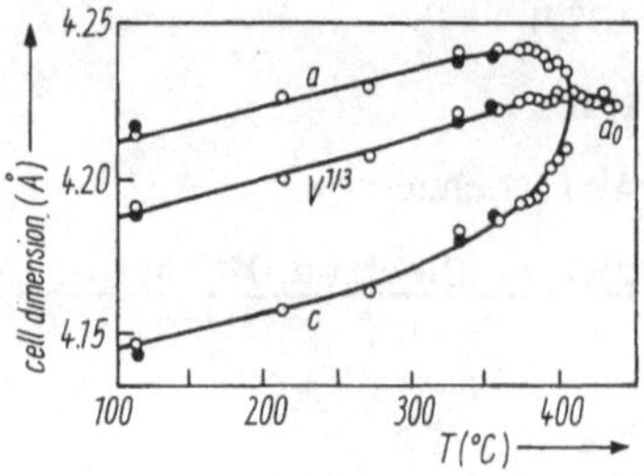

Fig. 2. Plots of the cell dimensions of $Mn_6N_{5.32}$ against temperature. a and c lattice parameters of the tetragonal phase, a_o lattice parameter of the cubic phase, $V^{1/3}$ cubic root of the unit cell volume of the tetragonal phase. o on specimen heating, • on specimen cooling

distributed randomly among the octahedral interstices. Fig. 1b (see on the photo pages before the Short Notes part) is an electron diffraction pattern at 430 °C of the same specimen film as that in Fig. 1a. The pattern could be interpreted by the NaCl type of structure with a random distribution of N atoms among the octahedral interstices.

In order to investigate the phase transition from tetragonal to cubic, lattice parameter measurements were made on $Mn_6N_{5.32}$ at various temperatures up to 440 °C. For this purpose, a polycrystalline aluminum film was attached to the specimen film as a reference material and superimposed diffraction patterns were taken. Neither composition changes due to nitrogen evolution nor specimen oxidation were observed during the high temperature measurement. Fig. 2 shows the result, where a and c represent the lattice parameters of the tetragonal phase, a_o the lattice parameter of the cubic one and $V^{1/3}$ the cubic root of the unit cell volume of the tetragonal phase. Open circles represent the values obtained on the heating cycle and solid ones those on the cooling cycle. The accuracy of the measurement is better than 0.05%. It can be seen that the phase transition occurs at 410 °C, whereas the unit cell volume changes continuously within the experimental error. Other specimens with compositions $Mn_6N_{5.21}$ and $Mn_6N_{5.65}$ exhibited the phase transition at 415 and 425 °C, respectively. It was also found that the transition temperature does not depend on whether the specimen is heating or cooling. Thus, we may say that the observed phase transition is of the second order or of the first order very close to second order.

The physical properties of Mn_6N_5 have not been studied probably because of the difficulty in specimen preparation. Accordingly, we cannot discuss the observed phase transition in connection with physical property changes. However, the phase transition cannot be of the order-disorder type, since no indication for the ordering of nitrogen atoms has been detected. It is suggested that the phase transition may be related to the magnetic transition, because other compounds in the Mn-N system such as Mn_4N /3/ and Mn_2N /4/ have been known to be magnetic materials.

References

/1/ F. LIHL, P. ETTMAYER, and A. KUTZELNIGG, Z. Metallk. 53, 715 (1962).
/2/ S. NAGAKURA and N. ŌTSUKA, J. Phys. Soc. Japan 39, 1047 (1975).
/3/ W.J. TAKEI, G. SHIRANE, and B.C. FRAZER, Phys. Rev. 119, 112 (1960).
/4/ M. MEKATA, J. HARUNA, and H. TAKEI, J. Phys. Soc. Japan 25, 234 (1968).

(Received July 29, 1977)

phys. stat. sol. (a) 43, K131 (1977)
Subject classification: 10.2 and 20.1; 22.6

Sektion Physik der Friedrich-Schiller-Universität Jena[1)]

Connection between Lattice Deformation and Coloration Pattern in Synthetic Quartz Monocrystals

By

J. HÄRTWIG, B. DIETRICH, R. BÖTTGER, and G. LÜBKE

Lang topographs of synthetic quartz monocrystals often show besides the dislocation contrasts fringes perpendicular to the growth direction /1 to 3/. After irradiating the samples with hard X-rays a pattern of smoky coloration fringes appears which are also perpendicular to the growth direction [0001]. Their absorption in the optical range is due to the colour centres which arise from substitutional aluminium impurity ions after irradiation /4 to 7/. Compared with the perfect areas of the crystal the intensity of the reflected beam is changed in the areas with coloration fringes. This is caused by the lattice deformation which originates in different concentrations of aluminium ions. Härtwig and Dietz /3/ showed that the maxima of reflected intensity lie on positions where the optical absorption changes strongly. In regions which are strongly coloured the lattice is expanded and the essential components of the displacement vector are parallel to the growth direction [0001].

The starting point for the further investigations was according to /3/ the assumption that the optical absorption A(z) of the coloured crystal and the relative change of the lattice parameter $\Delta c(z)/c \approx d\vec{u}(z)/dz$ are proportional to each other ($\vec{u}$ displacement vector, c lattice parameter in the [0001] direction, z axis parallel to the [0001] direction).

To check this assumption by means of the appearing reflected intensity of X-rays further X-ray investigations with a double crystal diffractometer (Bragg-Laue position, $CuK_{\alpha 1}$-radiation, $\mu t \approx 3$) and measurements of the optical absorption were carried out.

The intensity of the reflected beam is under special conditions (Z is constant within the Borrmann fan a function of μt and the strain efficiency parameter Z /8/,

1) DDR-69 Jena, Max-Wien-Platz 1, DDR.

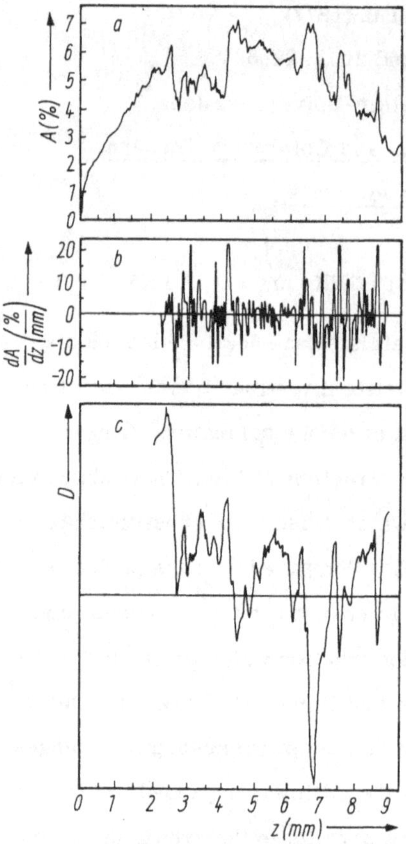

Fig. 1. a) Photometric absorption curve A(z); b) graphical differentiated absorption curve dA(z)/dz; c) transmittance curve of the topograph in Fig. 3

which is dependent on the displacement vector $\vec{u}$ through the second derivative of $\vec{u} \cdot \vec{g}$ ($\vec{g}$ reflection vector).

Under the present conditions /3/ one may assume that $Z \sim d^2(\vec{u} \cdot \vec{g})/dz^2$. Consequently dA(z)/dz and $d^2(\vec{u} \cdot \vec{g})/dz^2$ ought to be proportional to each other if $A(z) \sim d\vec{u}(z)/dz$ and if one takes into consideration the non-linear connection between the intensity and the parameter Z /8, 9/ (in the general case this is only possible by numerical methods). Therefore a qualitative correspondence of the contrast in the X-ray topograph and the first derivative of the optical absorption may be expected.

Fig. 1a shows the photometric absorption curve A(z) of an investigated crystal (Fig. 2, see on the photo pages before the Short Notes part). This curve has been differentiated graphically (Fig. 1b). Fig. 1c shows the transmittance curve of the topograph (Fig. 3, see on the photo pages before the Short Notes part) measured with the double crystal diffractometer. Comparing Fig. 1b and c one can see a very good qualitative accordance of the two curves in the positions of maxima and minima, so that the above assumption is justified experimentally. The quantitative deviations between both curves originate in the limited resolving power and in measuring errors during the photometric determination of the crystal and the topograph and in the necessity of considering the exact connection between intensity and Z.

So it is possible to state that the optical absorption A(z) and the relative change of the lattice parameter are proportional to each other.

The investigations were carried out within the contractual research for the VEB Carl Zeiss Jena.

References

/1/ A.R. LANG and V.F. MIUSCOV, J. appl. Phys. 38, 2477 (1967).

/2/ A.C.MCLAREN, C.F.OSBORNE, and L.A. SAUNDERS, phys. stat. sol. (a) 4, 235 (1971).

/3/ J. HÄRTWIG and R. DIETZ, Kristall und Technik 8, 855 (1976).

/4/ M.C.N. O'BRIEN and H.H.L. PRYCE, Rep. Bristol Conf. Defects in Crystalline Solids, Physical Society, London 1955 (p. 88).

/5/ E.W.J. MITCHELL and E.G.S. PAIGE, Phil. Mag. 46, 1353 (1955).

/6/ A.J. COKEN, J. chem. Phys. 25, 908 (1956); J. Phys. Chem. Solids 13, 321 (1960).

/7/ C.S. BROWN and L.A. THOMAS, J. Phys. Chem. Solids 13, 337 (1960).

/8/ Y. ANDO and N. KATO, J. appl. Cryst. 3, 74 (1970).

/9/ L.I. DATSENKO and E.N. KISLOVSKII, phys. stat. sol. (a) 25, 551 (1974).

(Received August 5, 1977)

phys. stat. sol. (a) 43, K135 (1977)
Subject classification: 1.6; 1.5; 2; 22.6

Research Institute for Technical Physics, Hungarian Academy of Sciences, Budapest[1)]

Induced Crystallization of Amorphous Silicon Oxide

By

M. NÉMETH-SALLAY, Á. BARNA, A. LŐRINCZY, and I.C. SZÉP

Depending on the chemical properties of silicon, an amorphous oxide layer can be formed on its surface during thermal oxidation. In the presence of impurities, some crystallized spots appear besides the amorphous oxide in the grown layer /1, 2/. Crystalline oxide structures are also observed when the silicon surface has been sand-blasted before oxidation; the crystalline phase can be identified as α-crystobalit /3/.

If proper surfaces are prepared and very clean surface treatment methods are used to grow silicon oxide layers, then they are always amorphous. In practice, however, collodion lac is often used to protect the surface of Si-SiO_2 structures during storage. After removing the lac, some structural changes are found on the oxide layer surface. Those changes have been studied at room temperature by the electron diffraction method.

The oxide layers thermally grown on silicon always have an amorphous structure. If the oxide surface is not protected by a lac layer, no structural changes occur for as long as a year or so. On the other hand, in the layers protected by collodion lac, the appearance of a crystalline phase can always be detected. This appearance seems to be independent of the age of the oxide, but it certainly depends on the length of time the surface was protected. ED pictures of silicon oxide are shown on Fig. 1 (see on the photo pages before the Short Notes part), a) after removing the collodion lac and b) when it has not been protected. The latter one shows a definite crystalline character which cannot be identified as some modification of SiO_2. The phenomenon depends neither on the origin of the silicon nor on its electrical properties.

To investigate the crystalline phase and the mechanism of the crystallization, we drew our attention to different protective materials and solvents. We found that only

1) Ujpest 1, P.O. Box 76, H-1325 Budapest, Hungary.

the solvents are the cause of the phase change. A large number of them have been tested (deionized water, alcohols, ethers, alkylchlorides, etc.) and it was found that the crystallization is connected, first of all, to the effect of ethyl ether, but the mechanism has not been explained yet. Some other solvents show no or very little effect. The electron diffraction method in reflection gives well detectable lines for ethyl ether treated specimens with lattice parameter of 4.2 to 4.4; 3.2 to 3.3; 2.85 to 2.95; 2.5 to 2.6 Å.

To study the dependence of the crystallization rate, some other experiments were carried out. 25 to 100 Å thick layers were grown at 900 °C and 800 to 1000 Å thick layers were prepared at 1100 °C without and with heat treatment in dry nitrogen. The samples were placed into a saturator where the effect of solvent vapour could be studied at room temperature. In the case of ethyl ether crystallization could be detected after 1/2 to 1 h treatment of the thin oxide layer. As to the N_2 annealing, the same effect could be achieved but the necessary time was about 20 times longer than that for an unannealed oxide. Moreover, the thicker the oxide layer, the longer time was needed for the appearance of crystallization. E.g. for the thick oxide layers 25 to 100 h were necessary to get crystallization whether they had been annealed or not in the as-grown and N_2 annealed oxide layers.

It can be concluded from our preliminary results that a low temperature crystallization process exists in silicon oxide layers and that it can be related to the solvent effect. In order to find out whether the structural properties of oxides can strongly affect their electrical properties, some other studies have been carried out using C-V, SIMS, and TSC methods. The detailed results of our investigation will be published later.

References

/1/ M.M. ATALLA, E. TANNENBAUM, and E.J. SCHEIBNER, Bell. Syst. techn. J. 38, 749 (1959).

/2/ S.W. JUG, Jr., R.E. MORRISON, and J.E. SANDOR, J. Electrochem. Soc. 109, 221 (1962).

/3/ A.G. REVESZ, Dissertation, Budapest 1967.

(Received August 12, 1977)

phys. stat. sol. (a) 43, K137 (1977)

Subject classification: 1.1; 1.4; 22

Rijksuniversitair Centrum Antwerpen[1)]

Electron Diffraction Evidence for a One-Dimensional Peierls Distortion in a Transition Metal Trichalcogenide: TaS_3 [2)]

By

G. VAN TENDELOO, J. VAN LANDUYT, and S. AMELINCKX[3)]

Unambigeous diffraction evidence for the occurrence of Peierls distortions is limited so far to very few cases. Two-dimensional systems such as the transition metal dichalcogenides (TaS_2, $TaSe_2$, VSe_2, ...) have been studied very intensely /1/. Also the one-dimensional organo-metallic systems as well as KCP have been the subject of detailed studies by means of many different techniques /2/.

The occurrence of "charge density waves" or "Peierls distortions" in transition metal trichalcogenides is suggested by the temperature behaviour of the transport properties as studied by Rouxel and collaborators /3/. The structure of one of these compounds, $TaSe_3$, is shown schematically in Fig. 1; it consists of linear chains of trigonal prisms ($TaSe_6$) resulting in the formation of strings of tantalum atoms separated by triangular sheaths of selenium atoms. For such a structure it is to be expected that structural instabilities accompanying a metal-semiconductor transition might be associated with a one-dimensional distortion as originally proposed by Peierls /4/. These considerations have lead us to examine a number of transition metal trichalcogenides (as well as other compounds such as $ZrSe_3$ having related structures), by means of electron microscopy and electron diffraction in an attempt to find evidence for the structural features associated with "charge density waves" or "Peierls distortions".

In the temperature range explored as yet, we have so far found clear evidence for a Peierls' distortion in TaS_3 only, confirming independently the very recent finding of Sambongi et al. /5/.

---

1) Middelheimlaan 1, B-2020 Antwerp, Belgium.

2) Work performed under the auspices of the association R.U.C.A.-S.C.K.

3) Also at S.C.K.-C.E.N., Mol 2400, Belgium.

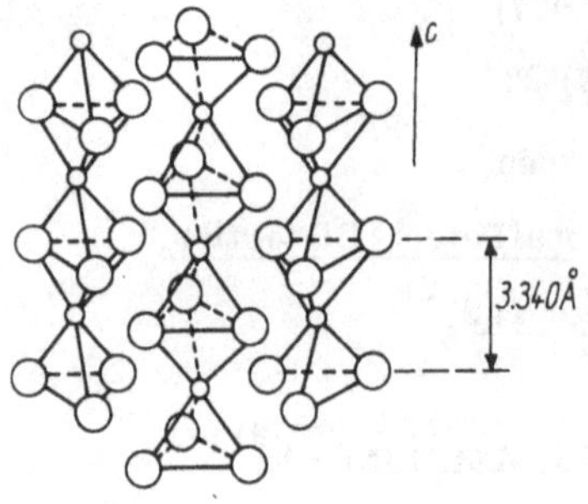

Fig. 1. Schematic representation of the structural features of a number of transition metal trichalcogenides

At and just below room temperature the diffraction pattern of TaS_3 exhibits very faint streaks of diffuse intensity parallel to the dense rows of Bragg reflections (Fig. 2a, see on the photopages before the Short Notes part). The streaks are at a distance of 1/4 of the interrow distance on both sides of the dense rows of Bragg spots; there are no visible streaks midway between these rows. On cooling the specimen holder to liquid nitrogen temperature, the streaks first develop reinforcements; at still lower temperatures these finally break up into weak, sharp, "superperiod" spots (Fig. 2b, see on the photo pages before the Short Notes part). Tilting experiments demonstrate that the diffuse streaks of Fig. 2a are the intersections of planes in reciprocal space with Ewald's sphere. These diffraction phenomena are in many respects similar to those observed in Cu_xNbS_2 $(0.8 < x < 1)$ /6/. We believe that the interpretation is similar as well, except that in the present case the superperiodicity is of different origin.

The structure of TaS_3 has not yet been determined; the orthorhombic lattice parameters have been measured however /9/: a = 36.804 Å, b = 15.173 Å, and c = = 3.340 Å. The similarity of the diffraction pattern with that of $TaSe_3$ suggests that the structure is based on the same features as those of $NbSe_3$ and $TaSe_3$ /3/. The diffraction phenomena in TaS_3 can then be interpreted as follows. On cooling, the linear chains become modulated with a period which is four times the period of the metal chains (i.e. of the height of a trigonal prism). The modulations in neighbouring chains are at first uncorrelated however. Such a structure leads in reciprocal space to planes of intensity perpendicular to the chains /7, 8/. The diffraction pattern (Fig. 2a) is consistent with such a geometry; the modulations presumably result from small displacements of the Ta-ions out of the centres of the trigonal prisms. At still lower temperatures the modulations in neighbouring chains become correlated; a two-dimensional superstructure then results (Fig. 2b). The superperiod spots give information on the wave vector of the modulation ($\vec{K} = 1/4\ \vec{c}^*$). The intensity of the superperiod spots and of the streaks increases relative to that of the Bragg spots for high order reflections, favouring an interpretation based on periodic distortions rather than on atomic ordering. In the present case the modulation wavelength is four times the distance between Ta-atoms in the undeformed structures as mentioned above. From Fig. 2b it is clear that the additional spots are almost invisible close to those rows of Bragg spots which are most strongly excited. This shows that these spots are not precisely in the same reciprocal plane as the Bragg reflections. Note also that the superperiod spots and the Bragg spots do not form a single rectangular array.

The c-parameter of the unmodulated structure being equal to the height of one trigonal prism, all Ta-atoms along a chain are equivalent. Moreover there are two different types of chains shifted one relative to the other over a distance c/2 along the chain direction. Along a given chain the modulation, with period 4c, can be formed with 4 different "phases". The superstructure therefore presumably consists of domains separated by translation interfaces.

Experiments are underway to verify whether or not similar phenomena take place in some of the other transition metal trichalcogenides; presumably at lower temperatures than in TaS_3. It is perhaps of interest to note that for the dichalcogenides the temperature at which the two-dimensional Peierls distortion sets in is higher in TaS_2 than for the related dichalcogenides.

We wish to thank Prof. Rouxel for useful discussions concerning the preparation of the compounds studied.

References

/1/ J. VAN LANDUYT, G. VAN TENDELOO, and S. AMELINCKX, phys. stat. sol. (a) 26, 359 (1974); 26, 585 (1974); 36, 775 (1976); 29, K11 (1975).

J.A. WILSON, F.J. DI SALVO, and S. MAHAJAN, Phys. Rev. Letters 32, 882 (1974); Adv. Phys. 24, 117 (1975).

C.B. SCRUBY, P.M. WILLIAMS, and G.S. PARRY, Phil. Mag. 31, 255 (1975).

P.M. WILLIAMS, G.S. PARRY, and C.B. SCRUBY, Phil. Mag. 29, 695 (1974).

/2/ For a recent review see, H.R. ZELLER, in Festkörperprobleme 13 (1973).

/3/ A. MEERSHAUT and J. ROUXEL, J. less-common Metals 39, 197 (1975).

J. CHAUSSY, P. HAEN, J.C. LAJAUNIAS, P. MONCEAU, G. WAYSAND,

A. WAINTEL, A. MEERSHAUT, P. MOLINE, and J. ROUXEL, Solid State Commun. 20, 759 (1976).

P. MONCEAU, N.P. ONG, A.M. PORTIS, A. MEERSHAUT, and J. ROUXEL, Phys. Rev. Letters 37, 602 (1976).

/4/ R.E. PEIERLS, Quantum Theory of Solids, Oxford Clarendon Press, 1975 (p. 108).

/5/ T. SAMBONGI, K. TSUTSUMI, Y. SHIOZAKI, M. YAMAMOTO, K. YAMAGA, and Y. ABE, Solid State Commun. 22, 729 (1977).

/6/ G. VAN TENDELOO, R. DE RIDDER, D. VAN DYCK, J. VAN LANDUYT, and S. AMELINCKX, phys. stat. sol. (a) 38, 185 (1976).

/7/ A. GUINIER, Théorie et Technique de la Radiocristallographie, Dunod, Paris 1964 (p. 508).

/8/ R. COMES, M. LAMBERT, H. LAUNOIS, and H.R. ZELLER, Phys. Rev. B 8, 571 (1973).

/9/ E. BJERKELUND, J.H. FERMOR, and A. KJEKSHUS, Acta chem. Scand. 20, 1836 (1966).

(Received September 6, 1977)

phys. stat. sol. (a) 43, K141 (1977)

Subject classification: 1.2 and 14.2; 18.1; 21; 21.1

Electrotechnical Laboratory, Tokyo[1)]

On the Relation between the Density of States in the Cubic State and Phase Transition Temperatures (T_c, T_m) of V-Zr-Nb Ternary Laves Compounds

By

Y. KIMURA

The cubic Laves compounds V_2(Hf, Zr) are known as high magnetic field conductors. The maximum upper critical field rises above 200 kOe at 4.2 K /1, 2/. All of these compounds undergo the martensitic transformation at low temperature. Zr-rich compounds and Hf-rich compounds undergo a cubic-to-hexagonal and a cubic-to-orthorhombic transformation, respectively /3 to 7/.

In this paper, we will report the relation between the density of states in the cubic state and the superconducting transition temperature as well as the relation between the density of states and the martensitic phase transition temperature.

V_2Zr was chosen as the mother alloy; because the highest increasing rate of the superconducting temperature per atomic percent is obtained for Nb when a part of Zr in V_2Zr is replaced by another transition metal among the V-based cubic Laves compounds and V_2Zr itself has the second highest superconducting transition temperature in the cubic Laves compounds.

This study was undertaken as a part of the program of the investigation on the superconducting properties of V_2(Hf, Zr) compounds.

We measured the susceptibility in the temperature range between 77 and 300 K and the specific heat between 70 and 150 K. The superconducting transition temperature, T_c, was measured by a standard four-contact technique.

The compounds were prepared by arc-melting of appropriate amounts of components on a water-cooled hearth in an argon atmosphere. They were turned over and remelted not less than five times to ensure homogeneity. The weight loss of samples on melting was negligible. V, Zr, and Nb used were of 99.9% purity. Samples for the susceptibility and resistivity measurements which were cut out from the ingots were sealed in silica tubes in a vacuum below $3x10^{-5}$ Torr and quenched into water after annealing for 72 h at 900 ^{o}C.

---

1) 5-4-1 Mukodai-machi, Tanashi-shi, Tokyo, Japan.

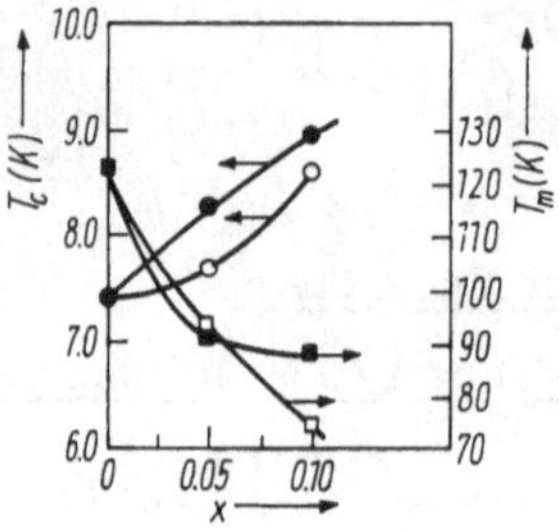

Fig. 1

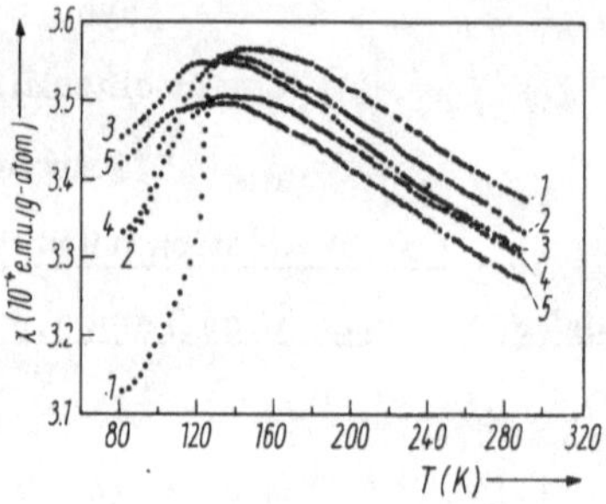

Fig. 2

Fig. 1. The dependence of the superconducting temperature and the martensitic transformation temperature on the concentration of Nb; ■, ● $V_{2-x}ZrNb_x$; □, ○ $V_2Zr_{1-x}Nb_x$

Fig. 2. The paramagnetic susceptibility χ of $V_2Zr_{1-x}Nb_x$ and $V_{2-x}ZrNb_x$ as a function of T; (1) V_2Zr, (2) $V_{1.95}Nb_{0.05}Zr$, (3) $V_{1.9}Nb_{0.1}Zr$, (4) $V_2Zr_{0.95}Nb_{0.05}$, (5) $V_2Zr_{0.9}Nb_{0.1}$

The X-ray investigation of crushed powder confirmed that these samples were of C-15 structure. If x is greater than 0.1 in $V_2Zr_{1-x}Nb_x$ and $V_{2-x}ZrNb_x$, these compounds are composed of two phases.

T_c and T_m measured are shown in Fig. 1. The open circles and squares are for $V_2Zr_{1-x}Nb_x$. The solid circles and squares are for $V_{2-x}ZrNb_x$. T_c and T_m have anticorrelation. The transition temperatures measured by the susceptibility are almost the same as those measured by the resistive method.

The values of $\Delta\chi/\chi = [\chi(170\text{ K}) - \chi(290\text{ K})]/\chi(170\text{ K})$ versus the superconducting transition temperature is plotted in Fig. 3. It is interesting that the value of $\Delta\chi/\chi$ is linearly dependent on the transition temperature.

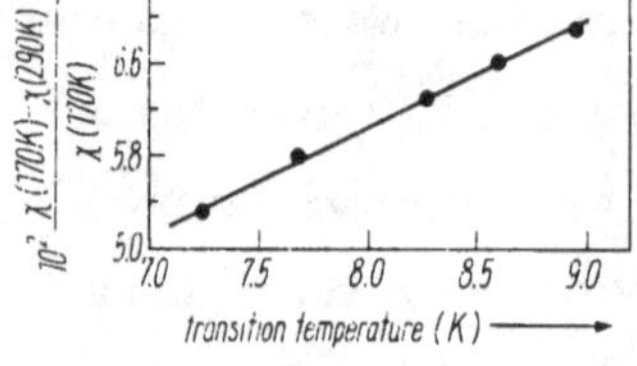

Fig. 3. $\Delta\chi/\chi$ as a function of T_c

Cohen et al. /8/ approximated an abruptly varying density of states near the Fermi level by

$$n(E) = N_o \quad \text{for} \quad E \geq 0 ,$$
$$\alpha N_o \quad \text{for} \quad E < 0$$

and determined the position of the Fermi level by equating the total number of occupied states at any temperature T to that at T = 0. Thus they derived a function $F_\alpha(T)$ defined as follows:

$$F_\alpha(T) = 1 - [F_\alpha(T)]^\alpha \exp(-T_o/T) ,$$

where α means the ratio of the s-density of states to the total density of states and T_o is the effective Fermi temperature of the d-band. According to their results, the susceptibility is written as follows:

$$\chi(T) = \mu_b^2 (1 - \alpha) N_o F_\alpha(T) + \chi_{orb} .$$

The orbital contribution to the susceptibility is given by the $F_\alpha(T) = 0$ intercept of χ. The density of states N_o is estimated from the slope of the solid line. We chose α and T_o so as to get the best fit for the resistance results. But there is some uncertainty about both values, because the temperature range of the resistance measurements in the cubic state is narrow and the residual resistance is very large.

The results are plotted in Fig. 4. Below 100 K a decrease in the effective density of states is indicated. The decrease is due to the transformation. Fig. 4 shows that χ_{orb} does not vary appreciably. The results of the

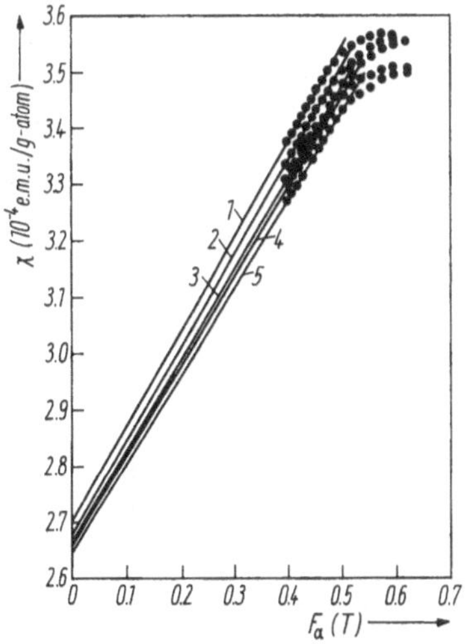

Fig. 4. The paramagnetic susceptibility of $V_2Zr_{1-x}Nb_x$ and $V_{2-x}ZrNb_x$ as a function of $F_\alpha(T)$ with $\alpha = 0.11$ and $T_o = 120$ K; (1) V_2Zr, (2) $V_{1.95}Nb_{0.05}Zr$, (3) $V_{1.9}Nb_{0.1}Zr$, (4) $V_2Zr_{0.95}Nb_{0.05}$, (5) $V_2Zr_{0.9}Nb_{0.1}$

Table 1

sample	T_c (K)	T_m (K)	χ_{orb}(10^{-4} e.m.u./g-atom)	N_o (states/eV atom)	H (J/K g-atom)
V_2Zr	7.45	123	2.71	6.5	2.27
$V_{1.95}Zr_{0.05}Nb$	8.27	91	2.70	6.3	0.3
$V_{1.90}Zr_{0.1}Nb$	8.96	88	2.67	6.2	0.01
$V_2Zr_{0.95}Nb_{0.05}$	7.68	93	2.66	6.2	0.35
$V_2Zr_{0.90}Nb_{0.1}$	8.60	74	2.60	5.9	0.03

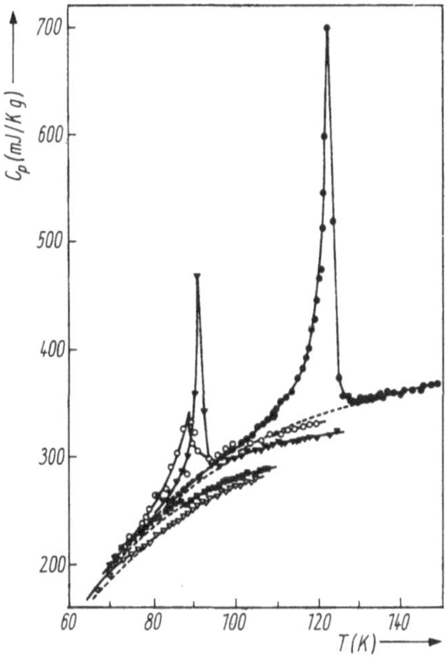

Fig. 5. The specific heat of $V_2Zr_{1-x}Nb_x$ and $V_{2-x}ZrNb_x$; ● V_2Zr, ○ $V_{1.95}Nb_{0.05}Zr$, ■ $V_{1.9}Nb_{0.1}Zr$, ▼ $V_2Zr_{0.95}Nb_{0.05}$, ▽ $V_2Zr_{0.9}Nb_{0.1}$

density of states obtained from the slopes are shown in Table 1. The density of states in the cubic state decreases as the concentration of Nb increases. But after transformation below T_m the density of states would increase with the increase of the Nb concentration corresponding to the increase of T_c. The same situations occurred in the V_2 (Hf, Zr) system where we could compare T_c with the electronic specific heat coefficient for some compounds. The V_2 (Hf, Zr) system is now under investigation.

The results for the specific heat are shown in Fig. 5 /9/. The transformation temperature T_m and the transformation energy ΔH which is estimated by substracting the background represented by the broken line in the figure decrease as the concentration of Nb increases. It is found that the phase transition is suppressed more and more as the density of states in the cubic state increases.

The extended results including the V_2 (Hf, Zr) system will be published in short time.

References

/1/ K. INOUE, K. TACHIKAWA, and Y. IWASA, Appl. Phys. Letters 18, 123 (1971).

/2/ K. INOUE and K. TACHIKAWA, Japan. J. appl. Phys. 12, 161 (1973).

/3/ J.E. DOHERTY and D.F. GIBBONS, phys. stat. sol. (b) 44, K5 (1971).

/4/ A.C. LAWSON and W.H. ZACHARIASEN, Phys. Letters A 38, 1 (1972).

/5/ A.C. LAWSON, Phys. Letters A 36, 8 (1971).

/6/ Y. KIMURA, Japan. J. appl. Phys. 16, 517 (1977).

/7/ D.E. MONCTON, Solid State Commun. 13, 1779 (1973).

/8/ R.W. COHEN, G.D. CODY, and J.J. HALORAN, Phys. Rev. Letters 19, 840 (1967).

/9/ Y. KIMURA, J. Phys. Soc. Japan 36, 306 (1974).

(Received June 28, 1977)

phys. stat. sol. (a) 43, K147 (1977)

Subject classification: 14.2; 8; 13; 21

University of Illinois at Urbana - Champaign[1)]

On Isotope Effects in $PdH(D)_x$

By

C.B. SATTERTHWAITE

Some time ago we proposed an explanation for the reverse isotope effect in the superconducting properties of $PdH(D)_x$ /1/ which involved a difference in electronic structure between PdH_x and PdD_x. Recent papers in this journal /2, 3/ are particularly relevant to the proposed explanation and provide strong support for it.

The difference in electronic structure between PdH_x and PdD_x, we believe, arises in the following way. When hydrogen is added to Pd /4/, occupying the octahedral interstitial sites Pd-H bonding states are formed which are hydridized electron states involving s-p states of Pd and s states of H. These states are well below the Fermi level and account for approximately 0.5 electrons per H atom. The remaining hydrogenic electrons enter the unfilled states in the Pd d-band and when those are filled, progressively fill the broad s-p hybridized band of Pd. Superconductivity appears only at concentrations of H which correspond to filling the s-p band well above the top of the d-band. The superconducting transition temperature rises steeply with increasing H concentration and thus appears to be critically dependent upon the degree of filling of the s-p band.

The argument for an electronic difference between the hydride and deuteride was based on the difference in zero point amplitude of H and D in PdH_x and PdD_x lattices of essentially identical lattice constant. Recent measurements of the lattice constants in these materials show a small difference, the PdH_x constant larger than the PdD_x constant by about 0.005 Å independent of x /5/ for $0.6 < x < 1.0$. Because this difference is small compared to the difference in amplitude, it does not alter the argument. From neutron scattering measurements /6/ root-mean-squared displacements of 0.23 and 0.20 Å were found for H and D, respectively, in the $PdH(D)_x$ lattices which are in approximate agreement with the harmonic approximation in which the ratio is $4\sqrt{2}$.

---

1) Urbana, Illinois 61801, USA.

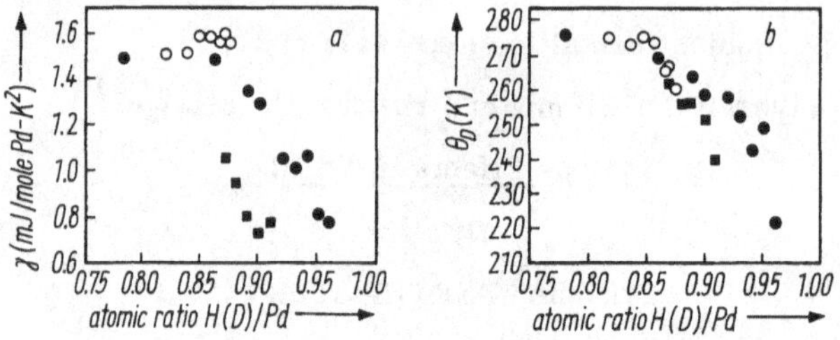

Fig. 1. a) Comparison of the electronic specific heat coefficient γ and b) comparison of Debye temperatures Θ_D for PdH_x and PdD_x; data taken from ● PdH_x /2/, ■ PdD_x /3/, o PdH_x /8/

As a result of the difference in amplitude the H atoms, on time average, will be in closer proximity to the surrounding Pd atoms than will the D atoms. Thus relative to the deuteride of the same concentrations one would expect for the hydride enhanced hybridization involving more electrons and resulting in stronger bonds with the Pd atoms leaving fewer electrons to fill the bands near the Fermi level.

For the same atomic ratio the hydride will have fewer electrons in the s-p band than the deuteride or for those properties dependent upon the electronic structure near the Fermi energy the deuteride of composition PdD_x should be comparable to the hydride of $PdH_{(x+\delta)}$. From the superconducting behavior we estimated δ to be approximately 0.05 /1/.

In /1/ we noted that in addition to accounting for the reverse isotope this model is consistent with a) the pressure dependence of T_c in PdH_x, b) significantly higher dissociation pressures, and c) smaller negative heats of absorption for deuterides than for hydrides at comparable temperature and compositions. More recently the Pd-H force constant has been found to be approximately 20% greater than that for Pd-D from analysis of neutron scattering results /7/ as predicted in /1/.

The specific heat studies of Wolf et al. show an isotope effect in both electronic and lattice specific heat which further supports our model for an electronic difference. The shift in Debye temperature Θ_D is consistent with a larger force constant in PdH_x than in PdD_x. And the electronic specific heat coefficient γ shows a shift such that $\gamma(PdH_{x+\delta}) \approx \gamma(PdD_x)$, where δ is approximately 0.05. Fig. 1a and b show these shifts in Θ_D and γ and show also that the work of /2/ is essentially in agreement with similar measurements of Mackliet et al. /8/ on PdH_x.

References

/1/ R.J. MILLER and C.B. SATTERTHWAITE, Phys. Rev. Letters 34, 144 (1975).

/2/ M. ZIMMERMANN, G. WOLF, and K. BOHMHAMMEL, phys. stat. sol. (a) 31, 511 (1975).

/3/ G. WOLF and M. ZIMMERMANN, phys. stat. sol. (a) 37, 485 (1976).

/4/ D.E. EASTMAN, J.K. CASHION, and A.C. SWITENDICK, Phys. Rev. Letters 27, 35 (1971).

/5/ J.E. SCHIRBER and B. MOROSIN, Phys. Rev. B 12, 117 (1975).

/6/ J.E. WORSHAM, Jr., M.K. WILKINSON, and C.G. SHULL, J. Phys. Chem. Solids 3, 303 (1957).

/7/ A. RAHMAN, K. SKÖLD, C. PELIZZARI, and S.K. SINHA, Phys. Rev. B 14, 3630 (1976).

/8/ C.A. MACKLIET, D.J. GILLESPIE, and A.I. SCHINDLER, J. Phys. Chem. Solids 37, 379 (1976).

(Received June 20, 1977)

phys. stat. sol. (a) 43, K151 (1977)

Subject classification: 20.1; 22.3

Department of Physics, Simon Fraser University, Burnaby[1)]

On Refractive Index Measurements in GaS and GaSe

By

T.A. MCMATH, G. JACKLE, and J.C. IRWIN

The fringe pattern that results from multiple internal reflections when white light is transmitted through a weakly absorbing parallel sided film provides an excellent method for determining the indices of refraction of the material concerned. For example, the present authors utilized the technique /1/ to measure the indices of refraction of GaSe and GaS and obtained results in good agreement with most of the available literature values /2 to 4/.

Recently, Akhundov et al. /5/ and Malik et al. /6/ have carried out similar experiments on the same compounds and have presented some quite different results. Both groups obtained interference curves exhibiting a marked heating effect and in addition Akhundov et al. /5/ reported a dependence of the dispersion on the incident light intensity. Finally Akhundov et al. /5/ observed strong resonances in the resultant dispersion curves for both materials in the spectral range 500 to 600 nm. Since similar effects were not reported in previous works /1, 3, 4/ additional experiments have been carried out in an effort to reproduce these observations.

Typical transmission curves for a GaS sample of thickness 7.4 μm and a GaSe sample of thickness 19 μm are shown in Fig. 1, along with the system response curve. The GaS crystals used in this work were cleaved from samples grown by a Bridgman technique and the GaSe crystals were obtained from both vapour transport and Bridgman grown samples. It is clear that no beating effect similar to that observed by Akhundov et al. /5/ and Malik et al. /6/ is evident in the curves of Fig. 1. Malik et al. /6/ attributed the beating to residual strains produced in the crystal during cooling after growth because of anisotropic thermal coefficients. They further stated that this anisotropy ruled out any scope for obtaining undeformed GaS crystals from growth in ampoules.

---

1) Burnaby, B.C., Canada V5A 1S6.

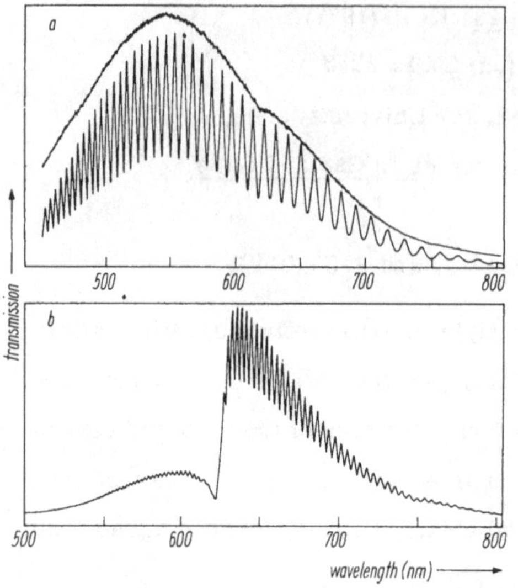

Fig. 1. Transmission spectra obtained for a) a GaS crystal and b) a GaSe sample at 300 K. The system response curve is also shown in a)

Beating effects have been observed with a few samples but their presence has always been correlated with a mechanical defect in the crystal, such as a step in the crystal surface or an air wedge separation between layers, or mechanical strain sometimes produced during mounting of the thinnest samples, which have little rigidity. The results of this work, however, indicate that it is certainly possible to obtain good interference patterns from GaS and GaSe crystals. If crystal anisotropy dictates that residual strains are always present in such crystals it is certain that their effects are much less pronounced than those observed by Malik et al. /6/.

Measurements to check the intensity dependence of the dispersion were made on a GaS sample of thickness 7.4 μm and a GaSe sample of thickness 5.5 μm. As described in /1/, white light from a tungsten filament projector lamp was incident on the samples, and the transmitted light was focussed on the entrance slit of a double grating spectrometer. The total unattenuated intensity incident on the sample from the source was about 70 mW/cm^2. The intensity was measured with a Spectra-physics 401C power meter and included all wavelengths between about 400 and 1100 nm. The resulting fringe pattern was investigated in the wavelength region from 450 to 650 nm. The measurements were repeated with the only change being the sequential insertion of neutral filters having optical densities of 0.3, 1.0, and 2.0 in the path of the incident light. The spectral lines from a Neon Geissler tube were used for wavelength calibration.

A portion of two of the resulting transmission patterns around the band edge of GaSe are shown in Fig. 2. The 2.0 OD filter was used to obtain curve A and curve B was obtained with the source unattenuated. No shift in the position or shape of the

Fig. 2. Transmission spectra in the vicinity of the band edge for a GaSe crystal. The incident light for curve A was attenuated by 100x from that of curve B. Both spectra were obtained at 300 K

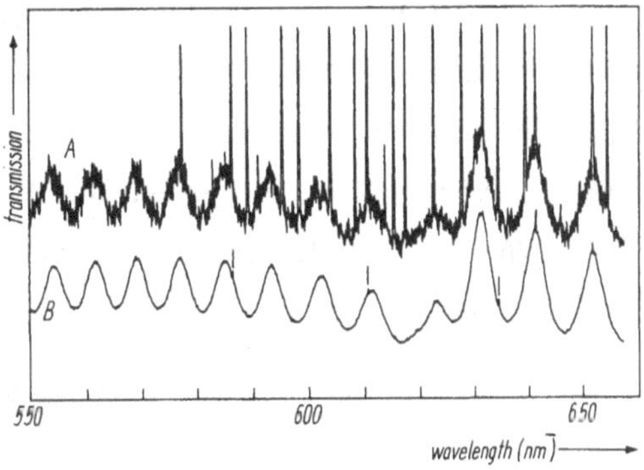

fringes is observable from these two patterns nor was there any discernable difference between any of the patterns obtained. Finally, the measurements were repeated with the sample mounted at the exit slit of the spectrometer. This reduced the total intensity incident on the sample by a factor of about 10^4 while maintaining the intensity at the wavelength of interest about the same. Once again, however, there were no detectable changes in the resultant fringe pattern. Thus for the range of light levels used in this experiment it must be concluded that the ordinary and extraordinary dispersion curves of GaS and GaSe are independent of incident light intensity.

It should also be noted that no evidence has been seen in any of our fringe patterns of the strong resonances observed by Akhundov et al. /5/. They observed such resonances in the dispersion of both GaS and GaSe in the vicinity of 570 nm. Such resonances would have produced drastic changes in the periodicity of the fringe patterns and should have been particularly visible with the thicker samples of GaS which are quite transparent at these wavelengths (for our thickest sample, 225 μm, the fringe spacing is about 0.25 nm).

Acknowledgements

The GaS crystals and some of the GaSe crystals were obtained from Dr. J.L. Brebner of the Université de Montreal. The financial support of the National Research Council of Canada is gratefully acknowledged.

References

/1/ J.A. MCMATH and J.C. IRWIN, phys. stat. sol. (a) 38, 731 (1976).

/2/ J.L. BREBNER and J.-A. DEVERIN, Helv. Phys. Acta 38, 650 (1965).

/3/ J.D. WASSCHER and J. DIELEMAN, Phys. Letters A 39, 279 (1972).

/4/ G.A. AKHUNDOV, S.A. MUSAEV, A.E. BAKHYSHEV, N.M. GASANLY, and L.G. MUSAEVA, Soviet Phys. - Semicond. 9, 94 (1975).

/5/ G.A. AKHUNDOV, L.G. MUSAEVA, and M.D. KHOMUTOVA, Soviet Phys.- Optics and Spectroscopy 39, 395 (1975).

/6/ A.I. MALIK, Z.D. KOVALYUK, and G.B. DELEVSKII, Soviet Phys. - Solid State 18, 420 (1975).

(Received August 26, 1977)

phys. stat. sol. (a) 43, K155 (1977)

Subject classification: 10.1; 21.1

Institute of Nuclear Research, Ukrainian Academy of Sciences, Kiev[1)]

Phenomenological Dependence of the Deformation Activation Volume upon the Deformation Value

By

S.D. MAMCHICH

Introduction It is known that the activation volume q in the expression for the thermo-activated plastic strain rate

$$\dot{\varepsilon} = A \exp\left[-\frac{U_o - q(\sigma - \sigma_i)}{kT}\right] \tag{1}$$

is dependent upon the total deformation value ε in general.[2)] However, the concrete form of this dependence is not given up to the present in the literature. Therefore an attempt to determine this dependence may be of definite importance.

Phenomenological approach If the deformation process can be described by the rate equation (1) then the relaxation of the tensile stress σ is as follows /1/:

$$\sigma - \sigma_i = \frac{U_o}{q} + \frac{kT}{q} \ln \frac{MkT}{Aq} - \frac{kT}{q} \ln (t + c) , \tag{2}$$

where M is a constant which includes the specimen geometry and the specimen and machine moduli, c an integration constant.

In the case of sufficiently long times when $t \gg c$

$$\sigma - \sigma_i = \frac{U_o}{q} + \frac{kT}{q} \ln \frac{MkT}{Aq} - \frac{kT}{q} \ln t . \tag{3}$$

On the other hand, the stress relaxation for $t \gg c$ may be described /2/ with satisfactory accuracy by the following expression:

$$\sigma = \sigma_1 - \frac{kT(\sigma_1 - \sigma_i)}{kT \ln(B/t_1) + U_o} \ln (t/t_1) , \tag{4}$$

---

1) Prospekt Nauki 119, Kiev-28, USSR.

2) A is a frequency factor including the entropy term, U_o the activation enthalpy, σ the applied stress, σ_i the athermal component due to long-range forces, k the Boltzmann constant, and T the absolute temperature /1/.

where σ_1 is the stress at the moment $t = t_1$ in the straight sections of the σ - ln t dependences (σ_1 can be obtained by extrapolation when the straight lines of these dependences are beginning at $t > t_1$), and B a coefficient weakly dependent on temperature and stress. The stress σ_i in expression (4) has the sense of a threshold stress.[3)]

Comparing the derivatives $\partial\sigma/\partial \ln t$ from (3) and (4), one receives

$$q = \frac{kT \ln(B/t_1) + U_0}{\sigma_1 - \sigma_i} \tag{5}$$

Since the total deformation ε is kept constant in the relaxation regime, it can be assumed

$$\frac{\partial q}{\partial(1/(\sigma_1 - \sigma_i))} = \frac{\partial q}{\partial(1/(\sigma_0 - \sigma_i))} = \frac{\partial q}{\partial(1/\varepsilon)} E , \tag{6}$$

where $\sigma_0 = \sigma\big|_{t=0}$, E the elastic modulus of the specimen. In this case it can be written

$$q = q_0 + \frac{a}{\varepsilon} , \tag{7}$$

where $a = (1/E)(kT \ln B + U_0 - kT f(t)\big|_{t=0})$, q_0 is an integration constant.

Discussion The dependence of the activation volume upon total deformation and temperature in the form (7) is in a satisfactory agreement with the experimental data found in /4/, used to define the temperature dependences of the activation volume deformation (Fig. 1). Furthermore, the strain rate equation in the form

$$\dot{\varepsilon} = A \exp\left[-\frac{U_0 - (q_0 + a/\varepsilon)(\sigma - \sigma_i)}{kT}\right] \tag{8}$$

is in qualitative compliance with the experimental fact of strain hardening.

The equation (8) describes logarithmic creep. In fact this expression can be written in the form

$$\exp\left(-\frac{\alpha}{\varepsilon}\right) d\varepsilon = D\, dt , \tag{9}$$

3) The athermic component σ_i in (1) acquires the sense of a threshold stress in this case as dislocations are pushed on by the external stress among the hard non-deformed precipitate formed in the soft matrix by impurity and alloying additions /3/.

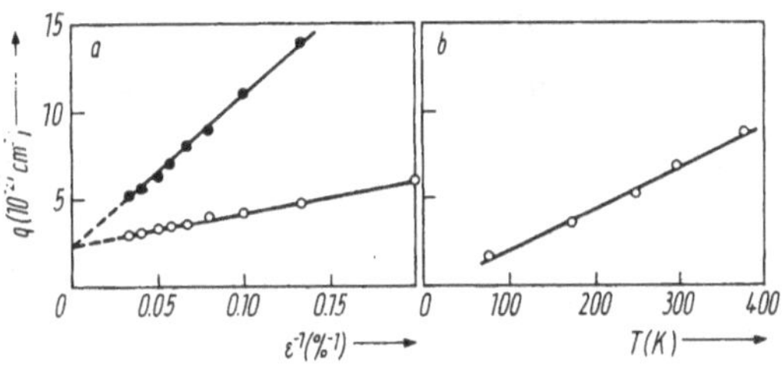

Fig. 1. Dependences of activation volume q upon a) deformation (Ni, ● 273 K, ○ 190 K) and b) temperature (Ni 15% Cr, ε = 10%) which are constructed according to the data of /4/

where

$$\alpha = \frac{a(\sigma - \sigma_i)}{kT}, \quad D = A \exp\left[-\frac{U_o - q_o(\sigma - \sigma_i)}{kT}\right].$$

The total deformation ε can be expressed as a sum of initial (elastic) ε_e and creep ε_p strains:

$$\varepsilon = \varepsilon_e + \varepsilon_p = \varepsilon_e(1 + \varepsilon_p/\varepsilon_e) . \tag{10}$$

In the case

$$\varepsilon_p \ll \varepsilon_e \tag{11}$$

expression (10) gives

$$\frac{1}{\varepsilon} \approx \frac{1}{\varepsilon_e}(1 - \varepsilon_p/\varepsilon_e) . \tag{12}$$

Substituting (12) in (9) we have

$$\exp\left[-\frac{\alpha}{\varepsilon_e}(1 - \frac{\varepsilon_p}{\varepsilon_e})\right] d\varepsilon_p = D\, dt . \tag{13}$$

After integration of this equation and substitution of the initial condition ($\varepsilon|_{t=0} = \varepsilon_e$) we obtain

$$\varepsilon = \varepsilon_e + \beta \ln(1 + \nu_c t) , \tag{14}$$

where

$$\beta = \frac{kT\,\varepsilon_e^2}{a(\sigma - \sigma_i)} ; \quad \nu_c = \frac{1}{\beta} A \exp\left[-\frac{U_o - (q_o + a/\varepsilon_e)(\sigma - \sigma_i)}{kT}\right] .$$

It must be emphasized that keeping of the condition (11) for receiving logarithmic creep from (8) reflects the experimental fact of the small deformation values of that type of creep comparing with the elongation which occurs after applying the stress /5/. That is why such inference of the logarithmic creep equation can be of independent importance.

References

/1/ F. GUIU and P.L. PRATT, phys. stat. sol. 6, 111 (1964).

/2/ S.D. MAMCHICH and V.N. DROBYASIN, phys. stat. sol. (a) 38, K5 (1976).

/3/ V.I. STARTSEV, V.YA. ILICHEV, and V.V. PUSTOVALOV, Plastichnost i prochnost metallov i splavov pri niskikh temperaturakh, Metallurgiya, Moscow 1975.

/4/ N. CLEMENT, C.R. Acad. Sci. (France) C278, 577 (1974).

/5/ D. MCLEAN, Mechanical Properties of Metals, John Wiley & Sons, New York/London 1962.

(Received July 1, 1977)

phys. stat. sol. (a) 43, K159 (1977)
Subject classification: 15; 13.1; 14.3; 22.3.1

Institute of Applied Physics, Academy of Sciences of the Moldavian SSR, Kishinev

The Thermoelectric Power of $Cd_3(As_xP_{1-x})_2$

By

S.I. RADAUTSAN, E.K. ARUSHANOV, L.N. LUKYANOVA, and A.N. NATEPROV

The semiconducting alloys $Cd_3(As_xP_{1-x})_2$ are of interest because of their high electron mobility, small effective mass, and the possibility of the existence of a band inversion similar to that found in CdHgTe alloys. Electrical /1/, optical /2 to 4/, magneto-optical /2/ properties of $Cd_3(As_xP_{1-x})_2$, and transport phenomena of $Cd_3(As_{0.7}P_{0.3})_2$ were investigated.

In this note experimental results on the magnetic field dependence of the thermoelectric power α and the Hall coefficient R of $Cd_3(As_xP_{1-x})_2$ as well as their band parameters are reported.

The crystals of $Cd_3(As_xP_{1-x})_2$ were grown by the Bridgman method. The evacuated quartz tube with stoichiometric quantities of the elements or with the required amount of the previously prepared Cd_3As_2 and Cd_3P_2 were pulled through a temperature gradient of 10 to 30 K/cm at the rate of 2 to 6 cm/d. The crystals were not specially doped. The composition of the samples was controlled by X-ray diffraction.

The electron concentrations and room temperature mobilities of the used samples were (1 to 3)x10^{24} m^{-3} and 0.6 to 1.3 m^2/Vs, respectively. The parameters of the typical samples are given in Tables 1 and 2.

In the investigated crystals in the range 77 to 300 K extrinsic conductivity is observed. The signs of R and α for all samples were negative. The magnetic field dependence of the Hall coefficient and thermoelectric power of $Cd_3(As_xP_{1-x})_2$ were investigated at 300 and 90 K in a magnetic field B up to 3.0 T. The value of R changes with magnetic field by not more than 15%, α increases in absolute value with B (Fig. 1, 2). The value of the high magnetic field thermoelectric power α_∞ has been calculated from the equation $\alpha_\infty = \alpha_0 + \Delta\alpha_\infty$, where α_0 is the zero magnetic field thermoelectric power. The value of $\Delta\alpha_\infty$ has been determined from saturation of $\Delta\alpha$ in magnetic field or from the dependence $B^2/(\alpha - \alpha_\infty)$ on B^2.

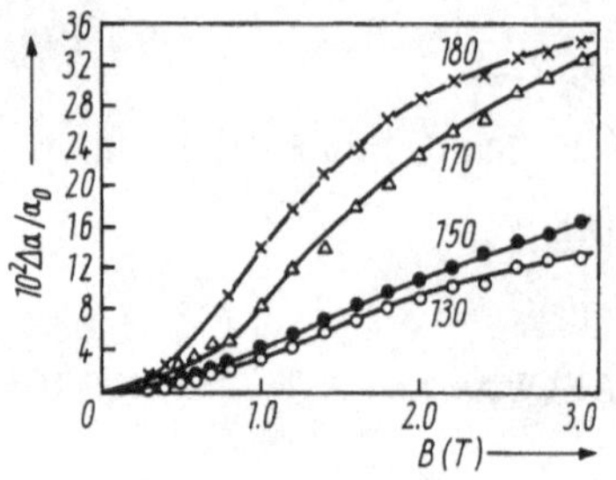

Fig. 1

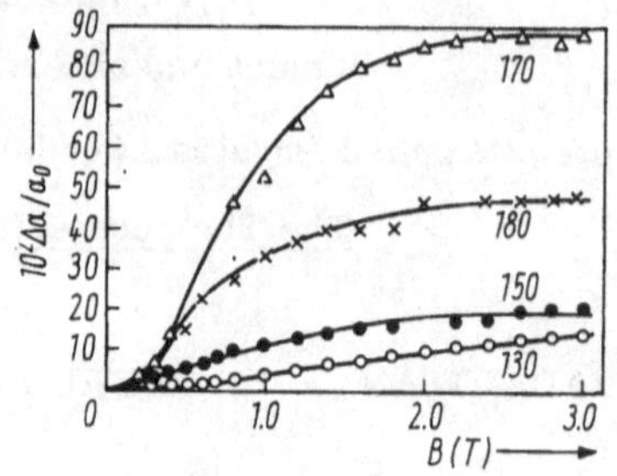

Fig. 2

Fig. 1. Magnetic field dependences of $\Delta\alpha/\alpha_0$ at 300 K

Fig. 2. Magnetic field dependences of $\Delta\alpha/\alpha_0$ at 90 K

The conduction band of Cd_3As_2 /7/, Cd_3P_2 /8/, and $Cd_3(As_{0.7}P_{0.3})_2$ /5/ is of Kane type. It is reasonable to assume a Kane model for the system of solid solutions $Cd_3(As_xP_{1-x})_2$ as done by Wagner et al. /2/, and Radoff and Bishop /3/.

The electron gas in the investigated samples is degenerate at 300 and 90 K (the reduced Fermi level η is greater than 5 (Tables 1 and 2)). Using the values of α_0, α_∞, the electron concentration n, and the energy gap E_g, the band parameters of $Cd_3(As_xP_{1-x})_2$ have been determined by the method described in /9/. It was assumed that at 0 K the value of E_g varies with composition according to the equation $E_g = 0.588 - 0.834\,x$ /3/, the temperature coefficient of the gap $\partial E_g/\partial T$ varies linearly from -0.1×10^{-4} eV/K in Cd_3As_2 /10/ to -3×10^{-4} eV/K in Cd_3P_2 /3/ for temperatures up to 300 K. The calculation was also done by using room temperature optical band gap variations as a function of composition of $Cd_3(As_xP_{1-x})_2$ alloys according to the data of /2/.

The value of P, used in the Kane-type band model, was calculated from the formula

$$P^2 = \frac{\hbar^2}{2}\left(\frac{1}{m_n} - \frac{1}{m_0}\right) E_g (E_g + \Delta)\left(E_g + \frac{2}{3}\Delta\right)$$

assuming a linear variation of Δ from 0.15 eV in Cd_3P_2 /11/ to 0.37 eV in Cd_3As_2 /11/, $\Delta \gg E_g$, $\Delta \ll E_g$ (Δ is the spin-orbit splitting), and in the investigated crystals P is respectively equal to 7.6×10^{-8} eVcm $\pm$ 18%; 8.2×10^{-8} eVcm $\pm$ 18%;

Table 1

Parameter values for $Cd_3(As_xP_{1-x})_2$ at T = 90 K

sample No.	x	n (10^{24} m^{-3})	α_o (10^{-6} V/K)	α_∞ (10^{-6} V/K)	$\frac{m^*}{m_o}$	$\frac{m_n}{m_o}$	η	γ	r	E_g(eV)
180	0.8	1.63	16.7	24.2	0.038	0.011	26.9	0.94	0.91	0.174
						0.006	35.2	1.03	1.0	0.085
170	0.7	2.97	14.3	26.7	0.062	0.022	23.5	0.68	0.88	0.20
						0.0005	31.8	0.8	1.0	0.004
150	0.5	2.0	24.2	28.4	0.051	0.023	20.6	1.07	0.79	0.26
						0.016	22.7	1.17	0.89	0.16
130	0.3	1.17	26.9	33.7	0.041	0.024	15.9	0.84	0.64	0.37
						0.023	16.2	0.89	0.69	0.32

The calculated values of the upper and lower lines in Tables 1 and 2 have been obtained using $E_g(x)$ according to the data of /2/ and /3/, respectively.

Table 2

Parameter values for $Cd_3(As_xP_{1-x})_2$ at T = 300 K

sample No.	x	n (10^{24} m^{-3})	α_o (10^{-6} V/K)	α_∞ (10^{-6} V/K)	$\frac{m^*}{m_o}$	$\frac{m_n}{m_o}$	η	γ	r	E_g(eV)
180	0.8	1.59	70.7	103.5	0.046	0.015	6.2	0.91	0.89	0.16
						0.010	6.5	0.95	0.93	0.10
170	0.7	2.7	56.2	82.9	0.053	0.016	8.04	0.94	0.92	0.18
						0.002	10.3	1.02	1.0	0.023
150	0.5	1.9	69.8	87.7	0.045	0.017	6.95	1.03	0.84	0.23
						0.011	7.8	1.13	0.94	0.13
130	0.3	1.27	84.2	106.9	0.041	0.023	5.1	0.87	0.69	0.33
						0.021	5.3	0.93	0.75	0.28

$6.7x10^{-8}$ eVcm ± 18% at 90 K and $7.7x10^{-8}$ eVcm ± 13%; $8.5x10^{-8}$ eVcm ± 13%; $7x10^{-8}$ eVcm at 300 K. The matrix element P relatively slightly varies with composition in the $Cd_3(As_xP_{1-x})_2$ system as in many semiconducting alloys.

Results of calculations of the density of states effective electron mass m^*/m_o, the effective electron mass at the bottom of the conduction band m_n/m_o, the nonparabolicity parameter γ ($\gamma = 3 \, d \lg m^* / d \lg n$), and the scattering factor r are given in Tables 1 and 2.

The calculated values excepting m_n/m_o are relatively slightly depending on the choice of the E_g value.

The value of the matrix element P and the scattering factor r in the investigated crystals is close to those determined for cadmium arsenide and cadmium phosphide /11/ and for composition x = 0.7 they are in agreement with the data of Blom et al. /5/.

In the investigated crystals r is close to 1 at 300 K. It may be supposed that scattering by optical lattice vibrations is the dominant one at room temperature. The possibility of this scattering is also supposed for the end point compounds of the system $Cd_3(As_xP_{1-x})_2$ /11/.

References

/1/ K. MASUMOTO and S. ISOMURA, Energy Conversion 10, 129 (1970).

/2/ R.I. WAGNER, E.D. PALLIK, and E.M. SWIGGARD, J. Phys. Chem. Solids, Suppl. 1, 471 (1971).

/3/ P.L. RADOFF and S.G. BISHOP, Phys. Rev. B 5, 442 (1972).

/4/ M. ZIVITZ and J.R. STEVENSON, Phys. Rev. B 10, 2457 (1974).

/5/ F.A. BLOM, K.J. YUNG, and M.J. GELTEN, Proc. XIII Internat. Conf. Phys. Semicond., Rome 1976.

/6/ E.K. ARUSHANOV and L.N. LUKYANOVA, Proc. All-Union Conf. Ternary Semiconductors and Their Application, Kishinev 1971 (p. 171).

/7/ F.A. BLOM and J. SCHRAMA, Phys. Letters A 30, 245 (1969).

/8/ S.I. RADAUTSAN, E.K. ARUSHANOV, and A.N. NATEPROV, phys. stat. sol. (a) 23, K59 (1974).

/9/ E.K. ARUSHANOV, A.N. NATEPROV, and G.P. THUIKO, Fiz. Tekh. Poluprov. 8, 2275 (1974).

/10/ M.J. AUBIN, Phys. Rev. B 9, 833 (1974).

/11/ S.I. RADAUTSAN, E.K. ARUSHANOV, A.N. NATEPROV, and G.P. THUIKO, Cadmium Phosphide and Arsenide, Kishinev 1976.

(Received July 26, 1977)

phys. stat. sol. (a) 43, K165 (1977)
Subject classification: 9; 21.1.1

Institute of Physics, Czechoslovak Academy of Sciences, Prague

Diffusion in the Force Field in Solid Solutions

By

V. PAIDAR

For diffusion in the force field $\vec{F}$ given by the potential V ($\vec{F} = -\nabla V$) the diffusion equation

$$\frac{\partial c}{\partial t} = D\nabla\left(\nabla c + \frac{c}{kT}\nabla V\right) \tag{1}$$

is commonly used /1/. In the usual notation c is the concentration, t time, T temperature, k is the Boltzmann constant, and D is the diffusion coefficient, which is assumed to be concentration independent. The aim of this note is to show the limits of applicability of (1) in solid solutions. For rather high concentrations a new equation for the motion of solute atoms in the force field $\vec{F}$ will be proposed.

We shall start from the generalized form of the first Fick's law stating that the flux $\vec{J}$ is proportional to the gradient of the difference of chemical potentials of both components in binary alloys. Following /2, 3/

$$\vec{J} = -M\nabla\frac{\partial f}{\partial c}, \tag{2}$$

where M is called mobility[1] and f is the density of free energy per atom. The "gradient-energy term" will not be considered here. Our treatment is restricted to the frame of the classical diffusion theory. In the potential V, f is composed of the configurational free energy f_o and cV ($f = f_o + cV$). Then the flux equation is

$$J = -M\nabla\left(\frac{\partial f_o}{\partial c} + V\right) = -\left(M\frac{\partial^2 f_o}{\partial c^2}\nabla c + M\nabla V\right), \tag{3}$$

where the coefficient $D = M(\partial^2 f_o/\partial c^2)$ is the diffusion coefficient. Using the continuity equation, the diffusion equation can be written in the form

1) Some authors defined mobility as B, M = Bc(1 - c), e.g. in /1/, but we shall use Cahn's notation here.

$$\frac{\partial c}{\partial t} = \nabla(D\nabla c + M\nabla V) \tag{4}$$

Both coefficients D, M may be generally functions of composition. However, if it is supposed (in the same way as in (1)) that D does not depend on position and, therefore, must be independent of composition in the considered composition range, the diffusion equation is

$$\frac{\partial c}{\partial t} = D\nabla\left(\nabla c + \frac{\nabla V}{\partial^2 f_o/\partial c^2}\right). \tag{5}$$

Let us consider a random binary substitutional solid solution with b.c.c. lattice. Taking into account the pair-wise interaction between atoms up to the second neighbours, the second derivative of the configurational free energy in the zeroth (Bragg-Williams-Gorsky) approximation will be (see e.g. /4/)

$$\frac{\partial^2 f_o}{\partial c^2} = W + kT\left(\frac{1}{c} + \frac{1}{1-c}\right), \tag{6}$$

where the constant $W = 8v_1 + 6v_2$ and v_1, v_2 are chemical interaction energies for the first and second neighbours, respectively. Inserting the expression (6) into (5) we get a complicated non-linear equation for concentration. However, considering small concentrations ($c \ll 1$), (6) can be rewritten as

$$\frac{\partial^2 f_o}{\partial c^2} = W + \frac{kT}{c} \tag{7}$$

and two less complicated (linear) equations are distinguished:

1) For $c \ll kT/W$, the constant term W in (7) can be neglected and the diffusion equation (5) takes the form of (1). This equation is strictly valid only for dilute ideal solid solutions ($c \to 0$, $W = 0$).

2) For $c \gg kT/W$, the second term in (6) is small and (5) can be approximated by

$$\frac{\partial c}{\partial t} = D\nabla\left(\nabla c + \frac{1}{W}\nabla V\right). \tag{8}$$

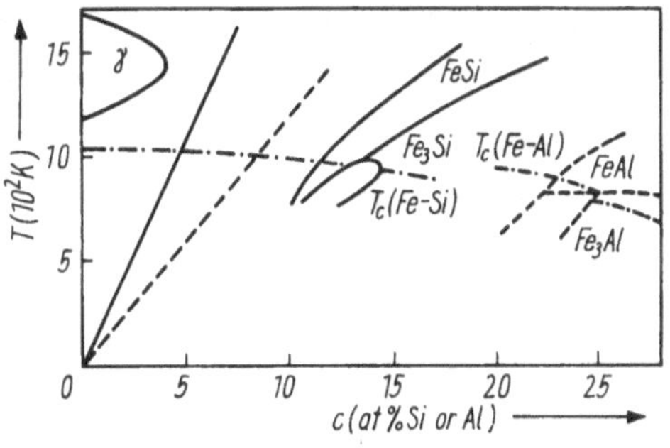

Fig. 1. The regions of applicability of the diffusion equations (1) (on the left from the straight line), and (8) (on the right). Full lines: Fe-Si alloys, dashed lines: Fe-Al alloys. The state diagrams of Fe-Si and of Fe-Al systems are taken from /7/ and /8/, respectively

As an example, let us compare Fe-Si and Fe-Al alloys. The energy parameters v_1, v_2 were determined in /5/ and /6/, respectively, yielding the values of $W = 2.2\times10^4$ k (3.04×10^{-19} J) and 1.2×10^4 k (1.65×10^{-19} J), respectively. The regions of applicability of the diffusion equations (1) and (8) are shown in the state diagram in Fig. 1, where the straight lines $c = kT/W$ are drawn for both alloys. While (1) is applicable for temperatures above 1000 K up to the concentration of about 10 at% Al in Fe-Al alloys, in the case of Fe-Si alloys it may be used only up to about 5 at% Si.

It must be emphasized that if the concentration dependence of the free energy (from a theoretical model or from experimental data for chemical potentials) is known, a relation between the diffusion coefficient D and mobility M can be derived. However, the concentration dependence of D (or M) cannot be determined in the frame of the phenomenological theory without a further assumption as, e.g., $M \sim c$ or M = const. Therefore, (8) is valid only if the second derivative of the free energy in (5) is approximately constant and if the diffusion coefficient changes very slowly with concentration in the considered composition range. If these two assumptions are not fulfilled, a more complicated (non-linear) diffusion equation (4) must be used in the composition range where (1) is not applicable.

References

/1/ P.G. SHEWMON, Diffusion in Solids, McGraw-Hill Publ. Co., New York 1963.

/2/ J.W. CAHN, Acta metall. 9, 795 (1961).

/3/ J.E. HILLIARD, Phase Transformations, ASM Seminar, 1970 (p. 497).

/4/ M.A. KRIVOGLAZ and A.A. SMIRNOV, The Theory of Order-Disorder in Alloys, GIFML, Moscow 1958 (in Russian); Mac Donald, London 1964.

/5/ G. INDEN and W. PITSCH, Z. Metallk. 63, 253 (1972).

/6/ P.S. RUDMAN, Acta metall. 8, 321 (1960).

/7/ G. SCHLATTE and W. PITSCH, Z. Metallk. 66, 660 (1975).

/8/ P.R. SWANN, W.R. DUFF, and R.M. FISHER, Metallurg. Trans. 3, 409 (1972).

(Received August 1, 1977)

phys. stat. sol. (a) 43, K169 (1977)

Subject classification: 14.3.2; 14.3.4

Department of Physics and Astronomy, University of North Carolina, Chapel Hill[1)]

Comment on the Small-Signal AC Response of Solid Electrolytes[2)]

By

D.R. FRANCESCHETTI and J.R. MACDONALD

In a recent interesting paper /1/ Kornyshev and Vorotyntsev (henceforth referred to as KV) presented an approximate analysis of the response of a solid electrolyte to small electrical perturbations. When this paper appeared we had recently completed an analysis /2, 3/ of a somewhat more general problem than that considered by KV, without making the approximations employed by them. Here we discuss the relationship of our result to that of KV and to earlier treatments /4 to 6/ of small-signal electrical response not mentioned by KV. We hope that this comparison will help in making clear the physical implications of the approximate perturbation approach used by KV for ac and transient response and by Itskovich and KV /7/ for steady-state dc behavior.

Although Itskovich and KV /7/ remark that some treatments of electrolyte with two mobile species of charge carrier are not directly reducible to the particular solid electrolyte case in which only one species of charge carrier is considered to be mobile, the small-signal theories of Macdonald /4, 5/, Friauf /6/, and Beaumont and Jacobs /8/ are not so limited and provide a single logically consistent treatment of both the two-mobile-carrier and one-mobile-carrier cases. These treatments are somewhat restricted in their applicability, however, by their neglect of compact layer effects through their use of the Chang and Jaffé /9/ boundary condition, which, for positive charge carriers reacting with a metal electrode placed to the left of the electrolyte, may be written

$$I_{pL} = -z_p e k_p (p_L - p_e) , \tag{1}$$

where I_{pL} is the current of charge carriers, z_p the valence number of the carriers, k_p a rate constant, p_L the concentration of charge carriers at the electrode surface, and p_e the equilibrium concentration of charge carriers.

1) Chapel Hill, North Carolina 27514, USA .

2) Work supported by U.S. National Science Foundation, Grant No. DMR 75-10739.

Our more recent analysis /2, 3/, however, made use of the boundary condition

$$I_{pL} = -z_p\, e \left[k_p(p_L - p_e) + (z_p e\eta/kT)\gamma_p p_e\right] , \tag{2}$$

which applies in the limit of small deviations from equilibrium to a wide range of first-order electrode reactions. Here p_L is evaluated a distance d from the electrode surface, η is the potential drop across the compact layer of thickness d and dielectric constant ε_1, and γ_p is a parameter (with the dimensions of k_p) which reflects the dependence of the electrode reaction rate on the overpotential η. (Our original treatment /3/ dealt with negative mobile charge carriers and defined η as the potential of the right-hand electrode minus that on the electrolyte side of the right-hand compact layer. Here, as in the treatment of KV, we consider positive charge carriers and define η as the potential of the left-hand electrode relative to that of the electrolyte side of the corresponding compact layer.) It is readily verified that for a system in which the static electric field is negligible in the absence of an applied static potential (as assumed in our treatment and that of KV) the Butler-Volmer, or Frumkin /10, 11/, expression for the rate of electrode reaction used by VR reduces to (2) with $\gamma_p = -k_p$.

Our result was derived for a slab of electrolyte between two identical metal electrodes. We let $R_\infty \equiv l/\sigma$ denote the limiting high frequency resistance and $C_g \equiv \varepsilon/4\pi l$ denote the geometric capacitance of the cell (l is the length of the electrolyte, not counting the compact layers, σ its conductivity, and ε its dielectric constant) and use a subscript "N" (i.e. normalized) to indicate an impedance or resistance expressed in units of R_∞ or a capacitance expressed in units of C_g. Our fundamental result may be written (for mobile positive charge carriers) as

$$Z_N = Z_{CJN} + Z'_N , \tag{3}$$

where /4 to 6/

$$Z_{CJN} = \frac{2 + (1 + i\Omega)r_p + 2i\Omega\gamma_1}{(1 + i\Omega)((1 + i\Omega)r_p + 2i\Omega\gamma_1)} \tag{4}$$

is the impedance of the cell as determined for Chang and Jaffé boundary conditions /3, 5/, and /3, 12/

$$Z_N' = \frac{(r_p + 2\gamma_1)\left[r_p(1 + i\Omega + \nu_p) + 2i\Omega\gamma_1\right]}{(C_{CN}(r_p(1 + i\Omega) + 2i\Omega\gamma_1) - \gamma_1 r_p \nu_p)(r_p(1 + i\Omega) + 2i\Omega\gamma_1)} \tag{5}$$

is a correction for compact layer effects. Here Ω is the angular frequency expressed in units of τ_D^{-1}, where $\tau_D \equiv \varepsilon/4\pi\sigma$ is the dielectric relaxation time; $\nu_p \equiv \gamma_p/k_p$, which is (-1) for the electrode kinetics assumed by KV; $r_p \equiv D_p k_p/1$, where D_p is the diffusion coefficient of the charge carriers; $C_{CN} \equiv \varepsilon_1 l_h/\varepsilon d$ is the normalized capacitance of the compact layer; and $\gamma_1 \equiv (M_p\Theta_1)\,\mathrm{ctnh}(M_p\Theta_1)$, where $\Theta_1^2 \equiv (1 + i\Omega)$, $M_p \equiv 1/2L_{Dp}$, and L_{Dp} is the effective Debye length for mobile positive carriers. Our results, (3) to (5), constitute an exact solution to the linearized equations of charge transport /1, 3/, subject to the boundary condition (2) (and the equivalent condition at the right-hand electrode) and entail no assumptions about the relative size of 1, L_{Dn}, d or r_p. The correction Z_N' is negligible in many cases but becomes potentially significant for $r_p \ll M_p$ and $M_p C_{CN}^{-1} = \varepsilon d/\varepsilon_1 L_{Dp} \approx 1$.

In their treatment /1/ KV consider the electrolyte to be in contact with a single polarizable electrode, obeying Frumkin ($\nu_p = -1$) delayed action kinetics, and require that at a distance L from the electrode $p = p_e$ and $\Phi = 0$, where Φ is the electrostatic potential. KV refer to L as the diffusion length. We feel that this terminology is unfortunate since the diffusion length concept is often associated with electrochemical cells in which there is hydrodynamic flow and frequently the presence of an indifferent or supporting electrolyte, conditions not usually encountered in the solid state. In dealing with solid electrolytes, one must in general consider all the material between two non-ohmic electrodes. It is readily shown /3/, however, that in the present case the impedance of a symmetrical cell expressed in units of R_∞ is equal to the impedance of an idealized half-cell, consisting of half the electrolyte bulk placed between a single polarizable electrode and an ohmic electrode (for which $p = p_e$ and $\Phi = 0$) expressed in units of its high-frequency-limiting resistance, which equals $R_\infty/2$. We are thus able to interpret L, as employed by KV, as 1/2.

KV explicitly assume $M_p \gg 1$, $C_{CN} \gg 1$, and $r_p \ll M_p$ in their treatment. Their result may be expressed as a normalized admittance $Y_N \equiv Z_N^{-1}$:

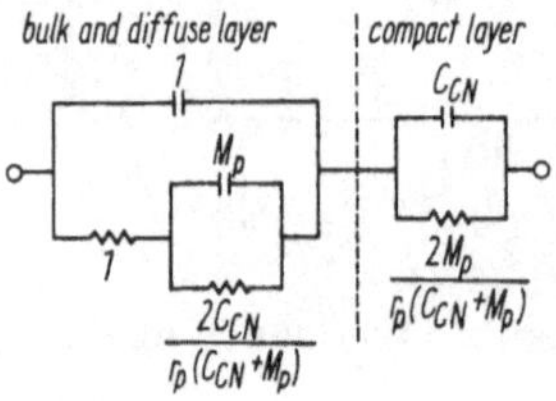

Fig. 1. Normalized equivalent circuit which reproduces Y_N (7) in the $\Omega \ll 1$ limit. Expressions for the resistive elements are given as normalized resistances, not conductances

$$Y_N = \frac{4M_p^2\Omega^2 + r_p(r_p + 2)(1 + M_pC_{CN}^{-1})^2}{(r_p + 2)^2(1 + M_pC_{CN}^{-1})^2 + 4M_p^2\Omega^2} + i\,\frac{4M_p^2\Omega^3 + 4M_p\Omega(1 + M_pC_{CN}^{-1})}{(r_p + 2)^2(1 + M_pC_{CN}^{-1})^2 + 4M_p^2\Omega^2} . \qquad (6)$$

(In the notation of KV, M_p, r_p, $M_pC_{CN}^{-1}$, and Ω are ε^{-1}, $2j^*$, γ, and Ωt_D, respectively.) Under the same assumptions concerning M_p, C_{CN}, and r_p and in the low frequency region ($\Omega \ll 1$), we obtain an expression for Y_N which in rationalized form is

$$Y_N = \frac{4M_p^2\Omega^2 + r_p(r_p + 2)(1 + M_pC_{CN}^{-1})^2}{(r_p + 2)^2(1 + M_pC_{CN}^{-1})^2 + 4M_p^2\Omega^2} + i\,\frac{4M_p\Omega(1 + M_pC_{CN}^{-1})}{(r_p + 2)^2(1 + M_pC_{CN}^{-1})^2 + 4M_p^2\Omega^2} , \qquad (7)$$

differing from (6) only in the absence of $4M_p^2\Omega^3$ in the numerator of the imaginary part. If higher order terms in Ω are retained until after the separation of real and imaginary parts of Y_N, (7) is obtained, but it should be noted that for $\Omega \ll 1$, $M_p^2\Omega^3 \ll M_p^2\Omega^2$ and should be neglected. We conclude that even under the restraints on M_p, C_{CN}, and r_p required by KV, their result is only valid for $\Omega \ll 1$, a limitation not stated by those authors, and that (7) provides an adequate expression for Y_N in that case. When $M_pC_{CN}^{-1} \ll 1$, this expression coincides in the small Ω region with the corresponding result of earlier treatments /3, 5, 6, 8/.

We present in Fig. 1 an equivalent electrical circuit which in the $\Omega \ll 1$ limit has the admittance given by (7). This admittance is readily transformed to yield an expression for the transient response to a small potential step, which should be valid for $t \gg \tau_D$. A discussion of equivalent circuit representations of the more general exact impedance of (3) to (5) has been included in a forthcoming publication /12/.

It should be noted that the above discussion applies only for a mobile species of charge which does not recombine with the oppositely charged immobile species present. An exact small-signal solution has been obtained /13/ for the intrinsic-extrinsic conduction problem with positive and negative species of arbitrary mobilities and valence numbers and arbitrary intrinsic and extrinsic generation and recombination rates, satisfying generalized Chang and Jaffé boundary conditions which incorporate the possible presence of specific ionic adsorption /14/. Work is in progress on extending this solution to include more general electrode kinetics.

References

/1/ A.A. KORNYSHEV and M.A. VOROTYNTSEV, phys. stat. sol. (a) 39, 573 (1977).

/2/ D.R. FRANCESCHETTI and J.R. MACDONALD, Bull. Amer. Phys. Soc. 22, 452 (1977).

/3/ D.R. FRANCESCHETTI and J.R. MACDONALD, J. Electroanal. Chem., to be published.

/4/ J.R. MACDONALD, Phys. Rev. 92, 4 (1953).

/5/ J.R. MACDONALD, J. chem. Phys. 58, 4982 (1973).

/6/ R.J. FRIAUF, J. chem. Phys. 22, 1329 (1954).

/7/ E.M. ITSKOVICH, A.A. KORNYSHEV, and M.A. VOROTYNTSEV, phys. stat. sol. (a) 39, 229 (1977).

/8/ J.H. BEAUMONT and P.W.M. JACOBS, J. Phys. Chem. Solids 28, 657 (1967).

/9/ H. CHANG and G. JAFFE, J. chem. Phys. 20, 1071 (1952).

/10/ K.J. VETTER, Electrochemical Kinetics, Academic Press, New York 1967 (p. 107).

/11/ P. DELAHAY, Double Layer and Electrode Kinetics, Wiley, New York 1965.

/12/ D.R. FRANCESCHETTI and J.R. MACDONALD, J. Electroanal. Chem., to be published.

/13/ J.R. MACDONALD, J. chem. Phys., to be published.

/14/ J.R. MACDONALD and P.W.M. JACOBS, J. Phys. Chem. Solids 37, 1117 (1976).

(Received July 8, 1977)

phys. stat. sol. (a) 43, K175 (1977)
Subject classification: 18.2; 10.1; 11; 21.1.1

Max-Planck-Institut für Metallforschung,
Institut für Werkstoffwissenschaften, Stuttgart

The Magnetic After-Effect Spectrum of α-Iron after Low Temperature Neutron Irradiation

By
W. MENSCH and J. DIEHL

Recently the relatively complicated magnetic after-effect (MAE) spectrum of α-iron after low-temperature particle irradiation and its variation with annealing has been re-examined by several groups. Whereas Walz et al. /1/ used electron irradiation with a relatively small dose, Dunleavy and Blythe /2/ studied the effect of a high-dose neutron irradiation (2×10^{18} n cm^{-2}). The measurements which are reported in this note were carried out after neutron irradiation, too, but with the much smaller dose of 3×10^{17} n cm^{-2} ($E_n \geq 0.1$ MeV). The total concentration of radiation induced defects is thus in between those of the two above-mentioned investigations. Furthermore, for the present measurements the type of the after-effects and the relaxation parameters can be evaluated in a rather detailed manner since in the authors' group internal friction measurements were performed on identically treated samples. Here we report on the observed MAE maxima and their annealing behaviour. A subsequent paper /3/ will deal with the correlation of these results with the corresponding internal friction measurements.

Polycrystalline samples (Johnson Matthey Co.) were purified by annealing in dry hydrogen (73 h at 1020 K) and some of them were doped afterwards with 160 at ppm of carbon. The samples were then irradiated in the Karlsruhe reactor FR 2 at $\leq$ 20 K /4/ and transferred to the cryostat for the MAE measurements at < 25 K. In performing the measurements the samples were heated in temperature/time steps of 2 K/30 min. The results are represented by isochronal relaxation curves. The relative decrease of the reluctivity $\Delta r(t_1, t_2, T_M) = 1/\chi_a(t_1) - 1/\chi_a(t_2)$ between $t_1 = 15$ s and $t_2 > t_1$ at the measuring temperature T_M is shown, where χ_a is the initial susceptibility. Fig. 1 presents results corresponding to the as-irradiated state of a purified sample (three different t_2 values); Fig. 2 represents results after annealing to several temperatures T_A. Essentially the same peaks are ob-

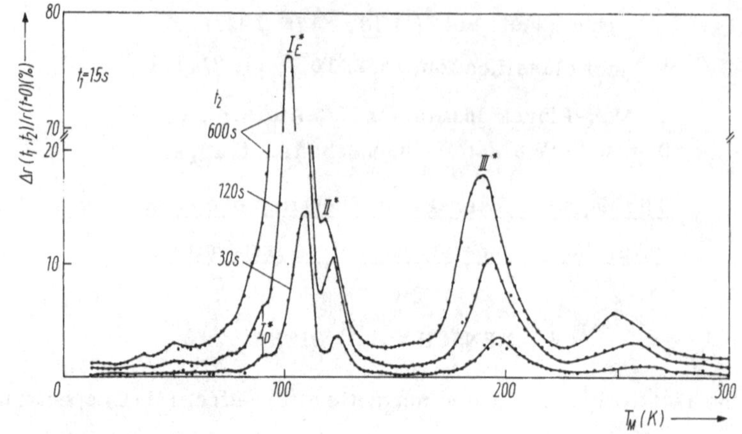

Fig. 1. MAE spectrum of purified iron after irradiation at ≤ 20 K with $3x10^{17}$ n cm^{-2}

served as in previous /5, 7/ and recent /2/ work. Peaks I_D^*, I_E^*, II^*, and III^* are labelled similarly as by the Grenoble group /6/. All these peaks anneal out rather rapidly at or slightly above the peak temperatures. In the as-irradiated state additional peaks are observed at 47 K (corresponding to the 52 K peak in /2/) and at 250 K (peak temperatures for t_2 = 600 s). During annealing a new peak appears at about 80 K. In this work it is named X and seems to correspond to Moser's peak II_B /5/ and to the peaks at 84 K in /1/ and at 90 K in /2/. A peak at 135 K, which is stable up to T_A = 220 K is revealed only after partial annealing of peak II^*. All the peaks mentioned here are found also in C-doped samples and presumably originate therefore from intrinsic relaxation processes.

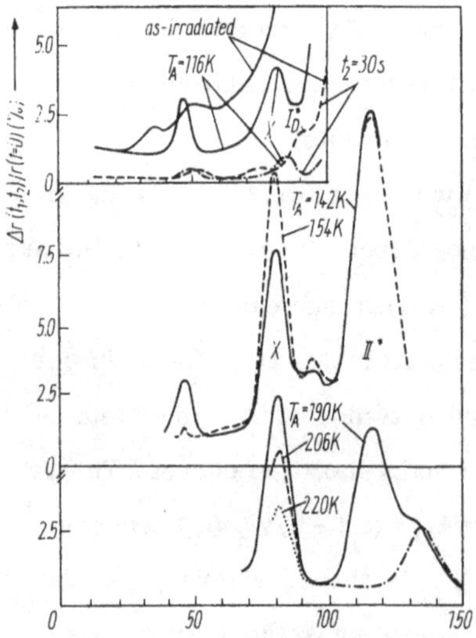

Fig. 2. MAE spectrum of purified iron, after n-irradiation and annealing up to various temperatures T_A. The two curves for t_2 = 30 s are shown for demonstrating the annealing of peak I_D^* and the different peak temperatures of peaks I_D^* and X. All other curves for t_2 = 600 s

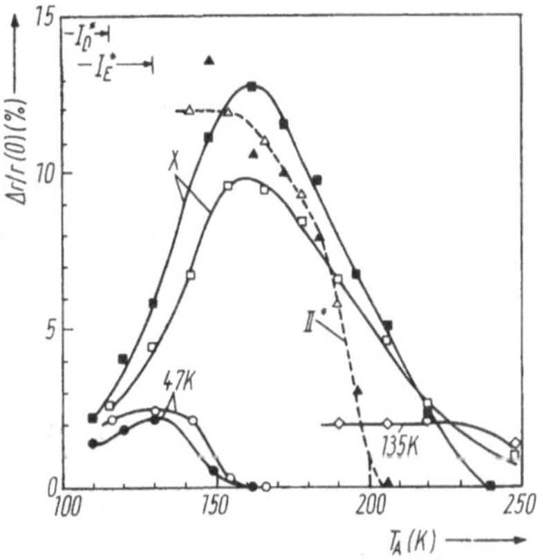

Fig. 3. Variation of the MAE peak heights for $t_1 = 15$ s, $t_2 = 600$ s as a function of annealing temperature T_A. Open symbols: purified iron; filled symbols: iron doped with 160 at ppm carbon

The variation of the peak heights with annealing (background subtracted) is shown in Fig. 3, for purified and for C-doped samples. In the upper left corner of Fig. 3 it is indicated by arrows that the I_D^* and I_E^* peaks have completely disappeared after $T_A = 116$ K and $T_A = 130$ K, respectively.

The present results appear to be in complete agreement with those after n-irradiation with a higher dose, reported in /2/. There are, however, some deviations from the observations of Walz et al. /1/ after e^--irradiation. A feature common to both types of irradiation is that during, or after, the annealing of the most prominent peak (I_E^*, 110 K in /1/), other peaks appear or increase in amplitude. Whereas in /1/ rather simple correlations are reported between the annealing and the formation of peaks at 110 K (corresponding to I_E^*) and at 84 K (corresponding to X) and between annealing of the 84 K peak and formation of a peak at 125 K (corresponding to II^*?), in the present measurements the sequence appears to be more complicated. The annealing of peak I_E^* apparently contributes simultaneously to the growth of the 47 K peak and to that of peak X, possibly also to the formation of peak II^*. Peak X is fully developed only after the 47 K peak has completely annealed out.

For the reasons mentioned in /1/ (monovacancies immobile up to above room temperature, see /8, 9/) we also assign all the present MAE peaks under discussion to the relaxation of interstitial atoms and/or of small clusters of interstitials with different configurations and different thermal stability.

Based on this general view it can easily be understood that, if the defect causing peak I_E^* anneals out by becoming mobile or by dissociation, more than one configuration of multiple interstitials are formed thus causing the growth of existing peaks or the formation of new ones. Furthermore, it can be explained that the annealing out of one cluster configuration (47 K peak) may contribute to the increase in concentration of another configuration (e.g. peak X). Finally, if there are such mutual interactions between the migration or dissociation and the formation of different interstitial cluster configurations, it seems to be quite plausible that e^--irradiated and n-irradiated samples behave somewhat differently because of the different concentrations and spatial distributions of the radiation induced defects.

Acknowledgements

We should like to thank D. Brunner and W. Decker for their continued cooperation in carrying out the experiments. We are also grateful to H.J. Blythe, H. Kronmüller, A. Seeger, F. Walz, and M. Weller for valuable discussions. The experiments were supported by Gesellschaft für Kernforschung GmbH, Karlsruhe.

References

/1/ F. WALZ, H. SCHREYER, and H. KRONMÜLLER, phys. stat. sol. (a) 43, K81 (1977).

/2/ L.H. DUNLEAVY and H.J. BLYTHE, phys. stat. sol. (a) 43, K87 (1977).

/3/ J. DIEHL, W. MENSCH, and M. WELLER, phys. stat. sol. (a) 43, K179 (1977).

/4/ W. DECKER, J. DIEHL, K. HAIN, H. KATHEDER, and CH. LEITZ, Kerntechn. 8, 257 (1966).

/5/ P. MOSER, Thesis, Grenoble 1966; Mém. sci. Rev. Métall. 63, 4 (1966).

/6/ V. HIVERT, R. PICHON, H. BILGER, P. BICHON, J. VERDONE, D. DAUTREPPE, and P. MOSER, J. Phys. Chem. Solids 31, 1843 (1970).

/7/ E. BALTHESEN, K. ISEBECK, and H. WENZL, phys. stat. sol. 8, 603 (1965).

/8/ H.-E. SCHAEFER, K. MAIER, M. WELLER, D. HERLACH, A. SEEGER, and J. DIEHL, to be published.

/9/ J. DIEHL, U. MERBOLD, and M. WELLER, to be published

(Received August 12, 1977)

phys. stat. sol. (a) 43, K179 (1977)
Subject classification: 12.1 and 18.2; 10.1; 11; 21.1.1

Max-Planck-Institut für Metallforschung,
Institut für Werkstoffwissenschaften, Stuttgart

Correlations between Magnetic After-Effect and Internal Friction in α-Iron after Low Temperature Neutron Irradiation

By

J. DIEHL, W. MENSCH, and M. WELLER

Studies of the magnetic after-effect (MAE) in high purity and carbon doped iron after low-temperature irradiation which were described in a preceding paper /1/ were accompanied by measurements of the internal friction (IF) on identically treated samples; these IF measurements are reported in detail elsewhere /2/. In this paper results of both experimental techniques obtained with purified samples will be compared in order to gain additional and more precise information on the relaxation processes of the radiation induced defects.

The samples (Johnson and Matthey Co.; purified in dry hydrogen at 1020 K for 73 h) were irradiated with $3x10^{17}$ n cm^{-2} at $\leq$ 20 K in the Karlsruhe reactor FR 2 and transferred cold to the measuring apparatus. The IF measurements were carried out in an inverted torsion pendulum at ν = 1.4 Hz. Several runs were made; in each run the measuring temperature T_M was increased in 2 K/1 min steps, from 80 K to a temperature T_A at which a 20 min isothermal annealing took place.

In Fig. 1 some typical results of the IF measurements are presented. Curve 1, corresponding to the as-irradiated state shows four maxima (labelled according to /3/) I_D^*, I_E^*, II^*, and III^*, in agreement with previous measurements /3, 4/. Curve 2, measured after annealing at T_A = 180 K, demonstrates that after the annealing out of peaks I_D^* and I_E^* (completed at T_A = 128 K) a small new peak, named X, appears. Furthermore it can be derived from curve 2, that peak II^* is apparently composed of two subpeaks, II_1^* and II_2^* (as indicated by the dotted and dashed curves, which refer to curve 1). Peak II_1^* anneals out very rapidly near the peak temperature (therefore not appearing in curve 2), whereas peak II_2^* is stable up to higher temperatures. Its peak temperature does not change with annealing. The shape and annealing behaviour of peak III^* suggests that it originates from rather complicated processes which will not be discussed here.

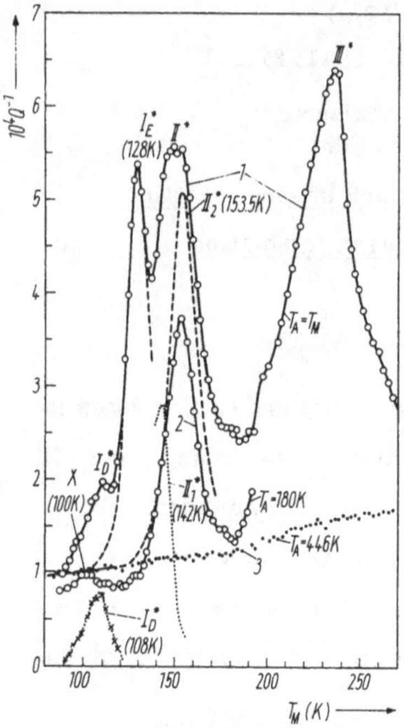

Fig. 1. Temperature dependence of the internal friction of purified iron after irradiation at ≦ 20 K with $3x10^{17}$ n cm^{-2}, as-irradiated and annealed at T_A

Curve 3 shows the background remaining after annealing at T_A = 446 K. Peaks I_D^* and I_E^* in curve 1 are separated from each other (dashed and dotted curves) by assuming that I_E^* is superimposed to this background and follows a slightly modified Debye curve (half-width enhanced by a factor of 1.2). A corresponding analysis was applied to the isochronal MAE curves of peaks I_D^* and I_E^*.

The correlation between MAE and IF peaks as it is suggested already by the labelling is examined more closely by two criteria:

(i) If corresponding IF and MAE peaks originate from the same relaxation process, the relaxation times involved in both types of measurements should follow an Arrhenius equation $\tau = \tau_o \exp(H/kT)$ with identical values of H and τ_o. The τ values belonging to the peak temperatures T_P, are $\tau(IF) = (2\pi\nu)^{-1}$ and $\tau(MAE) = (t_2 - t_1)/\ln(t_2/t_1)$, t_1 and t_2 being the two measuring times characterising each isochronal MAE curve (e.g. in /1/). In Fig. 2 $\log\tau$ is plotted versus T_P^{-1} for the observed IF peaks and for the corresponding peaks in several isochronal MAE curves from /1/. Straight lines are expected, and indeed obtained, in most cases. The H and τ_o values which due to the variation of τ by more than a factor 10^3 can be derived from these plots quite accurately are also given in Fig. 2.

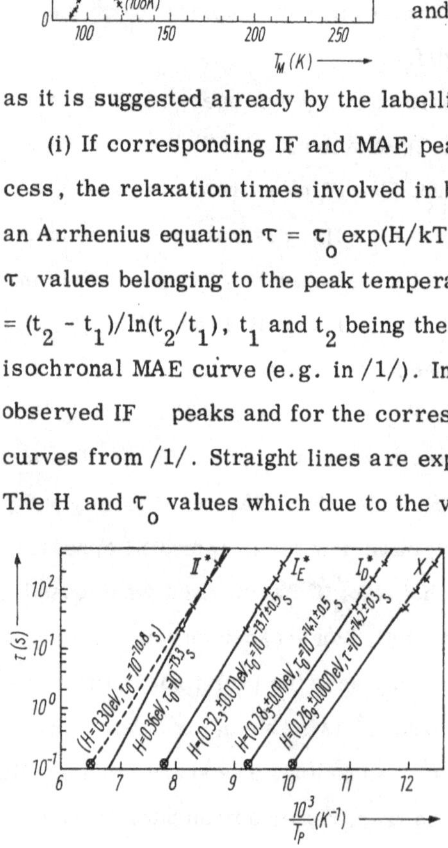

Fig. 2. Arrhenius plot for several MAE and IF peaks in purified iron after n-irradiation

Fig. 3. Variation of the height of IF peaks X and II_2^* (×) and of MAE peaks X and II^*(o) with annealing temperature T_A. (Peak heights are normalized to their maximum values)

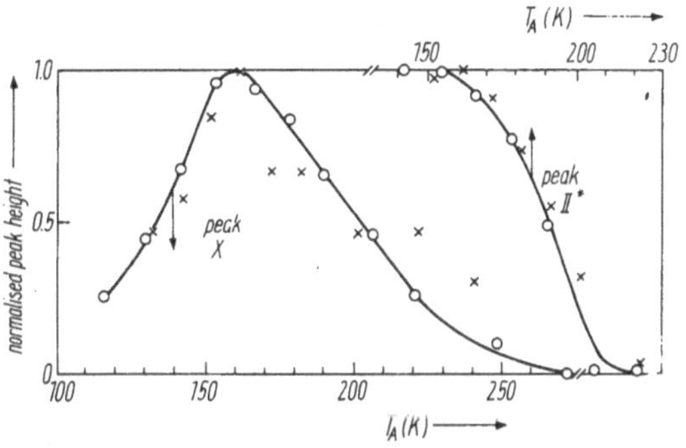

(ii) If MAE and IF peaks are caused by the same type of lattice defects, the temperature dependence of formation and annealing of IF and MAE peaks should be the same. Thus the variations of the peak heights (normalised to the maximum height) with annealing temperature T_A are plotted in the same diagram for IF and MAE peaks X and for peaks II_2^* (IF) and II^*(MAE) - both being called II^* in the following. The agreement is very good in both cases (in view of the fact that the extreme smallness of IF peak X does not allow higher precision). This criterion also applies to peaks I_D^* and I_E^* since they anneal out near the peak temperature in IF as well as in MAE.

Based mainly on Fig. 2 and 3 we arrive at the following results and conclusions: a) Since peaks X, I_D^*, I_E^*, and II^* fulfill criterion (ii), each of them is caused in MAE and IF by the same type of defects, which must be intrinsic ones because all these peaks appear in purified as well as in doped samples (see /1, 2/). b) All data points for peaks X and I_D^* fall on straight lines in the Arrhenius plot (criterion (i), Fig. 2). Thus with τ_0 values of about 10^{-14} s the corresponding relaxations are clearly due to the reorientation of rather simple defect configurations. c) Criterion (i) applies to peaks I_E^*, too, as established already by other authors /5/. However, the individual τ values in Fig. 2 and correspondingly also τ_0 are somewhat smaller than in these previous investigations /5/. This slight discrepancy may possibly be caused by annealing taking place during the measurements so that the peaks are virtually shifted to somewhat lower temperatures in the present measurements. d) Although IF and MAE peaks II^* coincide with respect to their annealing behavior and are therefore clearly correlated with each other, in the Arrhenius plot the IF peak deviates from the extrapolation of the straight line determined by the MAE data. This may indicate that the relaxation originates from a more complicated defect configuration (e.g. a larger cluster of interstitials), for which the relaxation parameters depend on the local stress field (being different in IF and MAE measurements), as it could be the case if more than one jump frequency contributes to the relaxation process.

For the MAE peaks at 47 and 135 K /1/ and for IF peak II_1^* no correlations between MAE and IF are reported here. It should be noted that this may simply be caused by an incomplete correspondence of both types of experiments. All the correlations established here are in agreement with or even support the view /1, 6/ that the low temperature relaxations indicate the reorientation of different configurations of interstitial atoms and/or of clustered interstitials. In addition close interstitial-vacancy pairs should be considered, especially with respect to peak I_D^*, as it was suggested in /3/.

Acknowledgements

We appreciate valuable discussions with W. Decker, H. Kronmüller, H.E. Schaefer, A. Seeger, and F. Walz. The experimental work was supported by Gesellschaft für Kernforschung, Karlsruhe.

References

/1/ W. MENSCH and J. DIEHL, phys. stat. sol. (a) 43, K175 (1977).

/2/ M. WELLER and J. DIEHL, Proc. 6th ICIFUAS, Tokyo 1977, in the press.

/3/ V. HIVERT, R. PICHON, H. BILGER, P. BICHON, J. VERDONE, D. DAUTREPPE, and P. MOSER, J. Phys. Chem. Solids 31, 1843 (1970).

/4/ M. WELLER and J. DIEHL, Proc. 5th ICIFUAS, Vol. I, Springer, Berlin/Heidelberg/New York 1975 (p. 352).

/5/ J. VERDONE, W. CHAMBRON, and P. MOSER, phys. stat. sol. (b) 61, K41 (1974).

/6/ F. WALZ, H. SCHREYER, and H. KRONMÜLLER, phys. stat. sol. (a) 43, K81 (1977).

(Received August 12, 1977)

phys. stat. sol. (a) 43, K183 (1977)
Subject classification: 1

Erewaner Staatliche Universität

Ein Dreistrahl-Röntgeninterferometer

Von
P.A. BESIRGANYAN, R.Z. GABRIELYAN und F.O. EIRAMSHYAN

Das von Bonse und Graeff /1/ beschriebene Fünfblock-Dreistrahl-Röntgeninterferometer arbeitet nur reflektierend (Bragg-Reflexion) und hat wegen der Vierfachreflexion nur eine schwache Lichtstärke. Dabei ist seine Anwendung sehr beschränkt: es ist nur für die NiK_{α_2}-Strahlung vorgesehen, deren Intensität in der Regel gering ist.

Hier wird ein Vierblock-Dreistrahl-Doppelinterferometer großer Lichtstärke vorgeschlagen, mit dem gleichzeitig Interferenzbilder entsprechend Bragg und Laue aufgenommen werden können. Das Interferometer arbeitet mit drei Reflexionen, deshalb ist die Lichtstärke größer als die des Interferometers /1/. Außerdem kann das hier vorgeschlagene Interferometer in einem ausreichend breiten Spektralbereich arbeiten.

Das Interferometer und der Strahlengang in ihm sind in Fig. 1 gezeigt. Das Interferometer is monolitisch, wurde also aus einem Silizium-Kristall gefertigt und hat vier Blöcke auf einer allgemeinen Basis. Im ersten und vierten Block liegen die $(10\bar{1})$- und $(1\bar{1}0)$-Ebenen symmetrisch hinsichtlich der Senkrechten der großen Oberfläche und bilden mit dieser die Winkel $\pm 30^{\circ}$, die $(0\bar{1}1)$-Ebenen aber sind senkrecht zu diesen großen Oberflächen. Um eine Laue-Reflexion zu erhalten, ist der vierte Block dünner als die übrigen. Im zweiten und dritten Block sind die $(01\bar{1})$-Ebenen parallel zur großen Oberfläche, die $(10\bar{1})$-und $(1\bar{1}0)$-Ebenen liegen aber symmetrisch zur Senkrechten auf der großen Oberfläche, wie das in Fig. 1 gezeigt ist. Die Blöcke I und IV, wie auch die Blöcke II und III, sind streng parallel zueinander. Das Interferometer ist für die NiK_{α_2}-Strahlung berechnet. Der Abstand a zwischen den Blöcken I und IV und der Abstand b zwischen den Blöcken II und III genügen der Bedingung

$$a : b = 1 : \sqrt{3}. \qquad (1)$$

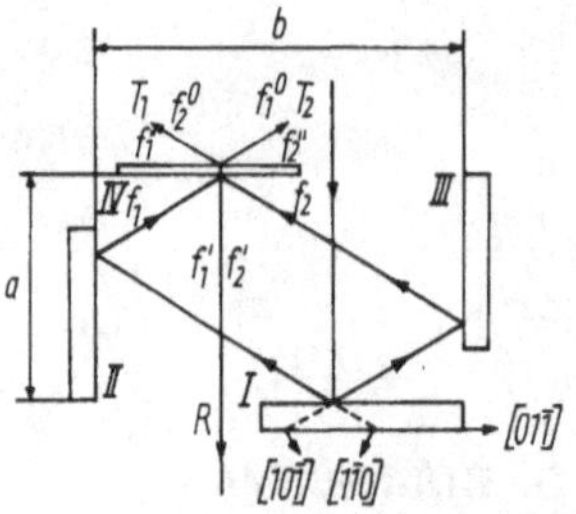

Fig. 1. Das Vierblock-Dreistrahlinterferometer

Der monochromatische Strahl der NiK_{α_2}-Strahlung fällt auf den ersten Block und wird von den $(\bar{1}01)$- und $(\bar{1}10)$-Ebenen reflektiert. Der von den $(\bar{1}01)$-Ebenen reflektierte Strahl fällt auf den zweiten Block, und nach der Reflexion von den $(\bar{1}10)$–Ebenen fällt er auf den vierten Block. Der von den $(\bar{1}01)$-Ebenen des ersten Blocks reflektierte Strahl fällt auf den dritten Block und nach der Reflexion von den $(01\bar{1})$-Ebenen auf den vierten Block. Bei strenger Erfüllung der Bedingung (1) überlagern sich die von den $(01\bar{1})$- und $(0\bar{1}1)$-Ebenen des zweiten und dritten Blocks reflektierten Strahlen ausreichend genau auf der Oberfläche des vierten Blocks, wo auch ein komplizierter Interferenzprozeß stattfindet. Der Strahl f_1, der auf den vierten Block fällt, wird gleichzeitig von den $(1\bar{1}0)$- und $(0\bar{1}1)$-Ebenen reflektiert, woraus die Strahlen f_1', f_1'' und f_1^o (siehe Fig. 1) resultieren. Der Strahl f_2 wird gleichzeitig von den $(10\bar{1})$- und $(01\bar{1})$-Ebenen reflektiert und bringt die Strahlen f_2', f_2'' und f_2^o hervor.

Somit findet eine Interferenz zwischen den nach Bragg reflektierten Strahlen f_1' und f_2' und den nach Laue reflektierten Strahlen f_1'', f_2'', f_1^o und f_2^o (die letzten beiden sind durchgehend) statt.

So haben wir faktisch also zwei Interferometer· ein Bragg-Bragg-Bragg-Interferometer (mit dem resultierenden Strahl R) und ein Bragg-Bragg-Laue-Interferometer (mit den resultierenden Strahlen T_1 und T_2).

Am interessantesten bei diesem Interferometer ist, daß hier an der Interferenz zwei Dreiwellen-Konfigurationen $(0\bar{1}1)$, $(1\bar{1}0)$ und $(01\bar{1})$, $(10\bar{1})$ teilnehmen; von Block zu Block wechseln sich diese Konfigurationen ab.

In der Tat, wie in Fig. 1 zu sehen ist, findet die Dreiwellenstreuung im ersten Block an den $(1\bar{1}0)$- und $(\bar{1}01)$-Ebenen statt, im zweiten und dritten Block aber an den $(0\bar{1}1)$- und $(1\bar{1}0)$-Ebenen. Im vierten Block ist dieses Bild komplizierter, hier wirken gleichzeitig die Konfigurationen $(01\bar{1})$, $(10\bar{1})$ und $(0\bar{1}1)$, $(1\bar{1}0)$, und es findet eine Interferenz zwischen den Wellen statt, die verschiedenen Konfigurationen der Dreiwellenstreuung angehören, während im Interferometer /1/ in allen Blöcken nur ein und dieselbe Konfiguration (101), (110) wirkt und die Interferenz in ihm nach der

Fig. 2. Die Lage der $(01\bar{1})$-, $(10\bar{1})$- und $(1\bar{1}0)$-Ebenen in den Blöcken des Interferometers

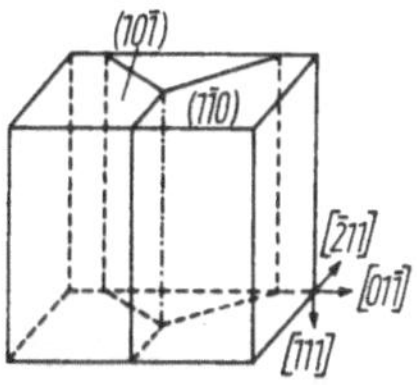

vierten Reflexion zwischen den Wellen stattfindet, die zu ein und derselben Konfiguration der Dreiwellenstreuung gehören. Somit hat die Benutzung von zwei Dreiwellen-Streukonfigurationen folgende Vorteile:

1. Die Zahl der Reflexionen wird verringert, wie auch die Zahl der Blöcke; die Interferenz ergibt sich nicht erst nach vier Reflexionen, sondern schon nach drei, wodurch die Lichtstärke des Gerätes bedeutend erhöht wird.

2. Es findet gleichzeitig eine Interferenz zwischen den nach Bragg und den nach Laue reflektierten Wellen statt, d.h. das von uns vorgeschlagene Interferometer besteht faktisch aus zwei Interferometern (Doppelinterferometer), was die Möglichkeit gibt, sehr interessante Untersuchungen auszuführen: von ein und demselben Kristallort gleichzeitig ein Interferenzbild nach Bragg und nach Laue zu erhalten und den Einfluß verschiedener Faktoren auf diese Bilder zu klären.

Sehr wichtig ist auch die folgende Tatsache. Die $(01\bar{1})$-, $(10\bar{1})$- und $(1\bar{1}0)$-Ebenen liegen symmetrisch, wie in Fig. 2 gezeigt ist, und haben eine bemerkenswerte Eigenschaft: wenn der Primärstrahl in einer von ihnen liegt, so liegen die reflektierten Strahlen in den zwei übrigen Ebenen. Es ergibt sich in gewissem Sinne ein geschlossenes System: der einfallende und die zwei reflektierten Strahlen liegen nur in diesen drei Ebenen und können zyklisch ausgetauscht werden. Mehr noch, wenn bei einer Wellenlängenänderung die Einfallsrichtung geändert werden muß, und sich damit auch die Reflexionsrichtung ändert, so verbleiben diese Strahlen wiederum in diesen Ebenen - sie verlassen sie nicht, wie man anhand elementarer Überlegungen zeigen kann.

Letztere Eigenschaft der Gruppe dieser Ebenen führt zu der Möglichkeit, den Spektralbereich für die Untersuchungen mit dem Interferometer zu erweitern (achromatisches Vierblock-Dreistrahlinterferometer), was man auf eine der folgenden Weisen verwirklichen kann:

1. Man läßt den Strahl unverändert und dreht das Interferometer um die Achse senkrecht zur Arbeitsoberfläche des zweiten und dritten Blocks des Interferometers,

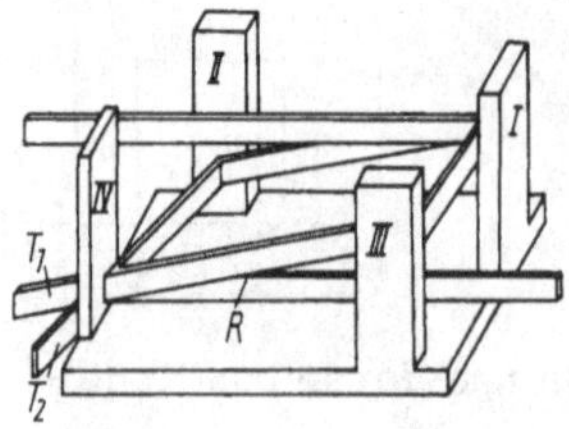

Fig. 3. Das achromatische Vierblock-Dreistrahl-Doppelinterferometer

bis die vorliegende Wellenlänge nicht mehr ausreicht für eine Dreiwellenstreuung bei der gegebenen Konfiguration der Reflexionsebenen.

2. Bei unbewegtem Interferometer dreht man den einfallenden Primärstrahl in der vertikalen Ebene um den Einfallspunkt im ersten Block bis die vorliegende Wellenlänge sich nicht mehr in der Dreiwellensituation befindet.

3. Um den einfallenden Strahl in der horizontalen Ebene bei unbewegtem Interferometer zu halten, müssen die Blöcke des Interferometers so geschnitten werden, daß die Reflexionsebenen des zweiten und dritten Blocks die vorherigen Orientierungen hätten, die Schnittlinie dieser Oberflächen des ersten und vierten Blocks mit dem einfallenden Strahl aber einen solchen Winkel ergeben würde, der die Dreiwellensituation für die gegebene Wellenlänge ausmachen würde.

Indem wir nach einer der aufgezeigten drei Methoden für den Strahlengang im achromatischen Interferometer vorgehen, erhalten wir das Bild, das in Fig. 3 gezeigt ist. Die Bezeichnung der Strahlen auf den Fig. 1 und 3 ist identisch.

Somit kann das vorgeschlagene achromatische Interferometer in einem ausreichend breiten Spektralintervall in Abhängigkeit von den Ausmaßen der Reflexionsebenen der Interferometerblöcke arbeiten.

Daher ist das vorgeschlagene Interferometer in seinen Anwendungen im Röntgenbereich im Prinzip frei von den oben aufgezeigten Beschränkungen.

Es muß auch bemerkt werden, daß die Interferenzbilder außerhalb der Richtung des Primärstrahls entstehen, was sehr bequem ist, besonders bei der Benutzung einer Synchrotron-Strahlung; beim Interferometer /1/ ist es fast unmöglich, Untersuchungen mit einer Synchrotron-Strahlung und überhaupt mit harter Strahlung auszuführen, weil der Interferenzstrahl in der Verlängerung des Primärstrahls entsteht.

Literatur

/1/ U. BONSE und W. GRAEFF, Acta cryst. A31, 254 (1975).

(Received August 16, 1977)

phys. stat. sol. (a) 43, K187 (1977)
Subject classification: 10.1; 21.1

Departamento de Física del Estado Sólido, Laboratorio de Microscopía Electronica,
Facultad de Ciencias Físicas, Universidad Complutense, Madrid

Relationship between Point Defect Clusters and the Temperature Dependence of Yield Stress in Fatigued Copper Polycrystals

By

J. VICTORIA, R. GONZÁLEZ, J. PIQUERAS, and L. BRÚ

Introduction The non-conservative motion of sliding dislocations with jogs is generally accepted as the most important source of the production of point defects during plastic deformation; by means of diffusion and collapse mechanisms, these defects can occasionally give rise to point defect clusters large enough to be observed in the electron microscope. The clusters act as thermally surmountable barriers for dislocations; they are, in large measure, responsible for the variation of yield stress or saturation stress in fatigue with temperature.

The variation of yield stress in fatigued copper polycrystals has previously been studied /1, 2/ and the different factors which influence the temperature dependence, such as degree of deformation, purity of the material, and grain size, have been analyzed. In this investigation the defect density in fatigued samples having different degrees of deformation, purity, and grain size have been studied by means of the electron microscope. Information regarding the factors which influence the formation of point defect clusters during fatigue is thus obtained and the results are related to the variation of the ratio $A = \sigma(293\ K)/\sigma(85\ K)$ between the yield stresses at 293 and 85 K found in /1, 2/.

Experimental method In the present work, two different kinds of copper were used. One, Marz copper, of 99.999% purity - including both metallic and other kinds of impurities - from Material Research Corporation, and the other of about 99.5% purity, hereinafter called LP (low purity) copper.

Part of the Marz copper samples were annealed for 3 h at 500 °C in a 10^{-5} Torr vacuum; the rest were annealed for 6 h at 800 °C. After this treatment the grain sizes obtained were 40 and 100 μm, respectively. The LP copper was annealed for three hours at 750 °C in the same vacuum and the grain size obtained was 40 μm.

The samples were then fatigued with a constant total strain amplitude, $\Delta\varepsilon_{total}$, in a W-type Hounsfield tensometer modified for fatigue tests.

After the fatigue, electron microscope specimens were prepared by means of conventional electropolishing techniques and observed at 200 kV with a Jeol 200 A electron microscope using the weak beam technique.

Results and discussion The mechanical treatment given to the different samples, as well as the corresponding density of defect clusters measured are summarized in Table 1. All samples were deformed with the same total strain amplitude, $\Delta\varepsilon_{total} = 0.040$.

Due to the relatively high strain amplitude used, a clearly defined structure of dislocation cells was already observed in the samples fatigued two cycles. As has previously been observed in single and polycrystals /3, 4/, the distribution of point defect clusters was markedly inhomogeneous, being mainly concentrated in those regions where the dislocation density is highest.

Table 1 shows the integral cluster densities in the zones which are almost free of dislocations. The relative size distribution functions (Fig. 1) were found to be exponential in nature, as has also been found in /3/. The mean diameter of the defect clusters was in all cases approximately 40 Å. The general feature of dislocation and cluster distributions does not depend, for analogous mechanical treatments, on the type of sample used.

Table 1

Data of the various samples used in this work

sample No.	copper	mean grain size (μm)	total number of cycles	yield stress σ (kp/mm^2)	integral density within the cells N (10^{15} cluster cm^{-3})
1	Marz	40	50	8.4	2.4
2	Marz	40	2	5.6	1.5
3	Marz	100	50	6.9	1.8
4	LP	40	50	5.0	1.4
5	LP	40	2	3.3	0.7

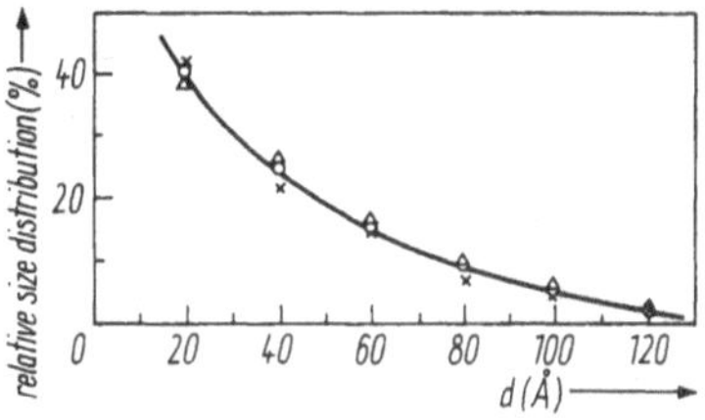

Fig. 1. Relative size distribution of point-defect clusters in the interior of the cells. × series No. 2; o No. 4; Δ No. 5

The high density of clusters, shown in Table 1, is characteristic of fatigued samples after the first cycles /4/. However, the degrees or kind of deformation - unidirectional, rapid hardening stage, or fatigue saturation - does not determine the density or final structure of defects since, as is shown in Table 1, these densities can vary with the purity and the grain size of the material. Both, purity and grain size can influence density and dislocation distribution during plastic deformation and consequently the number of point defect clusters produced.

Thompson et al. /5/ calculated the density of dislocations as a function of grain size in work hardening copper and established that the density of dislocations increases with decreasing grain size. In accordance with this increase in the dislocation density, a greater number of clusters should appear in copper of smaller grain size.

On the other hand, as the impurity content increases,-the motion of the mobile dislocations is hindered, at least in part of their length, by the solute atoms which act as obstacles to movement which can cause a smaller production of point defects and, consequently, of clusters.

Point defect clusters have a significant influence on the temperature dependence of yield stress in fatigued copper /3/ due to the interaction processes between mobile dislocation and clusters. It is shown in Table 1 that the density of clusters produced during fatigue depends on various factors which must, consequently, influence the dependence of yield stress on temperature. This influence should not attributed exclusively to the different density of point defect clusters since solute atoms, for example, influence the behavior of a metal during deformation.

In order to analyze the effect that point clusters have on the temperature dependence of yield stress from the electron microscope results presented in Table 1, we use the values of the ratio A between yield stresses at two temperatures, obtained in /1, 2/. These measurements, which were made with the same type of crystals as used in the present study, can be summarized for the purpose of the present discussion by the fact that, other experimental conditions being equal, the value of A increases with increasing number of cycles, purity, and grain size.

As in the above mentioned investigations the tensile yield stress of fatigue hardened specimens is subdivided into the thermal, σ^*, and athermal σ_μ, components. The ratio A can then be written as

$$A = \frac{\sigma^*(T_H) + \sigma_\mu}{\sigma^*(T_L) + \sigma_\mu} = \frac{N^{1/2} F(T_H, \dot{a}) + \alpha \mu b \varrho^{1/2}}{N^{1/2} F(T_L, \dot{a}) + \alpha \mu b \varrho^{1/2}} , \qquad (1)$$

where T_H and T_L represent the high and low temperatures, respectively, N the integral cluster density in the cell interior, ϱ is the dislocation density, b is the length of Burgers vector, μ is the shear modulus, and α is a constant. $F(T, \dot{a})$ is a function which depends on the nature of mobile dislocations and obstacles.

For each sample we define parameter C as

$$C = \frac{\sigma(T_H)}{N^{1/2}} , \qquad (2)$$

where the numerator represents the total yield stress at room temperature while the denominator relates to the thermal stress component since $\sigma^* = F(T, \dot{a})N^{1/2}$. In accordance with this, parameter C gives information on the relative value of the thermal component with respect to the total stress since the greater the relative value of σ^*, the smaller is C.

For samples 1 and 2, deformed with the same strain amplitude but different numbers of cycles, the values of C obtained from (2) and the experimental results are such that $C_1 > C_2$. If $C_1 > C_2$ it can be shown from (1) and (2) that must be $A_1 > A_2$ assuming that $F(T, \dot{a})$ is the same for both specimens. This conclusion agrees with the above-mentioned results of the temperature dependence of flow stress. The parameter C was obtained experimentally on the assumption that the main obstacles to dislocation motion are point defect clusters. These values of C agree qualitatively with the fact that A increases with increasing number of cycles. The same result is obtained when comparing samples 4 and 5 (of lower purity).

When comparing samples of different purities or of different grain sizes it can be concluded that the greater the ratio A, the greater the value of C, i.e. the lower the temperature dependence of the yield stress, the lower is the fraction of stress due to point defect clusters.

To compare the specimens with different grain sizes, the yield stress measured for the sample with the larger grain size was corrected in the form indicated by Carreker and Hibbard /6/.

References

/1/ J. VICTORIA, J. PIQUERAS, R. GONZÁLEZ, and L. BRÚ, phys. stat. sol. (a) 40, 167 (1977).

/2/ J. VICTORIA, R. GONZÁLEZ, and J. PIQUERAS, phys. stat. sol. (a) 40, K19 (1977).

/3/ J. PIQUERAS, J.C. GROSSKREUTZ, and W. FRANK, phys. stat. sol. (a) 11, 567 (1972).

/4/ R. GONZÁLEZ, J. PIQUERAS, and L. BRÚ, phys. stat. sol. (a) 29, 161 (1975).

/5/ A.W. THOMPSON, M.I. BASKES, and W.F. FLANAGAN, Acta metall. 21, 1017 (1973).

/6/ R.P. CARREKER and W.R. HIBBARD, Acta metall. 1, 654 (1953).

(Received August 22, 1977)

phys. stat. sol. (a) 43, K193 (1977)

Subject classification: 16; 20.1; 20.3; 22.9

Consiglio Nazionale delle Ricerche,
Laboratorio di Fotochimica e Radiazioni d'Alta Energia, Bologna

Photoconductive Properties of Fluorene Single Crystals

By

P. DI MARCO and G. GIRO

The literature reports few conflicting results on the photoconductivity of fluorene. Mukherjee claims that fluorene has no photoconductive properties /1/, while Donnini and Abetino showed a measure of fluorene activation energy of photoconductivity /2/. Bree and Zwarich pointed out that commercial fluorene contains many impurities very difficult to remove /3/. Since it is known that the impurity content strongly affects the photoconductivity /4/, it is reasonable to think that the reported discrepancies were probably due to a not carefully purified material. In order to test this hypothesis, photoconductivity measurements were carried out on fluorene crystals at different purity degrees.

Ultrapure fluorene (sample A) with an impurity content lower than 10 p.p.m. was obtained using a) treatment with maleic anhydride, in order to remove the anthracene, b) continuous chromatography, c) vacuum sublimation, d) zone refining (three times, 200 zone passes each) using a special multizone refiner, with an ingot length of about 90 cm, for an easier removal of benzo(f)indane /5/.

Fluorene crystals were grown from the melt using Bridgman's method, thin slices of about 1 mm thickness were cut perpendicularly to the c crystal axis and were placed between two conductive semi-transparent electrodes (SnO_2 on quartz) in a sandwich cell arrangement. The photoconductivity was measured at room temperature by irradiating the crystal, placed under N_2, with monochromatic light between 250 and 320 nm, by means of a 900 W Xenon lamp and a Hilgher & Watts monochromator. The high vapor pressure of fluorene hindered any measurement under high vacuum. The apparatus for light intensity and dc measurements are described elsewhere /6, 7/. Optical absorption and emission measurements of fluorene in solution and in crystalline state were performed using a Turner spectrofluorometer Spectro Model 210 and a Perkin Elmer 456 spectrophotometer. Photo-

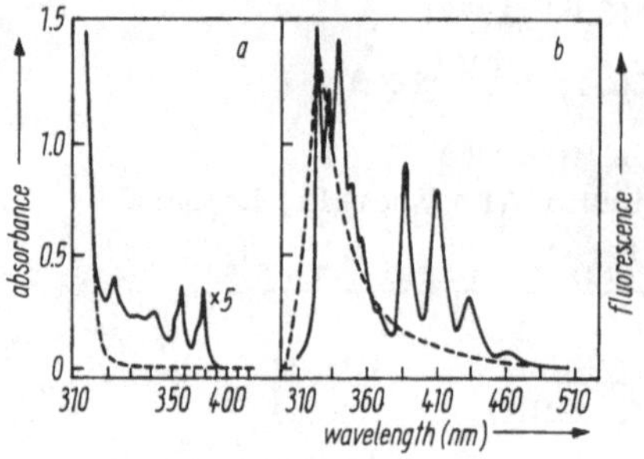

Fig. 1

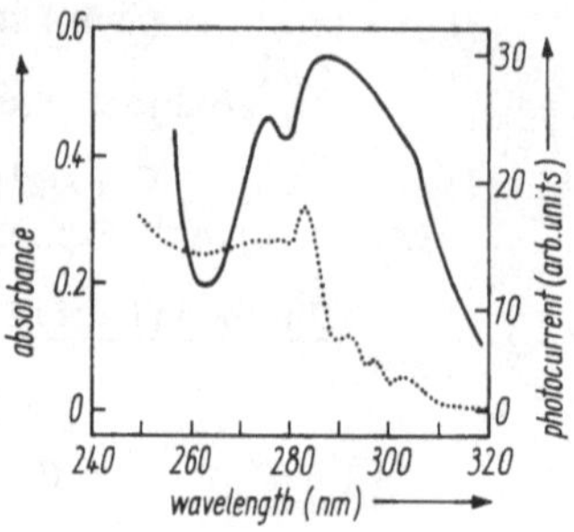

Fig. 2

Fig. 1. a) Tails of the optical absorption spectrum of fluorene: —— sample C, 1.2×10^{-2} M, in cyclohexane (impurity content: benzo(f)indane = 2.0x $\times 10^{-3}$ M/M; anthracene: 5.8×10^{-4} M/M); --- sample A, 1.6×10^{-2} M in cyclohexane.

b) Crystal fluorescence spectra of fluorene, —— sample C and --- sample A

Fig. 2. The dc action spectrum of ultrapure fluorene with the front electrode positively biased (——); optical absorption spectrum of solid fluorene, measured perpendicularly to the ab crystal section (---)

conductivity measurements were performed also on fluorene crystals doped with 1.2×10^{-4} M/M of benzo(f)indane (sample B), or purified with standard methods, i.e. continuous chromatography, vacuum sublimation and zone refining, using a BTL Multizone Refiner (sample C).

The results obtained from samples B and C agree well with the observation of Mukherjee; in fact no photocurrent signal was detected, despite the use of high electric field (up to 2×10^{4} V/cm). From Fig. 1, in which are reported the absorption and emission spectra of crystals purified with standard methods (full line), it is interesting to notice that anthracene and benzo(f)indane are not completely eliminated. Fig. 1 reports also the absorption and emission spectra of ultrapure fluorene (dotted line); in this case the amount of anthracene and benzo(f)indane is below 10 p.p.m. and the true fluorene fluorescence appears. The same crystals exhibit good photocurrent signals in the 250 to 320 nm range.

Fig. 2 shows the dc action spectrum of ultrapure fluorene, measured at $2x10^4$ V/cm, when the illuminated electrode is positively biased; the action spectrum is clearly symbatic with crystal absorption spectrum, therein reported.[1)]

The photocurrent signal measured at 303, 295, and 285 nm is proportional to the square of the applied voltages at high light intensity ($5x10^{13}$ photons/cm^2 s at 303 nm) and tends to be linear at lower light intensities. At the same wavelengths the photocurrent is directly proportional to the light intensity; these results indicate that at the highest light intensity used the photocurrent is not completely S.C.L. and at lower light intensities space charge effects become increasingly insignificant.

The symbatic relation between absorption and action spectrum suggests that in fluorene crystals, inthe 250 to 320 nm spectral region and under the conditions used in these experiments, the exciton-surface extrinsic charge generation mechanism is dominant. Intrinsic mechanisms of charge generation, like exciton autoionization or direct band gap transition, can be ruled out not only because they require an antibatic correlation between the absorption and the action spectrum /4/, but even from an evaluation of fluorene band gap. The energy of the lowest bound electron conduction band (E'_c) may be obtained from the following two relations /9/:

$$E'_c = I_g - 2P - E_a \ , \qquad (1)$$

$$P = I_g - I_c \ , \qquad (2)$$

where I_g is the ionization energy of the molecule in the gas phase, P is the polarization energy, E_a is the free molecule electron affinity, and I_c is the crystal ionization energy. The reported I_g and E_a values of fluorene are 8.48 eV /10/ and 0.34 eV /11/, respectively. Unfortunately, the crystal ionization energy of fluorene is not know and the determination of the polarization energy is not possible. In organic solid, however, where the Van der Waals forces are predominant, a relation exists between polarization energy and boiling temperature of a substance /12/. A P value lower than 1.6 eV can be inferred for fluorene crystals from the boiling point (293 to 295 ^oC), using the plot of the known P values for aromatic

1) The crystal absorption spectrum is in accordance with that measured by Bree and Zwarich /8/.

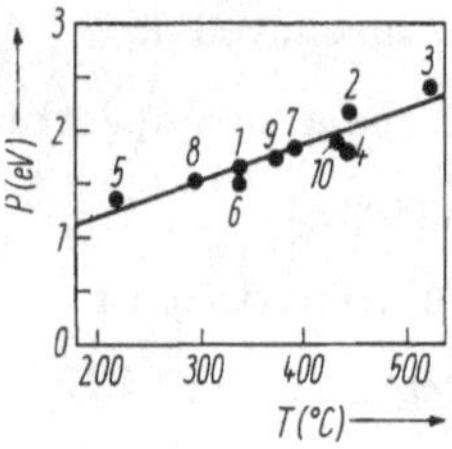

Fig. 3. Polarization energies of aromatic hydrocarbons plotted against their boiling points: 1 anthracene /13 to 16/; 2 chrisene /13, 16/; 3 coronene /17/; 4 benzanthracene /13, 16, 18/; 5 naphthalene /13, 16, 19/; 6 pentacene /20/; 7 perylene /14, 21/; 8 phenanthrene /13, 16/; 9 pyrene /13, 16/; 10 tetracene /13, 21 to 23/

hydrocarbons versus their boiling points (Fig. 3); therefore the lowest E'_c value can be set at 5.04 eV, which corresponds to 245 nm. This means that the onset of the intrinsic generation starts at or below 245 nm, beyond the spectral region considered in this work.

References

/1/ T.K. MUKHERJEE, J. phys. Chem. 74, 3006 (1970).

/2/ J.M. DONNINI and F. ABETINO, C.R. Acad. Sci. (France) 272, 433 (1971).

/3/ A. BREE and R. ZWARICH, Spectrochim. Acta 25 A, 713 (1961).

/4/ R.F. CHAIKEN and D.R. KEARNS, J. chem. Phys. 45, 3966 (1966).

/5/ P.G. DI MARCO and G. GIRO, to be published.

/6/ G. YGUERABIDE, Rev. sci. Instrum. 39, 1048 (1968).

/7/ G. GIRO and P.G. FUOCHI, Atti Accad. Lincei. 54, 627 (1973).

/8/ A. BREE and R. ZWARICH, J. chem. Phys. 51, 903 (1969).

/9/ D.M. HANSON, CRC Crit. Rev. Solid State Sci. 243 (1973).

/10/ A. STREITWIESER, J. Amer. Chem. Soc. 82, 4123 (1960).

/11/ G. BRIEGLEB, Angew. Chem. - Internat. Ed. 3, 617 (1964).

/12/ VAN ARKEL, in: Molecules and Crystals, Butterworths Sci. Publ., London 1949.

/13/ R.S. BECKER and E. CHEN, J. chem. Phys. 45, 2403 (1966).

/14/ M. BATLEY and L.E. LYONS, Mol. Cryst. liquid Cryst. 3, 357 (1968).

/15/ A.P. MARCHETTI and D.R. KEARNS, Mol. Cryst. liquid Cryst. 6, 299 (1970).

/16/ L.E. LYONS and G.C. MORRIS, J. Chem. Soc. 5192 (1960).

/17/ M. BATLEY and L.E. LYONS, unpublished.

/18/ N. KINOSHITA, Bull. Chem. Soc. Japan 35, 1609 (1962).

/19/ L.E. LYONS and J.C. MACKIE, Proc. Chem. Soc. 71 (1962).

/20/ E.A. SILINSH, A.I. BELKIND, D.R. BALODE, A.J. BISENIECE, V.V. GRECHOV, L.F. TAURE, M.V. KURIK, J.I. VERTZYMACHA, and I. BOK, phys. stat. sol. (a) 25, 339 (1974).

/21/ M. KOCHI, Y. HARADA, and H. INOKUCHI, Bull. Chem. Soc. Japan 43, 1690 (1970).

/22/ D.R. KEARNS and M. CALVIN, J. chem. Phys. 34, 2026 (1961).

/23/ H. INOKUCHI, Y. HARADA, and Y. MARUYAMA, 3rd Org. Cryst. Symp., Chicago (Ill.) 1965 (p. 29).

(Received September 6, 1977)

phys. stat. sol. (a) 43, K199 (1977)

Subject classification: 18.2; 22.8.2

Institute of Physics, Silesian University, Katowice[1)]

Frequency Dependence of Initial Magnetic Permeability for $HgCr_2Se_4$ near the Curie Temperature

By

J. JELONEK and M. SZCZYGIELSKI

The compound $HgCr_2Se_4$ has a normal spinel structure and belongs to the group of semiconducting ferromagnetics. The preparation, X-ray structure analysis, magnetic and electric properties of the compound investigated are published in /1 to 3/. The initial magnetic permeability of $HgCr_2Se_4$ was measured at frequencies 40 kHz to 50 MHz and temperatures 77 to 300 K by the resonance method /4 to 6/.

The powder samples were pressed to a toroidal shape. The measurements were carried out passing from paramagnetic to ferromagnetic state. The Curie temperature T_c of the investigated compound was 108 K.

The temperature dependences of real and imaginary parts of the initial permeability, μ_1 and μ_2, at various frequencies are shown in Fig. 1a and b. The μ_1

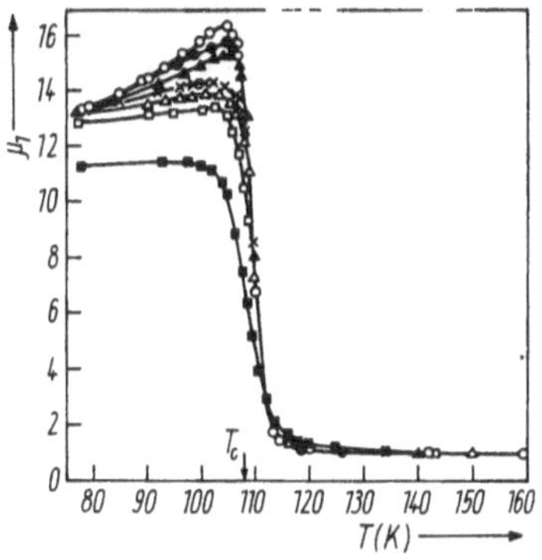

Fig. 1a

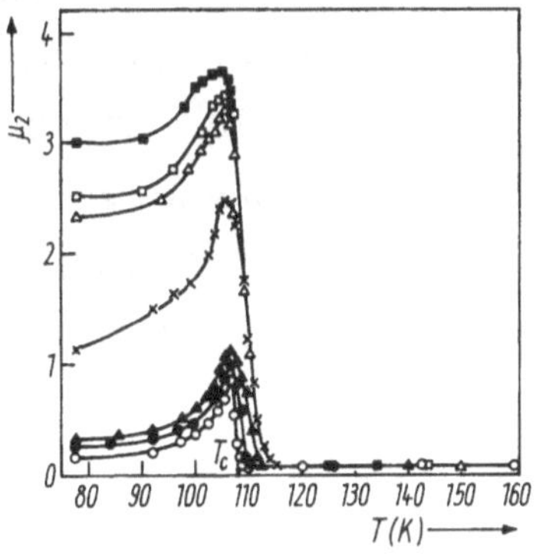

Fig. 1b

Fig. 1. Temperature dependence of a) the real component, μ_1, and b) the imaginary component, μ_2, of the initial permeability, at various frequencies:

○ 0.04; ● 0.4; ▲ 0.8; × 5.8; △ 15; □ 23; ■ 50 MHz

---

1) 40-007 Katowice, Poland.

component has a maximum which decreases with increasing frequency, whereas the μ_2 component (proportional to magnetic losses) has a maximum which increases with frequency. This bheaviour seems to suggest the presence of mobile magnetic regions in the specimen near T_c. These regions are excited by the weak alternating field to low frequency oscillations which are damped on increasing the measuring field frequency at constant temperature or due to the growing volume of the magnetic regions on decreasing the temperature at constant frequency. Near 80 K the sample exhibits a stable domain structure and the frequency dependence of the initial magnetic permeability corresponds to the typical courses for ferrites /7, 8/.

Damping of low frequency oscillations near T_c occurs in a wide frequency range which can be attributed to various local conditions in a given grain of the powder sample.

References

/1/ H. BRODA, D. KONOPKA, and I. KOZŁOWSKA, Acta Phys. Polon. A 42, 255 (1972).

/2/ S. KUBIAK, W. ZAREK, Z. DRZAZGA, J. KROK, and A. CHEŁKOWSKI, Acta Phys. Polon. A 45, 819 (1974).

/3/ I. KOZŁOWSKA, D. KONOPKA, M. JELONEK, J. HEIMAN, J. PIETKIEWICZ, and A. CHEŁKOWSKI, J. Solid State Chem. 14, 349 (1975).

/4/ J. BADY and R.J. FRANKLIN, IRE Nat. Convent. Rec. 15, Part 5, 135 (1957).

/5/ G.T. RADO, R.W. WRIGHT, and W.H. EMERSON, Phys. Rev. 80, 273 (1950).

/6/ A. GÓRAL, Rozprawy elektrotechniczne, Tom V, 1959 (p. 584).

/7/ J. SMIT and H.P.J. WIJN, Ferrites, Philips, Eindhoven 1959.

/8/ R.S. TEBBLE and D.J. CRAIK, Magnetic Materials, Wiley, New York 1969.

(Received June 21, 1977)

phys. stat. sol. (a) 43, K201 (1977)
Subject classification: 10.2 and 11; 22.1.2

Research Institute for Nuclear Physics, Polytechnical Institute, Tomsk

Lattice Location of ^{16}O Implanted in Si Crystals by Backscattering

By

I. CHERNOV, YU. KRYUCHKOV, A. MAMONTOV, N. SLAVIN, and YU. TIMOSHNIKOV

At present there are only few investigations on oxygen lattice location in crystals /1, 2/. The method used was based on channeling and nuclear reactions $^{18}O(p,\alpha)^{15}N$ /1/, $^{16}O(d,p)^{17}O$ /2/ which, however, have small cross-sections (1 mb/sr). For this reason the method has a low sensitivity.

Here we propose another method and discuss the results of an experimental investigation of the lattice location of ^{16}O implanted into a Si single crystal. The method is based on channeling of α-particles and their anomalous backscattering /3/. The cross-section of 15 to 25 MeV elastic α-scattering on light nuclei is strongly enhanced for large angles /4, 5/. For example, the cross-section of backscattering of 21.4 MeV α-particles on ^{16}O is 700 mb which is 1000 times larger than the cross-section of medium nuclei. The sensitivity of the method depends on the cross-sections of scattering on other admixed nuclei and on the matrix nuclei. Furthermore, it depends on beam intensity, exposure time, and the distribution of the admixture. For the targets Si and Ge, at a beam current of about 0.1 $\mu A/cm^2$ and at a solid angle of the detector of about 10^{-2} sr the minimum number of oxygen nuclei which can be determined is $5x10^{15}$ and $1x10^{14}$ nuclei/cm^2, respectively.

The measurements were done with the cyclotron of the Research Institute for Nuclear Physics of Tomsk Polytechnical Institute. Helium ions were accelerated to 21.4 MeV. The beam was collimated to a diameter of 1.2 mm and angular divergency of less than 0.03^o. The crystal was mounted on a double-axis goniometer which enables the target to be rotated in steps of 0.02^o. Backscattering particles were detected with a Si surface-barrier detector located at the scattering angle 175°. The beam current was measured with a Faraday cup. The target chamber was evacuated to $5x10^{-6}$ Torr.

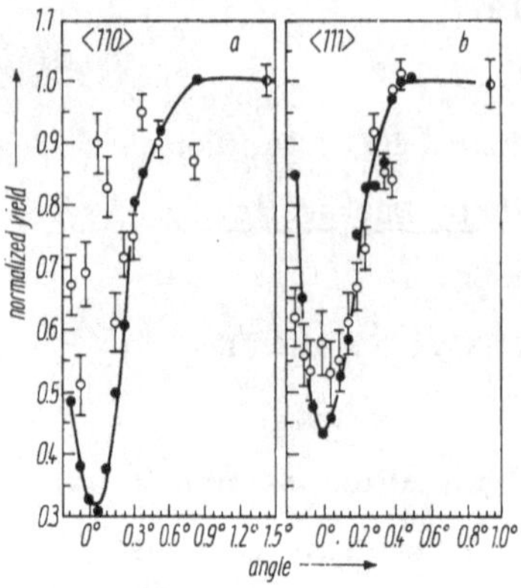

Fig. 1. Scattering yields of 21.4 MeV α-particles as a function of incidence angle relative to a) <110> and b) <111> directions for ^{16}O atoms (open circles) and Si atoms (solid curves). Implantation conditions: $1x10^{17}$ $^{16}O/cm^2$ at 150 keV, 350 °C

Small channeling angles (θ ≤ 0.3°) required high precision beam alignment with respect to the crystal orientation which was realized in two stages. At first the crystal was oriented with a gas laser and the quality of the orientation was checked by the blocking pattern recorded for 21.4 MeV α-particles. Then the target was scanned by the α-particle beam in two perpendicular planes and the minimum of the yield curve of backscattered particles was found.

The crystal targets were prepared from a cylindrical rod. After orientation by X-ray diffraction the rod was cut into thin disks perpendicular to two major crystal axes <110> and <111> each. The disks were polished and chemically etched. As a result, Si-targets approximately 8 mm diameter and 10 μm thick with relatively high perfection were obtained. Then the disks were implanted with $1x10^{16}$ to $1x10^{17}$ ions/cm^2 of ^{16}O. The implantation was carried out at 150 keV and 350 °C to reduce the accumulation of radiation damage.

Yields of backscattered α-particles from ^{16}O and ^{28}Si, normalized at random, are shown in Fig. 1 as functions of the incidence angle relative to the <110> and <111> axes. The narrow peak in the centre of the <110> curve indicates that part of ^{16}O must occupy interstitial sites in the centre of the <110> axial channel. At the same time part of the ^{16}O must be in substitutional positions to produce the dip under the peak. In the case of the <111> direction the angular distributions of the yields of scattered α-particles from ^{16}O and ^{28}Si are practically identical (Fig. 1b). Thus, oxygen atoms are not observed in this direction. Some increased α-particle yield from ^{16}O can perhaps be connected with the formation of oxygen surface layers.

Such a behaviour of angular scans indicates that part of implanted ^{16}O in silicon occupies substitutional sites and part occupies interstitial sites. Probably interstitial atoms are located at regular tetrahedral sites (see also /6/). Note that the group VI elements, Te /7/ and Se /8/, implanted in silicon are substitutional impurities. In our case part of the ^{16}O atoms occupies interstitial sites. This is perhaps connected with the high dose of implantation.

In such a way the method suggested is suitable for the lattice location of oxygen in single crystals, especially in crystals with $Z > 30$.

We wish to thank V. Sochoreva for preparing Si-targets and A. Komov and members of the cyclotron accelerator for their assistance.

References

/1/ P.P. MATYASH, N.A. SCAKUN, and N.D. DIKII, Zh. eksper. teor. Fiz., Pisma 19, 31 (1974).

/2/ H.-D. CARSTANJEN, Ion Beam Surface Layer Analysis, Vol. II, Ed. O. MEYER, G. LINKER, and F. KÖEPPELER, Plenum Press, New York/London 1976 (p. 497).

/3/ I. CHERNOV, B. KUZNETSOV, G. STARODUB, and A. YATIS, Radiochem. Radioanal. Letters 16, 41 (1973).

/4/ M. TAKEDA, S. KATO, and T. YAMANUKI, J. Phys. Soc. Japan 30, 56 (1971).

/5/ B. KUZNETSOV, R. OVSYANNIKOVA, and I. CHERNOV, Soviet Phys. J. nuclear Phys. 15, 681 (1972).

/6/ B. DOMEIJ, G. FLADDA, and N.G.E. JOHANSSON, Radiat. Eff. 6, 155 (1970).

/7/ L. ERIKSSON, J.A. DAVIES, N.G.E. JOHANSSON, and J.W. MAYER, J. appl. Phys. 40, 842 (1969).

/8/ O. MAYER, N.G.E. JOHANSSON, S.T. PICRAUX, and J.W. MAYER, Solid State Commun. 8, 529 (1970).

(Received September 8, 1977)

phys. stat. sol. (a) 43, K205 (1977)
Subject classification: 13.4; 10.2; 14.3; 22.1.1

Istituto di Fisica, Università di Bologna, Unità G.N.S.M. del C.N.R.

Dislocation Energy Levels in Ge

By

A. CAVALLINI, P. GONDI, and A. CASTALDINI

The electrical effects considered so far by various authors (reviewed in /1/) refer in general to dislocations introduced in Ge by deformations at T > 800 K.

In previous papers /2, 3/ we reported that it is possible to reduce the useful deformation temperatures down to ≈ 600 K by introducing large numbers of surface dislocation sources by means of suitable scratching. At these lower deformation temperatures there occur increases of the percentage of the screw dislocations /3/ and of widely dissociated dislocations /4/. In this paper we refer to Hall effect measurements from 20 to 300 K on such specimens deformed at 570 K; preliminary measurements /2/ were made from 75 to 300 K.

Before deformation Ge was n-type by Sb doping, donor concentration N_D = (1 to 5)x10^{13} cm^{-3}, lifetime 10^{-3} s, dislocation density 10^3 cm^{-2}. Dislocations were mainly introduced at 570 K in argon by bending, between two elliptical graphite plugs, rectangular {111} Ge slices (2x0.5)cm^2, 0.05 cm thick with the longer side in <110> direction orthogonal to the bending axis. The maximum stress in bending was 5x10^8 dyn/cm^2 . For deformation the specimens were scratched on one surface as for a diffraction grating: scratch distance 10^{-2} cm, scratches were parallel to the bending axis. Protective coatings appeared unnecessary at the deformation temperatures used.

Hall coefficient measurements from 20 to 300 K were done at 1 kG. To avoid dislocation density inhomogeneities we used for the Hall measurements small transverse specimens, 0.5x0.2x0.05 cm^3, cut from the samples used for bending in direction orthogonal to the longer side.

The dislocation density was evaluated by means of TEM observations; the uniformity of the dislocation distribution was controlled by means of etch pit counts; dislocation inhomogeneities appeared to be less than 30%.

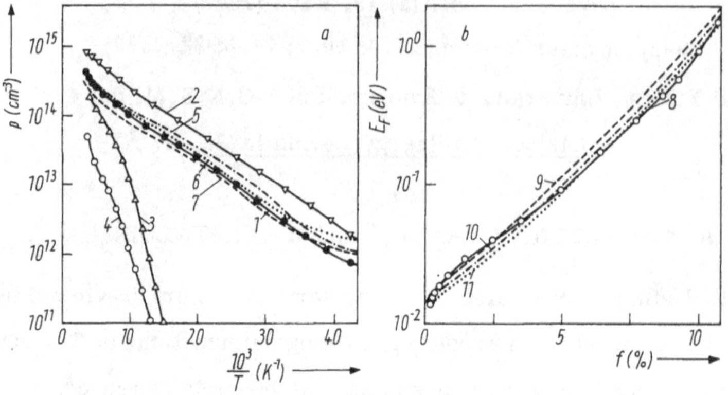

Fig. 1

The differences between the effects of the dislocations introduced at different deformation temperatures appear more evident in the specimens converted from n- to p-type.

For dislocations introduced at 570 K the dependence of the hole density, p, on T is shown by curves 1 and 2 in Fig. 1a: curves 3 and 4 show for comparison the behaviour of specimens deformed at T > 800 K. Data specifying the various diagrams are given in Table 1.

Table 1

Data relating to Fig. 1a and b

curve	1	2	3	4	5	6	7	8	9	10
origin.	ex.	ex.	exl.	ex.	th.	th.	th.	ex.[b)]	th.	th.
$N_D \times 10^{-13}$ (cm^{-3})	2.5[a)]	2.5[a)]	0.9[a)]	2.5[a)]	1.5[a)]	2.5[b)]	1.5[b)]	1.5[b)]	1.5[b)]	1.5[b)]
$N_d \times 10^{-8}$ (cm^{-2})	1[a)]	2[a)]	1.3[a)]	0.3[a)]	1.5[b)]	1.5[b)]	1.0[b)]	3.0[b)]	3.0[b)]	3.0[b)]
$E_o \times 10^2$ (eV)					2.5[b)]	2.5[b)]	2.5[b)]		3.0[b)]	2.5[b)]
T_d (K)	570[a)]	570[a)]	860[a)]	900[a)]				570[a)]		

N_D donor density; N_d dislocation density; E_o neutral dislocation level; T_d deformation temperature; ex. experimental data; th. calculated curve; ex[b)] E_F(f) corresponding to curve 2; exl from bibliography /1/ .a) Values measured; b) Values introduced into (1) for fitting

The occupation of the dislocation levels is conditioned by the electrostatic energy of the electrons on the dislocation and of the screening charges around them; Read considered the ionized donor effects /5/, Schröter and Labusch /6/ the screening due to the free carrier rearrangement. Particularly at high p, hole screening should be prevalent: therefore the experimental data are discussed on the basis of the Schröter-Labusch theory. According to these authors the average hole concentration corresponds to the quasi-Fermi potential:

$$E_F = E_o + \frac{e^2}{2\pi \varepsilon s} f \left[\log \frac{|f|\lambda}{s} + \frac{1}{4}\left(- \frac{N_D}{p} + 1 \right) \right] , \tag{1}$$

where $\lambda = \sqrt{kT\varepsilon / e^2 p}$, E_o neutral dislocation level, ε dielectric constant, s dangling bond distance, N_D donor concentration, $f = (N_D + p)/(N/s)$ the fraction occupied of the N_d/s sites per cm^3 offered by the N_d dislocations per cm^2. The theoretical trends of p are obtained by equating relation (1) to

$$E_F = kT \ln \frac{N_v T^{3/2}}{p} \tag{2}$$

and solving for p. As is known (2) gives the Fermi potential corresponding to the average occupation of the valence band for $N_v = 1.17 \times 10^{15}\ cm^{-3}\ K^{-3/2}$

The hole densities so calculated are drawn by dashed lines in Fig. 1a: they are given for different dislocation densities N_d and donor densities N_D so as to indicate the effects of changes of these parameters, i.e. the precision relating to the comparison with the experimental data. The different parameters are given in Table 1. In Fig. 1b values of E_F corresponding to curve 2 (experimental) are plotted as a function of f, for comparison with theoretical curves calculated with different E_o values.

The experimental data for the specimens deformed at 570 K may be related to the Schröter-Labusch theory for a neutral dislocation level $E_o = 0.025$ eV from the valence band: for the specimens deformed at T > 800 K application of (1) leads to a dislocation level at 0.09 eV and the Read treatment gives higher levels /1/. The differences between the experimental donor and dislocation densities and those introduced into (1) for fitting are not prejudicial. Questionable seems the fact that with some tens of specimens, with deformation temperatures from 550 to 600 K, we obtained curves always parallel to those shown in Fig. 1 whereas theoretical curves show relevant slope variations for small changes of the donor and dislocation densities.

With reference to the difference between the effects of the dislocations introduced at 570 K and at T > 800 K it may be worth recalling the results of EPR measurements reported in the literature /7, 8/, according to which in Si the concentration of unpaired electrons decreases by an order of magnitude with increasing deformation temperature. The same probably occurs in Ge and in this case low deformation temperatures may be necessary to make the effects of the unpaired electrons evident.

References

/1/ R. LABUSCH and R. SCHETTLER, phys. stat. sol. (a) 9, 455 (1972).

/2/ A. CAVALLINI and P. GONDI, Lettere Nuovo Cimento 10, 115 (1974).

/3/ A. CAVALLINI, P. GONDI, and F. SCHINTU, phys. stat. sol. (a) 27, 523 (1975).

/4/ A. CAVALLINI and P. GONDI, Lettere Nuovo Cimento 14, 222 (1975).

/5/ W.T. READ, Phil. Mag. 45, 775 (1954).

/6/ W. SCHRÖTER and R. LABUSCH, phys. stat. sol. 36, 539 (1969).

/7/ F.D. WÖHLER, H. ALEXANDER, and W. SANDER, J. Phys. Chem. Solids 31, 1381 (1970).

/8/ V.A. GRAZHULIS and YU.A. OSIPYAN, Zh. eksper. teor. Fiz. 58, 1259 (1970); 60, 1150 (1971).

(Received August 2, 1977)

Pre-printed Titles

Papers to be published
in "physica status solidi (a) - applied research"
Vol. 44, No. 1

M. WILKENS and P. RAPPS, Stuttgart

Electron Diffraction in Crystalline Specimens Containing a High Density of Lattice Defects (I)

S. AL-DALLAL, Bellevue

Epitaxial Growth of CdSe on ZnTe (111)

V.P. DOBRII, O.P. ERMOLAEV, and V.D. TKACHEV, Minsk

Investigation of Shallow Radiation-Defect Levels Produced in Ge by Fast-Neutron Irradiation

R. HEIDEMANN, D. JÄGER, and R. KASSING, Münster

Minority and Majority Carrier Mobilities Determined by Microwave Measurements

W. SCHNELLE and U. DILLNER, Jena

Thickness Dependence of Resistivity and Its Temperature Coefficient of Polycrystalline Bismuth Films

A.J. MORTON, Parkville, Victoria

Inversion Domains in γ-Brass Type Phases (III)

K. DE STROOPER, A. VAN ALBOOM, C. HENRIET-ISERENTANT, R. VANLEERBERGHE, and R. VANDENBERGHE, Gent

On the Cation Distribution and Magnetic Properties of Cobalt-Zinc Titanates

H. PFEIFFER, Jena

Growth Rate Dependence of the Growth-Induced Anisotropy in Garnets - A Simple Theoretical Model

M.E. VAN HULLE and W. MAENHOUT-VAN DER VORST, Gent

Space Charge Characteristics of Silver Halide Microcrystals

VIJENDER SHARMA and P.C. MATHUR, Delhi

Large-Signal Impedance Measurements and Power-Efficiency Consideration of n^+pp^+ GaAs IMPATT Diodes

A.K. SINGH, B.K. JAIN, and K. CHANDRA, Roorkee

Mössbauer Studies of Natural Goethite and Bog Iron Ore

H.J. HOFFMANN and W. SOHN, Karlsruhe

Analysis of Localized Levels in Semiconducting CdS from Generation-Recombination Noise Spectra

E.B. LEIKO, A. LUFT, and E.M. NADGORNII, Dresden and Chernogolovka

Slip Rosettes and Dislocation Structure Formed by Microindentations in Molybdenum Single Crystals

J.D. LIVINGSTON, Schenectady, New York

Grain Size in A-15 Reaction Layers

S. RADHAKRISHNA and V.S. SIVASANKAR, Madras

F_{Z1} Centres in Cesium Halides

N. RUSOVIĆ and H. WARLIMONT, Stuttgart

The Elastic Behaviour of β_2-NiAl Alloys

M. STUBIČAR, E. BABIĆ, D. SUBAŠIĆ, D. PAVUNA, and Ž. MAROHNIĆ, Zagreb

Phase Transformations during Isochronal Annealing of $Fe_{40}Ni_{40}B_{20}$ Glass

H.-J. KOS and D. NEUBERT, Berlin (West)

Two-Step Photoconductivity by Dislocations in Silicon

H. KNAUER, J. RICHTER, and P. SEIDEL, Jena

A Direct Calculation of the Resonance Tunneling in Metal-Insulator-Metal Tunnel Junctions

T.L. TANSLEY, North Ryde, New South Wales

A Composite Barrier Model for Thermally-Assisted Tunnelling in Heterojunctions

C. LAWTHER and K.V. ANAND, Durham and Canterbury

Physical Mechanism of Current Conduction and Light Emission in High-Resistivity ZnS:Mn Thin Films

V.Z. BENGUS and T.P. KOVALENKO, Kharkov

Mean Mobile Dislocation Velocity in LiF Crystals under Straining

J.J. BARA, A.T. PEDZIWIATR, Z.M. STADNIK, A. SZYTUŁA, J. TODOROVIČ, Z. TOMKOWICZ, and W. ZAREK, Cracow, Vinča, and Katowice

Investigations of Crystalline and Magnetic Properties of Nickel Ferrite-Aluminates

J.M. CORBETT and S.S. SHEININ, Waterloo, Ontario, and Edmonton, Alberta

A Determination of the Reflections Required to Describe Electron Diffraction Phenomena in Gold

S.R. HERD, Yorktown Heights, New York

TEM Observation on the Source of Perpendicular Anisotropy in Amorphous $GdCo(O_2)$

W. NOWAK, Dresden

Scattered-Light Investigations on Amorphous Systems

P. GONDI, C. PATUELLI, and R. TOGNASO, Bologna

Dislocation Source Activation by Load Impulses

D.L. SMITH, Pasadena, California

Recombination Currents in a P-N Heterojunction Diode

M. JURICH, K.-H. BERTHEL, and H.-J. ULLRICH, Dresden

Influence of Chemical Composition within the Range of Homogeneity on Phase Transition and Transition Temperature of V_3Si Single Crystals (I)

Papers to be published
in "physica status solidi (b) - basic research"
Vol. 84, No. 1

Review Article

G.N. RAO and A.K. SINGHVI, Kanpur
Dilute Impurity Hyperfine Fields

Original Papers

M.A.B. WHITAKER, Coleraine, Ulster
Calculation of Nuclear Double-Resonance Relaxation Rate

V.I. GREBENNIKOV, O.B. SOKOLOV, and E.A. TUROV, Sverdlovsk
Response of Metal Electrons to Local Fields of Holes Arising in X-Ray Processes

A.K. JONSCHER, London
A Many-Body Model of Dielectric Polarisation in Solids (II)

H.R. ANAND, West Lafayette, Indiana
Mössbauer Study of Dispersed Iron Ions in Alkali Halides

M. SCHWAB and H.G. KAHLE, Karlsruhe
Specific Heat of $TbPO_4$

V.S. LUBIMOV, E.F. MAKAROV, and S.P. IONOV, Moscow
The Magnetic Properties of Charge-Ordered Crystals

A. BIVAS, VU DUY PHACH, B. HÖNERLAGE, and J.B. GRUN, Strasbourg
Induced Absorption Due to Exciton-Biexciton Transitions in CuCl

E. PALANIYANDI and S. BALASUBRAMANIAN, Madurai
Polarizabilities of Donors in Elemental Semiconductors

M.N. SHARMA and A. KANT, Lucknow
A New Approach to Lattice Properties of Alkali Halide Crystals

G.V. IONOVA, S.P. IONOV, O.M. PACHEV, and E.F. MAKAROV, Moscow
Orbital-Charge Ordering in Crystals

L.P. MOKHRACHEVA, V.A. TSKHAI, and P.V. GELD, Sverdlovsk

Chemical Binding and Energy Band Structure of Vanadium Monocarbide

C.L. BRIANT and J.J. BURTON, Linden, New Jersey

Icosahedral Microclusters. A Possible Structural Unit in Amorphous Metals

A. RANFAGNI and G. VILLIANI, Firenze and Trento

Spin-Orbit-Allowed Tunneling Splitting in the T x ε_g Jahn-Teller Problem Splitting of the Zero Phonon Line

K. BÄRNER, Göttingen

Energy Bands and Electron Correlation Effects in Manganese Pnictides

M. FÄHNLE, Stuttgart

The Influence of Fast Neutron Irradiation on the Critical Current Densities in Nb_3Sn (I)

M. DHWAJ SINGH, Varanasi

The Effect of Internal Strain and Electrical Field of an Impurity on Electron-Phonon Interaction in n-Type Germanium and Silicon

J. LOOS, Prague

The Mean Field Phase Diagrams of YbIG

A.A. KORNYSHEV, A.I. RUBINSHTEIN and M.A. VOROTYNTSEV, Moscow

Image Potential near Dielectric-Plasma-Like Medium Interface

19. Magnetic Resonance

20. Optical Properties
    - 20.1 Spectra. Optical Constants (X-Ray Spectra See 20)
    - 20.2 Lasers
    - 20.3 Luminescence (X-Ray Spectra See 20)

21. Metals. Alloys
    - 21.1 Transition Metals (Group Sc to Zn) and Their Alloys
        - 21.1.1 Fe and Fe Alloys
    - 21.2 Alkali Metals
    - 21.3 Alkali-Earth Metals
    - 21.4 Rate-Earth Metals
    - 21.5 Actinides
    - 21.6 Noble Metals
    - 21.7 Semi-Metals (Semiconducting Alloys 22.7)

22. Semiconductors and Ionic Crystals
    - 22.1 Elements
        - 22.1.1 Germanium
        - 22.1.2 Silicon
        - 22.1.3 Group VI Elements
    - 22.2 III–V Compounds
        - 22.2.1 Arsenides
        - 22.2.2 Phosphides
        - 22.2.3 Antimonides
        - 22.2.4 Mixed Crystals
    - 22.3 IV–IV, II–V, and III–VI Compounds
        - 22.3.1 Mixed Crystals
    - 22.4 II–VI Compounds
        - 22.4.1 Sulphides
        - 22.4.2 Selenides
        - 22.4.3 Tellurides
        - 22.4.4 Mixed Crystals
    - 22.5 Halides
        - 22.5.1 Silver Halides
        - 22.5.2 Alkali Halides
        - 22.5.3 Alkali-Earth Halides
        - 22.5.4 Mixed Crystals
    - 22.6 Simple Oxides (CdO, ZnO See 22.4)
        - 22.6.1 Mixed Crystals
    - 22.7 Semiconducting Intermetallic Compounds
    - 22.8 Tertiary and Higher Compounds
        - 22.8.1 Salts of Oxy-Acids (Carbonates, Phosphates, Silicates Including Mica etc.)
        - 22.8.2 Spinels, Garnets, Ferrites
    - 22.9 Organic Semiconductors

23. Solidified Gases

Attention

The preceding classification scheme is closely akin to the scheme used by Physics Abstracts to classify the various fields of solid state physics. A paper will either be classified under one or more of the subsections or, if none of these are relevant, it will just appear under the general section heading (i. e. sections and subsections are equivalent in this sense).

The *subject classification* given at the head of each paper begins with the main subject matter of the paper, and this is followed, if necessary, by figures referring to subsidiary subject matter and the type of solids investigated.

The papers published in this issue are listed in the order of their subject classification at the end of the table of contents.

32914

Manuscripts and letters for physica status solidi (b) — basic research and physica status solidi (a) — applied research

should be sent to

DDR-102 Berlin, Neue Schönhauser Str. 20

or

Internationaler Buchversand GmbH, 1 Berlin 19, Eschenallee 21.

The object of "physica status solidi (b)" and "physica status solidi (a)" is to publish rapidly important and topical results concerning fundamental research in the field of solid state physics and aspects of application and experimental techniques of fundamental solid state research, respectively. The issues of "physica status solidi (b)" and "physica status solidi (a)" will be published on the 1st and the 16th day of each month, respectively. Either series will contain the following categories:

1. Review articles written by arrangement with the editors. They will be remunerated.
2. Original papers selected from those which have not been submitted for publication elsewhere, after being reviewed. (The latest date for acceptance of contributions is the 8th and the 23rd day of each month for "physica status solidi (b)" and "physica status solidi (a)", respectively, 50 days before the delivery of the journal.)
3. Short notes published in the off-set part of the journal. They are, if necessary after being reviewed, selected by the editorial staff. (The latest date for acceptance of manuscripts for this part is the 12th and the 27th day of each month for "physica status solidi (b)" and "physica status solidi (a)", respectively, 16 days before the delivery of the journal.)
4. Pre-printed titles of papers to be published in the next issue of both series.

Editorial Information

Manuscripts should be submitted preferentially in English. Besides, papers in German, French, and Russian are admitted. The final decision on the incorporation of the papers in the series (b) or (a) is taken by the editorial staff after consultation of the referee. Authors are asked to observe the following rules:

1. Manuscripts have to be written as concisely as possible. As most, 20 and 4 pages (28 lines per page; abstracts, tables, references, figures, and captions of figures included, where 3 figures are counted as about one page) are admitted for original papers and short notes, respectively. All manuscripts should be type-written on one side only. Two copies are required. Badly arranged manucripts are returned.
2. The author's name, initials, and institution should be given.
3. One abstract of about ten lines in English and one in German, French, or Russian are required for original papers.
4. Because of lack of space conclusions (if any) must not have the character of a summary (i. e. substantially a repetition of the contents of the abstract).
5. Formulae should not be written too small and with pencil. Separate lines for formulae are desirable. Indistinct writing will lead to a return of the manuscript.
6. Drawings should be enclosed on separate sheets of paper, not exceeding the maximum and minimum sizes of 20×30 and 10×15 cm^2, respectively. Photographs should be submitted, on white paper, only for review articles and original papers. Duplicates are required.
7. Captions for the figures should be submitted on separate sheets of paper, and special instructions with regard to the size of illustrations should be given on these.
8. References should be numbered in order of appearance and should be given, on a separate sheet, in the form used in physica status solidi.
9. Units, symbols, etc. should be used in accordance with the recommendations given by the IUPAP.
10. Each paper should contain a proposal for the subject classification according to the classification scheme appearing on the second and third cover of each issue.

The authors will be sent galley proofs of review articles and original papers. Proof reading of short notes and pre-printed titles will be done by the editorial staff.

For each original paper and short note a total of 75 reprints is available free of charge. The number of reprints free of charge per review article is 25. Requests for additional reprints (against payment) should be given together with the submitted manuscript. **The deadline for orders of additional reprints is the 15th and the 30th of each month** for "physica status solidi (b)" and "physica status solidi (a)", respectively, two weeks before the delivery of the journal.

FSC
www.fsc.org
MIX
Papier | Fördert gute Waldnutzung
FSC® C083411

Kolibri 360 GmbH
Ferdinand-Jühlke-Straße 7
99095 Erfurt, Deutschland
produktsicherheit@kolibri360.de